GLOBAL SHIFT
Second Edition

Peter Dicken is Professor of Economic Geography at the University of Manchester. His major research interests are global economic change and the spatial behaviour and strategies of transnational corporations. He is the author of numerous academic papers and research reports on foreign direct investment and on economic and industrial change in the UK, Europe and in East and South East Asia. The first edition of *Global Shift* (1986) is extensively used as a major text throughout the world. He is the co-author of *Location in Space: Theoretical Perspectives in Economic Geography* (3rd edition) (1990), Harper & Row, New York; and *Modern Western Society* (1981) Paul Chapman. He is currently actively involved in the work of the European Science Foundation. He has lectured and taught extensively in universities and colleges in Europe, the United States, Canada, Australia and South East Asia.

GLOBAL SHIFT

The Internationalization
of Economic Activity

Second Edition

Peter Dicken

University of Manchester

THE GUILFORD PRESS
New York   London

© 1992 Peter Dicken

Published by The Guilford Press
A Division of Guilford Publications, Inc.
72 Spring Street, New York, NY 10012

This book is printed on acid-free paper.

Last digit is print number: 9 8 7 6 5 4 3 2

Library of Congress Cataloging-in-Publication Data

Dicken, Peter.
   Global Shift: the internationalization of economic activity/
Peter Dicken. — 2nd ed.
     p. cm.
   Includes bibliographical references and index.
   ISBN 0–89862–488–6
   1. Industry — History — 20th century. I. Title.
HD2321.D53 1992
338.09′04 — dc20   91–28634
                                          CIP

To Valerie, Christopher and Michael
– and to my father who has seen the world change

Contents

PART TWO: PROCESSES OF GLOBAL SHIFT

PART FOUR: STRESSES AND STRAINS OF ADJUSTMENT
            TO GLOBAL SHIFT

Acknowledgements

The author and publishers would like to thank the following publishers who have kindly given their permission for the use of copyright material:

Association of American Geographers for Figures 4.6, 11.3 and 11.6

Australian Geographical Studies for Figure 9.7

Basil Blackwell for Table 5.3

Commission de Communautes Européenes and Gower Publishing Group for Figure 6.6

Croom Helm for Figure 11.5 and Table 5.1

HarperCollins Academic for Figure 4.8

Harvard Business School Press for Figures 4.10 and 5.6 reprinted with permission from Louis T. Wells, Jr., ed *The Product Life Cycle and International Trade*. Boston: Division of Research, Harvard Business School, 1972

Harvard Business School Press for Figure 7.4; reprinted by permission of Harvard Business School Press from *Managing Across Borders: The Transnational Solution* by Christopher A Bartlett and Sumantra Ghoshnal. Boston, MA, 1989. Copyright © 1989 by the President and Fellows of Harvard College.

Institute for International Economics for Table 8.9, reprinted with permission from William R. Cline *The Future of World Trade in Textiles and Apparel*, 1987.

Johns Hopkins University for Figure 13.3 from Kols, A. and Lewison, D. Migration population growth, and development. Population Reports, Series M, No. 7. Baltimore, Johns Hopkins University, Population Information Program, September–October 1983

Journal of International Business Studies for Table 5.2

Macmillan Ltd. and Holmes & Meier Inc. for Figure 7.7. Reprinted with permission from Peter J. Buckley and Mark Casson *The Future of the Multinational Enterprise* © 1986, Figure 2.7 Scanning of Scientific and Business Environment. 'A Long-run Theory of the Multinational Enterprise.'

Methuen and Co. for Figure 5.3

OECD for Figure 4.2 and Table 11.6

Oxford University Press, for Table 4.3. Reprinted from S. Hirsch, *Location of Industry and International Competitiveness* (1967), Clarendon Press. Reprinted by permission of Oxford University Press.

Pergamon Press for Table 9.3

Pion, London for Table 10.6

The Regents of the University of California for Figure 7.14 copyright 1986. Reprinted from the *California Management Review*, Vol. 28 No. 3; and Table 7.2 copyright 1989. Reprinted from the *California Management Review*, Vol. 32, No. 1

Stanford University Press and the Board of Trustees of Leland Stanford Junior University for Figure 6.8 © 1989. Reprinted from *Between Miti and the Market: Japanese Industrial Policy for High Technology* by Daniel I. Okimoto

John Wiley and Sons Inc. for Figure 4.9, reprinted with permission from D. M. Smith, *Industrial Location: an Industrial Geographical Analysis*, 2nd edition, 1981.

Preface

The first edition of *Global Shift*, published in 1986, was gratifyingly well received across a whole range of academic disciplines and at different academic levels, from introductory to graduate courses, according to previous background and current need. However if, as they say, 'a week is a long time in politics', six years is an eternity in studies of the global economy. Within those six years, enormous changes have occurred both in the nature and the extent of the globalization of economic activities. It is largely to incorporate these changes that this second edition has been prepared. Although the broad framework of this second edition is similar to that of the first edition the book has been almost totally rewritten. The focus of the book has been sharpened to an explicit concern with the internationalization and globalization of economic activities, rather than the more general subject of industrial change. The organization of the chapters and the links between them have been tightened to reflect this more specific focus.

The book is organized into four closely related, but distinct, parts, each of which is introduced by a brief prologue. Not only has all the empirical material been updated but also the emphasis given to certain key topics has been greatly increased. In particular, the following new elements have been introduced: (1) technology and technological change now have a chapter to themselves, with particular emphasis being placed upon the role of information technology and the emergence of flexible technologies of production; (2) there is specific treatment of the competitive strategies of business enterprises and of the increasingly intricate networks of inter-relationships between firms at the global scale, particularly through strategic alliances and subcontracting arrangements; (3) the increasingly pervasive issue of trade tensions, especially between North America, Europe and East and South East Asia, and the growing adoption of strategic trade policies are given more explicit attention; (4) in a closely related vein, there is specific discussion of the strengthening of regional integration in Western Europe (both the completion of the Single European Market and the opening up of Eastern Europe) and in North America (the Canada–United States Free Trade Agreement and the potential integration of Mexico; (5) an entirely new case study chapter on the internationalization of services (with particular emphasis on financial services) is included in place of the previous chapter on iron and steel. These various changes have been made without increasing the overall length of the book. It is surprising how many words seem to be unnecessary with the lapse of time!

The globalization of economic activity is a very complex phenomenon. The approach taken in this book is that it is primarily the outcome of the complex interaction between *transnational corporations* and *nation-states* set within the context of a *volatile technological environment*. These, I believe, are the three major forces – the primary generators – of global economic change. *Transnational corporations*, through their worldwide operations, and *national governments*, through their trade, foreign investment and industry policies as well as their macroeconomic measures, shape and reshape the global economic map. At the same time, revolutionary changes in the *technologies* of transport, communication, production and organization have facilitated the internationalization and globalization of production and trade of both goods and services. Few parts of the world are unaffected as transnational corporations restructure their operations on a global scale, as national governments attempt to build or preserve their own economic strength and as the pace of technological change quickens and as its nature changes. Such transformations are both geographically and sectorally uneven. We need to look not only at the *general* processes of change but also at their *specific* expression in different places and in different industries (both manufacturing and services). As far as the impact of global shifts is concerned, one of the most important is on employment. The question of 'where will the jobs come from?' faces people in all parts of the world.

My approach is deliberately broad-ranging. I have drawn on ideas and explanations from many fields of study – economics, economic geography, international business, industrial organization, political science – and on examples from all parts of the world. Much use is made of graphical material (maps, diagrams) and tables to capture in a more accessible manner the far-reaching changes which are occurring in the global economy. Previous knowledge, either of economics or of the other subject areas involved, is *not* assumed; the treatment throughout is *non-technical*. The book has been designed for use in several ways and in a variety of courses in different disciplines. Precisely *how* it is used will obviously depend on the background of the user and on the length and organization of the course. As far as possible, the text has been kept free of detailed references to literature – although quotations are used extensively – but each chapter ends with a guide to further reading and there is a substantial bibliography. In this way, the needs of readers at different levels can be catered for.

Thus, the book may be used in broad outline for an introductory course; as it stands for a middle-level course; or as the basis for a more advanced course through the use of the guides to further reading and the bibliography. The bibliography itself consists mainly of academic books and articles. But for such a rapidly changing, and topical, subject matter readers are urged regularly to consult such publications as *The Financial Times*, *The Wall Street Journal*, *The Economist*, *Business Week*, *Fortune*, *The Far Eastern Economic Review*, and the like, as well as the business pages of the 'quality' newspapers to keep abreast of current developments.

A book whose horizons and subject matter are global necessarily has to draw upon an enormous range of statistical data derived from a wide variety of sources. The most useful data sources are those produced by such international institutions as the United Nations, the World Bank, the GATT, as well as national statistical agencies. In this revision, the data used are the most recent available as of early 1991. Unfortunately, this does *not* mean that the figures on, say, production, trade or investment in different countries and different industries are for 1990! The world of official statistics just doesn't work with such rapidity and efficiency. There is always a substantial time-lag in published data sources and this differs from one official source to another. Hence, most of the data refer to varying dates between 1986 and 1989. That is what 'up-to-date' means!

For this second edition I owe enormous debts to very many people. Indeed, the extent to which this edition is an improvement on the first is very largely the result of the many positive and constructive comments and criticisms I have received from academic colleagues throughout the world. The discipline of presenting material on the subject of *Global Shift* to audiences in economics, economic geography, international business and international sociology in Europe, North America and Asia has been of enormous benefit in sharpening up my ideas. Particular thanks are due to Michael McDermott, Jakob Kol, Jim McConnell, Gunter Krumme, Nigel Thrift, Roger Lee, Lyn Collins, Gerry Bloomfield and Roger Hayter who commented in detail on the proposed changes to the first edition as well as to the many people who reviewed or commented on the first edition. As always, of course, they are totally absolved from all blame for the final outcome. They can, of course, take comfort from the fact that this edition is infinitely better than it would have been without their careful and constructive criticism. Closer to home I am particularly grateful to two colleagues at the University of Manchester, Jamie Peck and Adam Tickell, for providing such a stimulating intellectual (and social) environment on a day-to-day basis. Again, Nick Scarle of the Department of Geography has drawn all the illustrations with phenomenal skill and patience from the almost incomprehensible sketches and figures I provided for him. Jean Woodward has done a tremendous job in word processing a very messy and complex manuscript to a very tight deadline. It was a far bigger task than either of us realized. I am immensely grateful to both of them. Once again, however, my biggest debts are owed to Valerie, Christopher and Michael for continuing to provide such a wonderful mixture of love and laughter.

Peter Dicken
Manchester, May 1991

Chapter 1

Introduction

The world economy is changing in fundamental ways. The changes add up to a basic transition, a structural shift in international markets and in the production base of advanced countries. It will change how production is organized, where it occurs, and who plays what role in the process.
(Cohen and Zysman, 1987, p. 79)

The talk today is of the 'changing world economy'. I wish to argue that the world economy is not 'changing'; it has already changed – in its foundations and in its structure – and in all probability the change is irreversible.
(Drucker, 1986, p. 768)

Themes with variations

The notion that something fundamental is happening, or indeed has happened, to the world economy is now increasingly accepted. There are many different ways in which this 'new world' can be approached. In this book, two broad themes form the major threads of the argument. The major theme is that economic activity is becoming not only more *internationalized* but that, more significantly, it is becoming increasingly *globalized*. These terms are often used interchangeably although they are not synonymous. 'Internationalization' refers simply to the increasing geographical spread of economic activities across national boundaries; as such it is not a new phenomenon. 'Globalization' of economic activity is qualitatively different. It is a more advanced and complex form of internationalization which implies a degree of functional integration between internationally dispersed economic activities. Globalization is a much more recent phenomenon than internationalization; however, it is emerging as the norm in a growing range of economic activities.

The other theme woven through the book is that we live in a period of major economic change; an era of *turbulence* and *volatility* in which economic life in general is being restructured and reorganized both rapidly and fundamentally. These two themes are intimately connected in the sense that the pervasive internationalization, and growing globalization, of economic life ensure that changes originating in one part of the world are rapidly diffused to others. We live in a world of increasing complexity, interconnectedness and volatility; a world in which the lives and livelihoods of every one of us are bound up with processes operating at a global scale.

However, although we are often led to believe that the world is becoming increasingly homogenized economically (and perhaps even culturally) with the use of such labels as 'global village', 'global market-place' or 'global factory', we need to treat such all-embracing claims with some caution. The 'globalization' tag is too often applied very loosely and indiscriminately to imply a totally pervasive set of forces and changes with uniform effects on countries, regions and localities. There are, indeed, powerful forces of globalization at work – they are the central focus of this book – but we need to adopt a sensitive and discriminating approach to get beneath the hype and to lay bare the reality.

So, although the broad themes of globalization and turbulence will be the dominant themes throughout this book they will not be treated as monolithic or unvarying processes. Indeed, almost as important as the dominant themes themselves are the *variations* upon these themes, particularly the geographical variations. Change does not occur everywhere in the same way and at the same rate; the processes of globalization are not geographically uniform. The particular character of individual countries, of regions and even of localities interacts with the larger-scale general processes of change to produce quite specific outcomes. Reality is far more complex and messy than many of the grander themes and explanations tend to suggest.

A turbulent world

One would need to be a hermit to be totally unaware of at least some element of change and turbulence in economic life; to realize that what is happening in our own back yards is largely the product of forces operating at a much larger scale. Television news reports and specials, press headlines and the like constantly remind us of our uneasy present and precarious future. Turbulence and change are, of course, nothing new in human affairs. But there is no doubt that since the early to mid-1970s in particular the world economy, and its constituent parts, have been buffeted by extremely volatile forces.

During the 1970s and into the first half of the 1980s, unemployment in the Western industrialized countries soared to levels unknown since the world depression of the 1930s. Traditional industries contracted with accelerated speed; factories closed apace. The term 'deindustrialization' became common currency in both North America and Europe. During the latter half of the 1980s what had seemed to be an inexorable growth in unemployment abated somewhat and, indeed, there were substantial recoveries in some cases. But such recovery was extremely uneven geographically; many of the worst hit, older industrialized regions continued to face major employment problems. More generally, the shift in the composition of employment out of manufacturing into services gathered momentum. During the second half of the 1970s crippling financial debts became apparent in a number of developing countries, threatening not only to abort embryonic economic development and to multiply the already deep-seated welfare problems of the majority of the world's population but also to stretch to breaking point the political and social cohesion of entire countries. During the 1970s, too, trade tensions re-emerged both between the industrialized countries themselves and also between them and the so-called newly industrializing countries (NICs) of the Third World. Both Europe and the United States became locked in trade quarrels with Japan over the latter's penetration of their domestic markets in such products as automobiles and consumer electronics. In turn, Europe and the United States argued about mutual trade in steel and in agriculture. More generally, the Western industrialized nations accused the newly industrializing countries of flooding their markets with cheap imports of manufactured goods. Conversely the NICs themselves charged the older industrialized nations with restricting trade in manufactured goods and with failing to adjust to changing circumstances.

Most of these sources of turbulence continued through into the late 1980s and early 1990s, some in more muted, others in more amplified, form. The trade tensions between the Asian economies on the one hand and the United States and Europe on the other have, if anything, intensified. Although tariffs on trade have continued to fall there has been a dramatic increase in non-tariff barriers of various kinds. Trade conflicts have been made sharper by a number of other developments, such as the massive trade imbalances which affect, most notably, the United States on the one hand (with its huge trade deficit) and Japan on the other (with its massive trade surplus). Another is the creation of the Single European Market after 1992, with the fears it has engendered about the possible development of a more highly protected 'fortress Europe'. Together with the Canada–United States Free Trade Agreement this reinforces fears of a *de facto* division of the world economy into discrete trading blocs. All of these trade issues became entangled in the attempts to create a new multilateral trade agreement in the Uruguay Round of the GATT, which brought into sharp focus the problems of trading relationships between developed and developing countries in a whole variety of economic sectors, including services.

During the late 1980s and early 1990s four further sources of economic turbulence became more evident. First was the increasing separation, at a global scale, of what Peter Drucker (1986) calls the 'real economy' of the production and trade of goods and services and the 'symbol economy' of financial flows and transactions (capital movements, exchange rates and the like). The second new source of global economic turbulence appeared with amazing suddenness in late 1989 with the dramatic political changes in Eastern Europe and the beginnings of a transformation from the centrally planned command economies to market economies. At a stroke the whole nature and extent of the kind of 'European space' which had prevailed for more than half a century were thrown into question. Conversely, China's progression towards a more open economy, begun in 1979, was abruptly halted by the political massacres in Tiananmen Square in June 1989. Third, we were again reminded of the continuing political volatility of the Middle East with the Gulf war of early 1991, although this did not have the predicted deleterious effects on oil supplies. The fourth new source of volatility and uncertainty, which also became the focus of attention in the late 1980s, was the threat to the global environment created by global warming, via the greenhouse effect, and by the thinning of the atmospheric ozone layer.

Global interconnections: the increasing globalization of economic activities

The immediacy and longer-term impact of these major forces of turbulence are enormously enhanced by the growing interconnections between all parts of the world. The most significant development in the world economy during the past few decades has been the *increasing globalization of economic activities*. In one sense, of course, the internationalization of economic activities is nothing new. Some commodities have had an international character for centuries; an obvious example being the long-established trading patterns in spices and other exotic goods. The internationalization of economic activities was much enhanced by the onset and diffusion of industrialization from the eighteenth century onwards in Europe. Nevertheless until very recently the production process itself

> was primarily organized *within* national economies or parts of them. International trade . . . developed primarily as an exchange of raw materials and foodstuffs . . . [with] . . . products manufactured and finished in single national economies . . . *In terms of production, plant, firm and industry were essentially national phenomena.*
>
> (Hobsbawm, 1979, p. 313 – emphasis added)

The nature of the world economy has changed dramatically, however, especially since the 1950s. National boundaries no longer act as 'watertight' containers of the production process. Rather, they are more like sieves through which extensive leakage occurs. The implications are far reaching. Each one of us is now more fully involved in a *global* economic system than were our parents and grandparents. Few, if any, industries now have much 'natural protection' from international competition whereas in the past, of course, geographical distance created a strong insulating effect. Today, in contrast, fewer and fewer industries are oriented towards local, regional or even national markets. A growing number of economic activities have meaning only in a global context. Thus, whereas a hundred or more years ago only rare and exotic products and some basic raw materials were involved in truly international trade, today virtually everything one can think of is involved in long-distance movement. And because of the increasingly complex ways in which production is organized across national boundaries, rather than contained within them, the actual origin of individual products may be very difficult to ascertain.

Something of this increased global diversity of production can be gleaned simply from examining the labels on products. Many labels are geographically misleading, however, particularly in the case of products consisting of a large number of individual components, each of which may have been made in different countries. Generally, the labels signify the country of the final (assembly) stage of production. But where are such global products really made? Under such conditions what is a 'British' car, an 'American' computer, a 'Dutch' television or a 'German' camera? In today's global economy, some products can be regarded as having been made almost everywhere – or nowhere – such is the geographical complexity of some production processes.

What these developments imply is the emergence of a *new global division of labour* – a change in the geographical pattern of specialization at the global scale. Originally, as defined by the eighteenth-century political economist Adam Smith, the 'division of labour' referred simply to the specialization of workers in different parts of the production process. It had no explicitly geographical connotations at all. But quite early in the evolution of industrial economies the division of labour took on a geographical dimension. Some areas came to specialize in particular types of economic activity. Within the rapidly evolving industrial nations of Europe and the United States regional specialization – in iron and steel, shipbuilding, textiles, engineering and so on – became a characteristic feature. At the global scale the broad division of labour was between the industrial countries on the one hand, producing manufactured goods, and the non-industrialized countries on the other, whose major international function was to supply raw materials and agricultural products to the industrial nations and to act as a market for some manufactured goods. Such geographical specialization – structured around a *core* and a *periphery* – formed the underlying basis of much of the world's trade for many years.

This relatively simple pattern (although it was never quite as simple as the description above suggests) no longer applies. During the past few decades trade flows have become far more complex. The straightforward exchange between core and peripheral areas, based upon a broad division of labour, is being transformed into a highly complex, kaleidoscopic structure involving the *fragmentation* of many production processes and their *geographical relocation* on a global scale in ways which slice through national boundaries. In addition, we have seen the emergence of new centres of industrial production in the newly industrializing economies. Both old and new industries are involved in this re-sorting of the global jigsaw puzzle in ways which also reflect the development of technologies of transport and communications, of corporate organization and of the production process. The technology of production itself is undergoing substantial and far-reaching change as the emphasis on large-scale, mass production, assembly-line techniques is shifting to a more flexible production technology.

In effect, then, 'the traditional *international* economy of *traders* is giving way to a *world* economy of *international producers*' (Root, 1990, p. 7 – emphasis added). Just as we can identify a new international division of labour in production so, too, we can identify a 'new international financial system', based on rapidly emerging twenty-four-hour global transactions concentrated primarily in the three major financial centres of New York, London and Tokyo.

Focus and organization of the book

The focus in this book is on the *increasing globalization of economic activity* which is occurring within a dynamic framework of turbulence and volatility in the world economy. It is not an attempt to explain the comparative economic performance of individual nations, although such differential performance is clearly connected to the processes of internationalization and globalization.[1] Figure 1.1 presents a generalized picture of the production system. Note that the term 'production' is defined in very broad terms and refers to far more than just manufacturing. Virtually all aspects of the production system are becoming increasingly globalized. In the following chapters, attention will be focused on two major elements.

Particular attention is devoted to *manufacturing* activity for a number of reasons. It is in the manufacturing sector that major global shifts are especially apparent. It is in the manufacturing industries of the Western industrialized economies that the most severe problems of adjustment to changing world forces occur. It is to manufacturing industry that many of the developing countries look (rightly or wrongly) to stimulate the development of their economies. It is in the manufacturing sector that a small number of developing countries have recently emerged as world centres of production and trade, helping to shift the centre of gravity of the global economic system.

However, the globalization of manufacturing activity depends upon the globalization of other elements in the production system, notably in the *circulation* activities which connect the various parts of the production system together. An especially important development in the world economy in recent years has been the globalization of particular kinds of circulation activity – notably *business services*, such as banking and financial services, insurance, advertising, communications, hotels and the like – as part of an emerging global infrastructure. Indeed in a number of the older industrialized countries it is the services sector that is now the major dynamic element.

The purpose of this book, then, is to describe and explain the globalization of economic activity and to examine some of its implications for countries and regions. There are several ways of organizing the treatment of such a broad and complex subject. The major problem is that the processes involved are tightly interconnected and mutually interact with one another in intricate ways. We are not dealing with a linear sequence in which we can simply deal with one process at a time. There is no unique starting point. Unfortunately, in a book the chapters have to be arranged in a linear sequence. To try to get round this artificial constraint, Figure 1.2 shows how the individual sections and chapters of the book are organized. Its aim is to demonstrate

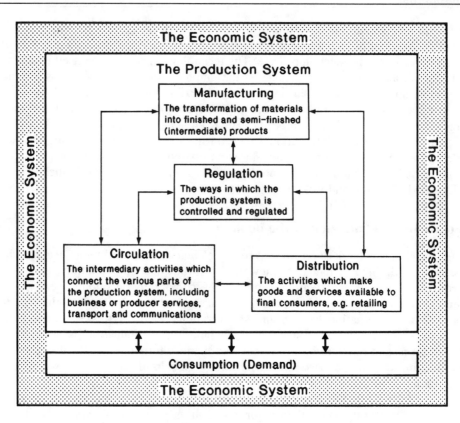

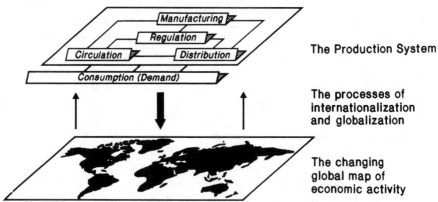

Figure 1.1 Basic elements of the production system (*Source*: based on Bailly, Maillat and Rey (1987), Figure 1)

both the sequential logic of the argument and the mutual relationships between individual chapters. Overall the book is organized into four related parts each of which begins with a Prologue that provides both the broader context of the chapters which follow and the ways in which they fit together.

Part One is primarily descriptive. It depicts in some detail both the aggregate and geographical trends in the internationalization of economic activity during recent

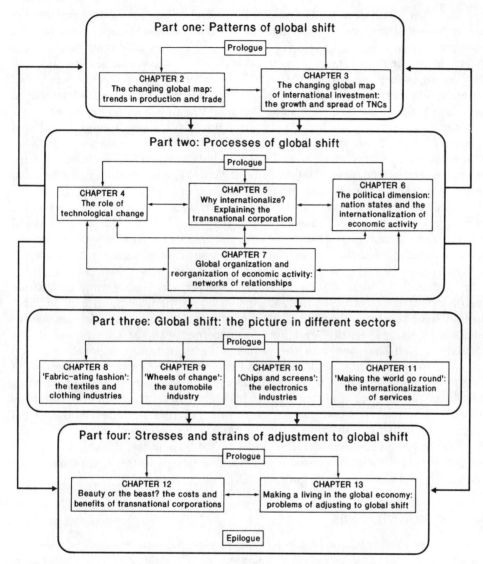

Figure 1.2 Relationships between the parts: the organization and structure of the book

decades. Chapter 2 is concerned with patterns of international production and international trade. Chapter 3 focuses upon patterns of international investment with particular reference to the spread of transnational corporations. Part Two is the explanatory core of the book. As Figure 1.2 indicates, this is organizationally the most complex in the sense that it deals with highly interconnected processes of change. There is no unique starting point here and no unique order in which the processes should be examined.

The overall logic of Part Two is that the globalization of economic activity is primarily the manifestation of the internationalization of capital as organized through business enterprises, of which the most significant is the *transnational corporation* (TNC). The TNC is arguably the most important single force creating global shifts in economic activity (Chapter 5). But its strategies and operations and, therefore, the resulting map of international production, trade and investment are much influenced by the forces of *technological change* (Chapter 4) and by the actions of *nation states* (Chapter 6). In the most general sense the globalization of economic activity can occur only if appropriate technologies exist both for overcoming geographical distance and for standardizing and fragmenting the production process. Nation states continue to play a highly significant role in shaping the globalization of economic activity both through trade policies (including the establishment of regional trading blocs) and through explicit policies to regulate and control TNCs. Chapter 7 focuses specifically on the complex relationships which have developed both within TNCs and also between independent and quasi-independent firms. Such networks of relationships are a crucial element in the global organization and reorganization of economic activity.

Whereas Part Two is concerned with general processes of globalization, Part Three examines in detail the globalization process in specific industries. The precise form of the internationalization of production, the form of global production systems and the manner in which global shifts have occurred vary substantially from one industry to another. Chapters 8 to 11, therefore, consist of case studies of individual industries – textiles and clothing, automobiles, electronics, and services – chosen to demonstrate some of this diversity of experience. Global shifts are particularly evident in these industries and have major employment implications for both the older and newer industrialized nations. Finally, Part Four is concerned with the stresses and strains created by global shift and with the problems facing national, regional and local economies. Chapter 12 assesses the costs and benefits of TNCs from the viewpoint of both host and home countries. Chapter 13 adopts a broader perspective and examines the problems of adjustment arising from global shifts in economic activity. The inter-connected problems facing economies occupying different positions in the global economy are explored, respectively, for the older industrialized countries, the newly industrializing countries and the less industrialized countries. Lastly, in a brief Epilogue, some questions are posed about possible futures in the light of the major themes addressed throughout the book.

Notes for further reading

1. Many writers have addressed the subject of comparative national economic performance. Among the most recent examples are Paul Kennedy's *The Rise and Fall of the Great Powers* (1987) and Michael Porter's *The Competitive Advantage of Nations* (1990).

PART ONE

Patterns
of
Global Shift

Prologue

The globalization of economic activity is evident in a number of ways. One indicator is the much increased geographical extent of the production of goods and services and the related flows of international trade. The second indicator is the explosion in the volume of international direct investment by transnational corporations and the increasing geographical extent and complexity of their activities. These two sets of indicators form the subject matter of the two chapters in Part One. In both cases attention is focused upon recent and current trends, specifically those occurring during the past two or three decades. However, it is impossible to understand the current situation without at least an outline knowledge of what has gone before. We need a baseline against which to measure current trends.

A historical perspective: evolution of a global economy

Although the development of a global economy was greatly accelerated by the process of industrialization beginning in the second half of the eighteenth century, its basis was established much earlier. This global system was a capitalist system which originated and developed in quite specific geographical locations.[1] The beginnings of a world economy were evident first in the expansion of trade during the period from 1450 to 1640, a period which Wallerstein has labelled the 'long sixteenth century'. Prior to, and even during, this period the pattern of trade was bimodal. On the one hand there was local trade – short distance, mainly concerned with basic necessities – the kind of trade focused upon the medieval market towns. On the other hand there was the much smaller volume of long-distance trade in luxury goods and rare items for a very tiny fraction of the population: the spice and fine cloth trade, for example, together with other exotic goods from distant parts of the world. During the late fifteenth and early sixteenth centuries, however, the geographical extent of trade increased dramatically as a result of the expansion of a small number of European maritime nations which came to form the core of an evolving world economy. By the middle of the seventeenth century economic leadership was centred on north-west Europe.

The development of a world trading system over a period of several centuries resulted in a *tripartite* geographical structure of core, semi-periphery and periphery. It also laid the foundations for a process which was to have even more far-reaching effects: industrialization. In turn, industrialization greatly accelerated the further expansion of world trade and further transformed its character. As the nineteenth century progressed the nature and geographical pattern of world trade changed to one in which the core (initially Britain) exported manufactured goods throughout the world and imported raw materials, especially from the colonies. Exports of textiles led the way, followed, in the second half of the nineteenth century, by heavy manufactured goods, such as iron and steel and also coal. In effect, a *new international division of labour* – a new pattern of geographical specialization – had emerged.

Progressively, the roles of core and periphery in the geographical division of labour became increasingly clearly defined. Industrial production became pre-eminently a core activity, a process reinforced by the political process of colonialization. Industrial goods were both traded between core nations and also exported to the periphery. Conversely, the periphery's role was a dual one. First, it supplied the core with primary commodities – raw materials for transformation into manufactured products in the core; foodstuffs to help feed the industrial nations. Second, it purchased manufactured goods from the core, particularly capital goods in the form of machinery and equipment. The relative fortunes of the core countries themselves waxed and waned. Most notably, the United States and Germany emerged to overtake the previously undisputed leader, Britain. By 1913 the United States was producing 36 per cent of total world industrial output while Britain's share had fallen to 14 per cent.

Of course, the process of development of the world economy is not a continuous, uninterrupted sequence of events. By its very nature it is a *discontinuous* process; periods of rapid growth of production and trade and geographical expansion are punctuated by periods of stagnation and recession. Such business cycles vary enormously in their frequency, duration and intensity. In recent years much attention has been focused upon the notion of long waves of economic development of roughly fifty years' duration, which are associated with a fundamental restructuring of economic activity at a global scale.

The emergence of transnational corporations

The evolution of a global economy based on increasingly extensive flows of international trade, and structured around the broad framework of a core and periphery, also involved the growth and spread of international investment. It involved, in other words, the early development of the transnational corporation, the institution which has subsequently come to dominate the global economy.[2] The development of companies with interests and activities located outside their home country was part and parcel of the early development of a global economy. Most students of economic history are familiar with the chartered trading companies which emerged in Europe from the fifteenth century onwards as an important part of the early evolution of the world economic system. Companies such as the East India Company, the Hudson's Bay Company and others created vast trading empires on the world scale. But, despite their worldwide extent, these trading companies were not really the true ancestors of today's TNCs. Apart from the fact that they were confined mostly to the colonial territories of their home countries their main *raison d'être* was trade and exchange not production.

In fact the first firms to engage in production outside their home country did not emerge until the second half of the nineteenth century, and then only hesitantly at first. But by the eve of the First World War there was substantial overseas manufacturing production by United States, British and continental European companies.

The growth of United States transnational activity and manufacturing in the late nineteenth and early twentieth centuries reflected the country's emergence as the world's major industrial nation. But in 1914 the major source of overseas investment was still the United Kingdom. In fact, the geographical spread of UK overseas manufacturing investment was considerably broader than that of US firms or of those from continental Europe. The UK pattern, of course, strongly reflected the nation's imperial position. Although US and UK transnational manufacturing investment was similar in several respects – both were investing heavily in food, chemicals and engineering industries – there were some differences which persisted for a long time. In particular, US transnational investments, from a very early stage, leaned towards the newer, more technologically sophisticated products in both producer and consumer goods. A good deal of UK transnational investment was in textiles. The early overseas manufacturing investments by continental European firms also displayed a distinctive industrial complexion; much of their investment was in chemicals and in electrical machinery. Between the First and Second World Wars, transnational manufacturing investment grew considerably, with especially rapid increases in the foreign network of United States TNCs. By 1939 the United States had become the major source of international investment in manufacturing as well as the world's leading industrial nation.

Post-1945: the shaping of a new global economic system

These broad contours of the global economic map persisted until the Second World War. Global production and trade were dominated by the old-established core economies of north-west Europe and the United States. Manufacturing production remained strongly concentrated in this industrialized core; a clear international division of labour was apparent. International direct investment by the rapidly developing transnational corporations was also dominated by firms in these leading core nations, particularly the United States and the United Kingdom. Such international investment at the global scale was most strongly concentrated in the developing countries which, on the eve of the Second World War, were host to two-thirds of total foreign direct investment.

This relatively stable and long-established structure was shattered by the Second World War, which devastated the global economy. As Scammell (1980, p. 2) observed:

> Across all historical studies of recent times the second world war cut its wide swathe, a great dividing line.

The world economic system which emerged after 1945 was in many ways a new beginning. It reflected both the new political realities of the postwar period – particularly the sharp division between East and West – and also the harsh economic and social experiences of the 1930s. The kinds of international economic institutions devised in the aftermath of war grew out of both these factors.[3] The major political division of the world after 1945 was essentially that between the West (led by the United States) and the East (the Soviet-dominated nations of Eastern Europe). Outside these two major power blocs was the so-called 'Third' World, a highly heterogeneous – but generally improverished – group of nations, many of them still at that time under colonial domination. The Third World was far from immune from the East–West confrontation. Both major powers made strenuous efforts to extend their spheres of influence, a process which had considerable implications for the subsequent pattern of global economic change.

The Soviet bloc drew clear boundaries around itself and its Eastern European satellites and created its own economic system, quite separate from the capitalist market economies of the West, at least initially. In the West the kind of economic order built after 1945 reflected the economic and political domination of the United States. Alone of all the major industrial nations, the United States emerged from the war strengthened rather than weakened. It had both the economic and technological capacity and also the political will to lead the way in building a new order. The institutional basis of this new order came into being formally at a conference at Bretton Woods in New Hampshire in 1944. The Bretton Woods conference brought into being two international financial institutions: the International Monetary Fund (IMF) and the International Bank for Reconstruction and Development (World Bank).[4]

The objective of the Bretton Woods system was to stabilize and regulate international financial transactions between nations on the basis of fixed currency exchange rates in which the US dollar played the central role. In this way it was hoped to provide the necessary financial 'lubricant' for a reconstructed world economy. The other major pillar of postwar international economic order was to be that of free trade. The view that the 'beggar-my-neighbour' protectionist policies of the 1930s should not be allowed to recur after the war was reflected in the establishment of another international institution in 1947: the General Agreement on Tariffs and Trade (GATT). The purpose of the GATT was to reduce tariff barriers and to prohibit other types of trade discrimination (see Chapter 6). Together, this triad of international bodies established in the immediate postwar period – the IMF, the World Bank and the GATT – formed the international institutional framework in which the rebuilt world economy evolved.

Although the world economy was rebuilt anew after the devastation of the Second World War such rebuilding did not take place on completely fresh, unbroken ground. The postwar economic 'architecture' did, indeed, point to a new world economic order but it was an order containing many traces of what had gone before. It is from this historical baseline, therefore, that recent global shifts in economic activity will be examined in the following two chapters. As these chapters will demonstrate, today's world is far more complex. The simple notion of core and periphery no longer captures the intricacy and variability of the global economy; the flows of international direct investment are increasingly diverse and interpenetrate in extremely complex ways.

Notes for further reading

1. For a superb historical survey of the evolution of the global economy, see Braudel's three-volume work (1984). Immanuel Wallerstein (1979) suggests a particular theoretical perspective on the evolution of a world system, based on the concepts of core, semi-periphery and periphery. A specifically historical–geographical perspective in this mould is provided by Knox and Agnew (1989). Bairoch (1982) presents a detailed quantitative survey of international industrialization between 1750 and 1980.
2. A broad survey of the evolution of international direct investment from the late nineteenth century is provided by Dunning (1983). Historical studies of the growth and spread of TNCs from major source countries are provided by Wilkins (1970, 1974), Vaupel and Curhan (1969) and Vernon (1971) for the United States; Stopford and Turner (1985) for the United Kingdom; Franko (1976) for continental Europe.
3. Excellent accounts of the politics of international economic relations in the postwar period are provided by Spero (1990) and Kennedy (1987) and of the international economy by Scammell (1980). MacBean and Snowden (1981) provide details of the major international financial and trade institutions which underpin the operation of the contemporary world economy.
4. The International Monetary Fund's primary purpose was to encourage international monetary co-operation among nations through a set of rules for world payments and currencies. Each member nation contributes to the fund (a quota) and voting rights are proportional to the size

of a nation's quota. A major function of the IMF has been to aid member states in temporary balance of payments difficulties. A country can obtain foreign exchange from the IMF in return for its own currency which is deposited with the IMF. A condition of such aid is IMF supervision or advice on the necessary corrective policies. The World Bank's role, as its full name suggests, is to facilitate development through capital investment. Its initial focus was Europe in the immediate postwar period. Subsequently, its attention shifted to the less developed economies.

Chapter 2

The Changing Global Map: Trends in Production and Trade

The roller-coaster: aggregate trends in global economic activity

The path of economic change is far from smooth; it is more like a roller-coaster. Sometimes the ride is relatively gentle with just minor ups and downs; at other times the ride is truly stomach-wrenching, with steep upward gradients being separated by vertiginous descents to what seem like bottomless depths. The experience of the world economy during the past four decades clearly bears this out.

The years immediately following 1945 were obviously ones of basic reconstruction of war-damaged economies throughout the world. It was to be expected that there would be considerable growth of production and trade during the 1950s as the world economy caught up after the deep recession of the 1930s and after the war itself. At the time it was felt that growth rates would then slacken in the 1960s. Not only did such slackening not occur but rates of growth reached unprecedented levels. Between 1948 and 1953 world trade increased at an average annual rate of 6.7 per cent; between 1958 and 1963 the rate had risen to 7.4 per cent, and between 1963 and 1968 it had accelerated further to 8.6 per cent.

Such growth rates grossly exceeded any previously experienced. A particularly important feature of the postwar period was that *trade increased more rapidly than production*, a clear indicator of the *increased internationalization* of economic activities and of the greater *interconnectedness* which have come to characterize the world economy. As Figure 2.1 shows, by 1988 total world exports were more than four times greater than in 1960; total world output was a little under three times greater than in 1960.

The period between the early 1950s and the early 1970s was one of almost continual growth in world production and trade, with only minor and short-lived interruptions. People began to believe that the roller-coaster days were over. Growth, in the Western industrialized economies at least, came to be an expectation rather than a hope even though, as we shall see, such growth was extremely variable from one country to another. The early 1970s delivered a severe shock; the long boom seemed to 'go bust', the 'golden age of growth' seemed, very suddenly, to become tarnished. Figure 2.2 shows that the roller-coaster was far from extinct. Growth rates of both production and trade declined with each successive decade.

The popular view is that the long economic boom was halted by the OPEC decision in 1973 to raise oil prices by 400 per cent. Certainly this was a massive shock to the system, but with hindsight it is possible to see that, rather than the entire picture, the oil price shock was merely the final piece in a jigsaw puzzle which had been taking shape since the 1960s. From about 1968 a number of important changes had been occurring which ultimately came together to throw the world economy into reverse. Commodity prices, other than oil, had been rising steeply. Labour costs in all the

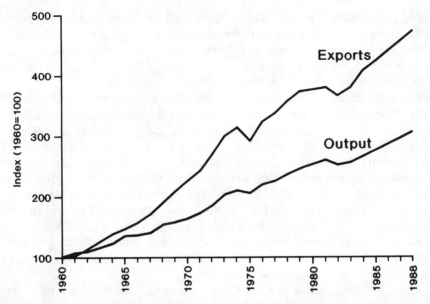

Figure 2.1 The growing interconnectedness of the world economy: the widening gap between exports and production (*Source*: GATT, *International Trade*, various issues)

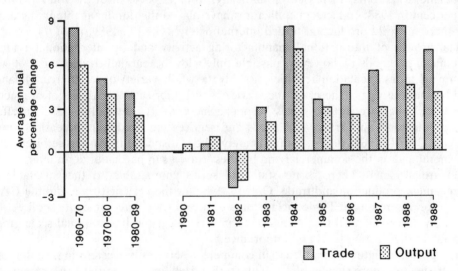

Figure 2.2 The roller-coaster of world merchandise production and trade, 1960–88 (*Source*: GATT (1990) *International Trade 1989–1990*, vol. II, Chart I.1)

industrialized nations had begun to accelerate as the level of wage settlements rose. The international monetary system created at Bretton Woods in 1944 became increasingly unstable as national currencies became more and more out of line with the fixed exchange rates. In 1971 the United States moved to a floating exchange rate, other countries followed and the Bretton Woods system was no more.

Thus the OPEC oil price increase was one – albeit the most spectacular – of a number of circumstances which came to afflict the world economy by the early 1970s. Arguably it was the biggest single factor in creating rampant price inflation, largely because so much of the world economic system had come to depend on oil as a cheap energy resource. A big rise in oil prices affected almost every product and service. But to attribute the world recession to a single cause is misleading.

Initially, some regarded the problems of the 1970s as being the result of 'an unusual bunching of unfortunate circumstances'. But as the recession deepened during the second half of the 1970s and into the early 1980s emphasis shifted towards the view that the changes were the result of more deep-seated and fundamental processes. A good deal of attention came to be focused on what is, in fact, a very old idea: that economic activity proceeds in a series of *long waves* each of which lasts for approximately fifty years.[1] During the 1980s, rates of growth were extremely variable, ranging from the negative growth rates of 1982 through to two years (1984 and 1988) when growth of world merchandise trade reached the levels of the 1960s once again. Overall, growth – albeit uneven growth – reappeared. However, the unevenness of such growth seemed to reflect the continuing difficulties of a world economy struggling to rediscover what had appeared to be a virtuous circle of growth.

The major economic sector driving the upsurge in both international production and especially international trade has been manufacturing industry. As Figure 2.3 shows, world trade in manufactures has generally outperformed world production in manufactures throughout the period from the 1960s. The share of manufactures in world merchandise trade has increased dramatically, from 52 per cent of the total in 1963 to 73 per cent in 1988. In fact, 'manufactures have played the dominant role in increasing the share of world production traded internationally' (GATT, 1989, p. 8).

The growth of international manufacturing activity and of international trade in manufactured goods is, however, possible only with the parallel development of what Figure 1.1 terms circulation activities, notably the whole variety of commercial, financial and business services. These are the services which lubricate the wheels of production and trade. They are in themselves becoming increasingly internationalized. In effect, 'the production – and trade – of goods and services are becoming increasingly interlinked' (GATT, 1989, p. 23). Unfortunately, the statistical data on the services sector in general and on the commercial and business services in particular at an international scale are abysmal. There is no statistical series comparable to that available for merchandise production and trade. Only in 1989, for the very first time, did the GATT Report on International Trade include data on world trade in commercial services. It is generally agreed that all the official data on services at the international scale grossly understate their actual level and importance.[2]

The growth of international trade in commercial services is accelerating. In the 1970s such trade grew more slowly than manufacturing trade but during the 1980s commercial services trade increased more rapidly. Table 2.1 shows that within the commercial services category it was the category 'Other private services' which experienced the greatest increase in share between 1970 and 1987. This oddly labelled category in fact includes virtually all the major business services (e.g. communications including telecommunications, financial services including banking, management, advertising, professional and technical services) and the ones of greatest significance in today's global economy. In 1970 this group of commercial services constituted 24 per cent of the total; by 1987 its share had grown to 40 per cent.

The growth of the financial sector has been especially marked:

World trade in goods is larger, much larger, than it has ever been before. And so is the 'invisible trade', the trade in services. Together, the two amount to around $2.5 trillion to $3 trillion a year. But the London Eurodollar market, in which the world's financial institutions borrow from and lend to each other, turns over $300 billion each working day, or $75 trillion a year, a volume at least twenty-five times that of world trade. In addition, there are the foreign exchange transactions in the world's main money centres, in which one currency is traded against another. These run around $150 billion a day or about $35 trillion a year – twelve times the world-wide trade in goods and services.

(Drucker, 1986, pp. 782–3)

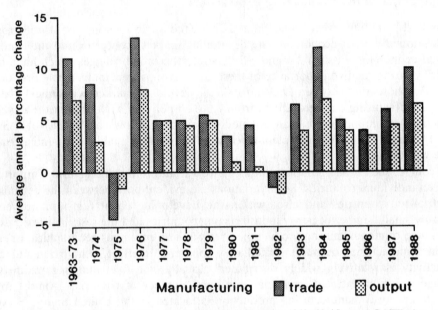

Figure 2.3 Growth of world manufacturing production and trade, 1963–88 (*Source*: GATT, *International Trade*, various issues)

Table 2.1 The changing composition of international trade in commercial services, 1970–87

	Percentage share in exports of commercial services		Average annual change in value (per cent)		
	1970	1987	1970–9	1980–7	1970–87
Shipment	22	13	15.0	1.5	9.5
Passenger services	5	6	19.0	7.5	14.0
Port services	12	11	21.5	0.0	12.0
Travel	29	30	18.0	6.5	13.0
Other private services and income	24	40	20.5	6.5	14.5
Commercial services	100	100	19.0	5.0	13.0

(*Source*: GATT (1989) *International Trade 1988–1989*, Table 21)

The changing global map of economic activity

These aggregate figures on international production and international trade are important in their own right as indicators of a changing global economy and, especially, of its increased interdependencies. But aggregate figures at the global scale mask very significant geographical variations. Changes in the production of goods and services and the trade in products and services have varied enormously from place to place on the world economic map. The globalization of economic activity is extremely uneven in both time and space; the geography of global economic change is exceptionally complex. In trying to unravel this complexity let us first look at production.

Global shifts in manufacturing production

Let us start by taking a very broad cut and divide the world economy into three groups: the developed market economies, the developing market economies and the centrally planned economies. Figures 2.4 and 2.5 illustrate the broad changes which have been occurring in the relative importance of these three groups as manufacturing producers. Figure 2.4 shows that between 1953 and 1985 the developed marked economies' share of world manufacturing output declined from 72 to 64 per cent while that of the developing market economies more than doubled, albeit from a low base level, to 11.3 per cent. The share of the centrally planned economies remained stable at around a quarter of the world total.

Of course, such broad grouping masks major variations in manufacturing importance between individual countries. Such variations are evident both between the established industrialized countries and also, within the developing country group, between a relatively small number of newly industrializing countries and the rest. Figure 2.5 is the global map of manufacturing production for 1986. The enormous geographical unevenness in manufacturing output is immediately apparent. In effect, although manufacturing activity is relatively widely distributed globally, the overwhelming majority of production is concentrated in a relatively small number of countries. Almost three-quarters of world manufacturing production is located in the United States, Western Europe and Japan. Nevertheless, several developing countries have recently emerged as important centres of manufacturing production.

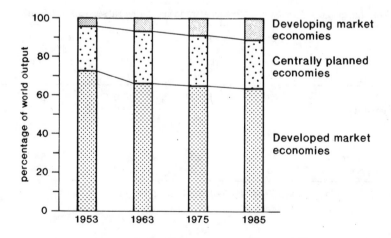

Figure 2.4 The changing distribution of world manufacturing output by major economic group, 1953–85 (*Source*: based on UNIDO (1986) World industry: a statistical review, 1985, *Industry and Development*, Vol. 18, Figure I)

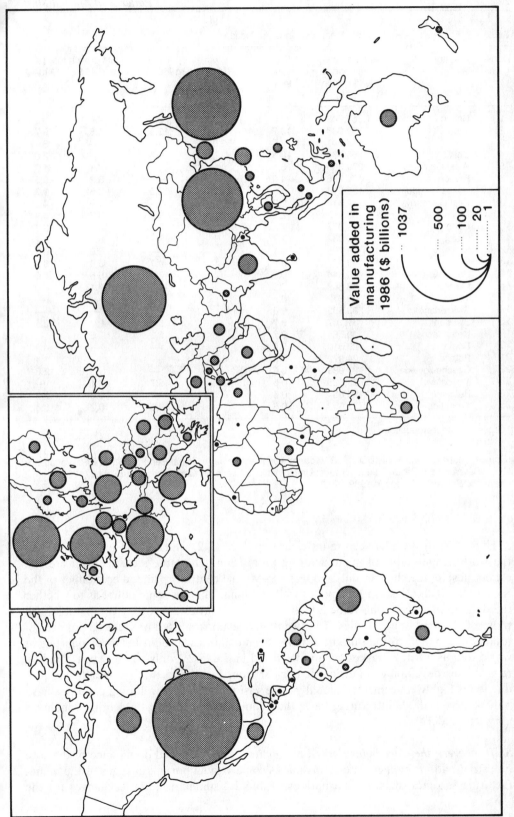

Value added in
manufacturing
1986 ($ billions)

1037
500
100
20
1

Figure 2.5 The global map of manufacturing production, 1986 (*Source:* UNIDO (1988) *Industry and Development: Global Report, 1988/1989,* UNIDO, Vienna)

Table 2.2 The world 'league table' of manufacturing production

Rank	Country	Manufacturing value added (US$ million 1986)	% of world total	Cumulative (%)	Average annual growth rate (%) 1960–70	1970–81	1980–7
1	United States	1,037,243	24.0		5.3	2.9	3.9
2	Japan	591,038	13.7		13.6	6.5	6.7
3	USSR	516,741	12.2	50.0	—	—	3.7*
4	China	456,434	10.5		—	—	12.6
5	West Germany	279,365	6.5		5.4	2.1	1.0
6	France	174,286	4.0		7.8	3.2	−0.5
7	United Kingdom	152,214	3.5	74.6	3.3	−0.5	1.3
8	East Germany	94,813	2.2		—	—	4.0*
9	Italy	93,512	2.2		8.0	3.7	0.9
10	Canada	79,077	1.8		6.8	3.2	3.6
11	Brazil	74,032	1.7		—	8.7	1.2
12	Spain	48,795	1.1		—	6.0	0.4
13	India	38,311	0.9		4.7	5.0	8.3
14	South Korea	36,644	0.9		17.6	15.6	10.6
15	Mexico	33,869	0.8		10.1	7.1	0.0
16	Taiwan	33,812	0.8		—	—	—
17	Switzerland	33,692	0.8		—	—	—
18	Sweden	31,107	0.7		5.9	0.7	2.5
19	Netherlands	30,564	0.7		6.6	2.6	—
20	Romania	26,810	0.6		—	—	4.8*
21	Poland	24,972	0.6	90.4	—	—	1.4*
22	Czechoslovakia	23,983	0.6		—	—	2.8
23	Yugoslavia	23,825	0.6		5.7	7.1	—
24	Belgium	23,654	0.6		6.2	3.0	2.3
25	Argentina	23,533	0.6		5.6	0.7	0.0

* Gross industrial production
— No data

(*Source*: based on data in UNIDO (1988) *Industry and Development: Global Report 1988–1989*; World Bank *World Development Report*, various issues; United Nations Economic Commission for Europe (1990) *Economic Survey of Europe in 1989/1990*)

These features can be seen more clearly in Table 2.2, which is the upper section of the world league table of manufacturing production. The table includes what are still called, despite the changes initiated in 1989–90, the centrally planned economies of the Soviet Union and Eastern Europe, as well as China. There is some doubt as to whether the data on manufacturing value added for these countries are strictly comparable with those of the market economies. The twenty-five countries listed produce 93 per cent of total world manufactured output; a mere seven countries account for three-quarters of world output. Of the market economies listed in Table 2.2 only six or seven could be regarded as developing countries. It is clear from Figure 2.5 that manufacturing production is highly concentrated globally. The vast majority of developing countries have only a very small manufacturing base; the 'manufacturing tail' of the world economy is very long indeed.

Shifts between the older industrialized economies   The continued dominance of the core industrial nations in aggregate terms masks some very important changes in the manufacturing fortunes of individual nations. Table 2.3 summarizes these changes for the

Table 2.3 Shifts between the older industrialized countries: changes in manufacturing production

	Rank*		Share of world manufacturing output (%)*		Average annual rates of change (%)		
	1963	1987	1963	1987	1960–70	1970–81	1980–7
United States	1	1	40.3	24.0	5.3	2.9	3.9
Japan	5	2	5.5	19.4	13.6	6.5	6.7
West Germany	2	3	9.7	10.1	5.4	2.1	1.0
France	4	4	6.3	5.4	7.8	3.2	−0.5
United Kingdom	3	5	6.5	3.3	3.3	−0.5	1.3
Italy	6	6	3.4	2.2	8.0	3.7	0.9
Canada	7	7	3.0	1.8	6.8	3.2	3.6

* Market economies only

(*Source*: World Bank, *World Development Report*, various issues; UNIDO, *Industry and Development: Global Report*, various issues)

seven leading manufacturing nations between 1963 and 1987. Three major elements of change are apparent:

1. The substantial decline in the United States' share of world manufacturing production: in 1963 it produced 40 per cent of world output; in 1987 its share had declined to 24 per cent. Its annual growth rate has fluctuated substantially but with some recovery in the 1980s.
2. The uneven manufacturing performance of the Western European economies: here the decline of the United Kingdom as a leading manufacturing nation continued; its share of world manufacturing production falling to 3.3 per cent and its world rank to fifth among the market economies and seventh overall. During both the 1960s and 1970s the United Kingdom's manufacturing performance was substantially poorer than its major competitors although, like the United States, there were some signs of recovery in the later 1980s. The other leading European manufacturing nations experienced fluctuating fortunes, as Table 2.3 reveals. Even West Germany has not performed quite as convincingly in the 1980s as the earlier experience suggested.
3. By far the most spectacular manufacturing performance of all the major industrialized nations in the postwar period has been that of Japan. As the pioneer manufacturing nation, the United Kingdom, continued its long decline, a new sun arose above the horizon. In 1963 Japan ranked fifth in the world league table; by 1987 it had risen to second place on the basis of extremely high annual growth rates. During the 1960s, manufacturing growth in Japan averaged 13.6 per cent, two and a half times greater than that of the United States and four times greater than the United Kindom. Even though Japan's growth rates fell to half of that level in succeeding periods, such growth continued to be very much greater than that of all the other established industrialized countries. By 1987, therefore, Japan's share of world manufacturing production had increased from 5.5 per cent in 1963 to 19.4 per cent.

Thus, within the core industrial nations there has been a pronounced shift in the centre of gravity of manufacturing production. Although the United States retained its leadership, its dominance was much reduced and its position increasingly challenged by the spectacular manufacturing growth of Japan. The differential manufacturing performance of the three major industrial regions during the 1980s is shown in Figure 2.6.

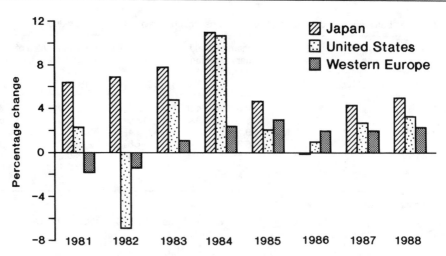

Figure 2.6 The differential manufacturing performance of the United States, Western Europe and Japan during the 1980s: annual changes in manufacturing value-added (*Source*: based on UNIDO (1987) *Industry and Development: Global Report, 1987*, Figures IV, V, VI)

Apart from 1986, when in the depths of the global recession Western Europe's manufacturing output grew more rapidly than that of either the United States or Japan, the region's performance *vis-à-vis* its two major competitors was poor. Even though Japan's growth rate has slackened since 1984, it remains substantial, and in 1988 it exceeded 4 per cent. Similarly, from a trough in 1982, the United States staged a dramatic recovery through the mid-1980s to expand to nearly 3 per cent in 1988. But in Western Europe growth was only 2 per cent in that year and throughout the 1980s scarcely exceeded that figure.

Shifts within the developing market economies   The developing market economies constitute an extremely varied group of countries but, for most of them, manufacturing remains relatively unimportant, as Figure 2.5 revealed. In fact, most of the rapid growth in manufacturing production has occurred in what the World Bank terms 'the middle income group' of developing countries.

Figure 2.7 shows the very substantial variations in manufacturing growth which occurred during the 1980s within the developing world. Although growth fluctuated over the period (the roller-coaster effect at a regional scale), the rates of manufacturing growth were highest of all in South East Asia and lowest in tropical Africa. Growth rates in Latin America and in the Indian subcontinent (both important manufacturing locations in absolute terms) were between these two extremes but with variability in manufacturing growth being much greater in Latin America. In fact only a relatively small number of developing countries – the so-called newly industrializing countries (NICs) – can be regarded as significant centres of manufacturing production on a world scale. The last few years have seen an enormous burgeoning of interest in the NICs, although precisely which countries are included in this category varies considerably. The number of nations identified as 'newly industrializing' varies from as few as eight to as many as twenty-three. Some writers exclude countries within Europe, others include several of the centrally planned economies of Eastern Europe (for example, Hungary,

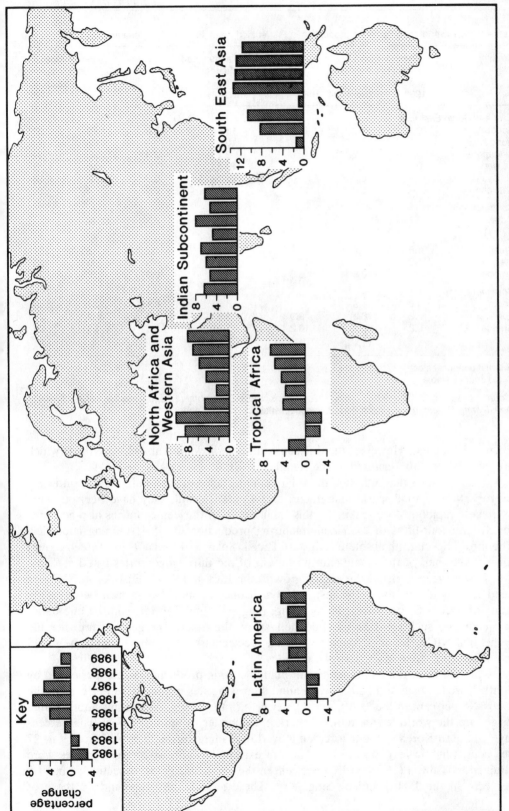

Figure 2.7 Growth rates in manufacturing in major regions of the developing world (*Source:* based on UNIDO (1988) *Industry and Development: Global Report 1988/1989*, Figures X–XIV)

Table 2.4 Growth of manufacturing production in the leading newly industrializing countries

Country	Share of world manufacturing output (%)*		Average annual rate of change (%)			Percentage of labour force in manufacturing	
	1963	1987	1960–70	1970–81	1980–7	1960	1980
East and South East Asia							
South Korea	0.1	1.2	17.6	15.6	10.6	9	27
Hong Kong	0.1	0.3	—	10.1	—	52	57
Singapore	0.1	0.2	13.0	9.7	3.3	23	39
Taiwan	0.1	1.1	16.3	13.5	7.5	16	32
Malaysia	—	0.2	—	11.1	6.3	12	19
Thailand	—	0.3	11.4	10.3	6.0	4	10
Southern Asia							
India	—	1.2	4.7	5.0	8.3	—	—
Latin America							
Brazil	1.6	2.4	—	8.7	1.2	15	24
Mexico	1.0	1.1	10.1	7.1	0.0	20	26
Argentina	—	0.8	5.6	0.7	0.0	36	34
Southern Europe							
Spain	0.9	1.9	—	6.0	0.4	31	40
Portugal	0.2	0.2	8.9	4.5	—	29	37
Greece	0.2	0.2	10.2	5.5	0.0	20	29
		11.1					

* Market economies only
— No data available

(*Source*: based on OECD (1979), Table 1; World Bank, *World Development Report*, various issues; UNIDO, *Industry and Development: Global Report*, various issues; Li (1988), Appendix 1)

Poland, Romania). However, the thirteen countries shown in Table 2.4 are those which are most commonly regarded as the significant NICs.

Table 2.4 shows that although the relative importance of each individual country (as a proportion of total world manufacturing output) is small, they have been growing extremely rapidly. More strikingly, this small group of thirteen countries also accounts for around four-fifths of total manufacturing production in all developing countries. The most important in absolute terms are Brazil, Spain and India. These three account for virtually half of the manufacturing output of the thirteen countries listed in Table 2.4. However, in terms of the *rate* of growth, the East and South East Asian NICs are head and shoulders above the rest. Here a distinction should be drawn between the 'gang of four' – South Korea, Hong Kong, Singapore and Taiwan – on the one hand, and countries such as Malaysia and Thailand on the other. The gang of four are the archetypal NICs; the ones which have risen spectacularly to prominence on the world manufacturing scene in the last two decades.

By far the most dramatic increase in manufacturing production was experienced by South Korea, which attained average annual growth rates of almost 18 per cent in the 1960s, 16 per cent in the 1970s, and 11 per cent in the 1980s. As a result South Korea surged up the world league table of market economies to tenth position. The percentage of South Korea's labour force employed in manufacturing increased from 9 to 27 per cent. Slightly less spectacular, but still extremely impressive, growth rates were attained by Taiwan (16.3 and 13.5 per cent in the 1960s and 1970s respectively, and 7.5 per cent in the 1980s) and by Singapore. These growth rates, and indeed those of the other NICs shown in Table 2.4, should be contrasted with those of the leading developed market economies discussed earlier (Table 2.3). It is especially notable that in the East and South East Asian NICs manufacturing growth rates remained at a high level throughout the 1970s and 1980s whereas those of the leading developed market economies fell to half or less of their 1960s levels. The growth performance of the non-Asian NICs was far more variable and less impressive.

In terms of manufacturing production, therefore, it is clear that there has been a considerable acceleration of growth in the global periphery. But it is also clear that such manufacturing growth is very unevenly distributed. A small group of developing market economies has begun to make a real impact on the world manufacturing scene, adding further to global shifts in the manufacturing system. Although the core industrialized countries continue to dominate, manufacturing production is no longer exclusively a core activity. Whereas in the first quarter of this century 95 per cent of world manufacturing production was concentrated into only ten countries, by 1986 twenty-five countries were responsible for the same proportion of world output.

The changing fabric of trade in manufactures

A distinctive feature of the postwar period has been the much more rapid growth of trade in manufactures than of production (see Figure 2.1). Indeed, manufacturing has come to account for an increasingly greater proportion of total exports in both developed and developing market economies, as Figure 2.8 reveals. In the developed market economies, manufactured exports increased from 70 per cent of the total in 1960 to 77 per cent in 1988. However, the shift was particularly strong in the developing market economies, where manufacturing counted for only 19.8 per cent of total exports in 1960 but for as much as 47.1 per cent in 1988. In fact, by the end of the 1970s, for the very first time, the value of manufactured products exported from the developing market economies exceeded that of food and raw materials. After 1973 exports of manufactured goods from the Third World grew at twice the rate of exports of raw materials. Without doubt the old international division of labour had been displaced.

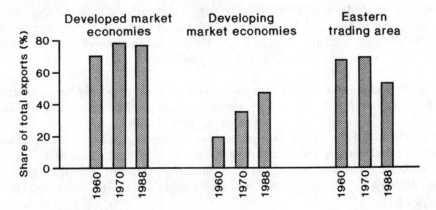

Figure 2.8 The share of manufactures in total exports by major economic group (*Source*: GATT, *International Trade*, various issues)

The changing geographical pattern of trade in manufactures is, not surprisingly, far more complicated than that of production, simply because trade consists of *flows* between areas. Theoretically every nation can trade with every other. In fact, trade flows in manufactures tend to be channelled into certain dominant routes. As in the case of production, the global pattern of trade in manufactures has changed a great deal – in some respects quite dramatically – during the past thirty years. In fact, it is through the lens of trade that global shifts in the world manufacturing system can be seen most clearly. The evolving pattern of trade in manufactures reflects the emergence of a truly *global* system of manufacturing.

Viewed in isolation international trade figures can be misleading. For example, other things being equal, international trade will be far more important relatively speaking for a small economy than for a larger one (an obvious example would be to contrast the United States with Singapore or Hong Kong). Similarly, although in general there is a relationship between manufacturing production and trade, the relationship is not exact. Some countries which are very significant as producers are less significant as exporters (again, the United States is an example as are Brazil and India). Figure 2.9 and Table 2.5 show the international pattern of manufacturing trade. Comparison with the production equivalents (Figure 2.5 and Table 2.2) reveals some interesting differences. First, the origins of manufactured exports are less concentrated than those of production. Eighty-three per cent of world manufactured exports originated from the top 25 countries compared with 93 per cent of manufacturing production. Second, no single nation dominates exports in manufactures to the extent that the United States dominates world manufacturing production. The most important exporter of manufactured goods in 1989 was the United States, with 11.8 per cent of the total (compared with its 24 per cent of manufacturing output). In fact several manufacturing nations occupy very different positions in the league tables of manufacturing production and manufacturing exports. For example, Hong Kong and Singapore are much higher in the export league table than in the production league table. Conversely, Brazil and India, both high-ranking producers of manufactures, are far less important as exporters of manufactures. Quite clearly, too, the economies of Eastern Europe and the Soviet Union are far less significant generators of manufactured exports than they are producers of manufactured goods.

Table 2.5 also reveals some interesting variations in the stability of the rankings over the ten-year period from 1978 to 1989. The top ten rankings were extremely stable with only very minor changes. In the lower part of the league table, however, there was very substantial change, with particularly large increases in the relative importance of the four East and South East Asian NICs, China and Mexico.

Shifts in the global network of trade in manufactures   It is not possible in a book of this kind to examine the fine detail of each nation's trading relationships in manufactured goods. Rather, it is more useful – and certainly more feasible – to examine trends both between and within the three major elements of the global economy: developed market economies, developing market economies and centrally planned economies. Within this framework attention can be drawn to particularly significant individual cases. The broad details of the changing flows of trade in manufactured goods are presented in Figure 2.10.[3]

The most striking feature of the network in 1963 and 1985 is the high degree of continuity and stability in the aggregate shares of exports of the three groups of

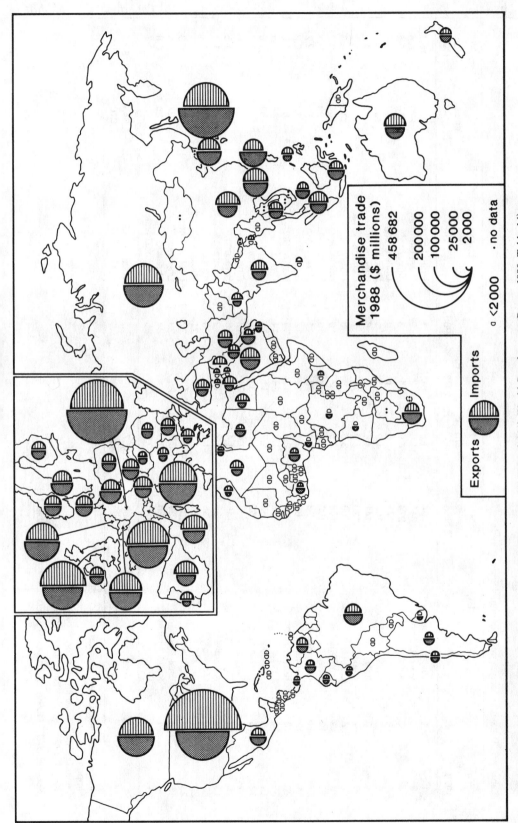

Figure 2.9 The global map of manufacturing trade (*Source:* World Bank (1990) *World Development Report 1990*, Table 14)

Merchandise trade
1988 ($ millions)

458 682
200000
100000
25000
2000

□ <2000   . no data

Exports   Imports

Table 2.5 The world 'league table' of manufacturing trade

Exports Rank 1978	Exports Rank 1989	Exports Country	Value ($ billion)	Share (%)	Imports Rank 1978	Imports Rank 1989	Imports Country	Value ($ billion)	Share (%)
1	1	United States	364	11.8	1	1	United States	493	15.4
2	2	West Germany	341	11.0	2	2	West Germany	270	8.4
3	3	Japan	274	8.9	4	3	Japan	210	6.5
4	4	France	179	5.8	5	4	United Kingdom	198	6.2
5	5	United Kingdom	152	4.9	3	5	France	193	6.0
6	6	Italy	141	4.6	6	6	Italy	153	4.8
9	7	Canada	120	3.9	10	7	Canada	120	3.7
7	8	USSR	109	3.5	8	8	USSR†	115	3.6
8	9	Netherlands	108	3.5	7	9	Netherlands	104	3.2
10	10	Belgium–Luxembourg	100	3.2	9	10	Belgium–Luxembourg	99	3.1
27	11	Hong Kong*	73	2.4	23	11	Hong Kong‡	72	2.2
21	12	Taiwan	66	2.1	12	12	Spain	72	2.2
20	13	Korea, Rep.	62	2.0	19	13	Korea, Rep.	61	1.9
33	14	China	52	1.7	29	14	China	58	1.8
14	15	Sweden	52	1.7	11	15	Switzerland	58	1.8
12	16	Switzerland	52	1.7	30	16	Taiwan	53	1.7
31	17	Singapore*	45	1.4	24	17	Singapore‡	50	1.6
18	18	Spain	45	1.4	12	18	Sweden	49	1.5
15	19	Australia	38	1.2	16	19	Australia	45	1.4
40	20	Mexico§	36	1.2	15	20	Austria	39	1.2
22	21	Brazil	34	1.1	34	21	Mexico§	34	1.1
23	22	Austria	32	1.0	17	22	East Germany†	27	0.8
16	23	East Germany	29	0.9	20	23	Denmark	27	0.8
—	24	Saudi Arabia	28	0.9	—	24	Thailand	25	0.8
24	25	Denmark	28	0.9	26	25	Czechoslovakia†	22	0.7
		Total	2,560	82.7			Total	2,645	82.4
		World	3,095	100.0			World	3,210	100.0

* Includes substantial re-exports.
† Imports f.o.b.
‡ Includes substantial imports for re-export.
§ Includes estimates of trade flows through processing zones.

(*Source:* GATT (1990) *International Trade 1989–1990*, Table 11)

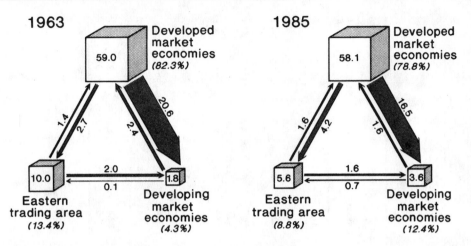

Figure 2.10 The network of world trade in manufactured goods, 1963 and 1985 (*Source*: GATT, *International Trade, 1966*, Table 26; *International Trade 1985–1986*, Table 24)

countries. Again, however, we need to look below these aggregates to see the finer details. When we do this we find evidence of substantial change, particularly in the rapid growth of manufactured exports from some developing market economies and also in the increased involvement of the Eastern Trading Area in the global trading system.[4]

Continued dominance of the developed market economies   Four-fifths of all world exports of manufactures were generated by the core economies of the global system in both 1963 and 1985. Roughly two-thirds of this trade is conducted within the core itself as trade between the developed market economies. Expressed in a different way, approximately half of total world trade in manufactures consists of trade between the core industrial economies. In fact, during the 1950s and the early 1960s it was largely the dynamic nature of trade between the developed market economies which drove the global manufacturing system. Since then, however, the dynamic has shifted.

The United States is the leading exporter of manufactures, followed by West Germany and Japan (Table 2.5). But their individual export performance differed substantially. Between 1963 and 1989 West Germany's share of the total fell from around 15 to 11 per cent. The United States' share fell from 17 per cent in 1963 to 12 per cent in 1989. The share of the pioneer manufacturing exporter, the United Kingdom, declined dramatically from 11 per cent of the world total in 1963 to around 5 per cent in 1989. The 'star performer' among the developed market economies was undoubtedly Japan: in 1963 it generated less than 6 per cent of total world exports; by 1989 its share had risen to 9 per cent.

At least as important as these aggregate figures are the more specific trading relationships between individual countries and groups of countries. Table 2.6 shows the major elements of the trade network in manufactures for the leading industrialized countries in 1973 and 1989. The table is read horizontally. For example, in 1989, 23.8 per cent of total US manufacturing exports went to Canada while 16.1 per cent of total manufacturing imports came from Canada. Unfortunately, data for 1973 were not available for

Table 2.6 The network of manufacturing trade of the leading industrialized countries, 1973 and 1989

		United States		Canada		Japan		Western Europe		Developing countries		Eastern Trading Area		Total trade balance in manufacturing	
		Exports	Imports	Exports	Imports	Exports	Imports	Exports	Imports	Exports	Imports	Exports	Imports	$ billions	% of exports
United States	1973			28.8	24.1	6.9	21.4	28.9	36.6	29.6	17.4	1.4	0.5	+0.19	0.4
	1989			23.8	16.1	9.3	25.9	29.4	23.4	31.8	30.8	1.9	3.4	−101.77	38.5
Canada	1973	82.6	76.8			0.8	5.3	8.6	14.0	6.2	2.9	0.2	0.8	−6.22	51.7
	1989	86.0	67.8			1.1	8.5	6.2	12.4	5.2	9.0	0.7	1.3	−23.56	34.1
West Germany	1973														
	1989	7.8	7.1	0.9	0.3	2.6	9.2	72.6	71.1	9.7	8.7	4.6	3.3	+114.52	37.9
France	1973														
	1989	7.5	8.9	1.0	0.4	2.0	5.6	67.2	73.3	18.1	7.3	2.9	2.1	−8.39	6.4
United Kingdom	1973														
	1989	13.5	11.6	2.0	1.1	2.4	7.9	59.1	67.3	17.9	10.2	1.9	1.4	−29.10	24.0
Italy	1973														
	1989	8.8	6.0	1.1	0.3	2.4	3.9	68.1	76.2	13.5	6.8	4.2	2.5	+30.96	24.8
EC	1973	8.1	8.9	1.3	0.7	1.4	3.2	67.1	79.6	14.2	4.4	4.6	2.6	+41.46	25.6
	1989	8.2		1.0		2.1		70.7		12.5		3.3		+78.56	8.8
Japan	1973	26.0	34.9	2.8	1.1			18.0	33.4	36.1	22.6	5.6	5.2	+24.70	71.4
	1989	34.5	28.7	2.5	1.1			20.9	30.5	32.4	31.5	4.5	7.0	+176.07	66.6

— Data not available

(*Source:* GATT (1981) *International Trade 1980–1981*; (1989) *International Trade 1988–1989*)

the individual European countries. Nevertheless, Table 2.6 reveals some significant relationships and trends.

One is the strength of regional relations in manufactured trade revealed in both North America and Western Europe. In the case of North America, Canada's immensely close ties to the US market increased on the export side but decreased slightly on the import side. In both 1973 and 1989, however, Canada was very highly dependent on the US market. The converse is far less true for the United States: the manufacturing trade network in North America is highly asymmetrical. The other clear regional relationship shown in the table is in Western Europe where the intensity of intra-regional manufacturing trade flows is very high indeed.

A second major feature of Table 2.6 is the trading position of Japan, which has become a major focus of controversy in recent years. In relation to the United States, Japan increased in importance as an exporter of manufactured goods but the United States' share of Japanese manufactured imports declined. A similar trend occurred in Japan–Western European trade relationships although to a lesser extent. However, it is the final columns of Table 2.6 which capture the most significant difference between the United States and Japan: their respective manufacturing trade balances. In 1989 the United States had a manufacturing trade deficit of $102 billion, equivalent to 39 per cent of its total manufactured exports. In the same year Japan had a manufacturing trade surplus of $176 billion, equivalent to 67 per cent of its total manufactured exports. In fact, in relative terms, Japan's position did not change greatly whereas the United States moved from a small surplus in 1973 to a massive deficit in 1989.

Third, Table 2.6 shows the relations between the leading industrialized countries and developing countries. Here the major contrast is between the United States and Japan on the one hand and Western Europe on the other. Both the United States and Japan are far more heavily engaged in manufacturing trade with the developing country group as a whole than is Western Europe. In 1989, 32.4 per cent of Japanese manufactured exports went to developing countries while 31.5 per cent of Japan's manufactured imports were from the developing countries. Between 1978 and 1989 developing countries declined as a destination for Japanese exports (from 36.1 to 32.4 per cent) but increased as a source of manufactured imports (from 22.6 to 31.5 per cent). In the United States' case, the importance of developing countries as manufacturing trade partners increased on both counts. The share of US manufactured exports going to developing countries increased from 29.6 to 31.8 per cent. But the importance of developing countries as a source of manufactured imports into the United States increased very dramatically indeed, from 17.4 to 30.8 per cent. This leads us on to shift the focus of discussion to the developing market economies themselves.

Acceleration of exports from the newly industrializing economies   Overall, the developing market economies increased their share of world exports in manufactures from 4.3 per cent in 1963 to 12.4 per cent by the mid-1980s. In fact, the developing market economies as a group have consistently outperformed both the developed market economies and the Eastern European economies in their rate of growth of manufactured exports. Table 2.7 shows that, for the period from 1970 to 1986, manufactured exports from the developing market economies grew at an average annual rate of 19.7 per cent compared with 13.3 per cent for the developed market economies. At the same time, the importance of manufactured exports as a proportion of total exports

Table 2.7 The growing importance of manufactured exports from developing economies

	Average annual growth 1970–86		Manufactured exports as a percentage of total exports		
	Manufactured exports	Total exports	1970	1980	1986
Developed market economies	13.3	12.5	65.8	67.2	73.1
Developing market economies	19.7	13.8	24.7	37.2	55.0
South and East Asia	21.4	17.7	41.8	50.5	68.8
Central and South America	17.3	10.7	9.5	15.7	24.0
Africa	12.1	9.0	6.6	5.6	10.3
West Asia	20.7	13.4	4.3	3.4	11.6

(*Source*: based on UNCTAD (1989) *Handbook of International Trade and Development, 1988*, Table 3.2)

from developing countries increased dramatically, from 25 per cent in 1970 to 55 per cent in 1986.

However, such increases in the growth of manufactured exports and in the enhanced importance of manufactured exports in the total export package varied substantially between different parts of the developing world. The leading region within the developing world was undoubtedly East and South East Asia, where manufactured exports increased at an annual rate of about 22 per cent and where manufactured exports grew to almost 70 per cent of total exports by 1986. In West Asia and in developing Africa, manufacturing remains a very small component of total exports. Central and South America's position lies between these two extremes. In fact, within these groups there are substantial variations in the significance of manufactured exports. For the leading NICs in particular, manufactured exports can account for between two-thirds and nine-tenths of individual country exports.

The changing geographical destinations of developing country manufactures are shown in Table 2.8. Two features are especially evident. One is the decline in importance of the developing country group as a destination for developing country exports. For the group as a whole, the share of total manufactured exports with an intra-group destination fell from 31.8 per cent in 1970 to 24.3 per cent in 1986. Other developing countries were even less important as a market for manufactured exports from developing countries in East and South East Asia. A logical corollary is that a larger proportion of developing country manufactured exports went to developed market economies. Within the developing market economies, however, it was the United States which became overwhelmingly the key destination for manufactured exports. For the developing country group as a whole, the United States accounted for 29 per cent in 1970 but for 38 per cent in 1986. The increased importance of the United States was even greater for exports from Central and South America. Western Europe's importance as a market declined slightly and Japan's grew but from a very low base level. This concentration of manufactured exports on the United States is a point to which we shall return.

Table 2.8 Geographical destinations of manufactured exports from developing countries

		Source of exports Percentage of total		
		Developing economies	South and East Asia	Central and South America
Destination of exports				
Developing economies	1970	31.8	28.0	42.8
	1986	24.3	23.6	29.4
Developed market economies	1970	60.6	66.7	55.6
	1986	67.3	69.9	67.4
Western Europe	1970	23.1	18.6	19.0
	1986	18.9	17.5	12.0
United States	1970	28.6	35.6	32.3
	1986	38.1	40.2	50.1
Japan	1970	4.1	6.0	1.4
	1986	5.5	6.8	2.2
Eastern Europe/USSR	1970	7.9	3.8	1.2
	1986	3.5	0.7	1.4

(*Source*: based on UNCTAD (1989) *Handbook of International Trade and Development, 1988*, Table 3.2)

The growing significance of the East and South East Asian NICs In our earlier discussion of global shifts in manufacturing production the point was made that only a relatively small number of developing countries has been involved. These NICs, particularly those in East and South East Asia, have grown very impressively as manufacturing producers but their share of total world output is still very modest (see Table 2.4). It is their dramatic increase in importance as exporters of manufactures that has made the NICs so remarkable. Table 2.5 showed one aspect of this: the surge of several NICs up the world league table of exporters. We now need to look at this phenomenon a little more closely.

In the global reorganization of manufacturing trade the increased importance of Asia as an exporter of manufactures is unique in its magnitude. Asian countries generated more than three-quarters of all Third World manufactured exports compared with only around a third in the 1950s. Thus, Asia increasingly dominates manufacturing trade both to other developing market economies and also to the developed market economies. Such dominance is reflected in the spectacular growth of the four leading Asian NICs as exporters of manufactures. Indeed, the 'gang of four' – Hong Kong, Singapore, South Korea and Taiwan – stand out from the other NICs from Southern Europe and Latin America as exporters of manufactures. Their significance as exporters is much greater than their share of manufacturing production would suggest. The third column of Table 2.9 reflects this difference in terms of the ratio of each country's share of world manufactured exports to its share of world production. A value of 1.0 would indicate that the two shares are equal. The higher the ratio the more important the country is as an exporter of manufactures than as a producer and vice versa.

As Table 2.9 indicates, each of the four Asian NICs had high ratios. Conversely, the non-Asian NICs were far more significant as producers than as exporters, at least in

Table 2.9 Growth of manufactured exports from leading newly industrializing countries

Country	Share of world manufactured exports (%)		Ratio of share of world exports:share of world production	Average annual rate of growth of manufactured exports (%)		
	1963	1988	1988	1965–80	1980–6	1986–7
East and South East Asia						
South Korea	0.01	2.1	1.8	27.3	13.1	36.2
Hong Kong	0.80	2.2	7.3	9.5	10.7	36.8
Singapore	0.40	1.4	7.0	4.8	6.1	27.5
Taiwan	0.20	2.1	1.9			34.6
	1.41	7.8				
Malaysia	—	—		4.4	10.2	29.2
Thailand	—	—		8.5	9.2	32.7
Southern Asia						
India	—	0.5	0.4	3.7	3.8	19.4
Latin America						
Brazil	0.05	1.2	0.5	9.4	4.3	17.3
Mexico	0.20	1.1	1.0	7.7	7.7	28.6
	0.25	2.3				
Southern Europe						
Spain	0.30	1.4	0.9	14.2	6.4	24.9
Portugal	0.30	—	—	3.4	11.0	26.5
Greece	0.04	—	—	12.0	4.6	5.3
	0.64					

— No data available

(*Source*: World Bank, *World Development Report*, various issues; GATT *International Trade*, various issues)

relative terms. The difference between export and production is especially marked in the case of the 'city states' of Hong Kong and Singapore. Table 2.9 also shows the extent to which the Asian four increasingly dominated total manufacturing exports from the leading NICs. Their combined share grew from 1.4 per cent of world exports in 1963 to 7.8 per cent in 1988. In fact, these four relatively small countries in East and South East Asia account for two-thirds of the total manufactured exports from all developing countries, a remarkable degree of concentration. Such increased importance was a reflection of their very high annual rates of growth in manufacturing exports; in particular the relative importance of both South Korea and Taiwan as manufacturing exporters has grown enormously.

A particularly striking characteristic of the manufactured exports of the four leading Asian NICs is their high degree of geographical concentration. Table 2.10 shows the main features of their trade network in 1989. A number of points are worthy of mention. First, all four NICs direct a roughly similar proportion of their manufactured exports to Western Europe (between 15 and 19 per cent). Second, apart from Singapore, a similar proportion of their manufactured exports goes to other developing countries. Third, the proportion of their exports going to Japan is relatively small, apart from the case of South Korea. Fourth, by far the most important destination of these countries' manufactured exports is the United States, ranging from 27 per cent of Hong Kong's exports to 39 per cent of Taiwan's. Fifth, the positions of the United States and Japan are reversed in the case of imports. The United States is a less

Table 2.10 The network of manufacturing trade of the four leading Asian NICs, 1989

	Hong Kong	South Korea	Singapore	Taiwan
United States				
Exports	27.0	35.2	30.5	38.6
Imports	7.5	25.6	19.3	21.7
Canada				
Exports	2.1	3.2	1.1	2.8
Imports	0.4	1.5	0.5	1.3
Japan				
Exports	5.6	18.6	5.2	10.2
Imports	18.3	41.6	28.5	42.5
Western Europe				
Exports	19.2	14.9	17.9	17.6
Imports	12.9	17.1	19.0	19.3
Developing Countries				
Exports	18.2	22.3	40.1	27.3
Imports	24.2	10.4	29.2	13.9
Eastern Trading Area				
Exports	25.3	—	2.0	0.2
Imports	35.9	—	2.4	0.7
Total trade balance in manufacturing				
$ billion	+5.81	+18.28	−4.53	+26.10
% of exports	8.7	31.6	14.2	42.7

— No data available

(*Source*: GATT (1990) *International Trade, 1989–1990*)

important source of manufactured imports into the four Asian NICs. In contrast, Japan is overwhelmingly the dominant source of such imports, particularly for South Korea and Taiwan, where more than 40 per cent of total manufactured imports originate from Japan. Sixth, the Eastern Trading Area is important as a trading partner for Hong Kong, and almost all of this is accounted for by trade with China. In fact, Taiwan also has substantial unofficial trade with China but this does not fully appear in the official GATT statistics. Lastly, three of the four Asian NICs have a substantial positive trade balance; only Singapore has a deficit in its manufactured trade.

NIC import penetration of industrialized country markets This discussion of the very high geographical concentration of manufactured exports from the East and South East Asian NICs leads us to a broader consideration of the extent of import penetration of developed country markets. At the global scale one of the most controversial aspects of the spectacular export growth of NICs in general and of the Asian NICs in particular is their growing penetration of the domestic markets of the developed market economies as well as their success in competing with industrialized nations elsewhere. In Chapter 13 we shall look at the economic and social implications of such increased penetration, particularly for employment in the developed market economies. At this point our concern is with the scale and nature of such import penetration.

Figure 2.11 shows the marked changes which occurred between 1964 and 1985 in the trade between the NICs on the one hand and the developed market economies as a whole (the OECD category in Figure 2.11) and the United States.[5] Until the early

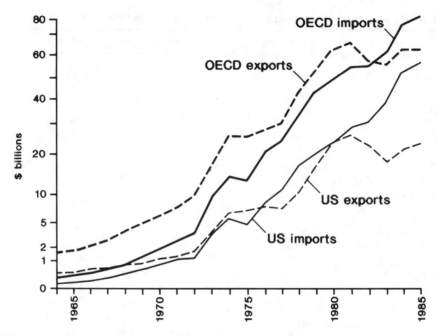

Figure 2.11 The shifting balance of manufactured trade between the NICs and the developed market economies (*Source*: OECD (1988), Figure 1)

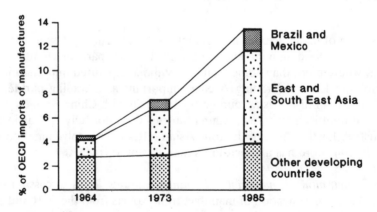

Figure 2.12 Shares of newly industrializing countries in manufactured imports to OECD countries, 1964–85 (*Source*: OECD (1988), Table A.2.1)

1980s the developed economies as a group had a trade surplus with the NICs. After 1982, however, the surplus was transformed into a deficit. Most of the explanation for this lies in the massive increase in exports to the United States from the NICs and the fall in United States exports to the NICs. Figure 2.12 reveals the dominance of the four East and South East Asian NICs in the growth of import penetration of the developed market economies between 1964 and 1985. The developing economies as a whole increased their share of manufactured exports into the OECD countries from 4.4 to

Table 2.11 NIC import penetration of leading developed market economies, 1964 and 1985

	Percentage of total NIC-manufactured imports to OECD countries		NIC percentage of country's total manufactured imports			
	1964	1985	1964	% from Asian NICs	1985	% from Asian NICs
United States	43.5	66.5	4.9	72.9	20.8	76.1
Japan	3.2	7.3	1.6	72.4	18.1	90.1
Australia	2.7	2.4	1.2	100.0	10.2	91.3
Canada	3.6	4.5	0.7	91.0	5.7	77.7
United Kingdom	25.9	4.4	5.0	96.2	4.7	92.6
West Germany	8.1	5.1	1.3	86.5	4.5	84.4
Italy	2.0	1.8	0.6	83.3	3.5	61.3
Sweden	2.0	0.8	0.7	100.0	3.2	89.2
France	1.1	1.9	0.2	60.0	2.4	87.5
Percentage of OECD total	92.1	94.7				

(*Source*: based on OECD (1988), Tables A.2.3–A.2.11)

13.4 per cent. The Asian NICs' share increased from 1.6 to 9.5 per cent. Thus, by 1985, 71 per cent of developing countries' manufactured imports to the OECD countries originated from just four East and South East Asian NICs. In comparison the combined share of Brazil and Mexico, even though they are much bigger industrial economies, was a mere 2 per cent in 1985.

Overall therefore six NICs had captured almost 10 per cent of all manufactured imports to the developed market economies by 1985. Although this was still a modest percentage the speed of growth from the early 1960s was truly spectacular. Between 1964 and 1973 OECD-manufactured imports from the six NICs increased by more than 30 per cent per year compared with an annual growth rate of 17 per cent for total manufactured imports into the OECD. Between 1978 and 1985, although the growth rates were lower, manufactured imports from the NICs still grew at double the annual rate of total imports (14 compared with 7.4 per cent).

The extent of import penetration was, however, extremely uneven between the industrialized countries (Table 2.11). In 1964, 44 per cent of NIC-manufactured imports into the OECD countries went to the United States and a further 26 per cent to the United Kingdom. By 1985 the picture had changed markedly. In that year 66.5 per cent of the NIC-manufactured imports to the OECD countries went to the United States alone and a further 7.3 per cent to Japan. The shares going to the four major European countries together totalled only 13.2 per cent. Looked at from a different perspective, roughly one-fifth of all manufactured imports into the United States and Japan in 1985 came from the NICs. The degree of NIC import penetration of the individual European economies was a good deal less than this, at between 2 and 5 per cent. Similarly, there was some variation in the relative importance of the four East and South East Asian NICs in each importing country's trade. They were especially important in the cases of Australia, the United Kingdom and Japan. In the case of the United States, imports from Brazil and Mexico are of considerable importance.

Thus, manufactured imports from NICs are strongly concentrated geographically into a small number of developed countries. Apart from the United States and Japan,

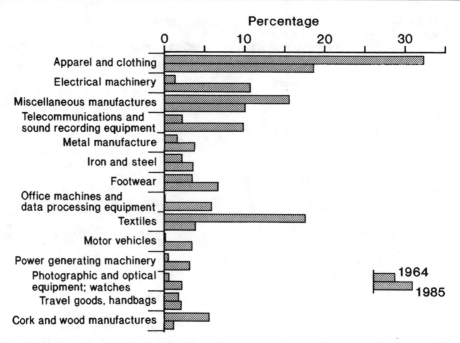

Figure 2.13 NIC imports into OECD countries by type of manufacturing industry, 1964 and 1985 (*Source*: OECD (1988), Table 2.2)

however, they still account for a relatively small share of individual countries' manufactured imports. A further significant characteristic of NIC-manufactured exports is their concentration into particular types of industry. However, as Figure 2.13 reveals, NIC-manufactured exports have not only become increasingly diverse but also there has been a clear shift in emphasis away from the traditional industries to more sophisticated sectors. In 1964, three sectors – apparel and clothing, miscellaneous manufactures and textiles – constituted 65 per cent of all NIC-manufactured imports into the developed market economies. By 1985 these same three sectors made up only 33 per cent of the total. Although apparel and clothing still dominate, with around 19 per cent of the total, this is a far cry from the 32 per cent share of twenty years earlier. Similarly, textiles imports declined as a share of the total, from 18 to 4 per cent. By the mid-1980s, in fact, it was such sectors as electrical machinery, telecommunications and sound recording equipment, office machines and data processing equipment and motor vehicles which had become particularly important. This would suggest that at least the leading NICs have moved substantially away from a heavy dependence on the low-skill, low-technology sectors.

A key characteristic of NIC exports, therefore, is their strongly selective nature, in terms of both geographical destination and sectoral concentration. Although the overall level of import penetration of the developed market economies is far from overwhelming, in some countries and in some sectors it is very high indeed. The sectors involved tend to be especially sensitive in both developed and developing market economies. Since they also tend to be some of the sectors in which global shifts in production and trade are most pronounced, we shall look at some of them in detail in the case studies of Part Three.

Increasing international involvement of the 'Eastern Trading Area' The term 'Eastern Trading Area' (ETA) is used by the GATT to refer to the countries of Eastern Europe and the USSR together with China and other centrally planned economies in Asia. They have more generally been known as the centrally planned economies but the changes initiated in Eastern Europe and the Soviet Union in 1989/90 make that term increasingly less applicable in those cases. Clearly, substantial changes are about to occur in the involvement of these countries in the global economy but it will be some time before the shape of the new order becomes clear. In 1990 the only country whose future international economic position was unambiguous was East Germany (the German Democratic Republic), which is now fully integrated with West Germany.

Prior to these changes, the international involvement of the ETA countries had remained fairly stable. In 1970, 12.4 per cent of the area's manufactured exports went to the developed market economies; in 1986 the figure was 12.7 per cent. There was a slight increase in the importance of developing market economies as an export destination, from 13.5 to 14.8 per cent. Over the same period internal trade decreased slightly: in 1970, 69.7 per cent of manufactured exports went to other ETA countries, in 1986 the figure was 66.4 per cent. In fact, East–West trade in manufactures is far more important for the ETA countries than it is for the developed market economies. Approximately half the ETA's exports to the West originated from the Soviet Union. Conversely, Western Europe is the major origin of manufactured imports into Eastern Europe. There is also a contrast between the two groups in the composition of trade. Manufactured goods account for 80 per cent of all Western exports to the East, whereas manufactures are only about 40 per cent of the ETA's exports to the West.[6]

The network of international trade in services

In the first section of this chapter we noted the significant development of international trade in commercial services, notably the circulation services which facilitate the operation of the global production system. To complete this brief survey of global trends – and particularly global shifts – in production and trade we need to look a little more closely at the network of international trade in services.[7]

Contrary to some popular opinion, a large service sector is not confined to the more advanced economies. Table 2.12 shows that services contribute a major share of gross domestic product (GDP) in all groups of economies, although the relative importance of services as a proportion of GDP certainly does vary directly with a country's income

Table 2.12 The contribution of the service sector to gross domestic product in developed and developing economies

| Country group | Percentage of GDP | | | | | |
| | Services | | Industry | | Agriculture | |
	1965	1987	1965	1987	1965	1987
Developing countries						
Low income	30	32	27	37	43	31
Lower-middle income	50	46	28	36	22	15
Upper-middle income	42	50	39	40	18	10
Developed market economies	54	61	40	35	5	3

(*Source*: based on World Bank (1989) *World Development Report 1989*, Table 3)

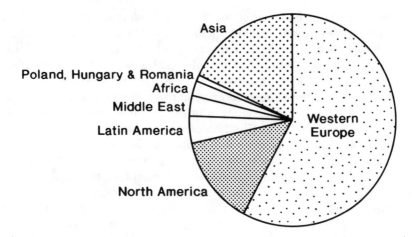

Figure 2.14 The regional composition of world commercial services trade, 1987 (*Source*: GATT (1989) *International Trade, 1988–1989*, Table 22)

level. However, what does differ is the *kind* of service activity involved. Much of the service sector in the developing countries is low-skill, low-technology private service activity (including wholesale and retail trade) whereas in the developed economies the relative importance of business-related services (including finance) is greater.

Many service activities are not 'tradable', that is, they have to be provided to customers on a face-to-face basis. Within the traded services category we can distinguish between those services related to transportation and what were termed in Table 2.1 'other private services', which include the major business-related services of finance, management, advertising and professional/technical services. It is the growth of international trade in these services (and international investment, too, as we shall see in the next chapter) which has been especially significant during the past decade or so.

Figure 2.14 shows that Western Europe was the dominant region in total commercial services trade (exports and imports) in 1987, accounting for 58 per cent of the world total, followed by Asia and North America. In terms of rates of growth during the 1980s, however, the most dynamic regions were North America and Asia. Table 2.13 is the commercial services equivalent of Table 2.5. A comparison of the two tables reveals a general similarity but with some significant differences. The United States is the leader in both cases. However, both the United Kingdom and France are more important as exporters of commercial services than of manufactured goods, while Japan and West Germany are less important as exporters of commercial services than they are as manufacturing exporters. Equally significant is the substantial increase in ranking of the leading East and South East Asian NICs as exporters of commercial services, a clear indication that these economies are now substantially more than mere assemblers of manufactured goods.

It is important, once again, to reiterate the intimate functional relationship which exists between the internationalization of services and the internationalization of manufacturing production and trade:

Increases in merchandise trade stimulate the expansion of such trade-related services as shipping, port services and merchandise insurance ... also ... the causation can run the

Table 2.13 The world 'league table' of commercial services trade

Rank 1970	Rank 1989	Exports	Value 1989 ($ billion)	Share (%)	Rank 1970	Rank 1989	Imports	Value 1989 ($ billion)	Share (%)
1	1	United States	102.5	15.7	5	1	Japan	80.5	12.0
3	2	France	67.1	10.3	1	2	United States	78.0	11.7
2	3	United Kingdom	47.3	7.2	2	3	West Germany	70.4	10.5
4	4	West Germany	45.4	7.0	4	4	France	52.1	7.8
6	5	Japan	39.4	6.0	3	5	United Kingdom	39.0	5.8
5	6	Italy	37.3	5.7	6	6	Italy	33.0	4.9
8	7	Spain	25.1	3.8	8	7	Netherlands	24.5	3.7
6	8	Netherlands	24.4	3.7	9	8	Belgium–Luxembourg	22.2	3.3
10	9	Belgium–Luxembourg	23.1	3.7	7	9	Canada	19.6	2.9
15	10	Austria	17.2	3.5	10	10	Sweden	14.9	2.2
12	11	Switzerland	15.5	2.6	—	11	Taiwan	14.0	2.1
9	12	Canada	13.1	2.4	15	12	Switzerland	13.7	2.0
—	13	Hong Kong	11.6	2.0	16	13	Spain	12.8	1.9
14	14	Sweden	11.6	1.8	11	14	Australia	12.4	1.9
27	15	Korea, Rep.	11.0	1.8	44	15	Saudi Arabia	10.5	1.6
22	16	Singapore	11.0	1.7	13	16	Norway	10.5	1.6
11	17	Norway	10.6	1.7	—	17	Hong Kong	10.1	1.5
13	18	Mexico	10.1	1.6	32	18	Korea, Rep.	9.5	1.4
16	19	Denmark	9.5	1.5	18	19	Austria	9.4	1.4
17	20	Australia	8.5	1.5	17	20	Denmark	8.7	1.3
26	21	Taiwan	7.5	1.3	20	21	Yugoslavia	8.2	1.2
33	22	Thailand	6.3	1.1	12	22	Mexico	7.6	1.1
—	23	Turkey	5.7	1.0	40	23	Singapore	6.5	1.0
18	24	Yugoslavia	5.4	0.9	26	24	Finland	6.2	0.9
20	25	Greece	4.9	0.8	—	25	Indonesia	5.2	0.8
		Total	571.1	87.5			Total	579.8	86.7
		World	653.0	100.0			World	669.0	100.0

— No data available

(*Source*: GATT (1990) *International Trade 1989–1990*, Table 12)

other way ... advances in these traditional services, as well as the availability of new services, can stimulate merchandise trade.

(GATT, 1989, p. 40)

But the relationship is not simply one between services trade and merchandise trade. Development of some kinds of services may also stimulate the growth of other services. These relationships become particularly significant components of the internationalization of business activity and, especially, the increasing pursuit of globalization strategies by business firms. These are all key issues to be explored throughout this book.

Conclusion: towards a multi-polar global economy

The world map of production and trade of the 1990s is substantially more complicated than that of forty or fifty years ago. Although there are clearly elements of continuity – particularly the persistent dominance of the developed market economies – some quite dramatic changes have occurred. For a quarter of a century after the end of the Second World War the global economy grew at an unprecedented rate. Growth rates were especially high in manufacturing industry as the manufacturing system itself became increasingly global in character. Such 'globalization' was reflected in the much faster growth of trade in manufactured goods than in production. World economic growth was savagely disrupted by the deep recession of the second half of the 1970s and early 1980s. By the second half of the 1980s, however, there had been a substantial, if uneven, economic recovery. In particular, the growth of international trade came to be driven more by the development of trade in commercial services than in manufacturing, a reflection of the increasingly close interconnections at a global scale between production and circulation activities.

There have been very substantial geographical variations in this pattern of growth. The world economic system continues to be dominated by a small group of core economies led by the United States. But although the United States remains the world's most important producer of manufactured goods its leadership has come under increasing challenge. Initially the major challenge was posed by West Germany; more recently Japan has emerged as the most rapidly growing manufacturing producer within the developed market economies. This is especially evident in the case of manufactured exports. Much of the United States' production is consumed domestically; the country is relatively less important as an exporter than as a producer.

One of the most striking developments of recent years, however, is the fact that manufacturing production is no longer almost exclusively a core-region activity as it had tended to be for the previous two hundred years. This is especially evident in the changing pattern of trade in certain manufactured goods. Although a handful of core economies still dominates international flows of goods and services the most spectacular growth rates – apart from that of Japan – have been achieved by the East and South East Asian NICs. From the viewpoint of North America and Western Europe the centre of gravity of the world manufacturing system has begun to shift towards the Pacific. A number of formerly peripheral countries now challenge the core of the world's manufacturing system. Certainly, in terms of rates of growth, the NICs have continued to perform substantially better than the developed market economies. Whereas the manufacturing production and trade of the developed market economies grew at much reduced rates in the second half of the 1970s and through much of the 1980s, that of the East and South East Asian NICs continued at high levels. As a result, NIC-

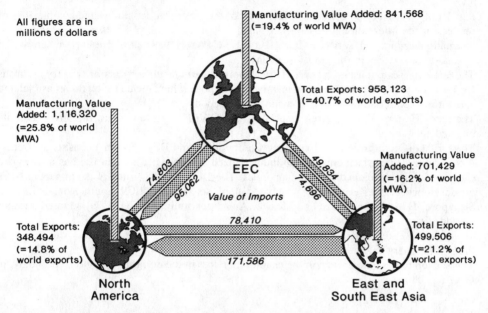

All figures are in millions of dollars

Manufacturing Value Added: 841,568 (=19.4% of world MVA)

Total Exports: 958,123 (=40.7% of world exports)

Manufacturing Value Added: 1,116,320 (=25.8% of world MVA)

Manufacturing Value Added: 701,429 (=16.2% of world MVA)

EEC

Value of Imports

74,803
95,062
49,834
74,696
78,410
171,586

Total Exports: 348,494 (=14.8% of world exports)

Total Exports: 499,506 (=21.2% of world exports)

North America

East and South East Asia

Figure 2.15 A multi-polar global economy – the 'triad' of economic power

manufactured products flowed increasingly into the developed market economies. Such import penetration has been highly selective, however, with a few sectors accounting for the bulk of NIC imports.

The fact remains that the actual extent of global shifts in economic activity is extremely uneven. Only a small number of developing countries have experienced substantial industrial growth; a good many are in deep financial difficulty whilst others are at, or even beyond, the margins of survival. Thus, although we can indeed think in terms of a new international division of labour, its extent is perhaps more limited than is sometimes claimed. What is clear, though, is that the relatively simple international division of labour organized around the three components of core, semi-periphery and periphery no longer exists. It is being replaced by a far more complex structure, what Storper and Walker (1984) called 'a mosaic of unevenness in a continuous state of flux'. More particularly the global economy is now 'multi-polar'. Within the mosaic of unevenness three clear regional blocs are evident: North America, the European Community and East and South East Asia (focused on Japan). This 'triad', to use Ohmae's (1985) term, sits astride the global economy like a modern three-legged Colossus. As Figure 2.15 shows, the three regions dominate global production and trade. Seventy-seven per cent of world exports are generated by them; 62 per cent of world manufacturing output is produced within them. They are the 'megamarkets' of today's global economy. As we shall see in the next chapter, they are also the dominant generators and recipients of international investment.

Notes for further reading

1. The nature of such long waves and, especially, the role of technological change is discussed in Chapter 4.

2. Part III of GATT (1989) contains a discussion of the statistical definitions and complexities of services in the international economy. Chapter 11 of this book presents a case study of the internationalization of services. Part II of UNCTAD (1988) also presents a substantial discussion of services in the global economy.
3. The latest figures published by GATT for export of *manufactures* by major country groups is for 1985. Later GATT data refer to *merchandise* exports which include such things as fuel and agricultural commodities as well as manufactured goods.
4. The term 'Eastern Trading Area' is used by the GATT in preference to the term 'centrally planned economies'.
5. The OECD (Organization for Economic Co-operation and Development) consists of twenty-four member states, including all the industrialized countries. It is often used as a surrogate category for the developed market economies. The OECD (1988) study of the impact of NICs on developed market economies was based upon six leading NICs: South Korea, Taiwan, Singapore, Hong Kong, Brazil and Mexico. An earlier study (OECD, 1979) adopted a rather broader definition of NICs.
6. Gwiazda (1982) reviews the detail of East–West economic relationships in the period prior to the revolutionary changes initiated in 1989.
7. This section draws extensively on the chapter 'Services in the domestic and global economy' in GATT (1989).

Chapter 3

The Changing Global Map of International Investment: The Growth and Spread of Transnational Corporations

Introduction: the definition and significance of transnational corporations

The changing global map of production and trade, described in Chapter 2, is one important indicator of the increasing internationalization of economic activity. A second major indicator, intimately related to the first, is the growth in the scale and complexity of international investment. Here, it is international or foreign *direct* investment, rather than portfolio investment, which is especially important.[1] The major channel of foreign direct investment is the transnational corporation (TNC). Indeed, the basic argument of this book is that the TNC is the single most important force creating global shifts in economic activity. The significance of the TNC, especially the very large global corporation, lies mainly in three basic characteristics:

1. its control of economic activities in more than one country;
2. its ability to take advantage of geographical differences between countries and regions in factor endowments (including government policies);
3. its geographical flexibility, that is, its ability to shift its resources and operations between locations at a global scale.

Hence, much of the changing shape of the global economic system is sculptured by the TNC through its decisions to invest or not to invest in particular geographical locations. It is moulded, too, by the resulting flows of materials, components and finished products as well as of technological and organizational expertise between geographically dispersed operations. Although the relative importance of TNCs varies considerably – from industry to industry, from country to country and between different parts of the same country – there are few parts of the world in which TNC influence, whether direct or indirect, is not important. In some cases, indeed, the influence of TNCs on an area's economic fortunes can be overwhelming.

Before looking at the growth of international investment and of TNCs in the global economy, it is important to be clear just what the term transnational corporation means. I prefer the term 'transnational corporation' to the more widely used term 'multinational corporation', simply because it is a more general, less restrictive, term. The term 'multinational corporation' suggests operations in a substantial number of countries whereas 'transnational corporation' simply implies operations in at least two countries, including the firm's home country. In effect, all multinational corporations are transnational corporations but not all transnational corporations are multinational corporations.

One way of defining a transnational corporation is in terms of its *ownership* of overseas assets and activities, where such ownership confers control over the overseas operation. This is the definition used in all national statistical sources, although the precise percentage of ownership used varies from country to country (this creates one of the many problems involved in making international comparisons of TNC activity).[2] We shall have to use this term during our discussion in this chapter of trends in international investment in the global economy. But it is important to appreciate that to measure a TNC's influence on this narrow definition of ownership of overseas activities is to take an unsatisfactorily narrow view of its geographical scope and influence.

As will become clear in subsequent chapters, there are many ways – direct and indirect, financial and non-financial – in which a business may be regarded as transnational in its activities, behaviour and influence on national and local economies. A more satisfactory and more comprehensive definition of a transnational corporation is that suggested by Cowling and Sugden (1987, p. 60):

> A transnational is the means of co-ordinating production from one centre of strategic decision making when this co-ordination takes a firm across national boundaries.

A broader definition of this kind is necessary to capture the increasing diversity in the forms of international involvement used by business firms. Many of these forms do not involve ownership or equity relationships but are, rather, various forms of collaboration between legally independent firms in different countries. We shall have a good deal to say about these in subsequent chapters. Unfortunately, there are no comprehensive, internationally comparable statistical data to match the broader definition of a transnational corporation. In trying to obtain an idea of the general growth and spread of TNCs we have to use the foreign direct investment data collected (very unevenly) by national governments.[3] However, we need to bear in mind the fact that these data massively understate the real level of TNC activity. They also tend to lag behind events. Even at the beginning of 1991, the latest available data for a range of countries on the stock of foreign direct investment (FDI) were for 1985!

We can only guess at the overall importance of TNCs in the world economy because of this serious absence of up-to-date and comprehensive figures. Even so, it is clear that, in relation to their numbers, TNCs are responsible for a disproportionate share of world employment, production and trade. Between one-fifth and one-quarter of total world production in the world's market economies is performed by TNCs. Such a large share of world production, together with the geographical extensiveness of their operations – their 'global reach' – also makes them an increasingly dominant force in world trade. An increasing proportion of world trade in manufactures is *intra-firm*, rather than inter-national trade. In other words, it is trade which takes place between parts of the same firm but across national boundaries (Figure 3.1). Unlike the kind of trade assumed in international trade theory, intra-firm trade does not take place on an 'arm's length' basis. It is, therefore, not subject to external market prices but to the internal decisions of TNCs.

In effect,

> international trade in manufactured goods looks less and less like the trade of basic economic models in which buyers and sellers interact freely with one another (in reasonably competitive markets) to establish the volume and prices of traded goods. It is increasingly

managed by multinational corporations as part of their systems of international production and distribution.

(Helleiner and Lavergne, 1979, p. 307)

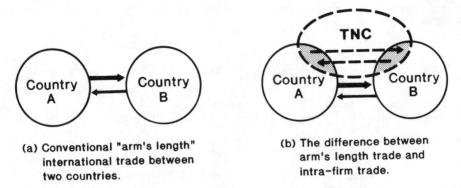

(a) Conventional "arm's length" international trade between two countries.

(b) The difference between arm's length trade and intra-firm trade.

Figure 3.1 'Arm's length' international trade and 'intra-firm' trade within a transnational corporation

Such trade may account for a very large share of a nation's exports and imports. In fact, very few countries collect trade statistics in such a way that intra-firm trade can be distinguished from total trade flows. However, such evidence as does exist is at least suggestive of the very important proportion of world trade which is carried on within the boundaries of TNCs. For example, more than 50 per cent of the total trade (exports and imports) of both the United States and Japan consists of trade conducted within TNCs. Possibly as much as four-fifths of the United Kingdom's manufactured exports are flows of intra-firm trade either within UK enterprises with foreign affiliates or within foreign-controlled enterprises with operations in the United Kingdom.[4]

The conventional image of the TNC tends to be narrow and stereotyped. Mention of the term 'transnational' or 'multinational' enterprise immediately evokes the picture of a gargantuan organization – an IBM or an ICI, a Unilever or a Philips, a General Electric or a Ford – whose activities encircle the globe and penetrate its remotest reaches. Such TNCs are, indeed, the dominant forces in the world economy. They are, as is so often pointed out, broadly equivalent in economic terms to some entire nations. As Benson and Lloyd (1983, p. 77) point out, 'of the 100 largest economic units in the world today, half are nation-states and the other half TNCs'. In fact, only perhaps 4 or 5 per cent of the total population of TNCs in the world can be regarded as truly *global* corporations. But, their sheer individual magnitude gives them a significance out of all proportion to their numbers. For example, some three-quarters of the United States' transnational activity is performed by fewer than 300 firms; four-fifths of UK direct investment abroad is in the hands of around 150 firms.

The United Nations Centre on Transnational Corporations (UNCTC) has identified a core of 600 transnational corporations in mining and manufacturing with annual sales of more than $1 billion in 1985. This 'billion dollar club' created more than one-fifth of the total industrial and agricultural production in the world's market economies. Of the 600, a mere 74 TNCs accounted for 50 per cent of the total sales. In addition to these 600 TNCs, the UNCTC also identified a further 365 major TNCs in business services. Table 3.1 lists some of the leading manufacturing TNCs in 1989. The table indicates the varied extent of transnationality even among the largest firms. For example, motor vehicle manufacturers figure prominently as major TNCs but they are not all equally

Table 3.1 'Vital statistics' of some of the world's leading transnational corporations in manufacturing, 1989

Corporation	Country of origin	Sales ($ million)	Total employment	Foreign employment as % of total*
General Motors	US	126,974	775,100	31
Ford	US	96,933	366,641	53
IBM	US	63,438	383,220	40
Toyota	Japan	60,444	91,790	20
General Electric	US	55,264	292,000	21
Hitachi	Japan	50,894	274,508	18
Matsushita	Japan	43,086	193,088	3
Daimler–Benz	Germany	40,616	368,226	19
Philip Morris	US	39,069	157,000	28
Fiat	Italy	36,740	286,294	18
Nissan	Japan	36,078	117,330	—
Unilever	UK/Netherlands	35,284	300,000	—
Du Pont	US	35,209	145,787	23
Samsung	South Korea	35,189	176,947	—
Volkswagen	Germany	34,746	250,616	34
Siemens	Germany	32,660	365,000	31
Toshiba	Japan	29,469	125,000	—
Nestlé	Switzerland	29,365	196,940	96
Renault	France	27,457	174,573	29
Philips	Netherlands	26,993	304,800	—
Honda	Japan	26,484	71,200	30
BASF	Germany	25,317	136,900	32
NEC	Japan	24,594	104,022	9
Hoechst	Germany	24,403	169,295	43
Peugeot–Citroën	France	24,091	159,100	—
BAT Industries	UK	23,529	311,917	86
Bayer	Germany	23,021	170,200	—
ICI	UK	21,889	133,800	52
Procter & Gamble	US	21,689	79,300	39
Mitsubishi Electric	Japan	21,213	85,723	6

* 1985
— Data unavailable

(*Source*: based on *Fortune*, 30 July 1990, p. 47; UNCTC (1988), Annex Table B.1)

involved in foreign production. Similar variation is apparent in other sectors such as chemicals and electrical production. Even so, the degree of foreign involvement by these giant firms is extremely high. In addition, during the postwar period the leading TNCs have become increasingly global in their operations.

In 1950 only three of the 315 largest TNCs (both US and non-US) had manufacturing subsidiaries in more than twenty countries. By 1975 forty-four TNCs from the United States alone had such an extensive geographical spread. Conversely, the number of large TNCs with subsidiaries in fewer than six countries declined dramatically. In 1950, 138 out of the 180 largest United States TNCs had subsidiaries in fewer than six countries; by 1975 only nine operated such a restricted international network. A similar trend was apparent among large UK firms. In 1950, 29 per cent had operations in more than six countries; by 1970 the figure was almost 60 per cent. Throughout the 1970s and the 1980s the foreign operations of most TNCs increased further in importance whilst the relative importance of their domestic (home country) activities declined. In other words, the degree of transnationality has been increasing, despite occasional interruptions. Much of the increase in transnationality during the 1980s seems to have been due to the rapid expansion of smaller and medium-sized TNCs, whereas those giant corporations which had already developed globally extensive networks were inclined to consolidate their existing operations.

Despite the massive and growing geographical extent of the global TNCs it is important to emphasize once again that they form only a very small percentage of the total number of TNCs. The majority are far less extensively spread geographically. It is important to recognize that not all TNCs are massive enterprises with enormously extensive geographical operations. There is a whole spectrum of TNCs of differing sizes and degrees of transnational involvement. Some are, indeed, long-established, highly experienced transnational firms whose operations are truly global. However, many operate in only a small number of countries. Many of these are the new entrants to the TNC population. This theme of variety in the TNC species is an important one which should not be ignored in the general tendency to concentrate upon the truly global corporations.

Take-off: the postwar surge in transnational production

The general pattern of change

Although there was very considerable growth and spread of TNCs during the first half of the century, it was as nothing compared with their spectacular acceleration and proliferation in the postwar period, especially between the 1950s and the 1970s. Such TNC growth was an integral part of the general economic growth in the world economy which, as we have seen already, was unprecedented. If anything, the activities of the TNCs grew even faster than the world economy as a whole. During the 1960s in particular the foreign output of TNCs was growing at twice the rate of growth of world gross national product and 40 per cent faster than world exports.

The notion that TNC activity accelerated dramatically in the decades after the Second World War is demonstrated graphically in Figure 3.2. The graph plots the average number of foreign manufacturing subsidiaries set up each year by the largest TNCs. The steep upward trend of each of the curves after 1945 is abundantly clear. Already by the 1946–52 period the average number of manufacturing subsidiaries being formed each year was 50 per cent greater than during the previous peak period (1920–9). After 1952 the growth rate was even more rapid, with a particularly large increase in the late 1950s. By 1965–7 the average number of manufacturing subsidiaries formed by the very large enterprises was more than ten times greater than between 1920 and 1929 and six and a half times greater than in the immediate postwar period.

In aggregate terms, there was some deceleration in the rate of growth of foreign direct investment between 1974 and 1983. Undoubtedly this was a reflection of the general slowdown in the world economy during the global recession. But even during that period, international direct investment remained more buoyant than domestic investment. During the 1980s there seems to have been a further acceleration in foreign direct investment. As Julius (1990, p. 60) points out, 'whereas in the 1960s FDI grew at twice the rate of GNP, in the 1980s it has grown more than four times as fast as GNP'.

This extremely rapid postwar proliferation of TNC manufacturing subsidiaries was driven initially by US firms. The United States emerged from the aftermath of war as the only major economy which was stronger rather than weaker. Such economic strength

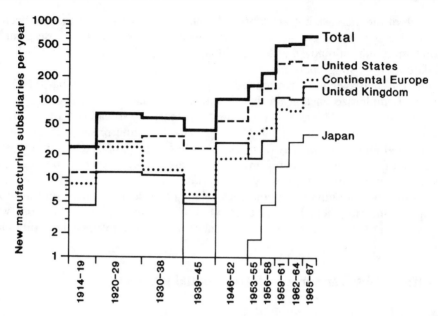

Figure 3.2 Annual rates of formation of overseas manufacturing subsidiaries by large transnational corporations, 1914–67 (*Source*: Vaupel and Curhan (1973), Tables 1.17.2, 3, 4, 5)

was reflected in a variety of ways, one of which was the spectacular growth of overseas manufacturing investment by US firms. Overall, transnational investment in manufacturing and services has continued to grow apace. However, some extremely important changes have become apparent in the composition of this investment. Figure 3.3 shows the growth in value of total overseas direct investment for the period from 1960 to 1985. Even allowing for the effects of inflation it is clear that transnational investment grew extremely rapidly. But the most significant feature of Figure 3.3 is the difference in relative growth rates for individual source nations and the consequent shifts in the relative shares of world foreign direct investment held by the leading source nations (Table 3.2).

For most of its history, world foreign direct investment has been overwhelmingly dominated by TNCs from the United States, the United Kingdom and one or two continental European countries. From the 1950s to the mid-1970s, US firms accounted for between 40 and 50 per cent of the world total. In 1960 US and UK TNCs made up two-thirds of the world total. But although TNCs from both countries have continued to invest heavily overseas other countries' outward investment has increased more rapidly, as Figure 3.3 shows. By 1985 the combined US–UK share of the world total had fallen to less than half. Conversely, the West German share of the total increased from 1.2 to 8.4 per cent while Japan's share grew even more sharply, from 0.7 to 11.7 per cent. From being a very minor player in terms of foreign direct investment in 1960 and not especially important in 1975, Japan had surged up the league table to third place by 1985. Indeed, in 1988 Japan ranked first in terms of the annual *flow* of outward investment for the first time.

Table 3.2 also reflects what might well be the harbinger of an important new development: the appearance of TNCs from some developing countries. There are

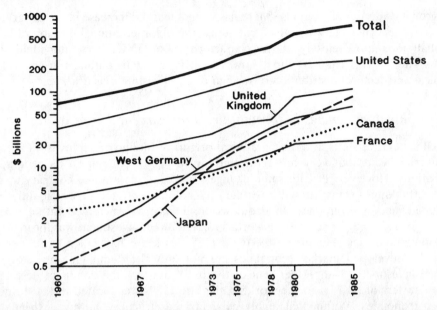

Figure 3.3 Growth of foreign direct investment in the world economy (*Source*: UNCTC (1983) *Salient Features and Trends in Foreign Direct Investment*, Annex II, Table 2; UNCTC (1988), Table I.2)

Table 3.2 Foreign direct investment in the world economy: the changing relative importance of leading source nations, 1960–85

Country of origin	Percentage of world total of outward direct investment		
	1960	1975	1985
United States	47.1	44.0	35.1
United Kingdom	18.3	13.1	14.7
Japan	0.7	5.7	11.7
West Germany	1.2	6.5	8.4
Switzerland	3.4	8.0	6.4
Netherlands	10.3	7.1	6.1
Canada	3.7	3.7	5.1
France	6.1	3.8	3.0
Italy	1.6	1.2	1.7
Sweden	0.6	1.7	1.3
Developed market economies	99.0	97.7	97.2
Developing market economies	1.0	2.3	2.7
World total	100.0	100.0	100.0

(*Source*: based on UNCTC (1988), Table I.2)

indications of the emergence of a considerable number of TNCs from a small number of developing countries, most obviously some of the NICs. Although quantitatively still small in comparison with the volume of TNC activity from the established source nations their share of world foreign direct investment increased from 1 per cent in 1960 to 2.7 per cent in 1985. Again, therefore, we see clear signs that the relatively simplistic division of the global economy has disappeared. The world's population of TNCs is not only growing very rapidly but also there has been a marked increase in the geographical diversity of its origins in ways which cut across the old international division of labour. We shall look more closely at the characteristics of TNCs from individual source nations later in this chapter. First, however, it is useful to outline the general geographical and sectoral distribution of foreign direct investment as a whole.

Geographical tendencies in foreign direct investment

As well as relative shifts in the geographical origins of foreign direct investment (FDI), equally significant, but very different, shifts have been occurring in its geographical destinations. However, despite some changes in recent years – and contrary to much popular thinking – transnational investment is still concentrated predominantly in the developed market economies. Indeed, this dominance has increased rather than de- creased. In 1938, 66 per cent of the world's foreign direct investment was located in the developing countries (Dunning, 1983); by 1985 this share had fallen to 25 per cent. Thus, the developed market economies are not only the dominant source of trans- national investment, with 97 per cent of the total, but also the dominant destination. Three-quarters of all FDI in the world in the late 1980s was located in the developed market economies. Within that highly concentrated structure, however, the detailed distribution of FDI had changed dramatically by the 1980s. Two particularly important points need to be emphasized. First, in general terms,

> it was becoming clear that the distinction, more or less valid during the 1950s and 1960s, between countries playing mostly the role of home countries for international investment and countries playing mostly the role of host countries was becoming blurred. Among OECD countries, inward and outward flows tended to become on the average more balanced.
>
> (OECD, 1987, p. 12)

In other words, the degree of *cross-investment* between the major developed market economies has increased very substantially. The geographical structure of FDI has become far more complex in recent years, a further indication of intensified competi- tion within the global economy.

Second, the most significant development of all has been the dramatic change in the position of the United States as a *host* country to FDI. Table 3.3 gives a summary view of this change, particularly in comparison with Western Europe. In 1975 the inward investment stock in Europe was almost four times greater than that in the United States; by 1985, the stock of FDI in the United States exceeded that in Western

Table 3.3 The increased importance of the United States as a host country to foreign direct investment

	Percentage of world FDI inward investment stock	
	1975	1985
United States	11.2	29.0
Western Europe	40.8	28.9
Japan	0.6	1.0
Other developed market economies	23.1	17.1

(*Source*: based on UNCTC (1988), Table I.3)

Table 3.4 The changing balance of inward and outward direct investment

	Foreign direct investment ratios*	
	1975	1983
Western Europe	1.20	1.56
United States	4.48	1.66
Japan	10.65	12.32

* FDI ratio $= \dfrac{\text{Outward stock}}{\text{Inward stock}}$

(*Source*: calculated from Dunning and Cantwell (1987), Table B1)

Europe. Western Europe's share declined from 41 per cent in 1975 to 28.9 per cent in 1985; the United States' share increased from 11 to 29 per cent. For every leading investing country, the United States became significantly more important as firms in Europe and Japan reoriented the geographical focus of their overseas direct investments, a point we shall return to later. Table 3.3 shows that Western Europe's share of total inward investment fell quite substantially between 1975 and 1985. However, this situation is changing again as non-European companies attempt to gain a direct presence in the European Community stimulated by the implementation of the Single European Market in 1992 (see Chapter 6). Anecdotal evidence suggests a major upsurge in direct investment in Europe by both US and, especially, Japanese companies.

The US direct investment position has been transformed from one of being overwhelmingly a home country for FDI to one in which the ratio of outward to inward investment is almost in balance and much closer to that of Western Europe (Table 3.4). The same cannot be said of Japan. While Japanese outward investment has grown very rapidly, there has been only very limited growth of inward investment. The outward/inward ratio increased from an already exceptionally high level of 10.65 in 1975 to 12.32 in 1983. Whereas in 1985 Japan accounted for almost 12 per cent of total world outward direct investment, it was the host to only 1 per cent of world inward investment. Along with trade frictions, this huge imbalance in the Japanese inward/outward investment account is causing major concern among businesses and policy-makers in the West. These issues will be discussed in Chapter 12, which examines the costs and benefits of TNCs to host and home countries.

Within the developing market economies, too, FDI is unevenly distributed. Table 3.5 shows that in both 1975 and 1985 the largest volume of FDI in the developing world was in Latin America and the Caribbean (12 per cent of the world total and roughly 50 per cent of the developing countries' total). Within that region the major concentrations are in Brazil and Mexico. However, the most rapid *rates of increase* in inward foreign investment between 1975 and 1985 were in Asia, particularly in Malaysia, Singapore, Hong Kong and Thailand. Indeed, Asia's share of inward FDI increased by almost 50 per cent, from 5.3 per cent of the world total in 1975 to 7.8 per cent in 1985. East and South East Asia has become the fastest-growing concentration of FDI in the developing countries. Here, again, we see evidence of the region's emergence as an increasingly substantial economic force in the global economy. Conversely, Africa's importance as a host region for FDI has declined substantially, a reflection of its

Table 3.5 The changing distribution of foreign direct investment within the developing countries

| | 1975 | | 1985 | |
	% of world total	% of developing countries' total	% of world total	% of developing countries' total
Developing countries	24.9	100.0	25.0	100.0
Latin America and Caribbean	12.0	48.3	12.6	50.6
Asia	5.3	21.1	7.8	31.2
Africa	6.7	26.8	3.5	14.0
Other	0.9	3.7	1.0	4.2

(*Source*: based on UNCTC (1988), Table I.3)

increasing marginality in the global economy. Africa's share of world FDI fell from 6.7 per cent in 1975 to 3.5 per cent in 1985. Its share of the developing countries' total declined by half.

The relative importance of FDI to individual host nations – expressed as the foreign share of a country's employment or production – also varies a good deal, although there are no reliable and comprehensive statistics. Among most developed countries, for example, foreign penetration is especially high in Canada. It is easy to see why Canada has been described as 'the largest branch plant economy in the world'. More than four out of every ten workers and more than half of all manufacturing production are in foreign firms. Conversely, foreign penetration of the Japanese economy through direct investment remains minuscule: only about 0.5 per cent of Japanese domestic employment and 1 per cent of domestic assets are accounted for by foreign firms operating in Japan.

The position in most of the other developed market economies lies between these two extremes. In the United Kingdom, for example, approximately 14 per cent of manufacturing employment and 21 per cent of capital investment in manufacturing is in foreign-owned firms. Foreign penetration of the US economy is much smaller than this, despite the popular view to the contrary. Roughly 4 per cent of US domestic employment and 8 per cent of capital investment is in foreign-owned firms.[5] Almost a third of FDI in the United States is by UK firms and 16 per cent by Japanese firms. In terms of employment in foreign firms in the United States, the United Kingdom accounts for 22 per cent, Canada for 20 per cent, continental Europe for 40 per cent and Japan for a mere 7 per cent.

Similar variation in the relative extent of foreign penetration is evident among developing countries. In Latin America foreign firms are especially important to the economies of Brazil and Mexico. Within Asia foreign firms are far more important to the economies of Singapore – well over half of all manufacturing employment and four-fifths of production – and Malaysia than to India or South Korea. In the latter case, only about one-tenth of the country's manufacturing sector is in the hands of foreign firms.

Finally, we should make brief mention of the importance of FDI in the centrally planned economies of Eastern Europe and the Soviet Union. Because of rigid restrictions on the inflow of foreign capital and the laws forbidding foreign ownership of, or equity participation in, state-owned enterprises, FDI in these economies has been negligible. However, foreign firms have long participated in the centrally planned economies through a variety of collaborative arrangements. A probable result of the

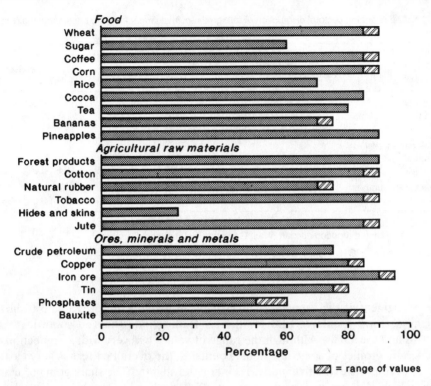

Figure 3.4 The control of commodity marketing channels by transnational corporations (*Source*: based on Clairmonte and Cavanagh (1983), p. 456)

recent political changes in Eastern Europe will be a large-scale inflow of FDI during the next few years. Already such activity is apparent in certain sectors, notably automobiles (see Chapter 9). It will be especially interesting to see whether the opening of investment opportunities in Eastern Europe diverts the attention of TNCs away from parts of the developing world and generates a major geographical reorientation of FDI flows.

Sectoral tendencies in foreign direct investment

Just as transnational investment as a whole is unevenly distributed geographically so, too, it tends to be concentrated rather more in some types of economic activity than in others. Historically, the major proportion of FDI was concentrated in the *natural resource-based sectors*: the mining of geographically localized minerals and ores, the operation of large-scale plantations for the production of commercial foodstuffs for export, and the like. TNCs are still heavily involved in such commodity sectors although in some cases national governments have acquired control of their natural resource operations. Even so, TNCs often remain in control of the commodities' marketing channels, as Figure 3.4 demonstrates.

Foreign direct investment in the extractive industries remains extremely important. In several instances, however, the emphasis has shifted towards other sectors. Initially the shift was particularly strong in *manufacturing* industry; in recent years it has been certain of the *service* sectors which have begun to attract a growing share of foreign

Table 3.6 The changing sectoral distribution of foreign direct investment from the major industrialized countries

| | Percentage of outward foreign direct investment* | | | | | |
| | Extractive | | Manufacturing | | Services | |
	1975	1985	1975	1985	1975	1985
Total for six industrialized countries[†]	22.9	25.7	45.7	35.1	29.2	36.9
Range	4.1/46.5	3.8/55.4	32.4/59.5	22.2/46.2	14.7/41.9	22.1/48.3
In developed market economies	21.8	23.7	46.2	36.0	31.3	38.9
Range	2.1/48.9	3.0/57.9	19.1/63.0	22.0/50.9	12.5/56.0	19.8/62.0
In developing countries	23.0	29.9	43.0	31.8	31.2	36.6
Range	12.6/34.1	9.9/46.4	18.2/64.1	20.4/57.7	19.1/60.6	33.2/43.4

* Totals do not sum to 100 because of residual categories
[†] United States, United Kingdom, Japan, West Germany, Netherlands, Canada

(*Source*: based on UNCTC (1988), Table V.4)

direct investment at the expense of investment in manufacturing. Table 3.6 shows the changing sectoral profile of FDI for the six leading industrialized countries between the 1970s and the 1980s. Although the range of values was very wide, a reflection of the idiosyncratic profiles of some individual countries, the overall pattern is very clear. The share of FDI in the extractive industries increased slightly. The share going to manufacturing, which was the dominant sector in 1975, declined. The share of total foreign investment going to the services sector increased in every case. By 1985 it was the service sector, and not the manufacturing sector, which contained the largest share of the total. However, 'it would be misleading to think of services FDI as displacing industrial FDI. Rather, what one observes is the growth of a global economy in which an increasing share of both goods and services are being produced by TNCs' (UNCTC, 1988, p. 372).

Such investment was not distributed evenly across the industries within the manufacturing and service sectors. Three broad types of manufacturing industry appear to have an especially large TNC involvement.[6]

1. *Technologically more advanced sectors* – for example pharmaceuticals, computers, scientific instruments, electronics, synthetic fibres.
2. *Large-volume, medium-technology consumer goods industries* – for example motor vehicles, tyres, televisions, refrigerators.
3. *Mass-production consumer goods industries supplying branded products* – for example cigarettes, soft drinks, toilet preparations, breakfast cereals.

Within the *services* sector it is predominantly those activities involved in trading, in banking, finance and insurance and in other business and commercial services (such as advertising, accountancy, legal services, real estate) in which foreign investment is especially high. Table 3.7 shows this pattern of concentration quite clearly but also illustrates the considerable variation in emphasis between individual countries. The years to which the figures apply differ because of variations in national data collection. There is inconsistency, too, in the precise classification of services from one country to another. Nevertheless the very strong concentration of FDI in the banking, finance and

Table 3.7 Foreign direct investment in service industries

Host country	Trading (%)	Banking, finance, insurance (%)	Transport and communications (%)	Other services (%)
United States (1986)	41	47	2	3
United Kingdom (1984)	24	43	2	2
Japan (1986)	44	35	3	15
West Germany (1985)	36	53	3	5
France (1983)	30	56	1	9
Italy (1985)	12	65	4	19
Netherlands (1984)	42	25	3	28
Canada (1984)	27	56	—	17
Brazil (1985)	18	66	1	15
Mexico (1981)	33	58	2	1
South Korea (1986)	—	13	4	64
Singapore (1981)	32	58	7	1
Taiwan (1985)	2	20	5	63
Malaysia (1984)	17	64	—	17
Thailand (1987)	39	16	9	8

— Data unavailable

(*Source*: based on GATT (1989) *International Trade 1988–1989*, Vol. 1, Appendix Table 14)

insurance sector in particular is abundantly clear from Table 3.7. But, equally clearly, there are significant national variations as well.

Home country influence: the national characteristics of transnational corporations

Quite clearly, TNC activity as a whole tends to be unevenly distributed, both geographically and sectorally. However, there are significant differences in the characteristics of TNCs from individual source nations. Despite a widespread popular view, TNCs are *not* all the same. Although, as profit-seeking enterprises operating within the capitalist market system they do, indeed, have some common characteristics, they are far from homogeneous. There are many reasons for such differences but one of the most important is, without doubt, the influence of the TNC's *home* environment and its political, social, cultural and economic characteristics. In the following sections of this chapter we look, briefly, at the broad trends and characteristics of FDI from four geographical sources: the United States, Western Europe (specifically the United Kingdom and West Germany), Japan and, finally, the newly industrializing countries.

TNCs from North America

United States TNCs: the rise and fall of the 'American Challenge'?[7] The phrase 'American Challenge' is derived from the (at the time) very influential book by the French writer, Jean-Jacques Servan-Schreiber (1968). He made the now infamous prediction (1968, p. 3) that: 'Fifteen years from now it is quite possible that the world's third greatest industrial power, just after the United States and Russia, will not be Europe, but American industry in Europe.' Of course, by 1983 – the year implied in

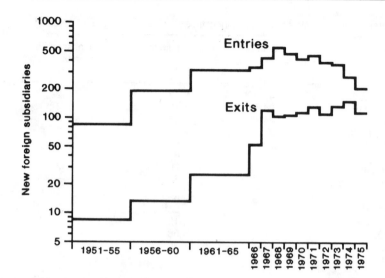

Figure 3.5 United States TNCs: entry and exit of foreign manufacturing subsidiaries, 1951–75 (*Source*: J. P. Curhan, W. H. Davidson and R. Suri (1977) *Tracing the Multinationals: A Sourcebook on US Enterprises*, Tables 2.2.4, 3.4.2)

the prediction – American industry in Europe was very important but it was certainly *not* the third most significant industrial power and is unlikely ever to be so. By the late 1970s, in fact, some writers were beginning to paint a different picture, suggesting that the expansion of United States TNCs had ended.

The same controversy continued into the second half of the 1980s. In 1986 *The Economist* printed a report headed 'American multinationals: The urge to go home', and asserted that 'American multinationals are turning inward'. Two years later, in 1988, the same journal printed an article headed 'America still buys the world', and reported that US foreign direct investment had continued to grow, over a twenty-year period, at a much faster rate than the American economy as a whole. Hence, reports of the decline of the United States TNC, like the reported death of Mark Twain, have been greatly exaggerated.

Nevertheless, change has occurred in the relative position of United States TNCs in the world economy. They no longer dominate as they threatened to do in the mid-1960s even though they still account for more than 40 per cent of the world total of overseas direct investment. For the very largest TNCs, at least, the late 1960s marked the apex of new foreign subsidiary formation. Figure 3.5 shows that whereas the rate of growth accelerated during the 1950s and early 1960s the peak was reached in 1968. Since that date, the rate of new subsidiary formation has fallen substantially. One would, of course, expect some slackening after the onset of world recession in the mid-1970s but it is clear from Figure 3.5 that the slow-down was already under way before this. Similarly, there had been a steady increase in the 'exit' of foreign subsidiaries from the population of United States TNCs. In the early 1950s the ratio between the establishment of new foreign subsidiaries and the exit of existing foreign subsidiaries was roughly 10:1. By the early 1970s, the ratio was 1.8:1.[8]

Figures such as these must be placed in their proper perspective, however. They refer only to the very large TNCs. The rate of formation of new subsidiaries by such

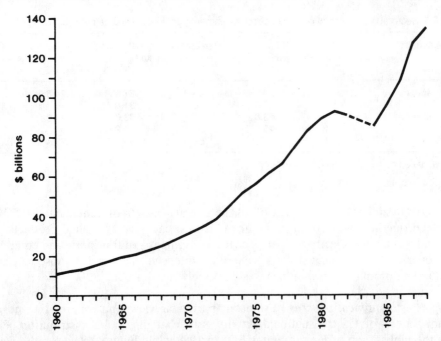

Figure 3.6 The growth of United States overseas direct investment in manufacturing, 1960–88 (*Source*: US Department of Commerce, *Survey of Current Business*, various issues)

firms is only a partial measure of TNC growth. Once an extensive network of foreign subsidiaries is established – as it was by United States TNCs as early as the 1950s and 1960s – the emphasis of subsequent growth shifts to the expansion of these existing subsidiaries and to the restructuring of the foreign network rather than on the creation of new plants. In addition, there has been very rapid growth in the overseas expansion of small and medium-sized US firms, many becoming transnational in their operations for the first time. Hence, US overseas manufacturing investment has indeed continued to grow (Figure 3.6). Although the United States' share of the world foreign investment total has fallen it should be remembered that the world total itself has expanded dramatically. In absolute, if not in relative terms, United States TNCs are more significant today than they were in the past. They constitute the largest and most extensive network of international production facilities in the world. The – not unreasonable – attention being given to the rapid growth of TNCs and of FDI from other countries (particularly Japan) should not let us forget this fact.

Sectoral distribution   In terms of sectoral distribution, United States FDI has shown some substantial changes in recent years, particularly between overseas investment in manufacturing and in services. As Table 3.8 shows, whereas in 1975, manufacturing was almost twice as important as services, by 1985 they were almost equal in relative importance. There has been a major and rapid growth of United States TNCs in the services sector, particularly in banking, financial and other business services.

Within manufacturing, United States TNCs correspond very closely to the three categories outlined earlier: the technologically more advanced sectors, large-volume medium-technology consumer goods and the mass production of branded consumer

Table 3.8 Sectoral distribution of overseas direct investment from leading source nations

Source country	Extractive		Percentage of country total Manufacturing		Services	
	1975	1985	1975	1985	1975	1985
United States	26.4	23.1	45.0	37.9	24.3	33.7
United Kingdom	11.1	33.3	59.5	31.8	29.4	34.8
West Germany	4.1	3.8	48.3	43.0	41.9	48.3
Japan	28.1	15.5	32.4	29.2	36.2	51.8
Canada	21.1	22.9	50.5	46.2	28.4	30.9

(*Source*: based on UNCTC (1988), Table V.4)

goods. Approximately 63 per cent of all the overseas assets of United States TNCs in manufacturing in 1988 were in just four industries: chemical and allied products, non-electrical machinery, electronic and electrical equipment and transportation equipment. Their overseas involvement in 'less technology-intensive' sectors and those less amenable to mass production techniques was extremely small.

Changing geographical patterns From our particular viewpoint, one of the most important aspects of US transnational activity is its changing global distribution. Figure 3.7 is the global map of United States FDI in 1988 while Figure 3.8 shows the general growth trends in the value of US overseas manufacturing investment between 1960 and 1988. Two geographical regions experienced faster than average rates of growth: Europe and the Far East (including Japan). The lowest growth rate of all was in Canada, with the result that by 1964 Europe had overtaken Canada as the most important location for US foreign manufacturing investment. It will be interesting to see what effect the Canada–United States Free Trade Agreement has on levels of US investment in Canada.

By 1980 more than 50 per cent of the total was located in Europe. The geographical distribution of US manufacturing investment within Europe has itself undergone a very marked change since the 1950s, as Table 3.9 reveals. Over the thirty-year period the biggest relative shift has been away from the United Kingdom and towards continental Europe. In 1956 almost 60 per cent of all US manufacturing investment in Europe was located in the United Kingdom, some six times more than in either West Germany or France. By 1988 the pattern had changed substantially. The United Kingdom's share had fallen to 28 per cent while that of West Germany had increased to 21 per cent. There is also evidence of considerably increased US manufacturing investment in the more peripheral countries of Europe, such as Spain and Ireland.

There is no doubt that the major factor in this geographical reorientation was the formation of the European Economic Community in the late 1950s. We shall look more closely at the European Community (EC), as it is known today, in Chapter 6. Suffice it at this stage to observe that the formation of the EC diverted manufacturing investment from the United Kingdom and diminished the country's role as an initial port of entry to Europe for US manufacturers. Table 3.10 provides support for this view. It shows that the gap between the growth of US manufacturing investment in the United Kingdom and in the EC widened in the period after the Community was formed and then narrowed again after the United Kingdom finally joined the Community in 1973. But it is also evident from Table 3.10 that US manufacturing investment was growing

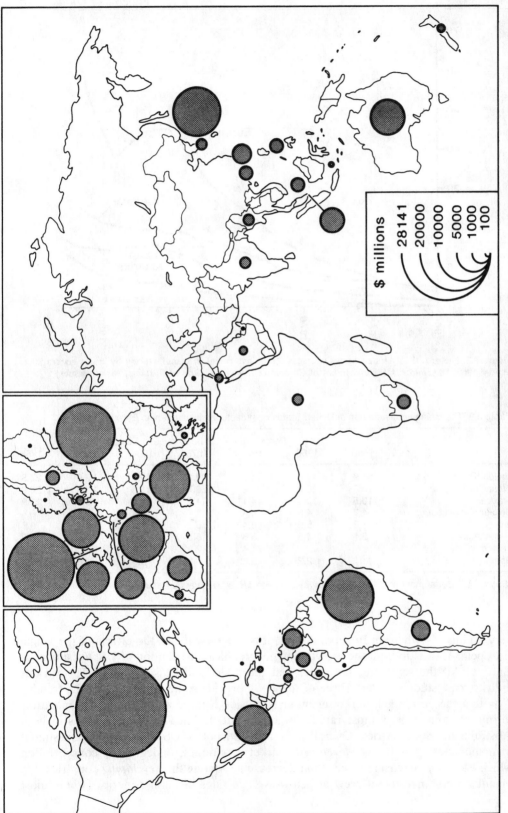

Figure 3.7 The distribution of United States overseas direct investment in manufacturing, 1988 (*Source:* US Department of Commerce, *Survey of Current Business,* August 1989, Table 13)

$ millions

28141
20000
10000
5000
1000
100

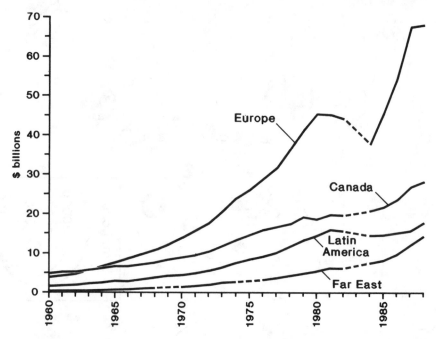

Figure 3.8 The growth of United States overseas direct investment in manufacturing by major geographical region, 1960–88 (*Source*: US Department of Commerce, *Survey of Current Business*, various issues)

Table 3.9 The changing distribution of United States manufacturing investment in Europe, 1956–88

	Percentage of US book value in Europe						
	1956	1966	1971	1975	1980	1984	1988
United Kingdom	57.9	41.6	35.3	29.1	30.6	29.0	27.8
West Germany	10.5	20.2	21.2	20.7	21.3	21.4	20.9
France	10.5	13.5	14.1	14.9	13.1	9.6	11.9
Belgium/Luxembourg	4.7	4.5	6.4	7.7	7.8	7.1	6.4
Italy	3.2	5.6	6.4	6.5	7.4	7.5	9.7
Netherlands	2.1	4.5	5.8	6.5	6.9	7.7	8.9
Ireland			1.4	2.3	3.8	8.5	6.1
Spain	1.1	2.2	2.6	3.8	3.9	3.3	3.9

(*Source*: US Department of Commerce, *Survey of Current Business*, various issues)

very rapidly elsewhere in Europe. As yet, it is too early to see the effects of the Single European Market in both the growth and geographical distribution of US investment in Europe. All the signs are, however, that the effect will be substantial. The various industry case studies in Part Three of this book provide some indications of the trends.

Within the developed market economies outside Europe and Canada, US manufacturing investment grew most rapidly in Japan but declined in relative terms in both Australia and South Africa. Overall some four-fifths of all US transnational investment in manufacturing is in the *developed* market economies, a degree of concentration which has been increasing rather than decreasing. Among the *developing* countries, US manufacturing investment grew at below-average rates in Latin America (where most

Table 3.10 Growth of United States manufacturing investment in Europe, 1950–79

	Total	UK	Percentage change EEC 6	Other Europe
Entire 1950–79 period	14.0	11.3	15.9	16.8
Period to EEC formation (1950–8)	13.0	12.2	15.0	8.7
Periods following formation:				
5 years (1958–63)	17.9	15.0	21.1	20.6
10 years (1958–68)	16.0	11.8	18.7	25.6
21 years (1958–79)	14.3	10.9	16.2	20.0
Period following UK entry (1973–9)	12.1	10.5	11.9	16.4

(*Source*: O. G. Whichard (1981) Trends in the United States direct investment position abroad, 1950–1979, US Department of Commerce, *Survey of Current Business*, Vol. 61, no. 2, Table 5)

of the investment was concentrated in Argentina, Brazil, Mexico and Venezuela) but at much faster rates in the Far East. The changing relative importance of Latin America and the Far East as host regions to US overseas manufacturing investment is especially interesting. Historically, Latin America accounted for most of this investment in the developing countries: almost 80 per cent in 1968 compared with 23 per cent in the Far East. But by 1988 these shares had changed to 72 and 57 per cent respectively.

Within Asia, however, the geographical balance changed considerably. During the 1950s most of the investment in the Asian region was located in India, the Philippines and Indonesia. Together these accounted for 90 per cent of the region's total. By 1988, however, this share had fallen to a mere 18 per cent as the geographical focus of US transnational investment shifted strongly towards such countries as Singapore, Hong Kong, South Korea and Taiwan. These four countries alone were host to 68 per cent of the total US manufacturing investment in the Far East in 1988 while Malaysia and Thailand accounted for a further 14 per cent. By 1988, therefore, four-fifths of US manufacturing investment in the region was in these six East and South East Asian NICs. Of the six, by far the largest concentrations were in Singapore (32 per cent) and Taiwan (18.5 per cent).

Employment in United States' overseas affiliates Statistics which describe the financial value of transnational activity are rather abstract notions for most people. What do the figures mean in 'real' terms such as employment? Usually such information is not available but in the United States' case statistics on employment in the foreign affiliates of US companies are available. Overall employment in the foreign affiliates of United States TNCs grew from 2,425,616 in 1966 to 4,079,700 in 1987, an increase of 68.2 per cent, although there was a decline of 8 per cent between 1982 and 1987. The detailed geography of these employment changes is shown in Figure 3.9. In some respects, the employment trends closely parallel the investment trends looked at earlier but in other cases there are important differences.

Within the developed market economies the biggest relative change occurred in Canada, where employment in US manufacturing affiliates declined by almost 10 per cent. As a result, Canada's share of the total fell from 21.4 per cent in 1966 to 11.5 per cent in 1987. The European share fell from 51 to 44 per cent (whereas the investment trend was almost exactly the opposite). Within Europe, the United Kingdom's share fell from 42 to 29 per cent although it remained the largest single concentration of US

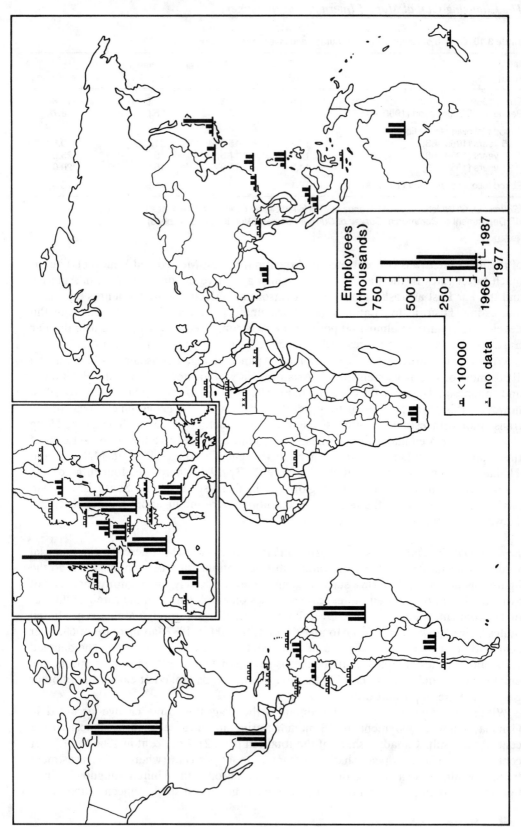

Figure 3.9 Employment in overseas manufacturing plants of United States TNCs, 1966–87 (*Source:* N. G. Howenstine (1982) Growth of United States multinational companies, 1966–1977, *Survey of Current Business*, Vol. 62, no. 4, Table 6; *Survey of Current Business* (1989), Vol. 79, no. 6, Table 8)

manufacturing employment in Europe. The fastest rates of employment increase in US firms in Europe were in the semi-peripheral regions. For example, US employment in Spain grew from 32,163 in 1966 to 136,200 in 1987; in Ireland from 5,694 to 31,100. Among the developed economies, the fastest rate of US overseas employment growth occurred in Japan: in 1966 a mere 18,628 were employed in US affiliates; by 1987 the numbers had soared to 222,400.

Most significantly, however, it is within the *developing* countries that the highest rates of US employment growth have been occurring. Overall, US manufacturing employment in the developing countries grew at almost five times the rate of such employment in the developed economies (184 compared with 39 per cent). As a result, the developing countries' share of the total increased from 20 to 34 per cent. This was in marked contrast to the trend in investment, where the developing countries' share remained virtually unchanged. Presumably it reflects a difference in the labour intensity of US manufacturing investment in the developing countries compared with that in the developed countries. But as Figure 3.9 shows, the incidence of US manufacturing employment in the developing countries is very uneven. The largest concentration was in Brazil, where the level increased from 115,255 to 378,900. In 1966, 29 per cent of US manufacturing employment in Latin America was located in Brazil; by 1977 this had increased to 40 per cent. The other major concentration of US employment in that region is in Mexico. Indeed, the Mexican employment growth rate was substantially higher than that of Brazil. Again, however, it was in the East and South East Asian countries that the most spectacular employment growth rates occurred though, of course, much of this was from a low starting point. For example, in 1966 United States TNCs employed a mere 1,750 manufacturing workers in Malaysia, 1,232 in Singapore and 4,804 in Taiwan. By 1987 Malaysia's employment in US manufacturing firms had grown to 54,000, Singapore's to 38,400 and Taiwan's to 49,100. Each experienced rates of increase well in excess of 1,000 per cent – in the cases of Malaysia and Singapore the increase was around 3,000 per cent.

Although United States TNCs no longer dominate the world economy as much as they did in the 1950s and early 1960s they remain, as a group, the largest single element in the world's TNC population. Their relative decline should not obscure the fact that they continued to grow in absolute terms throughout the 1970s and 1980s. But their global distribution showed signs of change. United States overseas manufacturing investment remained predominantly in the developed market economies but employment in United States TNCs grew more rapidly in the developing countries. In terms both of investment and employment, the two traditional locations for United States TNCs – Canada and the United Kingdom – declined in relative importance. At the same time, Europe remained the dominant focus with about half the world total of both investment and employment. Within Europe there were clear signs of a shift of emphasis – though still modest – towards the semi-peripheral countries, notably Spain. Outside Europe the most spectacular growth of US manufacturing investment and, especially, of employment was in Brazil and Mexico and, most of all, in the Asian NICs: Singapore, South Korea, Malaysia and Taiwan. Quite clearly, United States TNCs are continuing to play a very substantial role in the kind of global economic shifts which we identified in Chapter 2.

Canadian TNCs[9]　In terms of foreign direct investment, Canada is usually perceived as a host country: the largest branch plant economy in the world as it has sometimes been

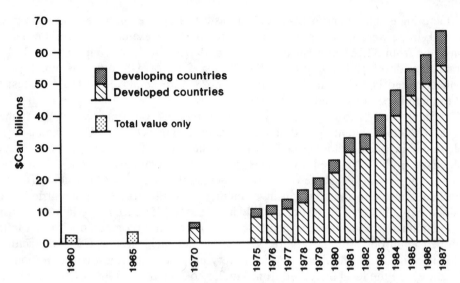

Figure 3.10 Growth of overseas direct investment by Canadian transnational corporations, 1960–87 (*Source*: Statistics Canada, *Canada's International Investment Position*, various issues)

called. It is certainly true that inward investment exceeds outward direct investment by a ratio of roughly 2:1. But Canadian TNCs are an important element of the global TNC population. As Table 3.2 revealed, Canada is the origin of around 5 per cent of the world's total international direct investment, a considerable increase in its world share. Its TNCs are not inconsiderable players on the world scene although they tend to attract less attention than TNCs from the United States, Japan or Europe.

Figure 3.10 shows the overall trend in the growth of total overseas direct investment by Canadian companies from 1960 (separate figures for manufacturing were not available). The graph also shows, for the period after 1970, the growth of investment in developed and developing countries. In general, Canadian investment in developed countries grew substantially faster than Canadian investment in developing countries. As a result, the share of overseas investment in the developed country group increased from 75 per cent in 1970 to 84 per cent in 1984.

Sectoral and geographical distribution   Sectorally, Canadian overseas direct investment is heavily concentrated in manufacturing. As a glance back to Table 3.8 reveals, almost half of the total in 1985 was in manufacturing, a higher level of sectoral concentration than any of the other leading source countries. Geographically, too, Canadian investment is very strongly concentrated not only in developed countries as a whole (as noted above) but, more specifically, in the United States (Figure 3.11). Two-thirds of Canadian overseas direct investment is located in the United States, the largest single concentration of any major overseas investing country. A further 16 per cent is located in Europe (of which more than half is located in the United Kingdom), 9 per cent in Latin America and less than 8 per cent in the Far East (including Japan).

A key question for future geographical patterns of Canadian overseas investment is the likely effect of the two major regional trading agreements: the Canada–United States Free Trade Agreement and the Single European Market. Both, especially the

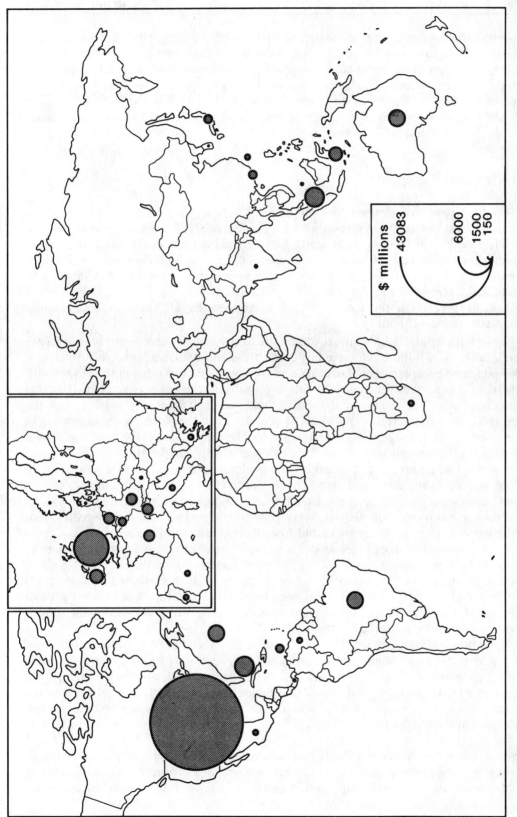

Figure 3.11 Geographical distribution of overseas investment by Canadian TNCs, 1987 (*Source:* Statistics Canada (1990) *Canada's International Investment Position, 1987*, Table 3)

$ millions

43083

6000
1500
150

former, could have a very substantial impact. It will be especially interesting to see whether Canadian firms decide that, in the absence of trade barriers, they can serve the US market from their home base. Canada's economic relationship with Europe is far less developed. Will the implementation of the single market after 1992 stimulate an increased flow of Canadian direct investment to Europe? If so, will the existing concentration on the United Kingdom be maintained or will more Canadian investment be directed to continental Europe?

TNCs from Europe

United Kingdom TNCs: retreat from the Empire[10]   As a group United Kingdom TNCs share at least one common feature with United States TNCs: both have experienced a relative decline in their share of world transnational activity in the last two or three decades. But this needs to be kept in perspective. The fact that the United Kingdom's share of a rapidly growing world total of transnational investment fell in no way indicates a diminution in the absolute importance of overseas investment by United Kingdom TNCs. On the contrary, such investment by UK firms has accelerated markedly in recent years.

Figure 3.12 shows the overall growth of UK overseas direct investment between 1962 and 1987. It is difficult to disentangle the trends in overseas direct investment in manufacturing and non-manufacturing activities because of changes in the coverage of the data. Up to and including 1981 the data excluded the key sectors of oil, banking and insurance. The inclusion of these activities after 1981 clearly inflates both the overall figures and makes it impossible to compare, say, the trends in manufacturing as opposed to non-manufacturing investment over the whole period.

However, the trend throughout the 1970s was clearly one in which the most rapid growth in foreign investment by British companies was in manufacturing industries. In 1965 manufacturing accounted for 50 per cent of total UK foreign direct investment; by 1981 it accounted for 63 per cent. British manufacturing companies raised the level of their overseas production activities very substantially indeed and there was a major relative shift away from domestic, and towards overseas, production. Many UK firms seem to concentrate more on production overseas than on exporting from a UK base to foreign markets. During the 1970s, for example, international production by UK firms was more than twice as great as their exports. In so far as this reflected a relative shift of manufacturing activity by UK firms away from the United Kingdom to other parts of the world it is a most important element in the general global shift in economic activity with which we are concerned in this book.

Like their US counterparts, United Kingdom TNCs tend to be drawn disproportionately from the larger domestic enterprises. The UK economy is, in any case, far more dominated by very large firms than other developed market economies. Again, however, it is important to emphasize that transnational activity is not confined solely to the largest firms. Increasingly, smaller and medium-sized enterprises are becoming transnational in their activities.

Sectoral distribution   Although UK manufacturing firms are especially active overseas investors, it is clear from Figure 3.12 that overseas investment in non-manufacturing activities is also extremely important. Reference back to Table 3.8 shows that over-

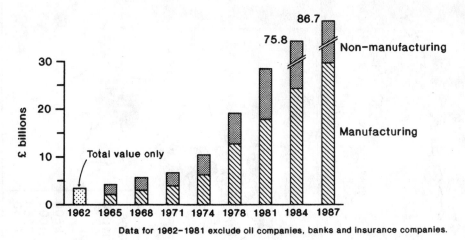

Figure 3.12 Overseas direct investment by United Kingdom TNCs, 1962–87 (*Source*: Business Monitor MA4: *Census of Overseas Assets* (1985); Business Monitor MO4: *Census of Overseas Assets* (1989))

seas investment in services has increased substantially. Within the services sector UK foreign investment in banking, finance and insurance is especially important (Table 3.7). Within manufacturing, United Kingdom TNCs have tended to be rather more heavily involved in lower-technology manufacturing sectors than United States TNCs. Whereas almost two-thirds of US transnational manufacturing investment was in the 'more technology-intensive' sectors of chemicals, mechanical and instrument engineering, electrical and electronic engineering and transportation equipment, the comparable figure for UK transnational investment was only 47 per cent. Apart from chemicals and, to a lesser extent, electrical engineering, United Kingdom TNCs were most prominent in the food and drink industries. In fact two sectors – food, drink and tobacco, and chemicals – accounted for approximately half of all UK transnational manufacturing investment.

Changing geographical orientation United Kingdom TNCs as a whole display a particularly extensive geographical spread of operations. The detailed geographical distribution of UK transnational manufacturing investment shown in Figure 3.13 confirms this impression. But the map also shows that the bulk of the investment was concentrated into certain geographical areas. Three areas – North America, Western Europe and Australia – contained more than four-fifths of the total. Outside these 'core' areas, UK manufacturing investment is spread extensively but thinly. Africa's share of the total (excluding South Africa) was a mere 2.2 per cent. Of this, the main concentrations were in Nigeria, Zimbabwe and Kenya.

Only 6.5 per cent of the United Kingdom's overseas manufacturing investment is located in the Far East (compared with 10.6 per cent of US manufacturing investment). This relatively low level of direct investment by British firms in the world's most dynamic economic region is surprising, given the long-standing historical connections between the United Kingdom and many of the countries of the region. Within the Far East, Singapore is now by far the biggest concentration of UK manufacturing investment (38 per cent of the regional total), followed by Hong Kong (19 per cent), India (16 per cent) and Malaysia (13 per cent). In fact, most of the high level of UK

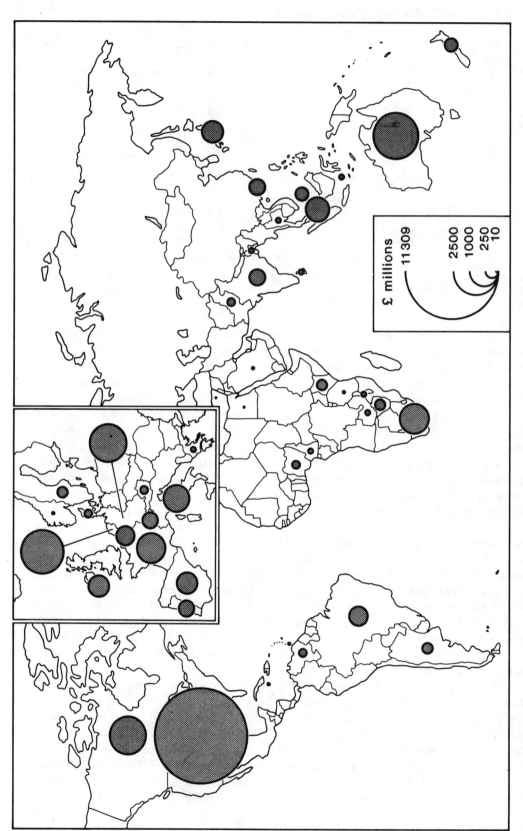

Figure 3.13 Geographical distribution of overseas investment by United Kingdom TNCs, 1987 (*Source:* Business Monitor MO4: *Census of Overseas Assets 1987* (1989))

£ millions

11309

2500
1000
250
10

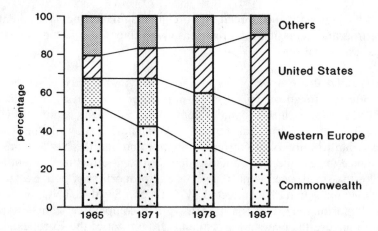

Figure 3.14 Changes in the geographical distribution of overseas manufacturing investment by United Kingdom TNCs, 1965–87 (*Source*: Business Monitor MA4/MO4: *Census of Overseas Assets*, various issues)

investment in Hong Kong is in non-manufacturing sectors. Within Latin America and the Caribbean, the investment was also spread thinly but with important concentrations in Brazil and, to a lesser extent, Argentina. However, only 3.3 per cent of the United Kingdom's overseas direct investment is in Latin America (compared with 13 per cent of the United States' investment).

Such a wide geographical spread is not too surprising given the United Kingdom's particular economic history and the legacy of its imperial past. In recent decades, however, its geographical composition has changed very substantially. At a broad geographical scale the major shift has been away from the Commonwealth countries and a reorientation towards Western Europe and the United States. Figure 3.14 shows this shift very clearly. In 1965 the Commonwealth countries – particularly the 'white' dominions – hosted more than half of all UK manufacturing investment overseas. By 1987 the Commonwealth share had fallen to less than a quarter. Conversely, the share located in Western Europe more than doubled. The United States' share increased even more dramatically, especially after the mid-1970s. By 1987 almost 40 per cent of UK overseas manufacturing investment was located in the United States.

To some extent, this massive geographical shift of emphasis in UK transnational investment towards Europe and the United States marked a return to the kind of geographical pattern of the early years of the present century. Most of the early manufacturing investments by British TNCs were located in the high-income countries of Europe and the United States. But from the 1920s to the 1960s the investment pendulum swung towards the Commonwealth and away from Europe and the United States. Since the 1960s, however, the investment pendulum has swung back again. In Europe the major focus of United Kingdom TNCs in manufacturing has been West Germany. However, the surge of British TNC involvement in the United States was very much a phenomenon of the second half of the 1970s. Among developing countries, UK manufacturing investment grew especially rapidly in Singapore, Malaysia, Hong Kong and Brazil. As yet, however, the actual level of such investment by UK firms in these NICs remains relatively small although there is no doubt that substantial growth is occurring. As in the case of United States TNCs, therefore, the falling relative share of the world foreign investment cake accounted for by UK firms masks significant actual growth and spread of foreign manufacturing investment. Without doubt, UK firms are increasing their overseas involvement and shifting substantial parts of their operations overseas.

TNCs from continental Europe   One of the reasons why Servan-Schreiber's (1968) prediction referred to above (p. 59) has not materialized has been the substantial growth of TNCs from each of the major continental European countries. Their degree of transnational activity varies substantially. As Table 3.2 showed, the leading European source countries are West Germany, Switzerland and the Netherlands, followed, at some distance, by France, Italy and Sweden. Overseas direct investment is especially significant for the smaller economies: a very large proportion of both Switzerland's and Sweden's economic activity is conducted by Swiss and Swedish TNCs in their overseas operations. Unfortunately, few comprehensive and comparable statistics are available to allow us to picture the overseas investments from most of the European countries.[11] The most comprehensive statistics are available for West Germany.

West German TNCs: leading edge of a 'European Challenge'[12]   The rapid growth of West German TNCs, particularly since the early 1960s, has constituted a major counter-challenge to American and British TNCs. Between 1950 and 1970 West German TNCs created at least one-third of the foreign manufacturing subsidiaries established by the largest European firms. West German transnational investment grew at a much faster rate than either American or British. Between 1960 and 1973, for example, the average annual increase in West German foreign direct investment was 23.6 per cent compared with 9.8 per cent for US foreign investment and 7.1 per cent for British. In 1970 West German overseas direct investment stood at DM18,447 million; by 1979 it had reached DM65,928 million; by 1987 DM157,912 million. Data are unavailable on the size distribution of West German TNCs but there is no doubt that the largest West German manufacturers are strongly transnational in their activities. A glance back to Table 3.1 shows that West German TNCs are strongly represented among the world's largest TNCs.

Sectoral and geographical patterns   In sectoral terms, German TNCs show a very marked concentration into the higher-technology sectors. In 1987 more than 70 per cent of all West German overseas manufacturing investment was in three sectors: chemicals (38 per cent), electrical engineering (17 per cent) and motor vehicles (16 per cent). Three of the six largest West German TNCs listed in Table 3.1 – Hoechst, BASF and Bayer – are in chemicals. Of the others, Volkswagen and Daimler–Benz are in motor vehicles while Siemens is among the world's largest electrical machinery and electronics manufacturers.

   The geographical distribution of West German transnational investment in manufacturing is shown in some detail in Figure 3.15. By far the largest single concentration in 1987 was in the United States. No less than 32 per cent of the total was located there. Europe, including the United Kingdom, contained 43 per cent of the total whilst about one-third was located within the EC. Altogether, more than four-fifths of all West German overseas manufacturing investment was located in the developed market economies. Within the EC, France and Belgium contained almost 40 per cent of West German manufacturing investment. The United Kingdom's share was extremely small – 10.3 per cent of the EC total – suggesting that West German firms prefer to serve the

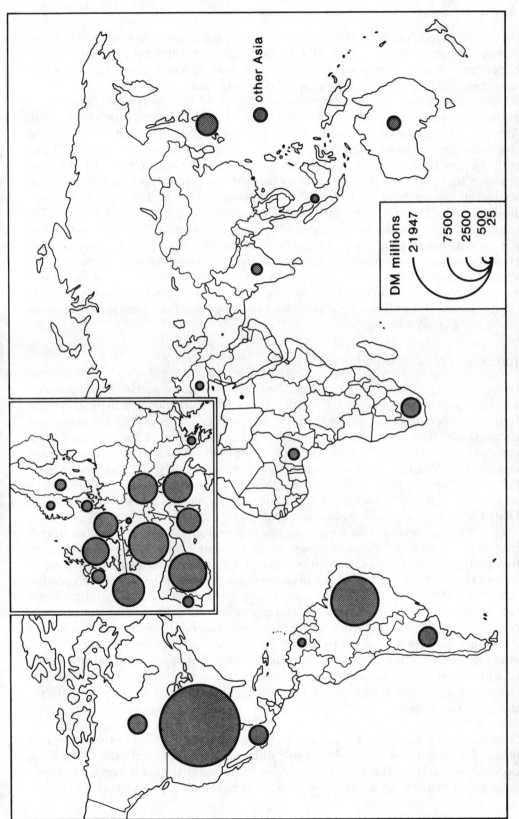

Figure 3.15 Geographical distribution of overseas manufacturing investment by West German TNCs, 1987 (*Source*: Deutsche Bundesbank Statische 3 (1989), Part 1, p. 23)

UK market through exports rather than by establishing a direct presence there. How-ever, a very substantial proportion of German investment within the EC – no less than 22 per cent – is located in Spain. Indeed, Spain's share of West German manufacturing investment in Europe is now only slightly lower than that of France.

Outside the developed market economies, Latin America – and, especially, Brazil – contains the lion's share of West German overseas manufacturing investment. Latin America, in fact, contains almost 90 per cent of all West German manufacturing investment in the developing countries. Of this huge share, Brazil contained 74 per cent. At least in aggregate terms, there has been relatively little West German manu-facturing investment in Asia. This is in marked contrast to American and British investment, as we have seen. Of course, individual German TNCs do have Asian production facilities but their aggregate importance is, so far, not great. Overall, the pattern of West German transnational investment in manufacturing in the late 1980s was the outcome of a marked reorientation since the 1960s. In the early 1960s the bulk of the investment was in other European countries. Since that time, West German TNCs have moved especially strongly into the United States as well as into Latin America. As other European TNCs have also moved with growing alacrity into the United States, Servan-Schreiber's 'American challenge' has been reversed – and on its own home ground too.

Japanese TNCs: a world apart?[13]

Perhaps more strongly than any others, Japanese TNCs demonstrate the importance of national origin in shaping the characteristics of transnational activities. As one Japanese writer observed, business enterprises are 'creatures of history, policy and circumstance'. The particular 'history, policy and circumstance' of Japan's recent econ-omic development have, indeed, produced a very distinctive form of transnational involvement. Much of the diversity and variety of the world's TNC population is, in fact, attributable to Japanese enterprises. Despite some convergence, Japanese TNCs as a group differ in a number of significant ways from TNCs originating from the United States, the United Kingdom and continental Europe.

One major difference is that Japanese transnational investment in manufacturing is relatively recent. Most of Japan's overseas investments before 1939 were in mining, transport and communications activities together with some investment in heavy indus-try and chemicals. Almost all of this investment was located in Japan's Asian colonies. After 1945 Japan's international economic strategy was based primarily on trade: on the development of exports and on the industrial self-sufficiency of the domestic economy. Before the late 1960s, the Japanese government operated a strongly restric-tive policy towards overseas investment by Japanese enterprises. In addition, the relatively low level of domestic labour costs provided little stimulus for Japanese firms to look overseas for cheaper labour locations. As a result, the level and growth of transnational investment by Japanese firms were very low indeed throughout the 1960s, as Figure 3.16 shows.

'Late start, but fastest growth' In the late 1960s, however, a number of developments in the Japanese economy combined to produce a spectacular take-off in Japanese overseas investment. The most important of these were: the relaxation of Japanese government restrictions on overseas investment and, indeed, its positive encourage-

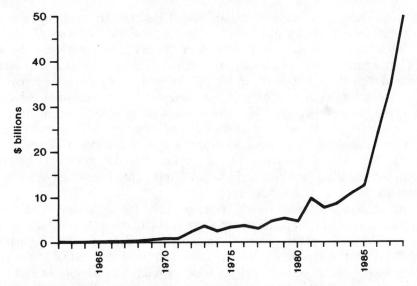

Figure 3.16 Growth of Japanese overseas direct investment, 1962–88 (*Source*: Ministry of International Trade and Industry, *Direct Overseas Investment from Japanese Companies*, various issues)

ment; the rapid increase in the value of the yen; the growing shortage and increasing cost of domestic labour; and shortages of indigenous natural resources. After taking off at the end of the 1960s and in the early 1970s, Japanese overseas investment surged ahead. Its growth was checked by the aftermath of the 1973 oil crisis but then accelerated again in the late 1970s, stimulated by a further revaluation of the yen. Again, there was a dip in the growth rate in the early 1980s but, since 1985 in particular, Japanese overseas direct investment has grown at a remarkable rate. 'Late start, but fastest growth' aptly summarizes the Japanese TNC experience. The result, as we have seen already, was that Japan's share of world transnational investment increased from 0.7 per cent in 1960 to 11.7 per cent in 1985 (Table 3.2). In terms of transnational investment, as in economic growth as a whole, the Rising Sun has been rising very fast indeed.

Some of the distinctive features of Japanese transnational activity are explicable in terms of two factors which are virtually unique to the Japanese situation. The first is the close involvement of the national government, the second is the key role played by the giant Japanese trading companies (the *sogo shosha*). We shall look more closely at the government aspect in Chapter 6 and at the *sogo shosha* in Chapter 11 when we examine the internationalization of business services. Here, our concern is to look at Japanese TNCs in the same way as we have looked at North American, British and European TNCs – in terms of size, sectoral involvement and geographical distribution. As we have noted several times in this chapter, transnational activity is generally associated with large enterprises. Even though there is much transnational investment by American, British and German firms over a wide size range, there is no doubt that large firms dominate. The Japanese case, in contrast, seems to be very different; a far larger proportion of small and medium-sized Japanese enterprises have operations overseas. This contrast with US firms in particular was especially marked in the 1970s.

Sectoral composition   The sectoral composition of Japanese transnational activity differs somewhat from that of the other major source nations. In terms of total transnational investment, manufacturing is considerably less important than the service sectors. Only a little over a quarter of Japanese overseas investment is in manufacturing compared with about a third of British and four-tenths of German overseas investment. The greater importance of overseas investment in service industries reflects a key attribute of the Japanese economy: its postwar emphasis on export trade as an economic strategy built upon an extensive international network of marketing institutions. The growth in importance of Japanese overseas investment in services has been especially marked. In 1960 only 22 per cent was in services (compared with 51 per cent in resource-based activities). By 1988 no less than 62 per cent of Japanese overseas direct investment was in the service sector.

Within manufacturing industry itself, Japanese TNCs again display an idiosyncratic pattern. Overseas investment is still less concentrated into the higher-technology sectors whereas these are the major focus of American and German TNCs in particular. Roughly 57 per cent of Japanese overseas manufacturing investment in 1988 was in the higher-technology sectors of chemicals, machinery, electrical equipment and vehicles compared with 63 per cent of US transnational investment and almost 80 per cent of West German investment. But this pattern is changing rapidly as Japanese overseas direct investment places less emphasis on the lower-technology sectors such as metals and textiles. The sectoral pattern is strongly differentiated geographically, however, as Figure 3.17 reveals. Outside Asia, where the sectoral profile of Japanese direct investment is relatively even (apart from major investments in mining), the sectoral distribution is very uneven but in rather different ways. Japanese FDI in both Europe and North America is strongly skewed towards the services sector. In Europe, especially, only 16 per cent of Japanese direct investment is in manufacturing whereas 76 per cent of the investment is in services. In North America manufacturing is relatively more important (32 per cent) but there, too, the major emphasis is on services (64 per cent).

In summary, Figure 3.17 displays two significant features. First, it shows the small amount of Japanese investment in virtually all manufacturing categories (apart from electrical machinery) in Europe compared with North America and even Latin America. Japanese FDI in Europe is overwhelmingly in the financial and commercial sectors. Second, there are significant differences between North America and Asia. In North America the dominant sectors for Japanese investment are real estate, banking and insurance and commerce. In Asia these sectors are relatively unimportant. The leading sector for Japanese investment in Asia is mining. In addition, the manufacturing profiles of the two host regions are substantially different, with Japanese investment in North America being more strongly concentrated in the more sophisticated manufacturing sectors.

Geographical patterns   It is clear that Japanese TNCs differ in a number of important respects from their American, British and German counterparts. We also find that their geographical distribution is rather different (Figure 3.18). At the broad geographical scale, 61 per cent of Japanese transnational investment is in the developing economies; in contrast, more than 80 per cent of overseas direct investment for the other leading source nations is concentrated in the developed market economies. But there is no doubt that the Japanese pattern is changing. In terms of manufacturing investment, there has been a very substantial shift towards North America (overwhelmingly the United States).[14] In

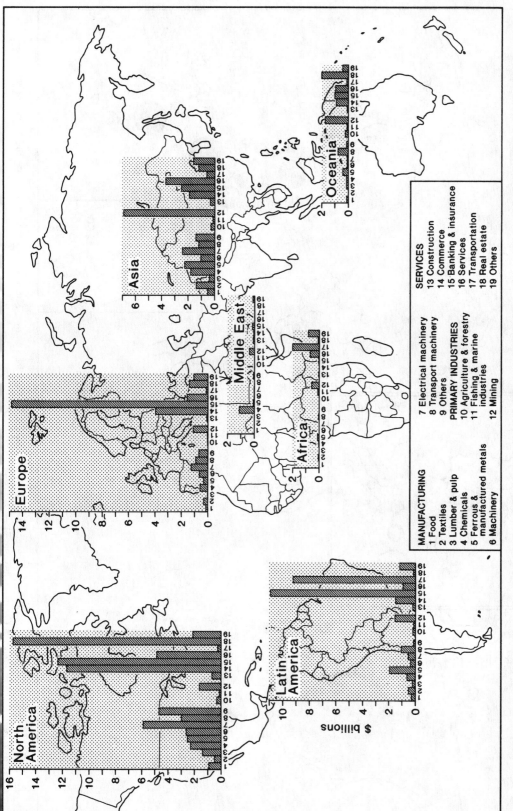

Figure 3.17 Sectoral profiles of Japanese overseas direct investment by major region, 1988 (*Source:* Ministry of International Trade and Industry (1989) *Direct Overseas Investment from Japanese Companies*, Table 4)

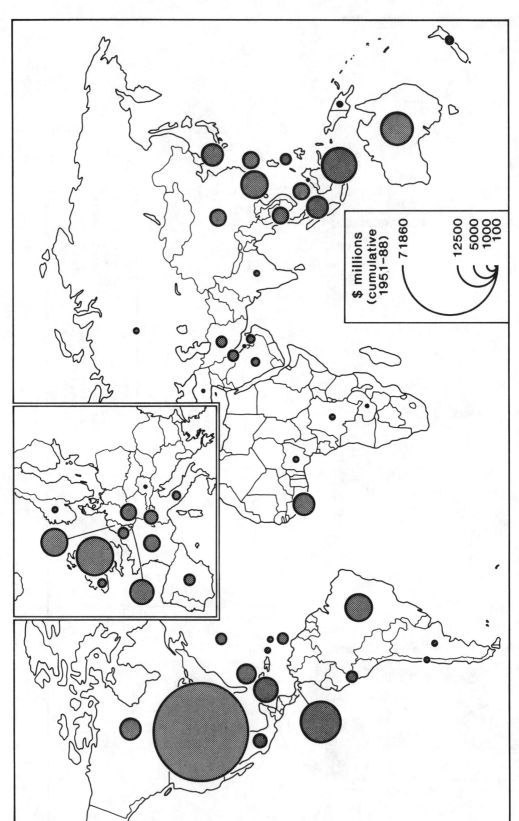

Figure 3.18 Geographical distribution of overseas direct investment by Japanese TNCs, 1988 (*Source:* Ministry of International Trade and Industry (1989) *Direct Overseas Investment from Japanese Companies*, Table 2)

Legend within figure:

$ millions (cumulative 1951–88)

71860

12500
5000
1000
100

1987, 48 per cent of Japanese overseas manufacturing investment was in North America compared with 25 per cent in Asia, the region which traditionally had been the dominant destination for Japanese manufacturing firms. Eleven per cent of Japanese manufacturing investment is in Latin America and only 10 per cent in Europe. Within Europe there is a very strong emphasis on the United Kingdom and, to a much smaller extent, West Germany.[15] The level of geographical concentration for all types of Japanese overseas direct investment is slightly different: among developed countries Europe is rather more significant (16 per cent) and North America rather less significant (40 per cent); among developing countries Latin America is more important (17 per cent) and Asia less important (17 per cent).

Quite clearly, Japanese FDI is in a considerable state of flux. Its present geographical distribution very much reflects the way in which such investment has evolved during the postwar period. According to Ozawa (1979, p. 12), 'Japanese overseas investments in the early 1950s were designed to help restore the Japanese economy as an exporter of manufactured goods and to secure overseas resources, circumstances similar to those of the pre-war years'.

In the mid-1950s Japanese firms began to take an interest in Latin America, especially Brazil. Investments were made in iron and steel, shipbuilding, machinery and textiles industries. By the early 1960s Japanese firms were looking increasingly towards neighbouring Asian countries. Much of the early investment in Asia was in non-manufacturing activities. Most of the manufacturing that did occur was designed to serve the local markets. In the late 1960s, however, the emphasis of Japanese investment in Asia shifted towards the labour-intensive manufacturing industries, particularly in Taiwan, Hong Kong, Singapore and Thailand.

The major take-off of Japanese transnational investment which occurred in the late 1960s and early 1970s consisted of two elements: a reinforcement of the labour-intensive activities in Asia and, to a lesser extent, in Latin America, together with a major drive into the US market for more sophisticated production. This reorientation towards North America and, to a lesser extent, Europe intensified during the late 1970s and through the 1980s. It reflected both the introduction of protectionist barriers against Japanese imports and the large-scale revaluation of the yen. These are developments which will be discussed again both in Chapter 6 and in the industry case studies of Part Three.

There is no doubt that, at present, Japanese transnational activity still differs quite substantially from that of the other major source nations. Such differences reflect the distinctive character of Japan's national economic and political situation. Japanese TNCs as a group seem to be rather different from those TNCs from the other developed market economies: in size, in sectoral composition and in their geographical distribution, as well as in their managerial practices. Whether these differences merely reflect the relative immaturity of Japanese TNCs is unclear as yet. It may well be that Japanese TNCs will come to resemble the American and European corporate giants. At present, though, Japanese transnational investment retains a very distinctive flavour.

A new wave? The emergence of TNCs from the newly industrializing countries[16]

Despite the changes which have been occurring in the relative importance of transnational investment from various source countries, the fact remains that a small number of industrialized nations generate almost all the transnational manufacturing investment in the world economy. Almost, but not quite. We are now becoming aware of a new wave: the spread of TNCs from the newly industrializing economies of East and South East Asia and Latin America. At a world scale the level of investment by TNCs from developing countries is minuscule (less than 3 per cent – see Table 3.2). But in certain parts of the world it is becoming very important. Most of the evidence, as yet, is extremely fragmentary, given the absence of official statistics for most countries. Almost certainly, however, the figures underestimate the true position. Developing country TNCs originate from a small number of countries, essentially the leading NICs, or countries such as India which have an important domestic manufacturing sector. In Asia the most important source nations are Hong Kong, South Korea, Taiwan, Singapore, Malaysia and India. The major Latin American generators of TNCs are Brazil, Argentina and Mexico. TNCs from the developing countries differ from core-country TNCs in a number of respects: their size, the kinds of manufacturing industry in which they operate, their ownership arrangements and their geographical distribution.

General characteristics   TNCs from the newly industrializing countries are, in general, very small compared with TNCs from the developed market economies. Invariably they have a very limited number of overseas affiliates and a limited geographical sphere of operations. But there are some important exceptions. The UNCTC found that seventeen of the 600 TNCs with sales of more than $1 billion in 1985 originated from developing countries. However, ten of these TNCs came from just three NICs: six from South Korea, two each from Brazil and Mexico. Of the seventeen 'billion dollar' TNCs from the newly industrializing countries on the UNCTC list, more than half were in the extractive and mineral processing industries (including petroleum refining). These were also mostly state-owned enterprises, often the outcome of the nationalization of pre-existing private companies. Among the other leading companies, however, were the very large South Korean firms engaged in machinery, electronics and motor vehicles (notably Samsung, Goldstar and Hyundai).

A good deal of Third World transnational activity, therefore, is in resource-based sectors and in distribution. Within manufacturing industry itself, the emphasis is overwhelmingly on relatively unsophisticated, low-technology products with high labour-intensity. In general, the kinds of manufacturing activity involved reflect the characteristics of the source country. Thus, many Indian overseas investments are in textiles, steel, food processing, machine tools, chemicals and electrical machinery. South Korean firms operate foreign plants in clothing, textile weaving, metal goods, plastic mouldings, paper, tyres and cement. Latin American firms manufacture machine tools, automobile parts, consumer durables and textiles in their overseas plants.

Although the general emphasis tends to be on technologically simple products and processes there are significant exceptions. There is a growing trend for some developing country TNCs to be moving 'up-market'. For example, recent Indian overseas investments have included plants in Singapore manufacturing minicomputers and precision tools and a sophisticated carbon black plant in Thailand. Similarly, the emphasis of overseas investment by Hong Kong firms has been shifting from textiles and clothing to the electronic sectors. A comparable tendency is apparent among TNCs from Taiwan and South Korea, as we shall see in Chapter 10. Nevertheless, developing country TNCs tend to be most heavily involved in those manufacturing sectors in which the developed country TNCs are least involved. In addition, their ownership characteristics tend to be different. Most developing country TNCs are engaged in joint ventures with indigenous entrepreneurs in their countries of operation. In comparison, most of the large TNCs from the industrialized countries tend to prefer majority or 100 per cent ownership of their overseas affiliates – although this relationship is no longer as clear as it was, as we shall see in subsequent chapters.

Geographically, too, developing country TNCs display a distinctive pattern. In particular they reveal a very strong 'neighbourhood' character with most of their foreign investments being located in the same general region as the source nation. In O'Brien's (1980, p. 304) words, 'geographical and cultural proximity play significant roles in determining the pattern . . . of investment. In the main, the Asian countries invest in Asia and the Latin Americans in Latin America.' A particularly clear example is the massive amount of manufacturing investment by Hong Kong firms in southern China.

Some case examples   Case studies of foreign direct investment from the leading Asian NICs – Hong Kong, South Korea, Taiwan and Singapore – help to compensate partly for the absence of more comprehensive data.

Hong Kong   Hong Kong is probably the major source of overseas investment from the East Asian NICs. Unfortunately, the Hong Kong government does not publish any relevant statistics. Chen (1981, 1983) showed that, in the early 1980s, the most important sectors in which Hong Kong firms were involved in overseas Asian locations were textiles and clothing followed by electronics. As we shall see in Chapter 8, many of the textiles and clothing investments were related to the quota restrictions of the Multi-Fibre Arrangement. More recently, vast quantities of manufacturing investment have flowed from Hong Kong to mainland China, particularly to the southern Chinese province of Guangdong. It has been estimated that Hong Kong businesses employ some two million workers in that region in thousands of small factories.

The Hong Kong Bank calculates that 75 per cent of Hong Kong's outward direct investment during the 1980s went to China. In 1989, Hong Kong invested almost $5 billion in Asia, a figure more than half that of Japan's investment in the region in the same year. Hong Kong firms are also investing in specific sectors, notably textiles, clothing and electronics in North America and Western Europe, but the bulk of the investment is in Asia. Undoubtedly some of this increased outward investment is related to the approach of 1997 and Hong Kong's reversion to Chinese rule. 'But the main reasons are more mundane: wage rates at home are rising, property prices are almost as high as Tokyo's, and inflation is nudging into double figures. No longer can Hong Kong's entrepreneurs make their millions from sweatshops in Kowloon; instead they must look to poorer parts of Asia' (*The Economist*, 8 December 1990).

South Korea   Euh and Min (1986) indicate that foreign direct investment by Korean firms began as early as 1968 with investments in Indonesia to procure timber for the Korean plywood industry. The first Korean overseas investment in manufacturing was the establishment of a food seasonings plant, also in Indonesia, in the early 1970s. But Korean overseas investment really did not begin to develop extensively until the late 1970s, partly in the light of government encouragement. Since then, the growth rate has been substantial overall and has accelerated during the late 1980s.

Table 3.11 reveals some significant features of Korean overseas investment in the mid-1980s. First, 51 per cent of the total investment was in mining and forestry.

Table 3.11 Overseas direct investment by companies from South Korea, 1984

Sector	Asia	North America	Europe	Latin America	Oceania	Total
			Percentage of total			
Mining	2.5	44.0	–	87.1	65.9	37.9
Forestry	33.3	0.7	–	0.02	28.3	13.0
Fishery	0.1	0.2	0.4	6.5	0.1	2.1
Manufacturing	36.3	10.9	2.7	2.5	3.1	16.3
Construction	3.3	5.7	–	–	–	7.8
Transportation	0.9	0.9	2.1	–	–	0.6
Trade	6.4	23.0	92.8	3.7	0.5	12.6
Real estate	16.9	1.4	2.0	0.3	0.6	4.5
Other	0.4	13.2	–	0.1	1.6	5.2
Total	100.0	100.0	100.0	100.0	100.0	100.0
Regional share	22.8	32.5	2.3	12.6	18.4	

(*Source*: based on Euh and Min (1986), Table II)

However, a further 16 per cent was in manufacturing and 13 per cent in trade-related activities. This sectoral pattern is a clear reflection of the major characteristics of the South Korean economy: a paucity of industrial raw materials and an export-oriented manufacturing base. Second, there is a clear geographical pattern both to Korean overseas investment as a whole and also to differences in sectoral emphasis. Slightly less than a quarter of total investment was located in Asia and a further 13 per cent in Latin America. But note that, in value terms, almost a third of the total was in North America. Much of this is explained by mining investment but there is also substantial Korean investment in North America in trade and in manufacturing. Korea's Latin American and Oceania investments are overwhelmingly in the mining sector. In contrast, the small volume of investment in Europe was almost entirely in trade-related activities. Within Asia, the investment emphasis was bi-modal, roughly a third being involved in both manufacturing and mining activities.

Within the manufacturing sector, most of Korea's overseas investments are in the production of labour-intensive standardized products. This is particularly the case in Asia. However, some of the very large Korean industrial conglomerates or *chaebol* – notably Samsung, Goldstar, Hyundai, Daewoo and the like – are investing overseas in more sophisticated high-technology processes in electronics (see Chapter 10) and in motor vehicles (see Chapter 9). Significantly, a growing number of these investments are in North America and Europe. Recent press reports suggest that foreign investment by these and other South Korean companies is growing very rapidly. In Asia there has been significant transfer of production in such industries as textiles, footwear and consumer electronics to offset declining domestic competitiveness.

Taiwan Taiwanese FDI was initiated as long ago as 1959 (Chen, 1986) when a cement plant was established in Malaysia. It was a further three years before the second recorded overseas investment by a Taiwanese firm. From the 1960s there was steady growth of FDI but it was not until the 1980s that the growth of such investment really began to accelerate. Although substantial Taiwanese investment has located in East and South East Asia, a surprisingly large proportion (54 per cent in 1985) was located in the United States. How far this reflects the inadequacies of the data is not clear. Within South East Asia, the largest concentration was in Indonesia. However, there is

Table 3.12 Overseas direct investment by companies from Taiwan, 1985

Sector	Percentage of total						
	Thailand	Malaysia	Singapore	Philippines	Indonesia	US	Total
Agriculture and fishery	3.4	0.2	–	–	0.9	–	2.3
Food, beverages	12.0	–	–	2.5	6.1	2.0	4.7
Textiles	7.9	1.8	7.6	–	29.4	–	5.3
Garment and footwear	–	–	6.1	0.6	–	–	0.8
Pulp and paper	36.4	–	–	–	43.9	0.4	7.5
Plastic and rubber	0.5	1.1	13.5	0.7	1.1	5.9	4.7
Chemicals	21.0	–	11.9	90.8	14.7	24.7	20.2
Non-metallic minerals	–	1.6	25.1	1.0	–	0.8	6.0
Basic metals and metal products	6.9	44.2	9.1	–	1.6	0.7	3.2
Electronic and electrical appliances	7.4	20.8	22.0	3.5	–	49.4	31.7
Trade	1.8	0.3	4.8	1.0	–	2.4	6.6
Others	2.7	30.0	–	–	2.3	13.7	7.0
Total	100.0	100.0	100.0	100.0	100.0	100.0	100.0
Country share	4.5	3.5	4.4	4.8	11.8	54.0	

(*Source*: based on Chen (1986), Tables 1 and 2)

also substantial Taiwanese investment in China which, for political reasons, does not appear in the official data.

Table 3.12 gives a partial picture of Taiwanese overseas direct investment in the mid-1980s for the South East Asian countries and for the United States. The table also shows the sectoral profile of the investment. In aggregate terms, the major concentrations are in electronic and electrical appliances and in chemicals. But the actual sectoral profile differs substantially both between the United States and the South East Asian countries and also between the South East Asian countries themselves.

Singapore The pattern of foreign direct investment by Singaporean firms differs substantially from that of South Korea and Taiwan although, unfortunately, Singapore does not publish data on outward investment (Lim and Teoh, 1986). As in the cases of the other East Asian NICs, Singapore firms began to invest abroad in the 1960s but it is only since the mid-1970s, and especially the 1980s, that such overseas investment has grown rapidly. A significant proportion of leading Singaporean companies have now become transnational although none is of the scale and diversity of the giant South Korean conglomerates.

Table 3.13 gives a partial picture of the geography and sectoral distribution of Singaporean overseas direct investment. The table is confined to Asia where, in fact, the bulk of Singaporean investment was located in the mid-1980s. Outside Asia most Singaporean direct investments are in trade-related activities. Table 3.13 also does not show investment in China although it seems that China is an increasingly important destination for investment by Singapore companies.

The most striking feature of Table 3.13 is the immensely strong neighbourhood characteristic of Singaporean direct investment within Asia. Two-thirds of the total investment was located in Malaysia and a further 24 per cent in Indonesia. Singapore's two closest neighbours thus contained almost 90 per cent of the total investment in Asia. The most important manufacturing sectors were food and beverage production,

Table 3.13 Direct investment by Singaporean firms in Asia

Sector	Percentage of total				
	Indonesia	Malaysia	Thailand	Hong Kong	Sri Lanka
Manufacturing	58	—	47	—	—
Food and beverages	—	36	—	27	80
Textiles and leather	—	11	16	17	7
Wood and paper	—	8	—	—	1
Chemicals and petroleum	—	11	2	1	3
Non-metal	—	14	13	—	—
Metal	—	11	—	51	—
Electrical and electronic	—	3	—	1	—
Transport equipment	—	6	—	—	—
Other manufacturing				1	1
Agriculture, forestry, mining	10	—	—	—	—
Services	32	—	22	—	8
Total	100	100	100	100	100
Country share	23.8	65.8	1.0	4.4	2.5

— No data

(*Source*: based on Lim and Teoh (1986), Table 4)

textiles and non-metal manufacturing industries. Unfortunately the lack of data on most of the non-manufacturing sectors in Table 3.13 makes further analysis impossible. This is a timely reminder to the reader of the problems of research in this area.

Summary   Overseas investment in manufacturing by TNCs from developing countries is still relatively limited on a world scale. But it is of rapidly increasing importance within the developing market economies themselves. Such investment differs substantially from that by TNCs from the developed market economies. The geographical pattern is heavily dominated by investment in other developing countries and especially in countries is close geographical or cultural proximity. Developing country TNCs tend to be small, to be involved in low technology sectors with relatively high labour-intensity and to have a close involvement with domestic firms in the countries in which they operate. However, a growing number of firms from the leading East Asian NICs in particular are beginning to invest substantially in the developed market economies of North America and, to a lesser extent so far, Europe.

Conclusion

The transnational corporation is the single most important force in the accelerated development of a global economic system. But it is not the simple homogeneous institution so often depicted; there is much variety between TNCs. An important source of such variety is the nature of the individual source nations themselves. Although there are considerable similarities between TNCs, especially the giant ones, whatever their national origin, there is no doubt that different source nations do create some important differences between TNCs. In the last four decades transnational activity has not only grown with unprecedented speed but also it has become increasingly diverse. Within an expanding world total of transnational investment the relative positions of the established source nations has changed whilst, at the same time, new sources have begun to emerge. The initial overwhelming dominance of the United

States has declined relatively. So, too, in relative terms, has the importance of TNC investment from the United Kingdom. The most rapid growth overall has been by West German and Japanese transnational investment. But there are also substantial signs of incipient TNC growth from some developing market economies. Such shifts in the geographical origins of TNCs mirror the shifting patterns of leadership and growth in the global manufacturing system as a whole.

Although most transnational investment originates from, and flows to, the core economies, the pattern of the investment flows has become increasingly complex. Today, there is a great deal of *interpenetration* of investment between national economies. This is shown particularly by the very rapid and recent growth of foreign investment in the United States itself. Even so, the sources of transnational investment remain geographically concentrated. Three of the four major source nations (the exception being Japan) are also the major hosts for foreign direct investment. A similar, though less intensive degree of geographical concentration is apparent within the developing economies. Here the largest concentrations of transnational investment are in Brazil and Mexico but the most rapid growth is in East and South East Asia. Sectorally, it is the technologically more sophisticated products, together with those based on mass-production technology and mass marketing, together with business services, which contain the most rapidly growing TNC activity. These are very broad generalizations, however: there are important exceptions to such geographical and sectoral tendencies.

A most important by-product of the global proliferation of TNCs in manufacturing industry in the last few decades has been the closely related development of TNCs in business services. The large banks, advertising agencies, hotel chains, car rental firms, insurance, legal and freight corporations, credit card companies, have followed the lead of their corporate clients in establishing a global network (Chapter 11). Hence, the internationalization of manufacturing and of manufacturing-related business services have become mutually reinforcing. Trends in one sector have helped to reinforce trends in the other and vice versa. *The result is an increasingly interlocking global system of transnational institutions.*

The internationalization – indeed globalization – of economic activities implied by the growth and spread of transnational corporations is further increasing the degree of interdependence and integration between national economies. Julius estimates, using very conservative assumptions, that by 1995 the global stock of foreign direct investment will be more than double the 1988 stock in real terms. As Julius (1990, p. 40) points out, 'we are in the middle of a decade (1985–95) when FDI gains its maturity as a major force in international economic integration. It is in this sense that quantitative increases in FDI flows have reached the threshold where they create a qualitatively different set of linkages among advanced economies.'

Notes for further reading

1. 'Direct investment' is defined as the investment by one firm in another with the intention of gaining a degree of control of that firm's operations. 'International' or 'foreign' direct investment is simply direct investment which occurs across national boundaries, that is, where a firm from one country buys a controlling investment in a firm in another country or where a firm from one country sets up a branch or subsidiary company in another country. 'Portfolio investment' refers to the situation in which firms purchase stock/shares in other

companies purely for financial purposes; that is, like any other investor, they build up portfolios of company shares. But such investments are not made to gain control.

2. Some of the issues involved in defining TNCs are explored in Cowling and Sugden (1987), UNCTC (1988), Julius (1990).

3. There is no single comprehensive world source of data on TNCs. Most individual nations publish figures on inward and outward direct investment but there are large variations in coverage and definition. The most comprehensive figures are those for the United States and are published regularly in the US Commerce Department's *Survey of Current Business*. Recent general surveys include OECD (1987), UNCTC (1988). Dunning and Cantwell (1987) have produced a comprehensive directory of international investment and production statistics.

4. Detailed discussions of intra-firm trade within transnational corporations are provided by Helleiner and Lavergne (1979), UNCTC (1988, chapter VI), Julius (1990).

5. Julius (1990), chapter 3, provides a useful summary of the relative significance of inward investment in the five leading industrial countries. Tolchin and Tolchin (1988), Graham and Krugman (1989) and Glickman and Woodward (1989) analyse the recent rapid growth of foreign direct investment in the United States; Young, Hood and Hamill (1988) examine foreign direct investment in the United Kingdom.

6. The classification is derived from Stopford, Dunning and Haberich (1981).

7. Overseas investment statistics for US firms are published annually in the Department of Commerce's *Survey of Current Business* together with commentary and analysis. The most detailed analyses are based upon periodic 'benchmark' surveys of the entire population of United States TNCs. Benchmark surveys have been published for 1966, 1977 and 1982. The 1982 survey is analysed in Whichard and Shea (1985).

8. Wilson (1978) in fact suggests that such trends simply reflect the 'maturing' of US transnational activity. This broader question of disinvestment is discussed in Chapter 7.

9. Statistics on Canadian overseas direct investment are published in Statistics Canada, *Canada's International Investment Position*.

10. Statistical data on the stock of overseas investment by United Kingdom TNCs is contained in the Business Monitor Series MO4: *Census of Overseas Assets*, published every three years.

11. Dunning and Cantwell (1987) and OECD (1987) summarize the available data for each of the European source countries.

12. West German transnational investment is less well documented (at least in English) than that of the other major source nations. Statistical data are published by the Deutsche Bundesbank.

13. Statistics on Japanese overseas investment are published annually by the Ministry of International Trade and Industry (MITI). Detailed analyses of overseas investments by Japanese TNCs are provided by Dicken (1988), Kojima (1990).

14. Tolchin and Tolchin (1988) and Glickman and Woodward (1989) discuss Japanese investments in the United States.

15. For a discussion of Japanese investment in Europe see Dunning (1986), Turner (1987).

16. There is now a small, but steadily growing, literature on transnational investment by firms from developing market economies. Buckley and Mirza (1986, 1988) on the 'Pacific Asia TNCs'; Chen (1986) on Taiwanese TNCs; Euh and Min (1986) on South Korean TNCs; Fong and Komaran (1985) and Lim and Teoh (1986) on Singaporean TNCs. Lall (1983), Wells (1983), UNCTC (1988) provide an overall review of developments.

PART TWO

Processes
of
Global Shift

Prologue

The two chapters of Part One outlined some of the major features of today's rapidly evolving global economy: the global shifts in production and trade in goods and services and in international investment. These broad patterns represent some of the major indicators of the internationalization and globalization of economic activity which are the central focus of this book. Chapters 2 and 3 were entirely descriptive. But what are the major processes which create the internationalization and globalization of economic activities? How can we begin to explain the patterns? The following four chapters, which make up Part Two of this book, address this question of *explanation*. They are concerned with the *processes*, rather than the patterns, of global shift. Before setting out the approach adopted in these chapters, however, we should be aware of the more important traditional explanations of international production, trade and investment.

Traditional explanations of international production, trade and investment

Many of the roots of modern explanations of international trade and production are traceable back to the ideas of the 'classical' economists of the late eighteenth and early nineteenth centuries. The most influential figures were Adam Smith, whose *Inquiry into the Nature and Causes of the Wealth of Nations* was published in 1776, and David Ricardo, whose book *On the Principles of Political Economy and Taxation* appeared in 1817. Smith developed the concept of *division of labour* as a key process in economic development. He also put forward the idea that an economy operates harmoniously without any overt control but through the guidance of the 'invisible hand' of the market. Ricardo introduced what is still probably the most fundamental concept not only in the theory of international trade but also in the trading policies of many nations: the principle of *comparative advantage*. The framework of classical economics, with its assumptions regarding the unfettered operation of markets and the nature of economic decision-making, also spawned *theories of the location of economic activity*. Indeed, the pioneer work in location theory (concerned with agriculture) was written by J. H. von Thünen in 1826, only a few years after Ricardo and very much in the same mould. From our viewpoint the most important of the early location theorists was Alfred Weber, whose work was published in Germany in 1909 and translated into English as the *Theory of the Location of Industries* in 1929. Although both trade theory and location theory have been greatly refined, extended and criticized over the years the influence of the pioneers on much current thinking remains strong, especially in the case of trade theory.

The principle of comparative advantage

Without doubt, the most basic concept in the whole of international trade theory is the principle of comparative advantage, first introduced by Ricardo in 1817. It also remains a major influence on much international trade policy and is therefore important on both counts in understanding the modern global economy. The principle of comparative advantage states that a country (or any geographical area) should specialize in producing and exporting those products in which it has a comparative, or relative cost, advantage compared with other countries and should import those goods in which it has a comparative disadvantage. Out of such specialization, it is argued, will accrue greater benefit for all.[1]

The principle of comparative advantage is at the heart of traditional attempts to explain geographical differences in production and trade. But, in itself, it says nothing about *why* such differences occur. The pioneering work in establishing the bases of comparative advantage was carried out by two Swedish economists: Eli Heckscher writing in 1919 and Bertil Ohlin whose *Interregional and International Trade*, published in 1933, refined and extended Heckscher's work. As a result their approach is generally referred to collectively as the Heckscher–Ohlin (H–O) theory of trade. As with the principle of comparative advantage the essence of the H–O theory is very simple. All products require a combination of different factors of production – natural resources (land and raw materials), a labour supply, capital in the form of money to buy the materials and machinery, technology, and so on. Different products vary in the precise combination in which these factors are used. Some industries, for example clothing, use a great deal of labour and are labour-intensive industries. Others, such as chemicals, employ relatively few workers but a vast amount of capital equipment and can be described as capital-intensive. Some industries occupy huge areas of land whilst others operate on tiny sites. Factors of production, of course, are very unevenly distributed geographically, particularly at an international scale. Some nations have an abundance of natural resources but a sparse population and, therefore, a limited labour force. Others have huge potential labour supplies but little capital, and so on.

The H–O approach is based upon such variations in the *factor endowments* of countries. Thus, a country (or region) will export those goods which use intensively the factors of production with which it is best endowed and import those goods which incorporate those factors of production with which it is poorly endowed. Although the H–O theory recognizes the existence of a number of production factors, for simplicity the theory was worked out using only two: capital and labour. In these terms, countries with an abundant supply of labour should export labour-intensive products while capital-abundant countries should export capital-intensive products.

There is a good deal of general substance in the H–O idea. Much of the spectacular growth of manufactured exports from the NICs has been in those industries which are, indeed, labour-intensive and therefore reflect that particular source of comparative advantage. But, as we shall show, it is not quite as simple as this in the real world.[2] The factor endowment theory of international trade attributes trade flows simply to differences in factor endowments. Yet much of world trade in manufactures, as we saw in

Chapter 2, actually occurs between countries with broadly *similar* factor endowments (the developed market economies).

Factor mobility and transport costs

Two particular assumptions of traditional trade theory are especially important. One relates to the *mobility of factors of production*. In trade theory, factors are assumed to be fixed and immobile geographically. An area has a particular endowment of production factors and this endowment forms the basis of the area's comparative advantage. Related to this assumption of the immobility of factors of production is a second assumption about *transport costs*. Trade theory specifically assumes that such costs are zero. In other words, despite its explicit concern with trade between areas, trade theory, in common with much other economic theory, remains curiously spaceless. In fact, neither of these two assumptions can be maintained in a study of the real world. Apart from land, factors of production are not geographically immobile and fixed in their location though there are differences in the degree of mobility of different factors. In general, for example, capital is far more mobile geographically than labour, while skilled labour tends to be more mobile than unskilled labour. Such differential factor mobility is an important element in the shifting global pattern of economic activity, as we shall see. More specifically, geographical distance imposes a cost on movement, whether of materials, finished products, people or less tangible things such as knowledge and information. Each of these differs in its degree of sensitivity to geographical distance. Within manufacturing industry itself, some materials and products are extremely costly to move in relation to their value whilst others are, relatively speaking, cheap to move. Either way, any explanation of the geographical distribution of economic activity at the global scale must incorporate the role of distance and transport costs.

Classical location theory

Historically, it has been the location theorists (a generally maverick band of scholars anxious to break out of the 'spaceless' world of economic theory) who have explored these issues.[3] Most of the leading location theorists have been more concerned with incorporating space into economic theory than with attempting to explain the actual location of economic activities. In addition, location theory is not as extensively developed as trade theory. But, as in the latter case, there are elements of location theory which are useful in helping us to understand the world about us. There are two broad bodies of location theory. One focuses on the costs of production as determining the location of industry; the other is concerned with the size and shape of a firm's market area. Attempts to integrate the two approaches have been largely unsuccessful although both approaches share a central interest in the role of geographical distance in shaping the location pattern of industry.[4]

The pioneer theorist of industrial location theory in general, and of the least-cost approach in particular, was Alfred Weber. Almost all subsequent work in industrial location theory has consisted of attempts to refine or elaborate upon Weber's approach. Weber's concern was to identify the optimal location for an individual firm (operating just one plant) given certain assumptions. On the basis of these assumptions, Weber envisaged the location of industry as being determined by two sets of factors:

1. *Primary or 'general' factors*: transport costs and labour costs.
2. *Secondary or 'local' factors*: forces of agglomeration or deglomeration.

Weber regarded transport costs as the initial determinant of industrial location. The primary aim was to minimize transport costs incurred in gathering together the necessary materials and transporting the finished product to the market. The nature of the materials used – whether they were weight-losing or weight-gaining in the process of production, whether they were geographically localized in their distribution or available more or less everywhere – was a most important consideration. Heavy materials, or those whose weight was reduced in the process of manufacture, would tend to pull the location of production to the source of the material. Manufacturing processes which added to the weight or bulk of the materials would tend to be located at the market.

Thus, Weber's basic argument was that a producer would locate primarily at the point of minimum total transport costs. But he recognized that the best location might well differ from this if other locational forces were more powerful. The major cause of such deviation, he suggested, would be the existence of geographical differences in labour costs. However, any savings in labour costs would have to more than offset the additional transport costs which would inevitably be incurred in locating at other than the minimum-transport-cost point.

Weber adopted a similar line of reasoning to determine the influence of his third location factor – agglomeration (external economies of scale). The spatial concentration of producers in a single location may well generate additional economies, particularly where producers are linked together in a functional manner. For example, in some industries closeness to suppliers and customers may be desirable or even obligatory. More generally, a cluster of economic activities makes possible the provision of a variety of services which might not be feasible for a single, isolated firm. Such external economies might also include an appropriate pool of labour. In this respect, labour and agglomeration factors may coincide in creating a strong locational pull. But in Weber's analysis, the location of production would shift from the minimum-transport-cost point only if the savings at the cheaper-labour or agglomeration location were greater than the additional transport costs which would be incurred.

Weber's emphasis on the importance of transport costs both for products and for factors of production added a significant dimension to the picture painted by the international trade theorists. At first sight it would appear that Weber's approach is most relevant for those economic activities in which transport costs are especially important. But this is a shrinking category as a result of technological changes in the transport media themselves and in the process of production (Chapter 4). As transport costs and raw material sources have come to exert less of a locational influence, Weber's emphasis on the locational attractions of labour locations has become increasingly relevant, particularly at the global scale.

Early attempts to explain international direct investment

In comparison with theories of international trade and industrial location early theoretical explanations of international direct investment were far less satisfactory.[5] For a considerable time, international direct investment was regarded as being simply one variant of international capital theory. In its most basic form this body of theory states that firms will place their investments in those locations where the financial return on the investment is greatest. There is, of course, much truth in this. But the early explanations of direct investment tended to equate such returns with geographical differences in interest rates on the financial investment. This was a reasonable assumption for portfolio investment but not for the kind of direct investment embodied in the transnational corporation. As Dunning (1973, p. 299) pointed out, 'unlike movements in portfolio capital . . . direct investment . . . involve[s] the transmission of other factor inputs than money capital, viz. entrepreneurship, technology and management expertise, and is as likely to be affected by the relative profitability of the use of these resources in different countries as that of money capital'. It was not until the early 1960s that a satisfactory body of theory concerned explicitly with foreign direct investment and the transnational corporation began to emerge (see Chapter 5).

The need for a new approach to explaining global shift

In unmodified form the traditional explanations of patterns of trade and of the location of production and investment are clearly inadequate to explain the complexities of the modern global economy. Nevertheless, they should not all be rejected out of hand. In particular, elements of both trade and location theory remain important to our understanding. Comparative advantage based upon different factor endowments *is* a most important element. Geographical distance reflected in movement costs *does* influence the spatial pattern of economic activities at the global scale. Differential labour costs and economies of spatial agglomeration *are* highly significant influences on global shifts in economic activity. But although we should continue to utilize some of the insights of these bodies of theory, we have to recognize their limitations. For example, both trade and location theory have tended to be *static* whereas we live in a *volatile and dynamic global system* in which technological change is rapid and endemic. Both trade and location theories have tended to assume very simple economic–geographical relationships and very simple decision-making processes, whereas the real world is made up of complex interactions between firms of widely varying sizes though increasingly dominated by massive transnational corporations. It is made up, too, of national governments and multinational political groupings each of which pursues its own, often widely differing, industrial and trading policies. These create important 'discontinuities' at national and regional boundaries and add greatly to the complexity of the global economic system. It is these organizational and institutional forces which are creating global shifts in economic activity and redrawing the global economic map. Comparative advantage is not simply 'given'; it is created and re-created by human action.

The major problem we face in explaining global shifts in economic activity and in trying to understand the processes of internationalization and globalization is that we are dealing with a very complex and highly interconnected set of processes. There is no unique point of entry; there are many possible approaches. The overall logic of Part Two is that the globalization of economic activity is the manifestation of the internationalization and globalization of capital in the form, primarily, of the transnational corporation. The TNC is undoubtedly the primary force shaping and reshaping the geography of the contemporary global economy. But it is misleading to regard TNCs as slicing their way at will through political boundaries and imposing their own corporate demands on supine and totally powerless governments. The situation is far more complex than this. Although the balance of power may well have shifted towards the TNC, this does not mean that its will always prevails. There are intricate bargaining relationships between TNCs and nation states. The extent to which TNCs can pursue and implement globally integrated strategies is strongly affected by the particular attitudes and behaviour of national governments, whether acting singly or within regional groupings like the European Community.

Hence, the current geographical structure of the global economy is the outcome of a dynamic and complex interaction between TNCs' competitive strategies on the one hand and national government strategies on the other. This interaction, which is sometimes co-operative and sometimes conflictual, is set within the context of a volatile technological environment. The four chapters of Part Two attempt to capture both the individual nature of these processes and also the interactions between them. The first (Chapter 4) examines those aspects of the technological environment which are most relevant to our understanding of the internationalization and globalization of economic activity. Chapter 5 explores some of the theoretical frameworks which have been put forward to explain the existence of TNCs, that is, why they engage in the international production of goods and services. Chapter 6 looks at the continuing role of the nation state in the global economy, again concentrating on those aspects of political decision-making which impinge most directly on the processes of globalization. Finally, Chapter 7 synthesizes these three distinct elements into a coherent treatment of how the global system of economic activities is organized and reorganized, with a particular emphasis on networks of relationships within and between firms.

Notes for further reading

1. All the standard texts on international economics contain explanations of the principle of comparative advantage as do most texts on international business. See, for example, Winters (1985), Czinkota, Rivoli and Ronkainen (1989), Daniels and Radebaugh (1989) and Root (1990).
2. See Dunning (1973) for a summary of criticisms of international trade theory.
3. Dicken and Lloyd (1990) provide a detailed account of location theories.
4. Several writers have pointed to the close link which exists between trade theory and location theory and urged their integration into a single body of theory. In 1933 Ohlin asserted that 'the theory of international trade is only a part of a general localisation theory' (p. vii). Following Ohlin's lead, some twenty years later Isard (1956, p. 207) stated that 'location and trade are as the two sides of the same coin'. He argued (ibid., p. 53) that 'location cannot be explained unless at the same time trade is accounted for and . . . trade cannot be explained without the simultaneous determination of locations'. Despite these and other pleas for integration between the two bodies of theory, however, relatively little progress has been made other than at the most general and abstract level.
5. General reviews of the early attempts to explain international direct investment are provided by Dunning (1973), Pitelis and Sugden (1991).

Chapter 4
The Role of Technological Change

Introduction: technology and economic change

Technological change is at the heart of the process of economic growth and economic development. As Joseph Schumpeter pointed out many years ago,

> the fundamental impulse that sets and keeps the capitalist engine in motion comes from the new consumers' goods, the new methods of production or transportation, the new markets, the new forces of industrial organization that capitalist enterprise creates.
>
> (Schumpeter, 1943, p. 83)

Technological change is the 'prime motor of capitalism'; the 'great growling engine of change' (Toffler, 1971); the 'fundamental force in shaping the patterns of transformation of the economy' (Freeman, 1988); the 'chronic disturber of comparative advantage' (Chesnais, 1986). Although technologies, in the form of inventions and innovations, originate in specific places, they are no longer confined to such places. Innovations spread or diffuse with great rapidity under current conditions. Indeed, one of the most significant sets of innovations is in the sphere of communications, which itself facilitates such technological diffusion.

Technology is, without doubt, one of the most important contributory factors underlying the internationalization and globalization of economic activity:

> It has long been understood that technological change, through its impact on the economics of production and on the flow of information, is a principal factor determining the structure of industry on a national scale. This has now become true on a global scale. Long-term technological trends and recent advances are reconfiguring the location, ownership, and management of various types of productive activity among countries and regions. The increasing ease with which technical and market knowledge, capital, physical artefacts, and managerial control can be extended around the globe has made possible the integration of economic activity in many widely separated locations. In doing so, technological advance has facilitated the rapid growth of the multinational corporation with subsidiaries in many countries but business strategies determined by headquarters in a single nation.
>
> (Brooks and Guile, 1987, p. 2)

However, in looking specifically at technology in this chapter, we need to beware of adopting a position of technological determinism. It is all too easy to be seduced by the notion that technology 'causes' a specific set of changes, makes particular structures and arrangements inevitable or that the path of technological change is linear and sequential. Technology in, and of, itself does not cause particular kinds of change. Technology is not independent or autonomous; it does not have a life of its own. *Technology is a social process*; it is created and adopted (or not) by human agency: individuals, organizations. The ways in which technologies are used – even their very creation – are conditioned by their social context. In effect, from the viewpoint taken here, this means the values and motivations of capitalist business enterprises operating within an intensely competitive system. Choices and uses of technologies, therefore, are influenced by the drive for profit, capital accumulation and investment, increased market share and so on. In one sense, technology is an *enabling* agent: it makes possible new structures, new organizational and geographical arrangements of economic activities, new products and new processes, while not making particular outcomes inevitable. But in certain circumstances technology may be more of an *imperative*. In a highly competitive environment, once a particular technology is in use by one firm, then its adoption by others may become virtually essential to ensure competitive survival. More generally, as Freeman (1982, p. 169) points out, for business firms 'not to innovate is to die'.

In this chapter we focus only on certain aspects of technology and technological change: those which specifically influence the processes of internationalization and globalization of economic activity. The chapter is divided into three major parts. First, some of the broad characteristics of technological change are discussed in order to identify the key technologies and their evolution over time. As in Part One of this book our focus is on the contemporary picture but with enough historical background to provide an understanding of what is happening today. Second, we focus on the 'space-shrinking' technologies of transport and communication which are obviously central to the processes of internationalization and globalization. Third, we look at technological changes in both products and processes and explore the extent to which totally new forms of production technology and organization are occurring.

Technological revolutions

A great deal of technological change passes unnoticed; it consists of the small-scale, progressive modification of products and processes. Freeman (1987) calls such changes *incremental innovations*.[1] More obvious are the *radical innovations*: discontinuous events which may drastically change existing products or processes. A single radical innovation will not, however, have a widespread effect on the economic system; 'its economic impact remains relatively small and localized unless a whole cluster of radical innovations are linked together in the rise of new industries and services, such as the synthetic materials industry or the semiconductor industry' (Freeman, 1987, p. 129).

More significant still are *changes of technology system*, which not only affect many parts of the economy but also may generate totally new industries. Freeman (1987) suggests that the following five 'generic' technologies have created such new technology systems:

1. information technology;
2. biotechnology;
3. materials technology;
4. energy technology;
5. space technology.

The fourth category of technological change identified by Freeman he terms *changes in the techno-economic paradigm*. These are the truly large-scale revolutionary changes which are:

the 'creative gales of destruction' that are at the heart of Schumpeter's long wave theory. They represent those new technology systems which have such pervasive effects on the economy as a whole that they change the 'style' of production and management throughout the system. The introduction of electric power or steam power or the electronic computer are examples of such deep-going transformations. A change of this kind carries with it many clusters of radical and incremental innovations, and may eventually embody several new technology systems. Not only does this fourth type of technological change lead to the emergence of a new range of products, services, systems and industries in its own right – it also affects directly or indirectly almost every other branch of the economy . . . the changes involved go beyond specific product or process technologies and affect the input cost structure and conditions of production and distribution throughout the system.

(Freeman, 1987, p. 130)

Long waves

The concept of long waves was briefly referred to earlier in both the Prologue to Part One and in our discussion of the roller-coaster of economic growth in Chapter 2. The notion that global economic growth occurs in a series of long waves of more or less fifty-year duration is generally associated with the work of the Russian economist N. D. Kondratiev in the 1920s, although he did not invent the idea.[2]

Figure 4.1 is a highly simplified picture of the kind of long-wave sequence commonly envisaged. Four complete K-waves are identified with the implication that we are entering a fifth. Each wave lasts for approximately fifty years and appears to be divided into four phases: prosperity, recession, depression, recovery. Each wave tends to be associated with particularly significant technological changes around which other innovations – in production, distribution and organization – cluster and ultimately spread through the economy. Such diffusion of technology stimulates economic growth and employment, though technology alone is not a sufficient cause of economic growth. As Rothwell (1981) points out, demographic, social, industrial, financial and other demand conditions have 'also to be right'. At some point, however, growth slackens:

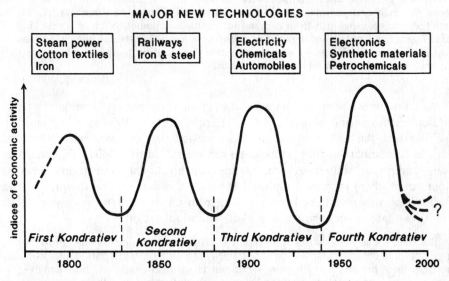

Figure 4.1 Long waves of economic activity and their associated major technologies

demand may become saturated or firms' profits squeezed through intensified competition. As a result, the level of new investment falls, firms strive to rationalize and restructure their operations and unemployment rises. Of course, a central assumption of the long-wave idea is that eventually the trough of the wave will be reached and economic activity will turn up again. A new sequence will be initiated on the basis of key technologies – some of which may be based on innovations which emerged during recession itself – and new investment opportunities.

Although there is some disagreement over the precise mechanisms and timing involved, it is generally agreed that major technological changes are associated with long waves. In fact, each of the waves is associated with changes in the techno-economic paradigm; with the culmination of a technology revolution, as one set of techno-economic practices is displaced by a new set. This is not a sudden process but one which occurs gradually and involves the ultimate 'crystallization' of a new paradigm. As Figure 4.1 shows, the first four K-waves were associated, respectively, with major technological revolutions in steam power, cotton textiles and iron; railways and iron and steel; electricity, chemicals and automobiles; and electronics, synthetic materials and petrochemicals.[3] Each successive K-wave has a specific geography; technological leadership in one wave is not necessarily maintained in succeeding waves. The technological leaders of K1 were Britain, France and Belgium. In K2 these were joined by Germany and the United States. K3 saw leadership firmly established in Germany and the United States although the other earlier leaders were still prominent and had been joined by Switzerland and the Netherlands. By K4 Japan, Sweden and the other industrialized countries were in the leadership group.

The fourth Kondratiev covers the period of economic growth and global shifts discussed in Chapters 2 and 3. Rothwell summarizes the evolution of industry during this period very succinctly:

> The post-war era has been characterized by the rapid growth of a bunch of 'new' industries based on new technological possibilities that emerged during the previous twenty years or more . . . The industries that emerged on a significant scale during the 1940s and 1950s – electronics, synthetic materials, solid-state devices, petro-chemicals, agro-chemicals, composite materials and pharmaceuticals – created rapidly growing new markets. At the same time there was a rapid growth in demand for capital equipment, often of a new kind. The wealth generated by the emergence of these new technology-based industries caused an associated boom in demand for consumer durables, leading to the rapid growth of the automobile and consumer white goods industries.
>
> (Rothwell, 1982, p. 364)

Table 4.1 outlines the major characteristics of the three phases identified by Rothwell. He describes the period between 1945 and approximately 1964 as a 'dynamic growth phase' in which the technological emphasis was on new products whose manufacture created large numbers of jobs. During the second half of the 1960s – the 'consolidation phase' – emphasis shifted to process and organizational innovations. Productivity (output per worker) increased rapidly but so too did demand, so that employment in manufacturing was maintained overall. But from the late 1960s and especially during the 1970s a phase of 'maturity and market saturation' set in:

> By this time the new industries were highly concentrated, production was centred on very large units and development was aimed primarily at process rationalization and productivity increase. Price became a much more significant factor in competition. Productivity growth (albeit at a lower rate than previously) outstripped demand growth in largely saturated

Table 4.1 The sequence of industrial evolution since 1945

1945 to approximately 1964 – dynamic growth phase	Mid-to-late 1960s – consolidation phase	Late 1960s to present – maturity and market saturation phase
Emergence of new industries based largely on new technological opportunities Production initially in small units	Increasing industrial concentration Highly dynamic economies	Industry highly concentrated Very large production units, often vertically integrated
Emphasis on product change and the introduction of many new products	Introduction of organizational innovations Increasing emphasis on process improvements Some major product changes, but based mainly on existing technology	Some product change, but emphasis mainly on production process rationalization Increasing organizational rationalization including foreign direct investment in areas of low labour cost Growing automation
Rapidly growing new markets Some market regeneration in traditional areas, e.g. textiles Competitive emphasis mainly on product availability and non-price factors	Markets still growing rapidly Competitive emphasis still mainly on non-price factors	Stagnating and replacement markets Where products are little differentiated the importance of price in competition is high
New employment generation (output growth greater than productivity growth)	Rapid productivity growth Output and productivity growth in rough balance (manufacturing employment roughly stable)	Productivity growth greater than output (demand) growth Rapidly growing manufacturing unemployment

(*Source*: based on Rothwell (1981), Table 1)

markets, and many jobs were lost. At the same time firms increasingly located production of mature product lines in areas of low labour cost, and further jobs were lost in the advanced economies.

(Rothwell, 1982, pp. 364–5)

Information technology: a key generic technology

The fifth Kondratiev cycle, which appears to have begun in the 1980s/1990s is associated primarily with the first of the five 'generic' technologies referred to above: *information technology* (IT). It, too, is likely to have a distinctive geography, in which Japan will be especially prominent, along with the United States, Germany and Sweden and, probably, some of the East and South East Asian NICs (Freeman and Perez, 1988).

Information technology[4] is, in Freeman's view, the new techno-economic paradigm around which the next wave of technological and economic changes will cluster. But, as Hall and Preston (1988, p. 30) point out, information technology in itself is nothing new: 'for thousands of years, since the first cave paintings and the invention of writing, humans have used tools and techniques to collect, generate and record data'. Consequently, they identify three main phases of information technology:

1. Simple pictorial representation and written language, evolving eventually into printing: its basic elements were paper, writing instruments, ink and printing presses.

2. Mechanical, electromechanical and early electronic technologies which developed during the late nineteenth and early twentieth centuries: the basic elements were the telephone, typewriter, gramophone/phonograph, camera, tabulating machine, radio and television.

3. Microelectronic technologies, which emerged only in the second half of the twentieth century: the basic elements are computers, robots and other information-handling production equipment, and office equipment (including facsimile machines).

Hall and Preston regard (1) as 'old IT'; (2) and (3) together as 'new IT'. They then employ a further term 'convergent IT' to refer to the newest advances of the 1970s and 1980s, whereby computers and telecommunications are integrated into a single system of information processing and exchange.

It is this quality of the *convergence* of two initially distinct technologies which is of the greatest importance for developments in today's (and tomorrow's) global economy. It is this kind of information technology which is most significant for the processes of internationalization and globalization of economic activities. When we use the term 'information technology' or 'IT' in this and the following chapters it is the 'convergent IT' which is involved. Figure 4.2 shows the nature of this convergence between communications technology, which is concerned with the transmission of information, and computer technology, which is concerned with the processing of information. As

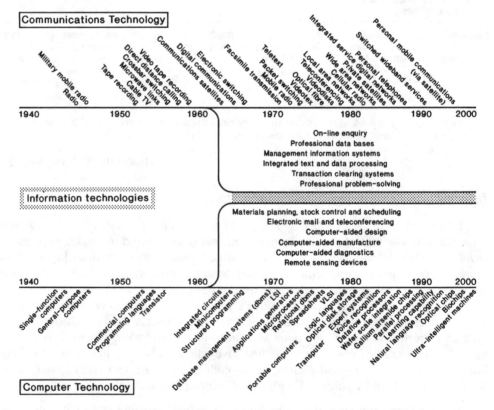

Figure 4.2 Information technology: the convergence of the technologies of communication and computers (*Source*: based on Freeman (1987), Figure 2)

the diagram indicates, it is not until the early 1960s that we can clearly identify convergent information technology.

The 'space-shrinking' technologies: transport and communications

A fundamental prerequisite of the evolution of international production and of the transnational corporation is the development of technologies which overcome the frictions of space and time. The most important of such *enabling* technologies – and the most obvious – are the technologies of transport and communications. Neither of these technologies can be regarded as the cause of international production or of the TNC; rather, they make such phenomena feasible. But without them, today's complex global economic system simply could not exist.

Indeed, both the geographical and organizational scale at which any human activity can occur is directly related to the available media of transport and communication. Similarly, the degree of geographical specialization – the spatial division of labour – is constrained by these media. Transport and communication technologies perform two distinct, though closely related and complementary roles. Transport systems are the means by which materials, products and other tangible entities (including people) are transferred from place to place. Communication systems are the means by which information is transmitted from place to place in the form of ideas, instructions, images, and so on. For most of human history, transport and communications were effectively one and the same. Prior to the invention of electric technology in the nineteenth century, information could move only at the same speed, and over the same distance, as the prevailing transport system would allow. Electric technology broke that link, making it increasingly necessary to treat transport and communication as separate, though intimately related, technologies. Developments in both have transformed our world, permitting unprecedented mobility of materials and products and a globalization of markets.[5]

Major developments in transport technology

In terms of the time it takes to get from one part of the world to another there is no doubt that the world has 'shrunk' dramatically (Figure 4.3). For most of human history, the speed and efficiency of transport were staggeringly low and the costs of overcoming the friction of distance prohibitively high. Movement over land was especially slow and difficult before the development of the railways. Indeed, even as late as the early nineteenth century, the means of transport were not really very different from those prevailing in biblical times. The major breakthrough came with two closely associated innovations: the application of steam power as a means of propulsion and the use of iron and steel for trains and railway tracks and for ocean-going vessels. These, coupled with the linking together of overland and ocean transport and the cutting of the canals at Suez and Panama, greatly telescoped geographical distance at a global scale. The railway and the steamship introduced a new, and much enlarged, scale of human activity. The flow of materials and products was enormously enhanced and the possibilities of geographical specialization were greatly stimulated. Such innovations were a major factor in the massive expansion in the global economic system during the nineteenth century.

The twentieth century, and especially the past few decades, has seen an accelera-

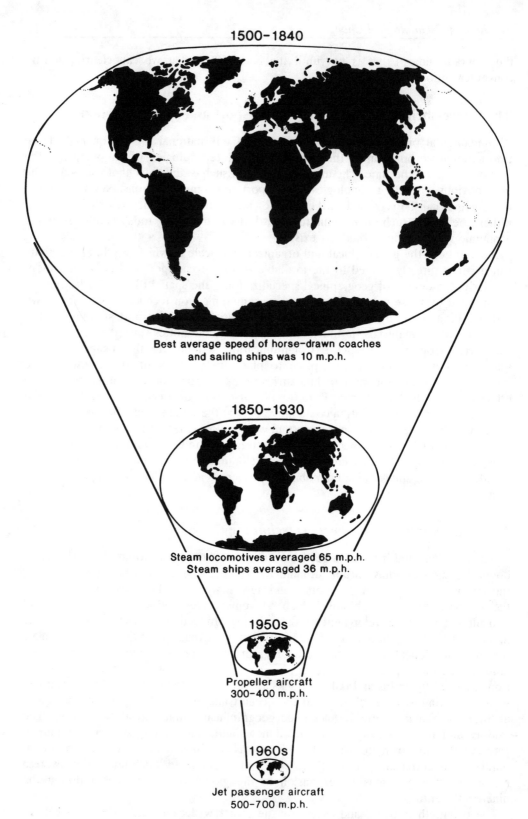

1500–1840

Best average speed of horse-drawn coaches
and sailing ships was 10 m.p.h.

1850–1930

Steam locomotives averaged 65 m.p.h.
Steam ships averaged 36 m.p.h.

1950s

Propeller aircraft
300–400 m.p.h.

1960s

Jet passenger aircraft
500–700 m.p.h.

Figure 4.3 Global shrinkage: the effect of changing transport technologies on 'real' distance (*Source*: based on McHale (1969), Figure 1)

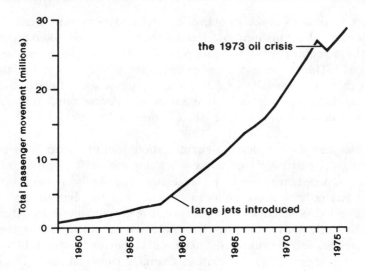

Figure 4.4 The 'take-off' of air traffic after the introduction of the jet aircraft (*Source*: based on Cherry (1978), Figure 3.16)

tion of this process of global shrinkage. In economic terms, the most important developments have been the introduction of commercial jet aircraft, the development of much larger ocean-going vessels (superfreighters) and the introduction of containerization, which greatly simplifies transhipment from one mode of transport to another and increases the security of shipments. Of these, it is the jet aircraft which has had the most pervasive influence, particularly in the development of the TNC. It is no coincidence that the take-off of TNC growth (see Figures 3.2 and 3.3) and the (more literal) take-off of commercial jets both occurred during the 1950s. As a consequence, in terms of time, New York is now closer to Tokyo than it was to Philadelphia in the days of the thirteen colonies. Figure 4.4 gives some indication of the effect of the introduction of large commercial jets on international air traffic from and to London. From the late 1950s the growth curve was virtually a straight line with a very steep upward slope.

Major developments in communications technology

Both the time and relative cost of transporting materials, products and people have fallen dramatically as the result of technological innovations in the transport media. However, such developments have depended, to a considerable degree, on parallel developments in communications technology. In the nineteenth century, for example, neither rail nor ocean transport could have developed as they did without the innovation of the electric telegraph and, later, the oceanic cable. Only with the ability to transmit information at great speed – for example to co-ordinate flows of commodities on a global scale – could the potential of the transport technologies be fully realized. Similarly, the far more complex global transport system of the present day depends fundamentally on telecommunications technology.

The communications media are, however, fundamentally significant in their own right. Indeed, as implied in our earlier discussion of the central role of information technology, communications technologies should now be regarded as the key technology transforming relationships at the global scale. As Henderson and Castells (1987, p. 6) observe: 'The new telecommunications technologies are the electronic highways of the informational age, equivalent to the role played by railway systems in the process of industrialization.' The communications technologies are significant for all economic activities but they are especially vital to those economic sectors and activities whose primary function is to collect, transform and transmit information; that is, the burgeoning business services, including finance (see Chapter 11).

Satellites, cables and faxes Global communications systems have been transformed radically during the past twenty or thirty years through a whole cluster of significant innovations in information technology (see Figure 4.2). Probably the most important catalyst to enhanced global communications has been the development of satellite technology. The use of satellites for commercial telecommunications dates only from 1965 when the Early Bird or Intelsat I satellite was launched. It was the first 'geostationary' satellite, located above the Atlantic Ocean and capable of carrying 240 telephone conversations or two television channels simultaneously. Since then the carrying capacity of the communications satellites has grown exponentially. Intelsat IV carried 6,000 simultaneous telephone conversations; Intelsat V carries 12,000 as well as television channels. According to the International Institute of Communications (1990) the 1989-launched satellite carries 120,000 bidirectional circuits for intercontinental communications. By the year 2000, the Intelsat system could be carrying 700,000 circuits.

Satellite technology, together with a whole host of other communications technologies, is making possible quite remarkable levels of global communication of conventional messsages and also the transmission of data. In this respect, the key element is the linking together of computer technologies with information-transmission technologies over vast distances. It has become possible for a message to be transmitted in one location and received in another on the other side of the world virtually simultaneously. Consequently,

> communications costs are becoming increasingly insensitive to distance. The crucial fact is the economics of satellite communication. Within the beam of a satellite it makes no difference to costs whether you are transmitting for five hundred miles or five thousand miles. The message goes from the earth station up twenty-two thousand three hundred miles to the satellite and down again twenty-two thousand three hundred miles. It makes no difference whether the two points on earth are close together or far apart . . . The important point about satellites is that their existence sets a limit on the extent to which costs are a function of distance. Many other technologies may compete with satellites, but in the end, satellite communication will ultimately be cheaper. Whatever that distance, it makes no difference how much further one communicates, the costs will be the same. Under those circumstances, the cost of access to any particular data base or information service becomes largely independent of its location. That does not make it free nor even necessarily cheap. There are costs for compiling the data and costs for manipulating it.
>
> (de Sola Pool, 1981, pp. 162–3)

Not only are transmission costs by satellite insensitive to distance but also the user costs have fallen dramatically. In the 1960s the annual cost of an Intelsat telephone circuit, giving a connection from one point on the earth's surface to any other, was more than $60,000. In the late 1980s the same facility cost only $9,000. Satellite communications are now being challenged by a new technology: optical fibre cables. Optical fibre systems have a very large carrying capacity, and transmit information at

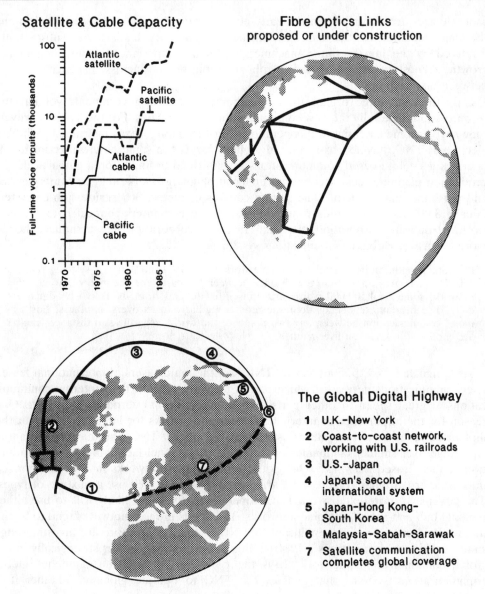

Figure 4.5 Developments in global telecommunications: satellites and optical fibre networks (*Sources*: based on *The Economist* (17 October 1987) Telecommunications Supplement, p. 23; Warf (1989), Figure 3; *Sunday Times* (6 December 1987), p. 79)

very high speed and with a high signal strength.[6] Figure 4.5 shows the rapid increase in satellite and cable capacity in the Atlantic and Pacific Ocean regions. It also shows the basic skeleton of the optical fibre communications system being constructed in the Pacific and the more comprehensive plans to build a 'global digital highway'. This will link the world's three major markets of North America, Western Europe and Japan, using a network of optical fibre cables capable of carrying 100,000 simultaneous messages. Within these very large-scale developments other smaller, but highly significant, changes are occurring. In the early postwar development of telecommunications the major medium was the telex. Within the last few years this has been substantially displaced by the fax (facsimile machine), which has experienced quite phenomenal growth. Almost two-thirds of the telephone traffic flowing across the Pacific is now between fax machines.

Only the very large organization, whether business or government, yet has the resources to utilize fully the new communications technologies. For the TNC, however, they have become essential to its operations.[7] For example, Texas Instruments, the US electronics TNC, has approximately fifty plants located in some nineteen countries. It operates a satellite-based communications system to co-ordinate, on a global scale, its production planning, cost accounting, financial planning, marketing, customer service and personnel management. The system consists of almost 300 remote job-entry terminals, 8,000 inquiry terminals and 140 distributed computers connected to the network. Mitsubishi Corporation, the giant Japanese corporation, has a similar, though more extensive, global communications system:

> the total length of its communication network lines, its central nerve system, reaches 450,000 kilometres, more than the distance between the earth and the moon. Through this network, some . . . 4,500,000 words are transmitted to and from its Tokyo headquarters daily. The headquarters sends domestic reports regularly to its overseas offices. Furthermore, communication between any two points of the world, regardless of distance, can be accomplished in less than five minutes.
>
> (Nakase, 1981, p. 85)

As Langdale (1989, p. 506) shows, TNCs are the major users of international leased telecommunications networks, which permit them to transmit their internal communications at great speed to other parts of their international corporate network: 'One reason for the importance of international leased networks for large TNCs is that the unit cost of a leased circuit falls as usage increases; major TNCs thus have a substantial advantage over smaller companies in being able to channel large volumes of information in their leased network.' The spatial form of such networks may be categorized into three broad types (Figure 4.6), although a TNC may well use a mixture of types. The larger and more extensive a TNC's operations, the more likely it is to use either regional hub-and-spoke or global networks. The nodes in these networks tend to be the major world cities and it is significant that NICs such as Singapore are investing huge sums in their communications infrastructures to position themselves strategically in the global communications network. It is the possession of such instantaneous global communications systems that enables the TNC to operate globally, whether it is engaged in manufacturing, resource exploitation or business services.

The mass communications media   Developments in the communications media have revolutionized the potential for large organizations to operate over vast geographical distances and, as such, have played a key role in facilitating the development of the TNC. But there is another sphere – that of the mass media – in which innovations have transformed the global economy and are facilitating the globalization of markets. Large business firms require large markets to sustain them; global firms need global markets. The existence of such markets obviously depends on income levels, but it depends, too, on potential customers becoming aware of the firm's offerings and being persuaded to purchase them. Even where consumer incomes are low the ground may be prepared for possible future ability to purchase by creating a desirable image. The mass media are

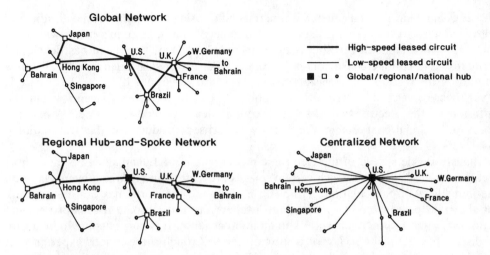

Figure 4.6 Major types of international leased telecommunications networks for a United States TNC (*Source*: Langdale (1989), Figure 1)

particularly powerful means both of spreading information and of persuasion, hence their vital importance to the advertising industry.

On a global scale, it is the electronic media – particularly radio and television – which are the most significant. In part, this is because of their vividness and sense of immediacy and involvement. But an important characteristic of the electronic media is that they make no demands on literacy, a demand which even the most primitive news-sheet makes. Access to the electronic media is largely governed by level of affluence. Not surprisingly, therefore, there are enormous geographical variations in the availability of radio and television. At one extreme, the United States had 2,043 radio receivers and 790 television receivers per thousand population in the mid-1980s. At the other extreme, India had a mere fifty-five radio receivers and four TV receivers per thousand. Whereas all the developed market economies have television networks, some developing countries do not. However, the electronic media are spreading very rapidly throughout the world.

Perhaps more than any other innovation in the mass media it has been the development of the transistor radio receiver which has had the most revolutionary effects, especially in developing countries. Not only is it portable but it is also relatively cheap. Largely because of this, the sequence of development of the mass media has been rather different in the developing countries:

> In the Western world, the pattern has been newspapers, radio, television. In Africa and much of Asia, the first contact the ordinary man has with any means of mass communication is the radio. It is the transistor which is bringing the people of remote villages and lonely settlements into contact with the flow of modern life.
>
> (Hachten, 1974, p. 99)

In such circumstances, the actual reach of the electronic media may be greater than the figures on sets per thousand population suggest. There is a great deal of collective listening and viewing in public places.

The electronic media transmit messages of all kinds. From our viewpoint, the most important point is that a very large proportion of these messages are commercial messages aimed at the consumer. Commercial advertising is a feature of most radio and television networks outside the centrally planned economies. Even in systems which are state controlled some advertising is often included. Indeed, 'television has developed primarily as a commercial medium . . . commercial advertising is carried by all but a handful of the world's . . . television systems' (Dizard, 1966, pp. 12, 13). Thus, there is a very close relationship between the global spread of advertising agencies and the diffusion of the electronic media.[8] The communications media, in effect, open the doors of national markets to the heavily advertised products of the transnational producers.

The worldwide spread of the electronic media has, in McLuhan's famous metaphor, created a *global village* in which certain images are shared and in which events take on the immediacy of participation. Although in one sense the world may not have shrunk for the rural peasant or the urban slum dweller with no adequate means of personal transport, it has undoubtedly shrunk in an indirect sense. It is now possible to be aware of distant places, of life-styles, of consumer goods through the vicarious experience of the electronic media. Hence the importance of these media to the TNC: they perform a vital role in creating a global shopping centre.

Summary   Technological developments in communications media have transformed space–time relationships between all parts of the world. Of course, not all places are equally affected. Consistent with the nature of the *time–space convergence* process as defined by Janelle is its inherent *geographical unevenness*. In general, the places which benefit most from innovations in the communications media are the 'important' places. New investments in communications technology are market-related; they go to where the returns are likely to be high. The cumulative effect is to reinforce both certain communications routes at the global scale and to enhance the significance of the nodes (cities/countries) on those routes. There is an additional factor which limits the universal spread of new communications technologies. In virtually all countries of the world, governments *regulate* the communications industries within their borders. Today, however, there is a strong trend towards the *deregulation* of telecommunications in several countries.

Technological changes in products and processes

When thinking of 'products' most of us tend to think of consumer products. But, for industry as a whole, most products are themselves intermediate in nature; they form the inputs to subsequent stages in the production process. Consequently, 'product and process innovation are inextricably interdependent . . . one firm's product is another's manufacturing equipment or material . . . the production process . . . [is] . . . a system of linked productive units' (Utterback, 1987, pp. 17, 18). This is a point we shall return to in Chapter 7. Bearing in mind this intricate relationship between product and process, however, it is useful to look at them separately in the first instance.

Product innovation and the product life-cycle

In an intensely competitive environment, the introduction of a continuous stream of new products is essential to a firm's profitability and, indeed, survival. As Casson (1983, p. 24) states: 'Long-run growth requires either a steady geographical expansion

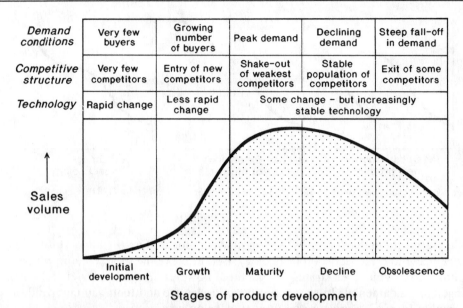

Demand conditions	Very few buyers	Growing number of buyers	Peak demand	Declining demand	Steep fall-off in demand
Competitive structure	Very few competitors	Entry of new competitors	Shake-out of weakest competitors	Stable population of competitors	Exit of some competitors
Technology	Rapid change	Less rapid change	Some change – but increasingly stable technology		

Sales volume

| Initial development | Growth | Maturity | Decline | Obsolescence |

Stages of product development

Figure 4.7 The product life-cycle

of the market area or the continuous innovation of new products. In the long run only product innovation can avoid the constraint imposed by the size of the world market for a given product.' The idea that the demand for a product will decline over time is captured in the concept of the product life-cycle.[9]

The essence of the product life-cycle is that the growth of sales of a product follows a systematic path from initial innovation through a series of stages: early development, growth, maturity and obsolescence (Figure 4.7). When a new product is first introduced on the market the total volume of sales tends to be low because customers' knowledge is limited and also they tend to be uncertain about the product's quality and reliability. Assuming that the new product gains a foothold in the market (and very many do not get beyond this initial stage) it then enters a phase of rapid growth as overall demand increases. Such growth is likely to have a ceiling, however: the product attains maturity in which demand levels out. Eventually, demand for the product will slacken as the product becomes obsolescent.

The kind of development path suggested by the product life-cycle concept clearly has very important implications for the growth of firms and for their profit levels. The product life-cycle implies that all products have a limited life; that obsolescence is inevitable. Of course, the rate at which the cycle proceeds will vary from one product to another. In some highly ephemeral products the cycle may run its course within a single year or even less, in others the cycle may be very lengthy. However, there is growing evidence that the general length of product cycles is tending to become shorter. Thus, in order to continue to grow and to make profits firms need to innovate on a regular basis (or to acquire innovations from other firms).

Figure 4.8 shows three ways in which a product's sales may be maintained or increased. One way is to introduce a new product as the existing one becomes obsolete so that 'overlapping' cycles occur (Figure 4.8a). An alternative is to find ways of extending the cycle for the existing product either by making minor modifications in the

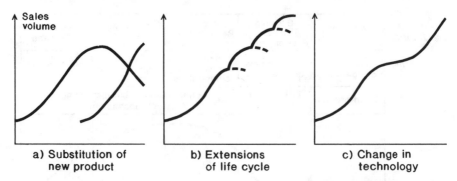

Figure 4.8 Some variations on the product life-cycle theme (*Source*: van Duijn (1983), Figure 2.3)

product itself to 'update' it or by finding new uses for it (Figure 4.8b). Third, changes may be made to the production technology itself to make the product more competitive (Figure 4.8c). Whichever strategy is pursued, however, innovation and technological change are fundamental. In so far as product cycles are shortening in many industries, this implies increasing pressure on firms to develop new products.

The production process and technology

In today's intensely competitive global environment product innovation alone is inadequate as a basis for a firm's survival and profitability. Firms must endeavour to operate the production process as efficiently as possible. Recent developments in technology – and, especially, in information technology – are having profound effects upon production processes in all economic sectors. Figure 4.9 shows that three major, and closely interrelated, decisions are involved in the production process: decisions relating to the *technique* to be adopted, the *scale* of production and the *location* of production. The technique decision concerns both the particular technology used and also the way in which the various inputs or factors of production are combined. It is almost always possible to vary the precise combination of, say, labour and capital according to their relative availability and cost. However, there are limits to such substitution of factors. Some production processes are intrinsically more capital-intensive than others and vice versa. Closely related to the question of technique is that of the scale of production. In general, the average cost of production tends to decline as the volume of production increases. The extent of such economies of scale varies considerably from one industry to another. They are much greater, for example, in automobile production than in the manufacture of fashion garments. Technique and scale are, themselves, closely related.

Table 4.2 identifies seven types of production process ranging from the manufacture of individual units to customers' requirements through to mechanized and automated mass production and continuous-flow processing. In each case, both the technique and the scale of production differ and so, too, does the type of labour required. Finally, both technique and scale are intimately linked to the question of location. Smith expresses this relationship very concisely:

> The choice of location cannot be considered in isolation from scale and technique. Different scales of operation may require different locations to give access to markets of different

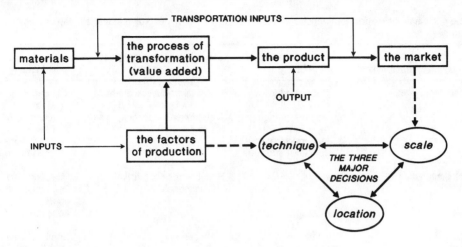

Figure 4.9 Basic elements in the production process (*Source*: Smith (1981), Figure 2.1)

Table 4.2 A classification of production processes

Type of process	Industry examples
1. Craft-type unit production to individual customer requirements	various
2. Craft-type batch production	aircraft; construction
3. Manual assembly	electronics assembly; garments
4. Mechanized assembly	automobiles; home appliances
5. Mechanized processing	textiles
6. Automated processing	pulp and paper mills; standardized metal working (e.g. ball bearings)
7. Continuous processing	petrochemicals; refining

(*Source*: based in part on Storper and Walker (1984), pp. 34–6)

sizes . . . Different techniques will favour different locations, as firms tend to gravitate toward cheap sources of the inputs required in the largest quantities, and location itself can influence the combination of inputs and hence the technique adopted.

(Smith, 1981, pp. 23–4)

Clearly, therefore, a firm which is seeking to reduce its production costs or to increase its efficiency and productivity can seek such economies at different points in the production process. It can attempt to purchase lower-cost inputs. In the case of material inputs this increasingly involves a shift to supplies in developing countries. In the case of labour, a relatively immobile factor of production, the search for lower costs may involve the physical relocation of production to a cheap labour location.

Production processes and the product life-cycle

We looked earlier at the product life-cycle as a way of understanding the drive for the continuous innovation and introduction of new products. It has been suggested by a number of writers that the nature of the production process itself also tends to vary

Table 4.3 Changes in the characteristics of the production process according to stage in the product life-cycle

| Production characteristics | Stage in the product cycle | | |
	Early	Growth	Maturity
Technology	1. Short production runs 2. Rapidly changing techniques 3. Dependence on specialist suppliers	1. Mass production methods gradually introduced 2. Variations in technique still frequent	1. Long runs and stable technology 2. Few innovations of importance
Capital intensity	Low	High because of high rate of obsolescence	High because of large quantity of specialized equipment
Critical labour requirements	Scientific and engineering	Management	Semi-skilled and unskilled labour
Industry structure	1. Entry is 'know-how' determined 2. Numerous firms providing specialist services	1. Growing number of firms 2. Many casualties and mergers 3. Growing vertical integration	1. Financial resources critical for entry 2. Number of firms declining

(*Source*: Hirsch (1967), Table II (1))

systematically according to stages in the product life-cycle (Figure 4.7). Each stage, it is argued, will tend to have different production process attributes: of technology, of capital intensity, of labour force characteristics and of industry structure – as Table 4.3 indicates. During the early stage of the cycle production technology tends to be volatile with frequent changes in product specification. Production tends to be in short runs or batches. There is a tendency to rely on specialist suppliers and subcontractors. Capital intensity is relatively low. The most important type of labour is scientific and engineering. Entry into the new industry is determined largely by 'know-how' rather than by financial resources.

By the time the product has progressed to the growth stage (assuming that it does) some important changes have occurred in the production process. The rapidly growing demand for the product (see Figure 4.7) permits the introduction of mass production and assembly-line techniques. Even though the technology may be evolving it is less volatile than in the earlier stage of the cycle. Capital intensity at the growth stage is considerably higher and the key type of labour is now managerial. The need is for administrative and marketing, rather than scientific, skills. In the growth stage also the number of firms engaged in the industry tends to be increasing but with a high casualty rate. There is a tendency towards increased vertical integration as firms seek to ensure the stability of component supplies and to exert greater influence over distribution of their product, often through acquisition and merger.

In the mature stage of the cycle demand has reached its peak (Figure 4.7) and the market is becoming saturated. The major emphasis is on reducing the costs of production through long production runs. By this stage the technology is stable with few important changes. Long runs require the installation of high-volume specialist equipment which, in turn, increases the capital intensity of the industry. Stable technology

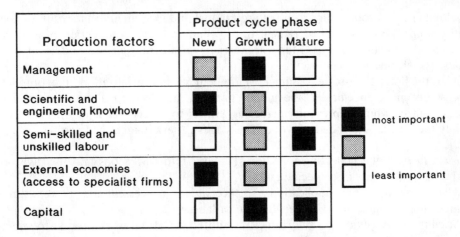

Figure 4.10 The relative importance of production factors at different stages of the product life-cycle (*Source*: Hirsch (1972), Chart 1, p. 41)

and the application of a finer division of labour alter the labour force focus. In the mature stage the emphasis is on semi-skilled and unskilled labour performing routine, repetitive tasks in a mechanized manner. Labour costs become an increasingly significant element in production costs. The major barrier to entry into the industry is finance. Acquisition and merger are important mechanisms of entry and exit.

According to this interpretation, therefore, each stage in the product life-cycle has a significant influence on the nature of the production process. Each stage has particular production characteristics. One of the most important of these is the way in which the relative importance of the major production factors changes (Figure 4.10). In general, as the cycle proceeds, the emphasis shifts from product-related technologies to process technologies and, in particular, to ways of minimizing production costs. In this respect, the relative importance of labour costs, especially of semi-skilled and unskilled labour, increases. More generally, different types of geographical location are relevant to different stages of the product cycle.

This view of systematic changes in the production process as a product matures is appealing and has some validity. There undoubtedly are important differences in the nature of the production process between a product in its very early stages of development and the same product in its maturity. But this linear, sequential notion of change in the production process is overly simplistic. At any stage, the production process may be 'rejuvenated' by technological innovation. There may not necessarily be a sequential sequence leading from small-scale production to standardized mass production. Taylor (1986, p. 753) is especially critical of the *technological determinism* which is implied in the product life-cycle: although 'there is an appealing logic . . . that derives from its simplicity. As the same time, however, simplicity is also its greatest weakness.'

Towards flexible production

This point leads us to consider the major recent developments that have been occurring in the technology of production processes and, particularly, those associated with the new techno-economic paradigm of information technology. Most technological developments in production processes are gradual and incremental: the result of 'learn-ing by doing'. But periods of radical transformation of the production process have occurred throughout history. Many argue that we are now in the middle of such a radical transformation.

Over the long time-scale of the development of industrialization, the production process or, more specifically, the labour process, has developed through a series of stages each of which represents increasing efforts to mechanize and to control more closely the nature and speed of work. The stages generally identified are:

1. *Manufacture*: the collecting together of labour into workshops and the division of the labour process into specific tasks.
2. *Machinofacture*: the application of mechanical processes and power through machinery in factories. Further division of labour.
3. *Scientific management* ('*Taylorism*'): the subjection of the work process to scientific study in the late nineteenth century (e.g. by F. W. Taylor). This enhanced the fineness of the division of labour into specific tasks together with increased control and supervision.
4. '*Fordism*': the development of assembly-line processes which controlled the pace of production.

Currently a major debate is raging as to what comes after Fordism.[10] Will it be a variant on Fordism, 'neo-Fordism', in which automated control systems are applied within a Fordist structure? Or will it be a totally new 'post-Fordism', in which the new technologies create quite different forms of production organization? It is a debate which stretches way beyond the bounds of technology and technological change into the realms of the social organization of production, of the ways in which the state regulates economic activity and the nature of consumption and markets. These are all issues which will be returned to in later chapters. Here our concern is specifically with technology.

The Fordist system is characterized by very large-scale production units using assembly-line manufacturing techniques and producing large volumes of standardized products for mass market consumption. It is a type of production especially character-istic of particular industrial sectors, notably automobiles. A number of writers now argue that this Fordist system of production (and its associated organizational struc-tures) is in 'crisis' and is being replaced by new modes of production.[11] The most important characteristic of the new system is *flexibility*: of the production process itself, of its organization within the factory and of the organization of relationships between customer and supplier firms (see Chapter 7).

The key to production flexibility lies in the use of information technologies in machines and operations. These permit more sophisticated control over the production process. With the increasing sophistication of automated processes and, especially, the new flexibility of electronically controlled technology, far-reaching changes in the process of production need not necessarily be associated with increased scale of produc-tion. Indeed, one of the major results of the new electronic and computer-aided production technology is that it permits rapid switching from one part of a process to another and allows the tailoring of production to the requirements of individual cus-tomers. 'Traditional' automation is geared to high-volume standardized production; the newer 'flexible manufacturing systems' are quite different:

Flexible automation's greatest potential for radical change lies in its capacity to manufacture goods cheaply in small volumes . . . In the past batch manufacturing required machines dedicated to a single task. These machines had to be either rebuilt or replaced at the time of product change. Flexible manufacturing brings a degree of diversity to manufacturing never before available. Different products can be made on the same line at will . . .

The strategic implications for the manufacturer are truly staggering. Under hard automation the greatest economies were realised only at the most massive scales. But flexible automation makes similar economies available at a wide range of scales. A flexible automation system can turn out a small batch or even a single copy of a product as efficiently as a production line designed to turn out a million identical items.

(Bylinsky, 1983, pp. 53–4)

Clearly, the potential of such flexible technologies is immense and their implications enormous for the nature and organization of economic activity at all geographical scales, from the local to the global. Some writers assert that flexible manufacturing is becoming the norm; the dominant style of production displacing Fordism. This is the 'post-Fordist' view which sees the hegemony of Fordism as being replaced by a new regime of flexible production and organization. However, not all agree with this diagnosis. Coombs and Jones (1989, p. 115) assert that 'it is premature to diagnose a global trend toward a unique and well-defined successor to Fordism as a paradigm for production organization'. They argue that it is more satisfactory to see current trends as consisting of a mix of three tendencies: neo-Fordist, neo-Taylorist and post-Fordist. Both Gertler (1988) and Sayer (1989) are also critical of the view which replaces Fordism with a system based on flexibility. They are not arguing that flexible technologies do not exist – clearly they do – but that they are not as universal and pervasive as is being claimed. Both Gertler and Sayer also argue that the Fordist system was never as all-pervasive as is sometimes claimed; that less rigid and smaller-scale production has always coexisted with mass production methods.

Nevertheless, there is no doubt that the application of information technologies does create new possibilities. Perez (1985, p. 447) identifies three major tendencies. First:

the trend towards information intensity rather than energy or materials intensity in production . . . [suggests that] . . . in product engineering, there would be a tendency to redesign existing goods to make them smaller, less energy-consuming, with less moving parts, more electronics and more software . . . In plant engineering, not only would energy-saving technology based on electronics be applied as a matter of good process design, but also materials-saving techniques.

Second, much enhanced flexibility of production, which

challenges the old best-practice concept of mass production in three central respects. High volume output of identical product is no longer the main route to high productivity, which can now be achieved for a diversified set of low-volume products. The 'minimum change' strategy in product development might no longer be necessary for cost effectiveness, as rapid technological change becomes much less costly and less risky. Market growth on the basis of 'homogeneous' demand is no longer essential, as the new technologies permit high profitability in catering to segmented markets and provide ample space for adapting production systems and output to specific local conditions and needs.

(*Ibid*. p. 449)

Third, arising from both increased automation and production flexibility is a potentially major impact on labour requirements in both volume of labour and type of labour required. At a global scale one possible outcome might be a reversal of offshore assembly in cheap labour locations and the reconcentration of such activities in highly automated plants in industrialized countries, closer to major markets.

A more realistic view of the current situation, therefore, would seem to be one which recognizes a continuing *diversity* of production processes and technologies but one in which the relative importance of specific processes is changing. Thus, we can find on the one hand a trend towards:

1. an *increasingly finer degree of specialization* in many production processes, enabling their fragmentation into a number of individual operations; and
2. an *increasing standardization and routinization* of these individual operations, enabling the use of semi-skilled and unskilled labour. This is especially apparent during the mature stage of a product's life-cycle.

On the other hand, we can identify a clear trend towards an *increasing flexibility in the production process*, which is altering the relationship between the scale and the cost of production, permitting smaller production runs, increasing product variety and changing the way production and the labour process are organized. The precise mix will vary from one economic sector to another, as we shall see in Part Three. In fact, it is a mistake to regard flexible manufacturing systems in a narrowly 'technological' light. Most of the benefits accrue not so much from the technology itself but from the organizational changes it involves. In other words, flexibility is more an 'organizational' property than a technical one (Bessant and Haywood, 1988).

Conclusion

The aim of this chapter has been to identify some of those key features of technological change which are most important in the internationalization and globalization of economic activity. Technological change is at the dynamic heart of economic growth and development; it is fundamental to the evolution of a global economic system. Much technological change is gradual and incremental, often unnoticed but nonetheless extremely significant. But there are periodic radical transformations of existing technologies – revolutionary developments in clusters of technologies – which dramatically alter not only products and processes in one industry but which also pervade the entire socio-economic system. These are the shifts in the techno-economic paradigm which seem to be associated with the long waves of economic change.

Currently, the major technological developments are those associated with the convergence of two initially distinct technologies – computer technology and communications technology – into a single, though complex strand: *information technology* (IT). IT is transforming both the technologies of transport and communication and also the technologies of products and processes. IT is capable of spreading into all sectors of the economy and to all types and sizes of organization but it is still the very large business organization, particularly the TNC, which is reaping the greatest benefits. Although the notion that we are shifting from a predominantly Fordist technological system to a post-Fordist system is far too sweeping and simplistic, it is nevertheless true that the spread of information technologies is generating fundamental change. The phenomenon of *flexibility* is extremely important in terms both of technologies of production and of the way production is organized. This is a topic we look at in some detail in Chapter 7.

In concluding this chaper, however, we should again remind ourselves that technological change, in itself, is not deterministic. We must not assume that a particular technology will lead inevitably and irrevocably to a particular outcome. More realistically:

> A frontier of new possibilities has been defined: a frontier which identifies the types of new products and services that can be made available. That frontier is itself a product of past choices . . . Specific choices within the frontier of technological possibilities are not the product of technological change; they are, rather, the product of those who make the choices within the frontier of possibilities. *Technology does not drive choice, choice drives technology.*
>
> (Borrus, quoted in Cohen and Zysman, 1987, p. 183, emphasis added)

Notes for further reading

1. The work of Christopher Freeman and his colleagues associated with the Science Policy Research Unit (SPRU) at the University of Sussex is especially relevant. Particularly useful discussions of technological change are provided in Freeman (1982, 1987), Freeman, Clark and Soete (1982), Perez (1985), Dosi *et al.* (1988).
2. There is a large literature on long waves. Kondratiev's original paper was republished in 1978. Major contributions were made by Schumpeter (1939), Mensch (1979), Mandel (1980). More recent important work includes Freeman, Clark and Soete (1982), van Duijn (1983), Freeman and Perez (1988), Hall and Preston (1988).
3. Freeman and Perez (1988), Table 3.1, provide a very detailed specification of each of the five K-waves.
4. Useful general introductions to information technology can be found in Forester (1985, 1987). More advanced treatments are provided by Freeman (1987), Hall and Preston (1988), Hepworth (1989).
5. Cherry (1978) and Hall and Preston (1988) provide comprehensive surveys of historical developments in communications technologies. Janelle (1969) and Abler (1975) discuss the concepts of cost–space and time–space convergence.
6. Moss (1987) and Warf (1989) discuss both satellite and optical fibre communications systems. See also International Institute of Communications (1990).
7. See Craig (1981), Schiller (1982), Bakis (1987), Hepworth (1989), and the examples given in Chapter 11 of this book. A more recent detailed study of the use of international leased networks by TNCs is provided by Langdale (1989).
8. A detailed study of US commercial involvement in Latin American television is provided by A. Wells (1972). A more recent study of the internationalization of television is provided by Negrine and Papathanassopoulos (1990).
9. The concept of the product life-cycle has been applied in a number of different ways. For its treatment within the marketing literature see Majaro (1982) and Paliwoda (1986). Van Duijn (1983) traces the historical evolution of the related concept of an industry life-cycle. The product life-cycle has also been employed to explain international trade and international production (see Chapter 5) as well as variations in regional growth. For a critique of this latter application see Storper (1985), Taylor (1986).
10. Some of the most useful contributions to this important debate are those by Blackburn, Coombs and Green (1985), Gertler (1988), Kaplinsky (1988), Schoenberger (1988a, 1989), Scott (1988), Coombs and Jones (1989), Sayer (1989), Amin and Robins (1990), Lovering (1990).
11. In the particular case of the automobile industry a recent study by Womack, Jones and Roos (1990) argues that 'mass' production is being superseded by 'lean' production. See Chapter 9 of this book.

Chapter 5

Why Internationalize? Explaining the Transnational Corporation

Introduction: approaches to explanation

Technological changes, particularly in the space-shrinking technologies of transport and communications, help to make possible the internationalization of economic activity and the development and geographical spread of transnational corporations. But they do not explain *why* business firms internationalize their activities by engaging in operations outside their countries of origin. For this we need to focus upon the TNC itself as an organizational form. It is clear that some of the explanations which have been offered to explain the growth of international production do not hold up when we look closely at the nature of transnational activity. For example, in the kind of world assumed by international trade theory transnational corporations simply could not exist. Traditional trade theory assumes that the factors of production – including capital – are geographically immobile and that there are no economies of scale in production. Both these assumptions are contravened by the very existence of transnational corporations. Similarly, the simple notion that capital will flow from areas of surplus to areas of deficit is not borne out by the fact that most transnational investment in the world economy is between capital-rich areas and only about a quarter of the total is located in the developing countries. We have to look elsewhere for an explanation.

The sheer variety of transnational activity, however, suggests that the search for an all-embracing explanation may well be illusory. Certainly there is no single, universally accepted, theoretical explanation of the TNC. Rather, there is a variety of competing explanations which operate at different scales of explanation and which derive from different ideological perspectives. The aim of this chapter is to present some of the major alternative explanations of the TNC and of the internationalization of economic activity.

The pursuit of global profits

The basis of each of the explanations is the fact that TNCs are essentially capitalist enterprises (a small number of TNCs are state-owned enterprises but they are in the minority). As such they must behave according to the basic 'rules' of capitalism. The most fundamental of these is the drive for *profit*, which is at the heart of all capitalist activity.

Business firms may well have a variety of motives other than profit, such as increasing their share of a market, becoming the industry leader or simply making the firm bigger. But in the long run none of these is more important than the pursuit of profit itself. A firm's annual rate of profit is the key barometer to its business 'health'; any firm which fails to make a profit at all over a period of time is likely to go out of business (unless 'rescued' by government or acquired by another firm). At best, therefore, firms must attempt to increase their profits; at worst, they must defend them. Of course, a capitalist market economy is a *competitive* economy. One firm's profit may be another firm's loss unless the whole system is growing sufficiently strongly to permit all firms to make a profit. Even so, some will make a larger profit than others. A key feature of today's world, of course, is that *competition is increasingly global* in its extent. Firms are no longer competing largely with national rivals but with firms from across the world. The pursuit is for *global profits*.

Expressed in the simplest terms, profit (P) is the difference between the revenue (R) which a firm receives from selling its products and the cost (C) of producing and distributing the firm's goods or services:

$$P = R - C$$

Obviously, therefore, profit can be increased either by increasing R or reducing C, or by a combination of the two. Our interest is in how this equation is worked out by firms operating at an international or even a global scale.

Scales of explanation: structure versus agency

A final general point to be made before we look at the various explanatory frameworks concerns the scale of explanation adopted. Consider the following two quotations:

> All rivers are unique, but the laws of fluid dynamics shape the flow of the Thames in the same way they shape the flow of the Hudson. Furthermore, the path to understanding unique events . . . lies through an investigation of *universal* laws of process.
> (Harvey, 1987, p. 371, emphasis added)

> Why then is the Thames not like the Hudson? Because of the very *specificities* of the history and geography of their respective formations.
> (Redfern, 1987, p. 414, emphasis added)

There is an explanatory tension, particularly evident in the social sciences, which is usually termed the 'structure–agency' dichotomy. Some explanations operate at the higher 'structural' level, offering a highly generalized explanation. Other explanations operate at the more specific level of 'human agency' and focus upon individual actions. For example, at the 'human agency' level international production would be regarded as the result of individual decisions by business enterprises to invest in particular locations outside the firm's home country. In one sense, all such decisions are unique. They are made by individuals or, more usually in large corporations, by groups of individuals on the basis of their perception of their firm's needs, and on the extent to which different locations meet such needs. But although at one level – that of the internal decisions of the business enterprise – international investment and location decisions are, indeed, unique, they are also part of a more general structure: the capitalist market system. From this perspective, therefore, the most appropriate level of explanation of the TNC, and of the growth of international production, is that of the system as a whole. The emphasis is on identifying 'universal laws of process'.

The following sections of this chapter reflect these differing theoretical approaches to the explanations of the TNC. No attempt is made to produce a single theoretical synthesis; each of the frameworks has its merits and its faults and should be judged accordingly. We begin by looking briefly at two theoretical frameworks which are in the more general 'structuralist' tradition: the circuits of capital and new international division of labour approaches. They are essentially macro-level approaches. We then turn to look in more detail at explanations which are couched at the micro-level of the business firm, notably the integrative 'eclectic' framework proposed by Dunning. We take a critical look at attempts to build sequential models of the evolution of TNCs and, finally, we explore some aspects of the strategic behaviour of TNCs, notably the tension which exists between forces of globalization and localization.

A macro-level approach: internationalization of the circuits of capital

One of the most useful attempts to explain the internationalization of economic activity at the general level is based upon the concept of the circuits of capital. This, in turn, is embedded within Marx's conceptualization of the capitalist system as a whole. From this perspective, the internationalization of economic activity and its major vehicle, the TNC, can be regarded simply as being part of

> a general trend of internationalisation inherent in the expansive nature of capitalism which tends to create a specifically capitalist world economy. Increasingly, this capitalist world economy is subject as a whole to the laws of motion of capitalism . . . international firms must be understood in terms of the internationalisation of capital and the accumulation of capital.
>
> (Radice, 1975, pp. 17, 18)

The 'laws of motion' of capitalism are derived primarily from the drive to enhance profits and to accumulate capital through increasing appropriation of surplus value from the process of production.[1] To Marxist writers the basis of the extraction of surplus value, or profit, is the exploitation of human labour power by capitalist firms which own the means of production. The internationalization of production, from this perspective, is the extension of the system of labour exploitation and class struggle to a global scale.[2]

The capitalist economic process can be envisaged as a continuous circuit, basically a very simple idea, as Figure 5.1 shows. Its essence is that 'the money at the end of the process is greater than that at the beginning and the value of the commodity produced is greater than the value of the commodities used as inputs' (Harvey, 1982, p. 69).

Thus, money (M) is used to purchase 'commodities' (C) in the form of raw materials and labour. These inputs are transformed in the process of production (P) and acquire increased value (C'). When exchanged for money (M') this increased value can be used to purchase a further round of inputs for the production process, and so the circuit proceeds.

The basic circuit of capital shown in Figure 5.1 can be expanded into three distinct circuits:

1. *The circuit of money capital*: as in Figure 5.1, the circuit of money capital appears as:

$$M - C \ldots P \ldots C' - M'$$

2. *The circuit of productive capital*:

$$P \ldots C' - M' - C' \ldots P'$$

3. *The circuit of commodity capital*:

$$C' - M' - C' \ldots P' \ldots C''$$

In fact, the three circuits are part of a completely *interconnected* whole:

$$\underbrace{M - C \ldots P \ldots \overbrace{C' - M' - \underbrace{C' \ldots P' \ldots C''}_{(3)}}^{(2)}}_{(1)}$$

From the viewpoint of our discussion, the important point is that each of these three circuits of capital has become *internationalized*. This is the argument particularly associated with the work of Christian Palloix (1975, 1977). He suggests that a clear historical sequence can be identified. The first of the three circuits to become internationalized, according to Palloix, was the circuit of commodity capital. This he equates with the development of world trade. The second circuit to become internationalized was that of money capital, as reflected in the flow of portfolio investment capital into overseas ventures. The circuit of productive capital, he argued, has more recently become internationalized. This is demonstrated by the massive growth of transnational corporations and of international production.

A major merit of this circuits of capital approach to the internationalization of economic activity is that it emphasizes the totally *interconnected* nature of finance, production and commodity trade. It is too easy to regard these as completely separate phenomena, yet – as Chapters 2 and 3 have shown – they are, indeed, components of a single system. The TNC itself is the clearest manifestation of this fact. The very large TNCs in particular display at the microscale the interlocking circuits of capital. They

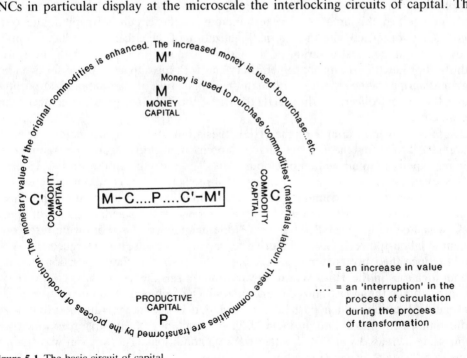

Figure 5.1 The basic circuit of capital

not only engage in international production but also generate a significant proportion of international trade within their own organizational boundaries and engage in sophisticated international financial transactions (including foreign currency dealings). The circuits of capital approach, by emphasizing the interconnection and interpenetration of the processes, also helps to capture something of the complexity of foreign direct investment flows, including the intricate cross-investments which we have observed to be occurring between the industrialized economies.

An explanatory approach based at this macro-level, therefore, helps to illuminate several basic features of the internationalization process. But it is insufficiently specific to deal with such questions as the precise form (geographical, organizational, sectoral) of transnational corporate activity. This is certainly not to deny the value of the circuits of capital approach even to non-Marxists, who can readily accept the broad framework it provides even if they do not necessarily subscribe to the entire explanatory package, particularly the overwhelming emphasis on class struggle and the exploitation of labour by capital. It serves to remind us that, in trying to explain the internationalization of economic activity, we are dealing with the workings of a dynamic capitalist market system.

Theories of the 'new international division of labour'

The economist Stephen Hymer who, as we shall see later, was a pioneer in developing theoretical explanations of the TNC, was probably the first to use the term 'new international division of labour' (NIDL) to explain the shift of production from the industrialized economies of the 'core' to the economies of the 'global periphery'. However, it was the publication of Fröbel, Heinrichs and Kreye's book, *The New International Division of Labour*, in English in 1980 which stimulated a considerable body of work in this area.[3] In the terminology of the circuits of capital, the NIDL theories are concerned with the internationalization of the circuit of productive capital. They deal with the restructuring of the global economy in general, with the shift of industrial production from core to periphery and with the impact of these shifts on both core and peripheral economies. Although the explanation of transnational corporations is not the major concern of the NIDL theorists, the TNC occupies the central role in the process.

There are several variants of the NIDL thesis but all assign the major role to the search by TNCs for cheap, controllable labour at a global scale. The emphasis is squarely upon the minimization of labour costs. The starting point of Fröbel, Heinrichs and Kreye's analysis was the appearance of three sets of pre-conditions: (1) developments in transport and communications technology; (2) developments in production process technology which permitted fragmentation and the standardization of specific tasks which could use unskilled labour; (3) the emergence of a 'worldwide reservoir of potential labour power'. Given the drive to maximize profits, it was argued that TNCs could reduce their production costs substantially by shifting their operations to cheap labour locations in the Third World. A variation on this theme emphasized a variety of 'push' factors in the industrialized economies which combined to stimulate firms to seek alternative locations. Such push factors included a fall in the rate of profit in the industrialized countries, and an increase in the militancy of labour forces which were demanding increased wages and better working conditions. The way out was again for firms to relocate to cheap labour sites overseas.

There is clearly some validity in this kind of explanation of some TNC behaviour. As we shall see in some of the industry case studies of Part Three (notably textiles, clothing and electronics), international relocation of assembly operations and the development of 'offshore' processing and sourcing by TNCs has been substantial. The NIDL theories have thrown valuable light on aspects of the internationalization process. But the view of the TNC and its strategies put forward in NIDL theories is excessively narrow and one-dimensional. As Elson (1988) has pointed out, even in industries such as textiles and clothing the availability of cheap labour is not usually the overwhelming locational consideration. Both Elson (1988) and Schoenberger (1988b) emphasize the major role of markets in TNC behaviour. More generally, the NIDL approach over-simplifies the variety of strategic options available to business organizations in general and to TNCs in particular. It also tends to underestimate the important roles played by national governments in influencing the internationalization of production, the subject of our attention in Chapter 6. Lastly, the NIDL theorists overstate the extent to which industrial production has been relocated to the global periphery. As we saw in Chapter 2 only a very small number of developing countries has developed a substantial degree of industrialization. Similarly, the evidence of Chapter 3 was that the overwhelming majority of foreign direct investment remains within the industrialized countries.

Micro-level approaches: the search for an integrative framework

An alternative approach to understanding the internationalization of economic activity through the TNC is to take a *firm-specific*, rather than a general system, view outlined in the preceding discussion. Instead of focusing upon the capitalist system as a whole, therefore, we now adopt a perspective based upon the capitalist firm. Before Stephen Hymer's pioneering study in 1960, there was no adequate theory of the TNC at the firm-specific level.[4] Since then numerous writers have contributed to this mode of explanation.[5] However, there has been a strong tendency to base such explanations on a stereotypical TNC – the very large, oligopolistic, mostly American enterprise – and theoretical conclusions have been extrapolated to apply to all TNCs. But TNCs from different source nations and of different sizes tend to differ considerably in their characteristics and behaviour. It is unwise to build an explanation of all TNCs on the study of just one type, no matter how important it may be. Any satisfactory explanation, therefore, needs to be sufficiently general to encompass the diversity of the world's TNC population.

Types of international production

As emphasized earlier in this chapter, international investment in production and distribution has to be seen within the context of the firm's attempts to increase or maintain profits in an increasingly competitive global environment. In this sense, international production by TNCs may be regarded as either offensive or defensive. Following our earlier breakdown of profit into revenue and cost components, we can classify overseas investments by business firms into two broad categories (although, as we shall see in Chapter 7, the boundary between them is becoming increasingly blurred):

1. *Market-oriented investments*. Most foreign direct investment in both manufacturing and service industries has been, and remains, of this type. Firms locate a facility in a

particular overseas market to serve that market directly. Often, the good or service produced overseas is virtually identical to that being produced in the firm's home country although there may well be modifications to suit the specific tastes or requirements of the local market. In effect, such specifically market-oriented invest-ment is a form of horizontal expansion across national boundaries.

2. *Supply- or cost-oriented investments*. Supply considerations are obviously the dominant motivation for firms in the natural resource industries. Such firms, of necessity, must locate at the sources of supply which, themselves, tend to be highly localized geographically. Often, such investments form the first element in a sequence of vertically integrated operations whose later stages (processing) may be located quite separately from the source of supply itself. In many cases – in line with the predictions of Weberian location theory – final processing occurs close to the market. Supply-oriented overseas investments have a very long history; many of the early transnational investments by American, British and continental European firms in the late nineteenth and early twentieth centuries were motivated by the desire to obtain and ensure raw material supplies. However, cost-oriented foreign investments by manufacturing firms are a relatively recent phenomenon and are very closely related to the kinds of technological developments discussed in Chapter 4.

These two broad types of transnational investment – market-oriented and cost-oriented – pose two rather different sets of questions. Why, in the case of market-oriented investment, is the market concerned not served by exports from the firm's domestic plants or by making licensing agreements with local firms in the foreign market? In the case of cost-oriented investment, why does not the firm simply purchase the necessary inputs from local suppliers rather than invest directly in foreign produc-tion facilities? As we shall see in Chapter 7, such contractual arrangements do, indeed, exist. But, equally, there is no doubt that cost-oriented investments are increasing in importance among TNCs even though market-oriented investments still account for the bulk of transnational investment in the world economy.

Dunning's 'eclectic' paradigm

John Dunning has put forward a framework which attempts to integrate various strands of explanation of international production. Dunning proposes a set of three general and interrelated principles which, he suggests, are fundamental to an understanding of international production. Because the three principles themselves are derived from a variety of theoretical approaches – the theory of the firm, organization theory, trade theory and location theory – Dunning labels his approach *eclectic*. It is this broad-ranging quality which is especially useful for our present purposes.

According to Dunning, a firm will engage in international production when each of the following three conditions are present:

1. a firm possesses certain specific advantages not possessed by competing firms of other nationalities (*ownership-specific advantages*); and
2. such advantages are most suitably exploited by the firm itself rather than by selling or leasing them to other firms. In other words, the firm *internalizes the use of its ownership-specific advantages*; and
3. it must be more profitable for the firm to exploit its assets in overseas, rather than in domestic, locations. In other words, *location-specific factors* play an important part,

in combination with the internalization of ownership-specific advantages, in determining whether or not – and where – overseas production occurs.

Let us look at each of these conditions in turn.

Ownership-specific advantages Ownership-specific advantages are assets which are internal to a firm. They are those

> which an enterprise may create for itself (e.g. certain types of knowledge, organization and human skills) or can purchase from other institutions, but over which, in so doing, it acquires some proprietary right of use. Such ownership-specific assets may take the form of a legally protected right, or of a commercial monopoly, or they may arise from the size, diversity or technical characteristics of firms.
>
> (Dunning, 1980, p. 9)

Stephen Hymer was the first to suggest that foreign direct investment could occur only if the investing firm possessed a particular advantage over indigenous firms. In his view, indigenous firms have a better understanding of the local business environment: the nature of the market, business customs and legislation. Foreign firms, in contrast, lack such knowledge, at least initially. To compete in the overseas market, therefore, foreign firms must possess some compensating advantage. Such advantages may be of many different types. The most obvious ones relate to size and market power. Large firms, in general, are in a better position to obtain their production inputs at favourable rates than smaller firms. They generally have better access to finance either from their own retained earnings or because of a better credit rating on the financial markets.

Technology – of production, of marketing and of organization in general – is a particularly important source of advantage. Technology, in the broadest sense of 'know-how', is an intangible asset easily transferable from one location to another. Caves (1971), for example, emphasizes the advantage which a particular brand image may give over lesser-known competitors. As he points out, a characteristic of many large firms, especially in consumer goods industries, is that they endeavour to *differentiate* their *products* from those of their competitors. This is done by making minor modifications to design or presentation and by reinforcing such differences through mass advertising. The practice of product differentiation is readily transferable from one market to another.

The Hymer and Caves approaches were based on the advantages of very large firms, generally those operating in predominantly oligopolistic industries. They were based particularly on observations of United States TNCs. But, as we have seen, not all TNCs follow this pattern. What sort of advantages do other firms possess? Table 5.1 uses Freeman's (1982) classification of firms based on their technology strategies to answer this question.[6] The first two categories – offensive and defensive firms – are the ones which obviously fit the conventional view of the TNC. Conversely, the last two categories – traditional and opportunist firms – tend not to be involved in overseas production for obvious reasons. But the two intermediate categories – imitative and dependent firms – do possess advantages which may be transferred overseas. Unlike the conventional image of the TNC, such firms cover the whole spectrum of sizes.

In the case of dependent firms, the stimulus to produce overseas may be related to the internationalization of their major domestic customers. By following overseas, the dependent firm retains its close ties with its dominant customer. At the same time, the dominant firm is assured of a known and reliable supply and avoids the possible risk of

Table 5.1 Principal advantages of successful firms and their applicability abroad

Firm characteristic	Principal source of advantage*	Size of firm	R & D spending	Transferable abroad?	Tied to firm?
Offensive	R & D capacity	Large	Heavy	Yes	Yes
Defensive	R & D capacity; applications and adaptation engineering; entrepreneurial skills	Large	Heavy	Yes	Usually
Imitative†	Production expertise – production engineering, quality control, applications and adaptations. Marketing expertise – product differentiation, brand names, marketing research, distribution	Large/small	Low	Yes	Sometimes
Dependent	Production engineering, applications and adaptation; production flexibility	Medium/small	Low/negligible	Yes	Yes
Traditional	Production craft skills; marketing association with traditionalism, regional image	Small	Negligible	No	Yes
Opportunist	Entrepreneurial skills	Small	Negligible	Not normally	Yes temporarily

* Note the principal source of advantage may be backed up by other advantages. For instance, among firms in the offensive and defensive categories, large size will facilitate economies of scale in production and marketing and various advantages in the finance area.
† Among large companies, skills may be either in production or marketing (supported by economies of scale, and advantages in finance, as above). Among small firms the advantages may accrue primarily from production.

(*Source:* Giddy and Young (1982), Table 4.4)

dealing with unknown indigenous firms. Thus, a transferable firm-specific advantage need not necessarily be related to large size. Small and medium-sized firms may well have a specific advantage which is exploitable in overseas locations although the nature of the advantage may be quite different from that possessed by the very large, research-intensive TNC. The kinds of advantage possessed by smaller, less technology-intensive firms include:

(a) Production experience and know-how in manufacturing certain items which are not in sufficient demand in foreign countries to support local production
(b) Skills in adapting machinery for low-volume production and in designing machinery to permit more flexible usage
(c) Imperfections in the market for secondhand machinery, arising from lack of knowledge relating to the availability of such machinery, that is, communications failure
(d) For the Third World multinationals most emphasis is placed on production rather than marketing expertise. Where marketing expertise is required, there is evidence of link-ups between developed and developing country firms, or indeed between developing-nation companies themselves.

(Giddy and Young, 1982, pp. 72–3)

Whatever the nature of the firm-specific advantage, however, we can assume that the branches and subsidiaries of TNCs do possess assets which are not available, on the same terms, to indigenous firms. In particular, TNC subsidiaries have access to the various resources of the parent company as a whole.

Internalization: bypassing the market[7] It is by no means inevitable that a firm must exploit its particular ownership-specific assets by investing directly overseas in production facilities. Its products may simply be exported at 'arm's length' through the usual trade channels. Alternatively, a technology may be licensed to indigenous firms in foreign countries in return for the payment of fees or royalties. Clearly, both of these alternatives are used extensively by both national and transnational enterprises. The main reason such alternatives may not be followed lies in the nature of the markets for materials, for intermediate goods and for finished products. In the miraculous world of neoclassical economics, markets are assumed to operate perfectly. If this were to be so then there would be no advantage to a firm in attempting to bypass the market. The global economic system would consist of a whole series of discrete transactions between independent buyers and sellers. But the world is not like this. *Markets are imperfect.* The greater the imperfection the greater will be the incentive for a firm to perform the function of the market itself by *internalizing* market transactions. The most obvious example of such internalization is vertical integration in which a firm decides to control either its own sources of supply or the destination of its outputs. In both cases, the functions of independent material suppliers or of wholesale and retail merchants are absorbed – internalized – within the firm.

The major incentive for a firm to internalize markets is *uncertainty*. The greater the degree of uncertainty – whether over the availability, price or quality of supplies or of the price obtainable for the firm's product – the greater the advantage for the firm to control these transactions. It is generally agreed that internalization is especially likely to occur in the case of knowledge. We have seen already that innovation and technological change are vital elements in a firm's ability to remain competitive and profitable. Many firms, especially large firms but also all those in high-technology industries, spend huge sums of money on research and development. To ensure a satisfactory return on such investment and to protect against predators, firms have a strong incentive to retain the technology and its use within their own boundaries. Rather than sell or lease the technology to another firm overseas the firm sets up its own production facilities and exploits its technological advantage directly. Thus,

> there is a special reason for believing that internalization of the knowledge market will generate a high degree of multinationality among firms. Because knowledge is a public good which is easily transmitted across national boundaries, its exploitation is logically an international operation; thus unless comparative advantage or other factors restrict production to a single country, internalization of knowledge will require each firm to operate a network of plants on a world-wide basis.
>
> (Buckley and Casson, 1976, p. 45)

One particularly important consequence of the internalization of markets by TNCs is its effect on prices. In external markets, prices are charged on an 'arm's-length' basis between independent sellers and buyers. In an internal market, on the other hand, transactions are between *related* parties – units of the *same* organization. The rules of the external market do not apply. The TNC itself sets the transfer prices of its goods and services within its own organizational boundaries.[8] Potentially, at least, this gives the TNC very considerable flexibility in setting its transfer prices to help achieve its overall goals.

The internal pricing of technology transferred between parent and overseas subsidiaries is especially amenable to manipulation because there may be no comparable external price. More generally, the ability to set its own internal prices – within the limits imposed by the vigilance of the tax authorities – enables the TNC to adjust transfer prices either upwards or downwards and, therefore, to influence the amount of tax or duties payable to national governments. For example, it would be in a TNC's interest to charge more for the goods and services supplied to its subsidiaries located in countries with high tax levels and vice versa. A similar incentive exists where governments restrict the amount of a subsidiary's profits which can be remitted out of the country. In general, the greater the differences in levels of corporate taxes, tariffs, duties, exchange rates, the greater will be the incentive for the TNC to manipulate its internal transfer prices. As Lall points out,

> it would appear that transfer pricing can be used most effectively by very large corporations with tightly exercised central control, sophisticated computational facilities and a wide experience of world conditions and of dealing with governments, and not by investors with limited overseas operations and a great deal of autonomy between different units.
>
> (Lall, 1973, p. 180)

Internalization of markets is, therefore, a most important element in explaining the existence of the TNC. There are, of course, limits; internalization carries costs as well as benefits, particularly the increased costs of communication between, and control of, the separate organizational units. But, as we have seen, the barrier of distance is now far less important than it was in the past, especially for the very large enterprise. However, the extent to which transactions or functions are internalized or externalized is in a continuous state of flux. In Chapter 7 we will look in some depth at processes of integration and disintegration by business firms and at the increasing proliferation of collaborative ventures and subcontracting relationships between different types of firm at the international scale.

Location-specific factors   The third element in the international production question is that of location itself and the extent to which economic and political conditions differ from place to place. Dunning (1980, p. 9) defines location-specific factors as 'those which are available, on the same terms, to all firms whatever their size and nationality, but which are specific in origin to particular locations and *have to be used in those locations*' (emphasis added). Thus, in the absence of more favourable locational conditions overseas, a firm would serve foreign markets by exports from a domestic base.

Variations in market size and composition   Several major types of location-specific factor are especially important in the context of international production. Spatial variations in the size and nature of markets are obviously of great significance. Figure 5.2 provides one approximate measure of market size – income as represented by gross national product (GNP) – and shows the enormous variation in income levels on a global scale. Per capita GNP in the OECD economies as a whole averaged $17,500 in 1988; in the lowest income group of developing countries, the average per capita income was a mere $320. In many cases, of course, income levels were very much lower than this. The largest geographical markets in terms of incomes, although not in terms of population, are obviously the United States and Western Europe. Such variations in GNP provide a crude indication of how the *level* of demand will vary from place to place across the world.

In addition, however, countries with different income levels will tend to have a different *structure* of demand. It has long been recognized that the type and mix of goods demanded tend to vary systematically with income. As incomes rise, so does the aggregate demand for goods and services. But such increased demand does not affect all products equally. In the economist's terminology, different products have different income elasticities of demand. (Income elasticity of demand is simply the way in which the quantity of a good demanded responds to changes in the incomes of consumers.) The basis of this relationship was first established in the 1850s by Ernst Engel. Figure 5.3 shows diagrammatically that each of the major types of economic good have different income elasticities of demand. One would expect countries with low income levels to spend a larger proportion of their income on primary products (basic necessities) and, conversely, countries with high income levels to spend a higher proportion of their income on 'higher-order' manufactured goods and services. Thus, countries with different per capita income levels will tend to differ greatly in both the magnitude and nature of their consumption patterns. Spatial variations in the size and composition of markets, therefore, constitute a most important location-specific factor.

The political dimension   A second factor is the influence of national governments. We devote the whole of the next chapter to this topic. In the present context, then, it is necessary simply to note the importance not only of government industry, foreign investment and trade policies but also of the general 'political climate' and national attitude towards foreign direct investments. As we shall see in the next chapter, government policies and actions constitute a most important source of market imperfection and have contributed enormously to the shape of international production in the world economy. Not unrelated to such national political differences are the less tangible – but no less significant – spatial variations in language and culture. A firm's perception of the *psychic distance* between potential locations for investment and its

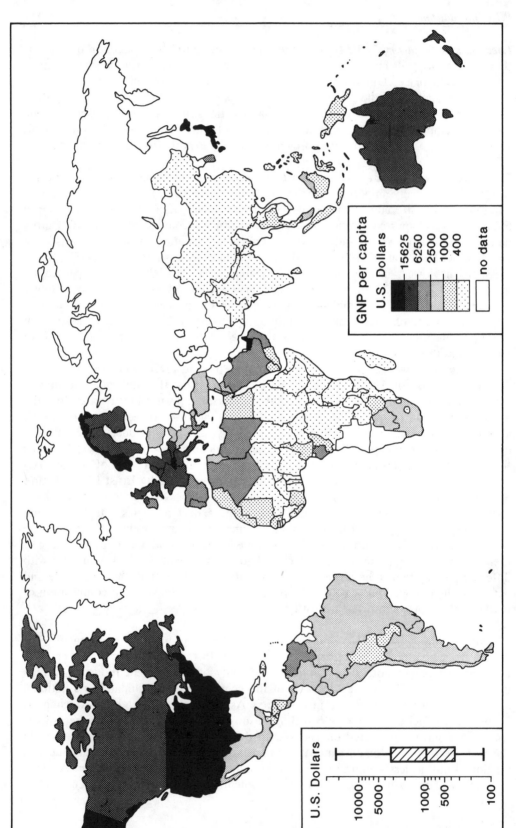

Figure 5.2 World variations in gross national product per capita (*Source:* World Bank (1990), Table 1)

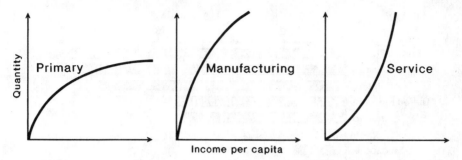

Figure 5.3 Relationship between income and the consumption of goods and services (*Source*: Hewings (1977), Figure 4.1)

home base may play a most important role in determining whether or not investment occurs. For example, much of the initial overseas investment by US firms was in countries of similar language and culture, particularly Canada and the United Kingdom. Similarly, we noted in Chapter 3 that many of the overseas investments by TNCs from the developing countries are located in countries with close linguistic and cultural affinities.

Spatial variations in production costs The third group of location-specific factors is that of variations in production costs. We have already had a good deal to say about the production process in Chapter 4. We noted, for example, that the relative import-ance of the various production factors tends to vary according to the stage in the product's life-cycle and, especially, with the maturity of the technology. More generally, the precise mix of production factors varies from one industry to another. In one sense, therefore, the key consideration is the relative importance of the individual factors in the firm's cost structure. But there is more to it than this. A particular factor of production may well be the most important element in a firm's total costs yet it may exert a negligible locational influence if its cost does not vary over space. If a factor costs the same everywhere it has a zero locational cost.[9]

Technological changes in production processes and in transport have evened out the significance of location for some of the traditionally important factors of production (for example, natural resources). Many now hold the view that, at least at the global scale, the single most important location-specific factor is labour. Labour is, of course, regarded as the primary variable in both the circuits of capital and new international division of labour theories discussed earlier in this chapter. As we have seen, labour – especially semi-skilled and unskilled labour – is especially important in the mature stage of the product cycle and has become increasingly important as the standardization and routinization of certain production processes have developed. Hence,

with the trend towards greater locational capability, labour moves to the forefront because of its degree of spatial differentiation. As capital develops its capability of locating more freely with respect to most commodity sources and markets, it can afford to be more attuned to labour force differences. Under the pressure of competition this becomes a necessity. The reasons for labour's persistent geographic distinctiveness lie in the unique nature of labour as a 'factor of production' – its embodiment in human beings.

(Storper and Walker, 1983, pp. 3–4)

The locational significance of labour as a production factor is partly reflected in geographical variations in *wage costs*. International differences in wage levels can be

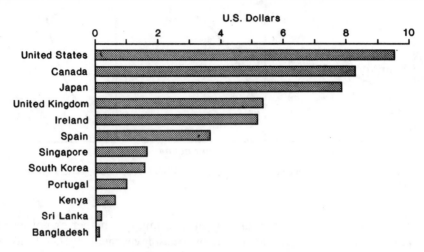

Figure 5.4 Hourly earnings in manufacturing industry in selected countries, 1985 (*Source*: United Nations (1988) *Statistical Yearbook 1985/86*, Table 24; International Monetary Fund (1986) *Yearbook of Financial Statistics*)

staggeringly wide, as Figure 5.4 shows, both between developed countries and, especially, between developed and developing countries. The latter earnings gap is especially great. For example, hourly earnings in manufacturing industry in the major NICs ranged from $3.66 in Spain down to less than $2.00 per hour in Singapore and South Korea. Earnings in less industrialized countries such as Kenya were around 62 cents per hour and were as low as 12 cents in Bangladesh. In comparison, hourly earnings in manufacturing in both the United States and Japan were between $7.90 and $9.50 per hour.

Spatial variations in wage costs are only a partial indication of the locational importance of labour as a production factor. The 'performance capacity' of labour varies enormously from place to place. For example, there are pronounced differences in *labour productivity*, which reflect a number of characteristics of the labour force including education, training, skill, motivation, as well as the kind of machinery in use. Another dimension of labour as a factor of production is its degree of 'controllability'. Largely because of historical circumstances, there are considerable geographical differences in the attitude of labour towards different kinds of production task.

A further significant dimension of labour as a factor of production is that it tends to be far *less mobile geographically*, particularly over great distances, than other factors such as capital or technology. In general, labour is strongly 'place-bound' although the strength of the bond varies a great deal between different types of labour. On average, male workers are more mobile than female workers, skilled workers are more mobile than unskilled workers. But there are exceptions to such generalizations. The 1960s, in particular, witnessed an enormous flow of migrant workers from regions of limited employment opportunities to the rapidly growing industrial regions of North America and Western Europe. Such flows do not, however, contradict the basic point that labour is a factor which is strongly differentiated spatially, particularly at the global scale.

Global variations in production costs are a significant element in the international investment-location decision. This is obviously the case for supply-oriented investments but it is also a critical consideration for market-oriented investments. In that case there is always a trade-off to be calculated between the benefits of market proximity on the one hand and locational variations in production costs on the other. But the problem is not merely one of variations in production costs at a single moment in time or even the obvious point that such costs change over time. A particularly important consideration is the uncertainty of the level of future production costs in different locations. One way of dealing with such uncertainty is for the TNC to locate similar plants in a variety of different locations and then to adopt a flexible system of production allocation between plants.[10] This is an issue we will return to in Chapter 7.

Integrating the three variables Within the general competitive pressures on firms to increase or sustain profits and to accumulate capital, the propensity to engage in international production is influenced by these three interrelated variables: ownership-specific advantages, internalization and location-specific factors. For international production to occur, all three conditions must be satisfied. Quite how they are satisfied will vary according to the type of investment involved, as Table 5.2 reveals. The table is self-explanatory but it is worth emphasizing that the scheme incorporates both of the major types of international production identified earlier: market-oriented investment and supply/cost-oriented investment. But it is important to point out that ownership-specific advantages and location-specific factors are not necessarily independent of each other. As Dunning has observed,

> today's ownership advantages of enterprises may be the inheritance of yesterday's country specific endowments. This is especially true of those to do with national resources and government policy. Further, such endowments may not give enterprises the same advantages at all points of the product and/or investment development cycle. It is possible that, as a new product becomes more standardised, ownership advantages will diminish or change . . . [also] . . . As an enterprise increases its degree of multinationality, the country specific characteristics of the home country become less, and that of other countries more, important in influencing its ownership advantages.
>
> (Dunning, 1979, pp. 283–4)

Table 5.3 outlines some of the links which exist between the ownership-specific advantages of firms and the location-specific characteristics of the firm's *home* country. It is this link which helps to explain the different characteristics of TNCs from different source nations. For example, the large domestic market and high level of technological sophistication of the US domestic economy have helped to produce the distinctive characteristics of United States TNCs. The lack of natural resources and the strong involvement of government and other institutions in technological and industrial affairs helps to explain the particular attributes of Japanese TNCs. Similarly, the specific characteristics of certain developing countries are reflected in the nature of their newly emerging TNCs. Thus, the particular geographical pattern of transnational activity will reflect the *combination* of ownership-specific advantages and the distribution of location-specific factors across both home and potential host countries.

The major value of Dunning's broad-based approach to the explanation of international production is that it can incorporate the *diversity* of transnational investment which is such a major feature of today's global economy. Hitherto, explanations were based upon a narrow, stereotype view of the TNC – the massive, oligopolistic enterprise in high-technology or research-intensive industries. Such explanations could not easily cope with, for example, much Japanese transnational investment or with the emergence

Table 5.2 Types of international production and their major determinants

Type of international production	Ownership-specific advantages	Location-specific advantages	Internalization advantages	Illustration of types of activity which favour TNCs
1. Resource-based	Capital, technology, access to markets	Possession of resources	To ensure stability of supply at right price. Control of markets	Oil, copper, tin, zinc, bauxite, bananas, pineapples, cocoa, tea
2. Import-substituting manufacturing	Capital, technology, management and organizational skills; surplus R & D and other capacity, economies of scale. Trade marks	Material and labour costs, markets, government policy (e.g. with respect to barrier to imports, investment incentives, etc.)	Wish to exploit technology advantages, high transaction or information costs, buyer uncertainty, etc.	Computers Pharmaceuticals Motor vehicles Cigarettes
3. Rationalized specialization: (a) of products (b) of processes	As above, but also access to markets	(a) Economies of product specialization and concentration (b) Low labour costs, incentives to local production by host governments	(a) As type 2 plus gains from interdependent activities (b) Economies of vertical integration	(a) Motor vehicles, electrical appliances, agricultural machinery (b) Consumer electronics, textiles and clothing, cameras, etc.
4. Trade and distribution	Products to distribute	Local markets. Need to be near customers. After-sales servicing, etc.	Need to ensure sales outlets and to protect company's name	A variety of goods, particularly those requiring close consumer contact
5. Ancillary services	Access to markets (in the case of other foreign investors)	Markets	Broadly as for types 2 and 4	Insurance, banking and consultancy services

(*Source*: Dunning (1980), Table 1)

Table 5.3 Links between selected ownership-specific advantages and the country-specific characteristics likely to generate and sustain them

Ownership-specific advantages	Country characteristics favouring such advantages
1. Size of firm (e.g. economies of scale, product diversification)	Large and standardized markets. Liberal attitude towards mergers, conglomerates, industrial concentration.
2. Management and organizational expertise	Availability of managerial manpower; educational and training facilities (e.g. business schools). Size of markets, etc. making for (1) above. Good R & D facilities.
3. Technological-based advantages	Government support of innovation. Availability of skilled manpower and in some cases of local materials.
4. Labour and/or mature, small-scale intensive technologies	Plentiful labour supplies; good technicians. Expertise of small-firm/consultancy operation.
5. Product differentiation, marketing economies	National markets with reasonably high incomes; high income elasticity of demand. Acceptance of advertising and other persuasive marketing methods. Consumer tastes and culture.
6. Access to (domestic) markets	Large markets. No government control on imports. Liberal attitude to exclusive dealing.
7. Access to, or knowledge about, natural resources	Local availability of resources encourages export of that knowledge and/or processing activities. Need for raw materials not available locally for domestic industry. Accumulated experience of expertise required for resource exploitation/processing.
8. Capital availability and financial expertise	Good and reliable capital markets and professional advice.
9. As it affects various advantages above	Role of government intervention and relationship with enterprises. Incentives to create advantages.

(*Source*: Dunning (1979), Table 6)

of TNCs from developing countries which are predominantly small and labour, rather than capital, intensive. Placed within a broader theoretical framework such apparently 'unconventional' TNCs can be readily accommodated. Dunning's self-styled eclectic theory has, however, been criticized as being merely 'a list of factors likely to be important in the explanation of the modern ... [TNC] ... rather than the explanation itself. Theoretical relations between the different factors too often remain untheorized' (Taylor and Thrift, 1986, p. 11). Nevertheless, seen as a pragmatic framework which attempts to integrate significant elements of other bodies of explanation, Dunning's approach is extremely useful as a conceptual structure within which specific cases can be examined.

TNC development as a sequential process

At the level of the *individual firm*, we can explain why international production occurs and why TNCs exist at all in terms of *specific* combinations of Dunning's three sets of conditions. But is there an identifiable *evolutionary sequence* of TNC development? Does the transition from a firm producing entirely for its domestic market to one engaged in overseas production follow a systematic development path? The answer to these questions is both 'yes' and 'no': yes, in the sense that certain common patterns of development are evident; no, in the sense that we should not expect all firms to follow the same sequence or in the sense that all firms will inevitably become TNCs.[11] It is useful to consider the broad path of TNC development, however, because it helps to give some sense of the dynamics of the processes involved.

Figure 5.5 illustrates the kind of sequence most commonly identified. It begins with

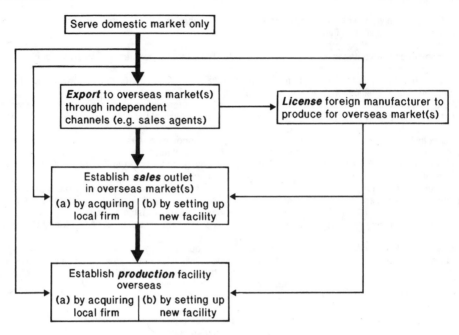

Figure 5.5 Paths of development in the evolution of a transnational corporation

the assumption that, initially, the firm is purely a domestic firm in terms both of production and of markets. In all national economies the majority of firms are of this type. However, the limits of the firm's domestic market may be reached and overseas markets may need to be penetrated to maintain growth and profitability. It is generally assumed that this is done initially through exports using the services of overseas sales agents. Such agents are, of course, independent of the exporting firm. However, the benefits of internalization which we discussed earlier may eventually stimulate the firm to exert closer control over its foreign sales by setting up sales outlets of its own in overseas locations. This may be achieved in one of two ways: by setting up an entirely new facility or by acquiring a local firm (possibly the previously used sales agency itself). Acquisition is, in fact, one of the most common methods of entry both to new product markets and to new geographical markets. It offers the attraction of an already functioning business compared with the more difficult, and possibly more risky, method of starting from scratch in an unfamiliar environment. Eventually the time may come – though not inevitably – when the need is felt for an actual production facility overseas. Again, this may be achieved through either acquisition or 'greenfield' investment.

Although there is a good deal of anecdotal evidence to support this sequence of development among firms which actually became TNCs there is nothing inevitable about the progression through each of the stages. Figure 5.5 shows a number of possible variations on the main theme. For example, a common means of involvement in overseas markets is through licensing agreements with local firms which are permitted to use a particular technology or manufacture a specific product for a defined market on payment of fees and royalties. In some cases, such agreements may form the basis for other forms of direct foreign involvement. Figure 5.5 also suggests that the general sequence may be 'short-circuited' at various points. Indeed, a firm may bypass the

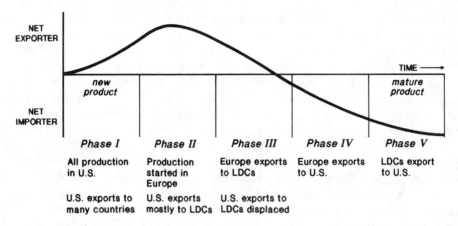

Figure 5.6 The product life-cycle and its suggested locational effects on United States production and trade (*Source*: Wells (1972), Figure 1, p. 15)

intermediate stages and set up overseas production facilities as a first step. One way in which this may occur is through the acquisition of another domestic firm which already has foreign operations. Thus, a firm may become transnational almost incidentally.

The product life-cycle explanation of sequential development

Developmental sequences such as this do not incorporate a specific *driving* mechanism. A whole variety of circumstances – both internal and external to the firm – may move the sequence along from one stage to the next depending upon the particular configuration of ownership-specific, location-specific and internalization factors. In the literature on international production, however, one particular driving mechanism has received most of the attention. The product life-cycle, which we discussed in Chapter 4 (see Figures 4.7, 4.8, 4.10), was specifically adapted as an explanation of the evolution of international production by Raymond Vernon in 1966.[12] Vernon's major contribution was to introduce an *explicitly locational dimension* into the product cycle concept, which in its original form had no spatial connotation at all. In fact, he argued, each phase of the product life-cycle has important locational implications in terms both of demand and of the production process itself.

Vernon developed the locational aspects of the product cycle in the particular context of the geographical spread of United States TNCs. Figure 5.6 reflects this US-oriented interpretation of the locational tendencies within the product cycle. His starting point was the assumption that producers in any particular market are more likely to be aware of the possibility of introducing new products in that market than producers located elsewhere. The kinds of new products introduced, therefore, would reflect the specific characteristics of the domestic market. In the United States case, according to Vernon, the high average-income level and high labour costs tended to encourage the development of new products which catered to high-income consumers and which were labour-saving (both for consumer and producer goods). Examples often quoted include vacuum cleaners, washing machines, drip-dry shirts and the like.

In explaining where the manufacture of these new products would take place, Vernon drew upon elements of location theory, especially the notion of external economies (what Alfred Weber had termed agglomeration economies). The unstandardized nature of the new product and its production technology created the need for rapid communication between producers, suppliers and customers. Such communication is obviously facilitated by close spatial juxtaposition between them. In this first phase of the product cycle, as Figure 5.6 shows, all production is assumed to be located in the United States and overseas demand is served by exports. But this situation is unlikely to last indefinitely. Vernon suggested that US firms manufacturing the new product would eventually set up production facilities in the overseas market either because they saw an opportunity to reduce production and distribution costs or because of a threat to their market position. Such a threat might come from local competitors or from government attempts to reduce imports through tariff and other trade barriers.

It follows from the nature of the product cycle model that the first overseas production of the product will occur in other high-income markets. In the specific case of US investment this tended to be Western Europe and Canada. The newly established foreign plants would come to serve their national markets and thus displace exports from the United States, which would be redirected to other areas in which production had not begun (phase II in Figure 5.6). Eventually, the production cost advantages of the newer overseas plants may lead the firm to export to other, third-country markets (phase III) and even to the United States itself (phase IV) from these plants. Finally, as the product becomes completely standardized, production may be shifted to low-cost locations in developing countries (phase V). It is interesting to note that when Vernon first suggested this possibility he regarded it as a 'bold projection'. At that time (the mid-1960s) there was little evidence of Third World export platforms serving European and US markets.

This locational interpretation of the product life-cycle became common currency, although even by the early 1970s Vernon himself was beginning to voice some doubts:

> The product cycle sequence may have seemed an efficient concept by which to describe the activities of US-controlled multinational enterprises during most of the decades of their existence. But by 1970, the concept was frequently exhibiting procrustean tendencies to discard or distort information in order to have the facts conform rather more nicely to the theory. For instance, the product cycle sequence relies heavily on the assumption that the special conditions of the US environment – especially factor costs and consumer tastes in the United States – will set in train a sequence that leads step by step to international investment. Though this may be an efficient way to look at enterprises in the US economy that are on the threshold of developing a foreign business, the model is losing some of its relevance for those enterprises that have long since acquired a global scanning capacity and a global habit of mind . . .
>
> By 1970, the product cycle model was beginning in some respects to be inadequate as a way of looking at the US-controlled multinational enterprise. The assumption of the product cycle model – that innovations were generally transmitted from the US market for production and marketing in overseas areas – was beginning to be challenged by illustrations that did not fit the pattern.
>
> (Vernon, 1971, pp. 107, 108)

For such reasons, Vernon shifted the emphasis of his product cycle model to that of the *oligopolistic* behaviour of TNCs and suggested that the development sequence consisted of three phases: innovation-based oligopolies; mature oligopolies; and senescent oligopolies. The innovation phase, he suggested, will still tend to be located in the country of origin though he broadened the scope. For example, European innovations and Japanese innovations would tend to be rather different from US innovations, perhaps emphasizing materials-saving considerations rather than labour-saving ones. In the second phase, however, locational developments will tend to reflect oligopolistic strategies, that is, 'the pricing and investment decisions of an oligopolist are taken with explicit regard for their effects on the decisions of the others in the industry' (Vernon, 1974, p. 97).

In seeking to maintain market stability firms will tend to react to match the actions of their major competitors. In terms of the location of international production it has been suggested that firms will tend to pursue a follow-the-leader strategy.[13] The move by one major firm to establish production facilities in a particular country is likely to be followed by the other major firms in the same industry. This may lead to a clustering of investment decisions in both time and space. In the mature oligopoly phase, firms exert their production and market power to prevent the entry of competitors. In Vernon's third phase of his revised product cycle model – senescent oligopoly – production locations are determined largely by geographical differences in costs. In this phase, the tendency to seek out low-cost locations at a global scale becomes especially strong.

How valid is the product life-cycle as an explanation of the locational evolution of TNCs? There is no doubt that a good deal of the initial overseas investment by US firms did fit the product cycle sequence quite well; but, as Vernon himself has pointed out,

> certain conditions of that period are gone. For one thing, the leading MNCs have now developed global networks of subsidiaries; for another, the US market is no longer unique among national markets either in size or factor cost configuration. It seems plausible that the product cycle will be less useful in explaining the relationship of the US economy to other advanced industrialised countries, and will lose some of its power in explaining the relationship of advanced industrialised countries to developing countries. *But strong traces of the sequence are likely to remain.*
>
> (Vernon, 1979, p. 265, emphasis added)

The 'traces' identified by Vernon include the innovating activities of smaller firms just beginning their overseas operations and the pattern of development of European and Japanese innovating firms. Most of all, he suggests, the emerging TNCs from the developing countries may follow a product cycle sequence in their investments in other developing countries. However, not all would agree with Vernon's view that 'strong traces of the sequence are likely to remain'.

It is clear that the product cycle can no longer explain the majority of international investment by TNCs. As these firms have become more complex globally it is unrealistic to assume a simple evolutionary sequence from the home country outwards. Even within strongly innovative TNCs the initial source of the innovation and of its production may be from any point in the firm's global network. In addition, as we saw in Chapter 3, much of the world's international investment is reciprocal or cross-investment between advanced industrial countries. Such investment cannot easily be explained in product cycle terms. Like all stage or cyclic models, the product cycle is *programmatic*: it can describe a general sequence but it says nothing about the length of each phase and it cannot predict the timing of the transition between one phase and the next. Having said this, however, there is some general value in such ideal-type models in emphasizing the dynamic nature of the major processes involved and in pointing to possible interrelationships between key elements. But their actual application to real-world circumstances must be time- and place-specific.

In the case of the product life-cycle model of the locational evolution of TNCs its specific context was that of the postwar expansion of large US enterprises. Its current relevance to TNCs in general is limited to the explanation of *some* – though not all – types of international production. As Giddy has observed,

> as an explanation of international business behaviour, the product cycle model has only limited explanatory power. It does describe the initial international expansion of many firms and it does show how demand for a product becomes more elastic over time and across countries. The multinational enterprise, however, has succeeded in developing a number of other strategies for surviving in overseas production and marketing. Hence, the product cycle model must now take its place as only one facet of the more general phenomenon of large international firms successfully applying a diversity of monopolistic advantages across national boundaries in order to internalize imperfectly competitive factor markets.
>
> (Giddy, 1978, p. 97)

The sequential model discussed above assumes a market-driven process. It fits the conventional view of the TNC as developing progressively, from serving markets through exports to direct investment in overseas production facilities. The broader definition of the TNC introduced earlier (see p. 48) allows for other possibilities. Recall that the essence of this broader definition is the *co-ordination* of international production from one centre of strategic decision making. It is quite possible for a firm to begin its transnational activities by co-ordinating production in overseas locations in order to serve the firm's domestic market rather than to serve overseas markets. This might subsequently evolve into a more elaborate transnational network of operation. We shall return to some of these alternative forms of transnational organization in Chapter 7.

Why internationalize? Competitive strategies of transnational corporations

Firms internationalize their operations for a host of different reasons, both general and specific. Such reasons may well change in detail over time but the fundamental reason – to enhance the firm's overall competitive position – remains largely unchanged. Developments in the technologies of production and in the space-shrinking technologies of transport and communication, discussed in Chapter 4, have greatly intensified competition at the global scale. Hence, the pressures on firms to operate globally have also intensified. The decision to internationalize is a major *strategic* decision. Although it shares the general characteristics of all strategic decisions, the very fact of seeking to operate across national boundaries gives the international strategic decision added complexity.

Some of the most influential ideas on a firm's competitive strategy have been developed by Michael Porter (1985, 1986). According to Porter, business firms seek to achieve a competitive advantage in their particular industry through the pursuit of one of three 'generic strategies' (Figure 5.7):

1. *Cost leadership*: being the lowest-cost producer of a good or service.
2. *Differentiation*: being different from competitors in some way or other.
3. *Focus*: applying either of these two strategies on a broad or a narrow front. The latter is generally known as a *niche* strategy and may apply to a specific geographic market, a particular segment of a production process or a particular type of customer.

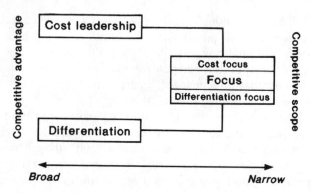

Figure 5.7 Porter's classification of 'generic' competitive strategies

Each of these generic competitive strategies applies to all firms regardless of the geographical extent of their operations. At the international or global scale, however, such competitive strategies take on an additional dimension: 'Industries may vary along a spectrum from multidomestic to global in their competitive scope' (Porter, 1986, p. 17). Figure 5.8 summarizes the basic features of industries at each end of this spectrum. Until the 1960s most TNCs invariably pursued a *multidomestic* (or nationally responsive) strategy in their international operations, whereby autonomous national subsidiaries served individual national markets. TNCs capitalized on their ownership-specific advantages and transferred intangible assets (through the internalization of markets for intermediate goods) to their overseas subsidiaries. This could be done very cheaply – effectively at marginal cost – once a network of overseas subsidiaries was

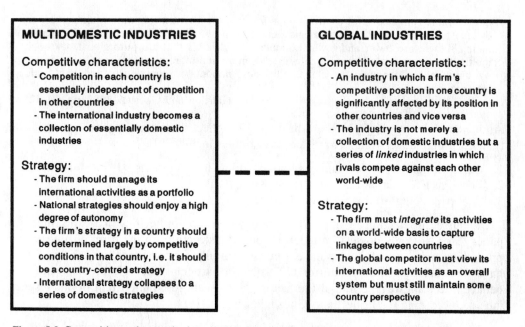

Figure 5.8 Competitive and strategic characteristics of multidomestic and global industries (*Source*: based on material in Porter (1986), chapter 1)

established (Doz, 1986). Consequently, these affiliates possessed a number of competitive advantages compared with purely domestic firms, enabling them to compete very effectively in specific national markets. As Figure 5.8 shows, the essence of a multidomestic or nationally responsive competitive strategy is an international network of commonly owned, but quasi-independent, operations. Each of these is responsive to the characteristics of individual national markets.

This kind of internationally competitive strategy is still very common among TNCs. Much foreign direct investment remains oriented towards national or regionally compact markets. But a growing number of TNCs, most notably (though not exclusively) the larger ones, have been developing *globally integrated* competitive strategies. Technological changes in transport and communications, in the production process and in the organization of production, together with international convergence in market characteristics, have increased the benefits to be gained from operating at very large scales. Increasingly, however, these are not so much economies of *scale in production* (given the development of the kinds of flexible production technologies discussed in Chapter 4) as economies of *scope and co-ordination*.

Dividing industries into these two polar opposite types is obviously a severe simplification of reality. It suggests two mutually exclusive strategic orientations: global integration on the one hand and national responsiveness on the other. In fact, each must contain elements of the other. Firms operating in so-called multidomestic industries must take account of global forces. Conversely, firms operating in global industries must be responsive to national and local differences. The intensification of global competition in a world which still retains a high degree of local differentiation creates, for all TNCs, an *internal tension* between *globalization* forces on the one hand and *localization* forces on the other.

In a world of

> a much more complex set of environmental forces . . . firms must respond simultaneously to diverse and often conflicting strategic needs . . . In the emerging international environment . . . there are fewer and fewer examples of industries that are pure global, textbook multinational, or classic international. Instead, more and more businesses are being driven by simultaneous demands for global efficiency, national responsiveness and worldwide learning.
>
> (Bartlett and Ghoshal, 1987, pp. 10, 12)

An increasing number of TNCs are beginning to reflect these simultaneous pressures in their corporate advertising. Three recent examples include:

1. 'the key to global performance is understanding local markets' (J. P. Morgan);
2. 'Local insight. Global outlook' (Hong Kong Bank);
3. 'the art of being local worldwide' (ABB).

Of course, in selecting a competitive strategy firms do not begin with a blank sheet of paper. They not only possess an inherited physical structure, in the form of offices and production units, but they also have a *history*; an inherited way of doing things. They have what Heenan and Perlmutter (1979) call a particular strategic predisposition to behave in certain ways. Some of the factors which seem to be important in this respect include, according to Bartlett and Ghoshal (1987, p. 14):

1. the influence of strong leaders who often leave an 'indelible impression' on their organizations;

2. home country culture and social systems;
3. the firm's specific history of internationalization.

Variables such as these help to explain much of the variety which exists in the strategic behaviour of TNCs, even within the same industry. We need to get away from the stereotype view of the TNC. Far too many people tend to regard all TNCs as being essentially the same. At one level, of course, there is some truth in this. All TNCs do share the common characteristics of being profit-seeking, primarily (though not exclusively) capitalist enterprises engaged in the search for profit in an intensely competitive environment. But the real picture is one of organizational and behavioural variety rather than uniformity. There is no such thing as a homogeneous transnational corporation which stamps an identical footprint on the economic landscape.

Conclusion

Internationalization is one potential strategy that is being used increasingly by business firms. It is one among a whole variety of strategies by which firms seek to maximize, increase, or at least sustain, their profitability. An internationalization strategy can contribute to both parts of the firm's profit equation. On the one hand, expansion and increased penetration of international markets through direct investment can increase revenues. On the other hand, location of production in countries where some, or all, of the factors of production can be acquired at lower relative cost contributes towards the minimization of the cost element in the profit equation.

Markets can be extended into new geographical areas, however, and lower-cost production locations can be utilized only if geographical distance can be overcome without excessive cost and if the dispersed operations can be co-ordinated and controlled. A prerequisite of international production and of the TNC, therefore, is the presence of enabling technologies – of transport, communication and organization. However, neither these technologies nor those in products and processes explain why firms choose to operate overseas. They are necessary but not sufficient elements in an explanation.

Explanation of the TNC and of the internationalization of economic activity it generates can be approached in a variety of ways and at a variety of levels. At the macro-scale of the capitalist system as a whole, an explanatory framework based upon the internationalization of the circuits of capital is especially useful because it emphasizes the complex interconnections between the three major circuits of money, productive and commodity capital. The very large TNCs can be regarded as a microcosm of the interlocking internationalized circuits of capital, operating, as they do, in all three circuits at the same time. The circuits of capital framework operates primarily at the macro-scale of the system as a whole, however. It is not really designed to deal with the more specific questions of the form of transnational corporate activity.

At the micro-level of explanation, Dunning's self-styled eclectic paradigm, based primarily on industrial organization and location theory, is very helpful. According to this approach, a firm will engage in international production if, and only if, three closely related conditions are satisfied: ownership-specific advantages, internalization and location-specific factors. This approach is sufficiently broad to incorporate the diversity of international production and, in particular, to explain the widely differing characteristics of TNCs of different sizes and from different source nations. But, in the final analysis, any explanation must be based upon the particular circumstances of time and place and on the type of economic activity involved. Dunning's eclectic approach is a useful framework in which such specific cases may be examined.

In this chapter, we have addressed the question why firms internationalize by examining a variety of explanatory approaches. Ultimately we need to develop an explanatory framework which combines both *general* elements of the way in which the system as a whole operates and also *specific* elements of the behaviour of particular firms. We need an approach which incorporates elements of both because all investment and location decisions, at whatever geographical scale, are a mixture of both generality and uniqueness. In Chapter 7 we shall look at how the process operates to create both the complex global structures of TNCs themselves and also the network of relationships with other firms of different sizes and types across the world. Before doing this, however, we turn to the other major type of institution in the global economy which plays a major role in the internationalization of economic activity: the nation state. Not only are states still significant actors in their own right in the world economy but also they have the capacity to encourage or to inhibit the global integration or nationally responsive competitive strategies of TNCs.

Notes for further reading

1. Harvey (1982) provides an extremely useful discussion of the capitalist system from a Marxist perspective. For a detailed treatment of international production as the process of internationlization of the various circuits of capital, see Palloix (1975, 1977), Jenkins (1984a). Barr (1981) adapts and expands the concept at the level of the individual enterprise.
2. For a series of contributions along these lines see the volume edited by Peet (1987).
3. Apart from Fröbel, Heinrichs and Kreye (1980), key contributions have been made by Jenkins (1984a), Schoenberger (1988b), Henderson (1989).
4. Hymer's 1960 study was his Ph.D. dissertation, which was eventually published posthumously in 1976. Dunning and Rugman (1985) assess the significance of Hymer's work on the theory of foreign direct investment, as do some of the contributions in Pitelis and Sugden (1991).
5. Key conributions include those by Vernon (1966, 1971, 1979), Caves (1971, 1982), Buckley and Casson (1976), Hymer (1976), Dunning (1977, 1983, 1988a). Useful reviews of the theoretical literature are provided in Pitelis and Sugden (1991).
6. This line of argument is developed by Giddy and Young (1982) in their discussion of 'conventional theory and unconventional multinationals'.
7. The initial concept of 'internalization' (though not the term itself) is generally attributed to Coase (1937). However, its specific application to the explanation of the TNC dates only from the 1970s and is associated primarily with the work of Buckley and Casson (1976) and Rugman (1981).
8. Transfer pricing is a particularly controversial issue in the relationship between TNCs and national governments, and is discussed in this context in Chapter 12.
9. The distinction between basic costs and locational costs was first drawn by Rawstron (1958) and subsequently developed by Smith (1981).
10. A formal solution of this problem is provided by De Meza and Van Der Ploeg (1987).
11. For a discussion of evolutionary models of firm – and especially TNC – development see Taylor and Thrift (1982, chapter 2). They argue that it is unrealistic to assume a single corporate development sequence and suggest that four different sequences can be identified. They also distinguish between the truly global corporation and the smaller TNC. Håkanson (1979) provides one example of an evolutionary sequence from domestic to international production.

12. Vernon's approach was a reaction to the inadequacies of existing theories of international trade and investment. His initial (1966) version of the product cycle model was subsequently modified (1971, 1974) to an explicitly oligopolistic interpretation. Hirsch (1967) and L. T. Wells (1972) both made significant contributions to the product cycle interpretation of international production and trade. In 1979 Vernon re-evaluated the product cycle concept in the light of changed circumstances. Critical appraisals of the product cycle explanation of international production can be found in Buckley and Casson (1976), Giddy (1978), Taylor (1986).

13. This particular hypothesis was investigated empirically by Knickerbocker (1973), using the Harvard data on the 187 largest United States TNCs for the period from 1948 to 1967. He found that 50 per cent of the foreign subsidiaries of firms in the same industry were located in the same overseas market within three-year clusters. Such behaviour he found to be especially common among the largest firms in each industry. For a discussion of this concept in the context of the international tyre industry see Rees (1978), Gwynne (1979).

Chapter 6

The Political Dimension: Nation States and the Internationalization of Economic Activity

Introduction: the changing position of the state in a global economy

For some three hundred years, from its emergence in the mid-seventeenth century, the nation state was regarded, rightly, as the dominant actor in international economic relationships. Historically, the state was the primary *regulator* of its national economic system. The world economy, quite legitimately, could be conceptualized as a set of interlocking *national* economies.[1] Trade and investment in the world economy were literally 'inter-national'. The legacy of this remains in the continuing collection of trade and foreign investment data within national statistical boxes, as Chapters 2 and 3 demonstrated. The major theme of this book, of course, is that the world has changed as economic activities have become not just increasingly internationalized but also increasingly globalized. Not only has the degree of interdependence and interconnection within the world economy increased dramatically but also the emergence of the transnational corporation has produced a rival to the nation state's traditional role as the dominant actor or institution in the world economy.

This has led some writers to assert the demise of the nation state as an important force in the world economy. Two brief quotations illustrate this viewpoint:

> the nation state is just about through as an economic unit.
>
> (Kindleberger, 1969, p. 207)

> a major feature of concentration and centralisation in late capitalism is its international scale, which *makes most nation states relatively insignificant elements* within the operation of a world economy dominated by a small number of companies which are larger and more wealthy than many individual states.
>
> (Johnston, 1982, p. 61, emphasis added)

In fact such statements greatly distort and oversimplify the real position. Despite the growing penetration of national economies by TNCs, the political institution of the nation state remains a most significant force in shaping the world economy. In a few cases – most obviously, though not exclusively, the centrally planned economies – it is the overwhelmingly dominant force. But even in the market economies the importance of political involvement in the economy should not be underestimated.

The state has, whether explicitly or implicitly, played an extremely important role in the process of industrialization in all countries. As far as today's NICs are concerned, for example, 'it is difficult to see how . . . [they] . . . could have achieved the unusual economic and export performance which distinguishes them as NICs without the crucial input of government policies' (Bradford, 1987, p. 313). The same is certainly true of Japan. Indeed, *all* governments intervene to varying degrees in the operation of the market and, therefore, help to shape different parts of the global economic map. It could be argued, in fact, that the world economy is today more, rather than less, politicized as the interdependence between countries has increased. Certainly, questions of trade imbalances, exchange rates and the like are as much political, as economic, phenomena.

Nation states operate within a world system of differential power relationships.[2] Expressed in the simplest terms, if the goals of business organizations are to achieve maximum (or at least satisfactory) profits, one of the goals of nation states is to maximize the material welfare of their societies. In an increasingly integrated and interdependent global economy nations are forced to *compete* with one another in a struggle to attain such goals. States compete to enhance their international trading position and to capture as large a share as possible of the gains from trade. They compete to attract productive investment to build up their national production base which, in turn, enhances their international competitive position.

The competitive basis of nation states is derived from a complex set of sources. On the one hand,

> the international competitiveness of national economies is built on the competitiveness of the firms which operate within, and export from, national boundaries. To a large extent, it is thus an expression of the dynamism of domestic firms, their capacity to invest and to innovate both as a consequence of their own R & D and of successful appropriation of technologies developed elsewhere.
>
> (Chesnais, 1986, p. 87)

On the other hand, the particular combination of conditions within nation states has an enormous influence on the competitive strengths of the firms located there. This is the essence of Porter's argument in his book *The Competitive Advantage of Nations* (1990). In his view:

> Competitive advantage is created and sustained through a highly localized process. Differences in national economic structures, values, cultures, institutions, and histories contribute profoundly to competitive success. The role of the home nation seems to be as strong as or stronger than ever. While globalization of competition might appear to make the nation less important, instead it seems to make it more so. With fewer impediments to trade to shelter uncompetitive domestic firms and industries, the home nation takes on growing significance because it is the source of the skills and technology that underpin competitive advantage.
>
> (Porter, 1990, p. 19)

Thus, the varying fortunes of national economies – and the state policies which underpin them – within the intensely competitive international environment undoubtedly influence the globalization of economic activity. It is certainly true that an individual nation state's degrees of economic freedom – its economic autonomy – are constrained, both by the actions of transnational corporations and of other nation states. But national governments, whether singly or collectively, continue to play a most important role in shaping the global economic map and, indeed, in either encouraging or inhibiting the global ambitions and strategies of business firms. National boundaries create significant differentials on the global economic surface. Political spaces are among the most important ways in which location-specific factors (of both supply and demand) are 'packaged'. Political boundaries create *discontinuities* of varying magnitudes in the flows of economic activities. Governments can modify (or even help to create or destroy) comparative advantage. Although the TNC may well be the single most important force creating global shifts in economic activity, it is not the only force involved. Both TNCs and nation states are interlocked within the complex processes of globalization. It is the outcome of this interaction that is producing the increasingly complex geography of the global economy.

State economic policies

The functions of the state are many and varied and impinge upon all aspects of life. The very role and behaviour of the state are subjects of sharp ideological disagreement. At one extreme is the view of the state as the 'handmaiden' of capitalism, whose primary functions are seen to serve the interests of capital rather than labour. At the other extreme the state may be seen as the neutral arbiter or referee between the various interest groups in a pluralist society. Such debates about the more fundamental role of the state are not pursued here. Rather, the aim is to focus on those actual policies of national governments which explicitly set out to influence the level and distribution of economic activity at an international scale. This is not quite as simple as it sounds because 'there is hardly a single policy issue without its international dimension' (Knight, in Carter, 1981, p. viii).

At one extreme, the *macroeconomic* policies pursued by governments to control domestic demand or to manage the money supply have extremely important implications for the distribution and redistribution of economic activity. At a more basic level, governments are generally the providers of the physical infrastructure of national economies – roads, railways, airports, seaports, telecommunications systems – without which private sector enterprises, whether domestic or transnational, could not operate. They are the providers, too, of the human infrastructure: in particular of an educated labour force as well as of sets of laws and regulations within which enterprises must operate. Between these two extremes of government involvement in the workings of the economy lie those policies whose explicit purpose is to influence the level, composition and distribution of production and trade. It is these policies which form the substance of this chapter.

National governments have at their disposal an extensive kit of tools with which to influence economic activity and investment within their own boundaries and to shape the composition and flow of trade and investment at the international scale. Such policies can be divided into two broad types: those whose objective is to *stimulate* economic activity; and those whose objective is to *regulate* it. Although often viewed separately, trade policies, policies towards foreign investment and industry policies overlap to a very considerable degree. The boundaries between them may be blurred. Each may reinforce – or counteract – the other. They may be employed as part of a deliberate, cohesive, all-embracing national economic strategy or, alternatively, individual policy measures may be implemented in an *ad hoc* fashion with little attempt at co-ordination. As we shall see, there is very wide variation in practice among the market economies, both developed and developing.

The first section sets out briefly and in general terms the major types of policies which governments may employ: policies towards trade, foreign investment and the support and regulation of industry, together with a brief introduction to what has become one of the most significant developments in the global economy in recent years: international economic integration. Subsequent sections develop these general policy areas in specific national contexts: the Western industrialized economies (North America and Western Europe), Japan and, lastly, the newly industrializing countries. Treatment of individual national policies is inevitably brief in this chapter. However, further discussion is contained in the industry case study chapters of Part Three. In those chapters, as well as in Chapter 12, the interactions between nation states and TNCs are spelled out more explicitly. In Chapter 13 particular policies of adjustment to global shifts are treated in some detail. The aim of the present chapter, then, is to provide a broad outline of policies relevant to the theme of this book rather than to attempt an exhaustive account of each country's national policies. Throughout this chapter, the focus is upon the world's market economies (both developed and developing). The centrally planned economies are not dealt with although, as many of them are currently moving towards incorporating the market into their economic management, the distinction is in any case becoming increasingly blurred.

Trade, foreign investment and industry policies: an overview

Trade policies

Of all the measures used by nation states to enhance their international competitive position, policies towards trade have the longest history. The shape of the emerging world economy of the seventeenth and eighteenth centuries was greatly influenced by the mercantilist policies[3] of the leading European nations. The successful challenge to Britain's industrial supremacy in the late nineteenth century by the United States and Germany was based on a strongly protectionist trade policy by these nations. The deep world recession of the 1930s was also characterized by national retreat behind trade barriers. Since the late 1970s a new wave of protectionism – *neo-mercantilism* – has swept the world economy as governments attempt to 'manage' trade in a variety of ways. Indeed, a whole new area of strategic trade policy has emerged in recent years, particularly in the United States, as we shall see later.[4]

The GATT framework   Trade policy is also unique in that it is set within an international institutional framework: the General Agreement on Tariffs and Trade (GATT). The GATT was one of the three international institutions formed in the aftermath of the Second World War which constituted the framework of the postwar world economy. Initially there were twenty-three signatories; today approximately a hundred nations belong fully to the GATT and a further thirty countries apply the GATT principles on a *de facto* basis. The workings of GATT are extremely complex and need not detain us here.[5] It is important, however, to appreciate its basic aims and objectives, which are:

1. the reduction of tariffs;
2. the prohibition of quantitative restrictions and other non-tariff barriers to trade;
3. the elimination of trade discrimination.

The GATT system is based essentially on *reciprocity* and *non-discrimination* between nations. A central element is the *most favoured nation principle*, by which a trade concession negotiated between two countries is also applied to other countries. Prior to the mid-1960s, GATT was most concerned with trade between the developed nations. As a result, widespread dissatisfaction emerged among the developing countries with the state of world trade in respect of their own commodities and products. This led to the establishment, in 1964, of UNCTAD, whose major role was to promote the trading interests of the developing nations. A particularly sensitive issue was the access of developing country exports to developed country markets. Pressure led, in 1965, to the

Policies towards imports	Policies towards exports
1 Tariffs	Financial and fiscal incentives to export producers
	Export credits and guarantees
2 Non-tariff barriers	Setting of export targets
Import quotas (eg 'voluntary export restraints'; 'orderly marketing agreements')	Operation of overseas export promotion agencies
Import licences	Establishment of Export Processing Zones and/or Free Trade Zones
Import deposit schemes	'Voluntary export restraint'
Import surcharges	Embargo on strategic exports
Anti-dumping measures	Exchange rate manipulation
Special labelling and packaging regulations	
Health & safety regulations	
Customs procedures and documentation requirements	
Subsidies to domestic producers of import-competing goods	
Countervailing duties on subsidised imports	
Local content requirements	
Government contracts awarded only to domestic producers	
Exchange rate manipulation	

Figure 6.1 The major types of trade policy used by national governments

adoption within GATT of a generalized system of preferences (GSP) under which exports of manufactured and semi-manufactured goods from developing countries would be granted preferential access to developed country markets. In fact, there were a number of exclusions from the GSP, of which one of the most important was textiles. As we shall see in Chapter 8, trade in textiles is strongly affected by special trading agreements. Nevertheless, the GSP did mark a major shift in international trade policy, particularly in manufacturing industry. As far as the developing countries were concerned it permitted non-reciprocal trading agreements to be made.

International trade negotiations within GATT have occurred periodically in a series of 'rounds' since 1947. So far there have been seven such mammoth and long-drawn-out rounds, the eighth being the Uruguay Round, which was scheduled to conclude at the end of 1990. Thus, national trade policies have to be seen within this international institutional context although this does not necessarily mean that nations invariably abide by the rules of GATT or respond to the urgings of UNCTAD. What it does mean is that there are international pressures on national governments in their pursuit of trade policies which are not present in the other policy areas discussed below. On

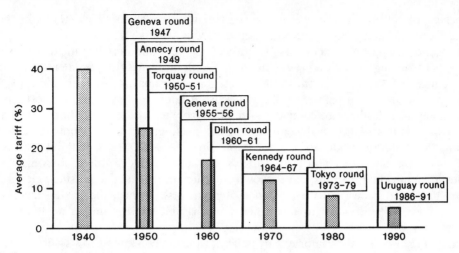

Figure 6.2 Reduction in tariffs and the sequence of GATT negotiating rounds

the other hand, the boundaries between trade policies and industry and foreign investment policies have become increasingly blurred, as we shall see.

Major types of trade policy Figure 6.1 summarizes the major types of trade policy pursued by national governments. In general, policies towards imports are regulatory whereas policies towards exports, with one or two exceptions, are stimulatory. Policies on imports fall into two distinct categories: tariffs, and non-tariff barriers (NTBs). *Tariffs* are, essentially, taxes levied on imports which increase the price to the domestic consumer and make imported goods less competitive (in price terms) than otherwise they would be. In general, the level of tariff increases with the stage of processing: tariffs tend to be lowest on basic raw materials and highest on finished goods. Although tariffs may be regarded simply as one means of raising revenue, their major use has been to protect domestic industries: either 'infant' industries in their early delicate stages of development or 'geriatric' industries struggling to survive in the face of external competition. Largely through the successive rounds of international negotiation in GATT, the general level of tariffs in the world economy has declined very substantially. Figure 6.2 shows that in 1940 the average tariff on goods was around 40 per cent; today the average tariff is around 5 per cent. Figure 6.2 also shows the eight GATT rounds within which tariff reductions and other trade measures have been negotiated.

Although, in general, tariffs have continued to decline, the period since the mid-1970s has seen a marked increase in the use of NTBs. Indeed, today NTBs may well be more important than tariffs in influencing the level and composition of trade between nation states. Certainly much of what has been termed the 'new protectionism' consists of the increased use of NTBs. As Figure 6.1 shows, such trade barriers can take a variety of forms. The most important type of NTB is the import quota, a limit on the quantity of a particular good which may be imported. In some cases, quotas are established as part of so-called 'orderly marketing agreements' or 'voluntary export restraints'. Such devices have become extremely important in each of the industry cases described in Part Three of this book. In addition to quotas, importers may be required to seek import licences, to pay deposits or be subject to import surcharges. Companies suspected of setting their prices in overseas markets at levels below those in their domestic markets may be subjected to anti-dumping penalties.

A second group of NTBs consists of various regulatory and bureaucratic devices such as special labelling or packaging regulations, health and safety requirements, customs procedures and documentation. A third category overlaps closely with some of the industry and foreign investment policies discussed later. For example, domestic producers may be subsidized to compete more effectively against imports; firms may have to comply with specific local content requirements; government contracts may be confined to domestic producers. Although the aims of GATT are to prohibit 'quantitative restrictions and other non-tariff barriers to trade' relatively little success has been achieved. Individual nations have shown a reluctance to abandon such means of controlling their trade (and their balance of payments). Hence, the greatly increased operation of NTBs has led to intensified tensions and stresses in the world trading system.

Figure 6.1 also lists the major export policies which may be pursued. Again, some of these may overlap with industry and foreign investment policies. In particular, financial and fiscal advantages may be granted to exporting firms. In addition, specific tax and tariff concessions may be applied: export earnings may be tax-free or taxed at a lower rate, tariffs may be waived or reduced on those imports which are essential for export activities. Governments may operate an export credit guarantee scheme, fix export targets and operate export promotion offices overseas. Particular geographical areas may be set aside as export processing or free trade zones. As we shall see, these have proliferated rapidly in many developing countries in recent years.

Most export policies are, of course, aimed at stimulating a nation's exports but there are some instances where policies are adopted which regulate exports. One example is the kind of 'orderly marketing arrangement' mentioned earlier, whereby the exporting nation 'voluntarily' restricts its exports of a good to the other nation (often the term 'voluntary' is hardly appropriate). Another example of export restriction occurs in the case of strategically or militarily sensitive items. Many countries operate such selective export embargoes. Finally, one measure common to both import and export policy is the manipulation of the nation's currency exchange rate. Devaluation of the currency makes exports cheaper and imports more expensive and vice versa. However, the ability of a government to manipulate its exchange rate in a controlled way in a world of floating and volatile currencies is very limited.

Bhagwati (1988) divides this proliferating array of non-tariff barriers into two broad types:

1. *'High track' restraints*, which bypass the GATT rule of law. These are the 'visibly and politically negotiated' restraints negotiated by trading partners: the voluntary export restraints and orderly marketing arrangements which have proliferated in a variety of industries, including textiles and clothing (see Chapter 8), automobiles (Chapter 9), consumer electronics (Chapter 10), steel, footwear, machine tools, etc.
2. *'Low track' restraints* which 'capture' and 'pervert' the GATT rules. Such measures include anti-dumping provisions and countervailing duties. These play 'legitimate roles in a free trade regime . . . but not if they are captured and misused as protectionist instruments' (Bhagwati, 1988, p. 48).

It has been estimated that NTBs affect more than a quarter of all industrialized country imports and are even more extensively used by developing countries.[6] Kostecki

(1987) identified 113 major voluntary export restraint agreements in operation in 1986–7, excluding the most comprehensive case of the Multi-Fibre Arrangement in textiles and clothing. Of the 113 cases, almost 90 per cent protected the EC, US and Canadian markets and almost two-thirds of the arrangements involved exports from Asia, primarily Japan and South Korea. Overall, Kostecki estimated that some 12 per cent of total world (non-fuel) trade in the mid-1980s was covered by export-restraint arrangements. Thus, whilst efforts continue within the GATT to maintain multilateral trade agreements, the trend towards managed trade on a bilateral basis is intensifying. Currently there is a whole series of smouldering trade disputes in a whole range of industries. Some of these disputes will be discussed in some detail at later stages of this chapter and in the case study chapters of Part Three.

Foreign investment policies

In a world of transnational corporations and of complex flows of investment at the international scale, national governments have a clear vested interest in the effects of foreign investment, whether positive or negative. From a national viewpoint, foreign investment is of two types: *outward* investment by domestic enterprises and *inward* investment by foreign enterprises. Few national governments operate a totally closed policy towards foreign investment although the degree of openness varies considerably. In general, the developed market economies tend to be less restrictive in their policies towards foreign investment than the developing market economies. One obvious reason is the fact that the developed economies are the major sources of, as well as the dominant destinations for, the world's foreign investment. There are, of course, some exceptions to this general pattern. Figure 6.3 summarizes the major types of national policy towards foreign investment.

Most national policies are concerned with inward investment though governments may well place restrictions on the export of capital for investment (for example, through the operation of exchange control regulations) or insist that proposed overseas investments be approved before they can take place. A far more extensive battery of policies exists in the case of inward investment. The policies listed in Figure 6.3 fall into four broad categories. The first category relates to the entry of foreign firms into a national economy. Governments may operate a 'screening' mechanism to attempt to filter out those investments which do not meet national objectives, either economic or political. Foreign firms may in fact be excluded entirely from particularly sensitive sectors of the economy or the degree of foreign penetration in a sector may be limited to a certain percentage share.

More generally, there may be a restriction on the extent to which individual firms may be owned or controlled by a foreign enterprise. Government may insist that only joint ventures involving indigenous capital may be permitted, possibly on a 51:49 per cent basis in favour of domestic firms. Another possibility is for government to require foreign firms to employ local personnel in managerial positions. Generally speaking, of course, foreign firms must comply with prevailing national codes of business conduct – be good corporate citizens – including those relating to the disclosure of information about the firm's activities. This latter point is frequently a major bone of contention between TNCs and national governments.

The second category of policies shown in Figure 6.3 relates to the operations of foreign firms. A particularly common requirement is for such firms to meet a certain level of 'local content' in their manufacturing activities. Such a requirement is designed

---

Policies relating to inward investment by foreign firms

- Government screening of investment proposals
- Exclusion of foreign firms from certain sectors or restriction on the extent of foreign involvement permitted
- Restriction on the degree of foreign ownership of domestic enterprises
- Insistence on involvement of local personnel in managerial positions
- Compliance with national codes of business conduct (including information disclosure)

- Insistence on a certain level of local content in the firm's activities
- Insistence on a minimum level of exports
- Requirements relating to the transfer of technology
- Locational restrictions on foreign investment

- Restrictions on the remittance of profits and/or capital abroad
- Level and methods of taxing profits for foreign firms

- Direct encouragement of foreign investment: competitive bidding via overseas promotional agencies and investment incentives

---

Policies relating to outward investment by domestic firms

- Restrictions on the export of capital (e.g. exchange control regulations)
- Necessity for government approval of overseas investment projects

---

Figure 6.3 Major types of national policy towards foreign direct investment

to increase the positive effects of a foreign investment on indigenous suppliers and to reduce the level of imported materials and components. Conversely, government may insist on a foreign firm exporting a specified proportion of its output. One of the major elements in the foreign investment 'package' is that of technology. As we shall see in Chapter 12, there is much dispute about the extent to which TNCs do, in fact, transfer technology beyond their own corporate boundaries. Governments may wish to stimulate such technological diffusion through regulatory or stimulative measures.

The third set of policies relates to government attitudes towards corporate profits and the transfer of capital. All governments are concerned to minimize the outflow of capital; on the other hand, TNCs invariably wish to remit at least part of their profits and capital abroad. Similarly, TNCs aim to minimize their liability to taxation; national governments wish to maximize their tax yield. Hence, variations in the restriction on the remittance of capital and profits, together with variations in the level and methods of taxing TNC profits, are extremely important.

The final category of policies towards foreign investment shown in Figure 6.3 aims to stimulate inward investment. Indeed, an increasingly common feature of today's world economy – of developed as well as of developing economies – is the scramble to attract foreign investment. Competitive bidding via overseas promotional agencies and invest-ment incentives has become endemic throughout the world. The important point about such international (and inter-regional) competition is that for certain types of invest-ment it is truly global in extent. For the cost-oriented transnational investments, for example, countries and localities in Europe or North America may well be in direct competition with locations in Asia and Latin America.

In the case of foreign direct investment there is no international body comparable to the GATT which provides a set of rules for international trade. The Uruguay Round negotiations of the GATT include, however, efforts to agree on a set of 'trade-related investment measures' (TRIMs). In effect some of the industrialized countries, led by the United States, wish to prohibit or restrict a number of the measures listed in Figure 6.3, notably local content rules, export performance requirements and the like. The advocates of TRIMs argue that such measures restrict or distort trade. The opponents, including many developing countries, see such measures as essential elements of their economic development strategies. They, in turn, wish to see a tightening of the regulations against the restrictive business practices of transnational corporations.

Industry policies

National policies towards trade or foreign investment are explicitly concerned with international or cross-border issues and are, therefore, most obviously relevant to our interest in the internationalization of economic activity. But there is a third policy area – industry policy – which, although essentially concerned with internal issues, also has broader international implications. Indeed, it is becoming increasingly apparent that the boundaries between trade policy and industry policy, in particular, are often extremely blurred. Figure 6.4 lists the major types of stimulatory and regulatory industry policies used by national governments.

The most obvious *stimulatory* measures are the various financial and fiscal incentives governments may offer to private sector firms. As Figure 6.4 shows, the financial measures most commonly used fall into two categories. On the one hand, governments may provide capital grants or loans to firms to supply part or all of the investment required for a particular productive venture. The other major financial, or rather fiscal, incentive employed by governments is that of tax concessions. Under this banner a whole variety of measures may be employed. For example, firms may be permitted to depreciate or write down their capital investment against tax at an accelerated rate, they may be granted tax reductions or even tax exemptions. Such concessions may be for a specified period: the so-called 'tax holiday'.

Governments may also use various types of employment and labour policy to help shape and encourage industrial activity within their boundaries. Most governments are concerned to stimulate employment and, therefore, to reduce unemployment. In pursuit of such aims, firms may be encouraged to increase their labour force by direct subsidy. Training may be paid for – or even provided in government establishments – to provide a labour force with appropriate skills.

National governments tend to be the largest individual customers in any economy for an enormous variety of goods and services (including, of course, defence). Thus government policies of procurement are extremely important. The award of large government contracts or, conversely, their withdrawal or cancellation, may make or break a private sector enterprise and have enormous employment repercussions.

Policies aimed at stimulating industrial activity

Investment incentives:
- Capital-related
- Tax-related

Labour policies
- subsidies
- training

State procurement policies

Technology policies

Small firm policies

Policies to encourage industrial restructuring

Policies to promote investment

Policies aimed at regulating industrial activity

State ownership of production assets

Merger and competition policies

Company legislation

Taxation policies

Labour regulation
- labour union legislation
- immigration policies

National technical and product standards

Environmental regulations

Health & safety regulations

Some or all of these policies may be applied either generally or, more commonly, selectively. Selectivity may be based upon several criteria, e.g.

1. Particular sectors of industry e.g.
(a) to bolster declining industries
(b) to stimulate new industries
(c) to preserve key strategic industries

2. Particular types of firm e.g.
(a) to encourage entrepreneurship and new firm formation
(b) to attract foreign firms
(c) to help domestic firms against foreign competition
(d) to encourage firms in import-substituting or export activities

3. Particular geographical areas e.g.
(a) economically depressed areas
(b) areas of 'growth potential' or of new settlement

Figure 6.4 Major types of national policy towards industry

The rapid and far-reaching technological developments discussed in Chapter 4 have led many governments to try to stimulate research and development in key sectors and to encourage technological collaboration between firms. The perceived need to stimulate entrepreneurial activity has produced a whole battery of policies to encourage small and medium-sized enterprises. Governments may also attempt to restructure firms – and even entire industries – to improve their international competitiveness.

Regulation of national industrial activity can also take a variety of forms, as the right-hand box of Figure 6.4 suggests. State ownership of productive assets is present in many countries although a current trend in many market economies is towards increased privatization. Entry into particular sectors may be discouraged through the operation of merger and competition policies, although, again, there is a current trend towards the *deregulation* of certain industries such as telecommunications and financial services. Company legislation in general is designed to regulate the ways in which companies can be formed and operate and may be reinforced by specific taxation policies. Regulation of the labour market may be pursued through the encouragement or discouragement of labour union activity or through immigration policies. Technical and product standards are usually defined in specific national terms as are health and safety regulations and environmental regulations.

As Figure 6.4 suggests, the various stimulatory and regulatory policies may be applied *generally* across the whole of a nation's industries or they may be applied *selectively*. Such selectivity may take a number of forms: particular sectors of industry, particular types of firms, particular geographical locations.

International economic integration: regional trading blocs

Within the last thirty years or so an additional element has been added to the politics of the world trading system: the emergence of regional trading blocs. Although not new, such groupings have proliferated in recent years. Table 6.1 lists the major trading blocs which exist at present among the world's economies. The postwar proliferation of regional trading blocs dates from the late 1950s and the early 1960s. A major stimulus was the formation of the European Community (EC) in 1957 followed by the European Free Trade Association (EFTA) in 1960. During the 1960s regional blocs were spawned in other parts of the world, especially in Latin America and Asia. Most recently Canada and the United States have signed a comprehensive Free Trade Agreement under which all trade barriers between the two countries will be abolished by 1998. Both the Canada–United States Free Trade Agreement and the completion of the Single European Market in 1992 will be discussed in later sections of this chapter. At this stage, the aim is merely to outline very briefly the general nature and characteristics of the various types of regional trading blocs.

As Table 6.1 shows, regional trading blocs come in a great variety of shapes and sizes. They also vary considerably in their degree of economic integration. Each type shown in Figure 6.5 represents an increasing degree and complexity of integration. The loosest form of integration is that of the *free trade area*, in which member nations remove trade restrictions between one another but maintain their individual trade policies towards non-member nations. A *customs union* is a free trade area between members together with the erection of a common external trade policy (tariffs and non-tariff barriers) towards others. Free trade areas and customs unions are concerned only with trade in commodities and goods. A *common market* goes a good way beyond this:

Table 6.1 Major regional trading blocs in the world economy

Regional bloc	Date of establishment	Current membership
European Community (EC)	1957*	Belgium, Denmark, France, Greece, Ireland, Italy, Luxembourg, Netherlands, Portugal, Spain, United Kingdom, Germany
European Free Trade Area (EFTA)	1960	Austria, Norway, Sweden, Switzerland
Canada – United States Free Trade Agreement	1989	Canada, United States
Latin American Integration Association (LAIA)	1960	Argentina, Bolivia, Brazil, Chile, Colombia, Ecuador, Mexico, Paraguay, Peru, Uruguay, Venezuela
Central American Common Market (CACM)	1960	Costa Rica, El Salvador, Guatemala, Honduras, Nicaragua
The Andean Group	1969	Bolivia, Colombia, Ecuador, Peru, Venezuela
Caribbean Community (CARICOM)	1973	Antigua, Bahamas, Barbados, Belize, Dominica, Grenada, Guyana, Jamaica, Montserrat, St Kitts-Nevis-Anguila, St Lucia, St Vincent, Trinidad & Tobago
Association of South East Asian Nations (ASEAN)	1967	Brunei, Indonesia, Malaysia, Philippines, Singapore, Thailand

*Denmark, Ireland, United Kingdom joined in 1973, Greece in 1981. Spain and Portugal became members in 1986.

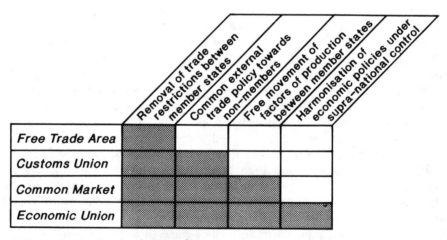

Figure 6.5 Types of regional economic integration

as well as removing trade barriers between member states and operating a common external tariff it also permits the free movement of factors of production (capital, labour, etc.) between its member states. The highest form of regional economic integration is the *economic union*, in which not only are internal trade barriers removed, a common external tariff operated and free factor movements permitted but also broader economic policies are harmonized and subject to supranational control. Thus, an economic union involves a much greater surrender of national sovereignty and not just the removal of trade restrictions.

Regional trading blocs are essentially *discriminatory* in nature. As such they go against the general principle of non-discrimination established by GATT, although provision for such integration was incorporated in the GATT agreement. Article XXIV of the GATT allows for the creation of free trade areas and customs unions, with certain specified provisos. Most of the blocs have a strongly defensive character; they represent an attempt to gain advantages of size in trade by creating large markets for their producers and protecting them at least partly from outside influence. As such the most important of the regional blocs – particularly the EC and potentially the Canada–United States FTA – have a very considerable influence on patterns of world trade. The establishment of a regional trading bloc may either *create* trade (where a country begins to import a product which it formerly produced itself) or *divert* trade (where a country begins to import a product from a member country which it formerly imported from elsewhere in the world).[7] It is the latter effect that creates apprehension among countries which are not members of a particular regional bloc. As we shall see in the following chapters, regional trading blocs also have a very substantial impact on the location of international investment.

Factors influencing the specific policy mix

We have now identified the major kinds of policies which national governments may employ to influence trade, foreign investment and industry. The precise mix of policies actually pursued by any specific nation depends upon a whole variety of factors, including:

1. *The nation's political and cultural complexion and the strength of institutions and interest groups.* In general, conservative governments are less inclined to pursue interventionist policies than liberal or socialist governments. However, much depends on the power of institutions and interest groups within the national economy: for example, business and financial interests, labour unions, environmental groups. The particular form of political structure – whether centralized or federal – will also be important. An especially significant factor, therefore, is the degree of consensus or conflict between institutions and interest groups, a situation in which the nation's history and culture will be significant influences.
2. *The size of the national economy, especially that of the domestic market.* This is especially relevant for trade policies: the larger the domestic market the less important relatively is external trade likely to be and vice versa. However, much will depend upon:
3. *The nation's resource endowment, both physical and human.* A weak natural resource endowment will necessitate the import of essential materials which, in turn, must be paid for by exports of other, usually manufactured, products. A strong endowment of particular types of raw material may well influence the kind of economic policy pursued. Similarly, the size, composition and skill level of a nation's potential labour force will constrain the kinds of policy which may be pursued.
4. *The nation's relative position in the world economy, including its level of economic*

development and degree of industrialization. A nation's 'degrees of freedom' depend very much on its relative position in the world economy and, in particular, on the extent of its dependence on external trade and investment. The kinds of industry and trade policies pursued will also tend to reflect the nation's degree of industrialization. The policy emphasis of established industrial nations will differ from that of nations in the process of industrializing.

One useful way of looking at broad differences in national policy position is to distinguish between two simplified models of government economic intervention: the developmental (or plan-rational) state and the regulatory (or market-rational) state. Johnson (1982) describes the differences between these two models as follows:

> A regulatory, or market-rational, state concerns itself with the forms and procedures – the rules, if you will – of economic competition, but it does not concern itself with substantive matters. For example, the United States government has many regulations concerning the antitrust implications of the size of firms, but it does not concern itself with what industries ought to exist and what industries are no longer needed. The developmental, or plan-rational, state, by contrast has as its dominant features precisely the setting of such substantive social and economic goals . . .
>
> The government will give greatest precedence to industrial policy, that is, to a concern with the structure of domestic industry and with promoting the structure that enhances the nation's international competitiveness. The very existence of an industrial policy implies a strategic, or goal-oriented, approach to the economy. On the other hand, the market-rational state usually will not even have an industrial policy (or, at any rate, will not recognize it as such). Instead, both its domestic and foreign economic policy, including its trade policy, will stress rules and reciprocal concessions (although perhaps influenced by some goals that are not industrially specific, goals such as price stability or full employment).
>
> (Johnson, 1982, pp. 19–20)

It is easier to find examples of developmental states in today's world than of pure regulatory states, although countries like the United States and the United Kingdom are more regulatory than developmental in their current policy stance. In broad terms, however, many of the current politico-economic disputes in the global economy are at least partly the reflection of a clash between these two modes of government involvement. In the following sections of this chapter, we look more closely at some specific national cases from the viewpoint of those policies which have the greatest influence on the internationalization and globalization of economic activity. Further discussion of national policies in specific industrial sectors appears in the case study chapters of Part Three.

Western Europe: the European Community[8]

National policies

Despite the establishment of the European Economic Community in 1957 and its subsequent enlargement in 1973, 1981 and 1986, the industrial policies of member states have remained essentially *national* policies. Although national perspectives and priorities are now having to change in the context of the completion of the Single European Market, traces of earlier policies will certainly persist. Historically, most of the continental European nations developed fairly centralized approaches to industrial policy. France is perhaps the clearest example of this, with its long history of strong state involvement that goes back to the seventeenth century.

France's industrial policy in the postwar period has been an integral part of the series of national economic, indicative, plans. A major component of French industrial policy has been the promotion of 'national champions' in key industrial sectors. In some cases, this was sought through state ownership of large-scale enterprises. In its desire for technological independence, especially from the United States, France invested massively, but selectively, in certain sectors. In the late 1960s and early 1970s, for example, four *grands programmes* were launched in the nuclear, aerospace, space technology and electronics industries. Subsequently, the focus was narrowed to high-growth sectors in energy conservation equipment, office information systems, robotics, biotechnology and electronics as well as to the problem sector of textiles.

The French government has also exerted considerable influence on its industry through its purchasing policies and through its control of the major financial institutions. Both powers were used extensively. For example, government monopoly of purchasing in the heavy electrical engineering sector was used to enforce reorganization of the industry. State control of financial institutions enabled the French government to steer investment funds to targeted firms and sectors. In a number of ways, therefore, including the encouragement of industrial mergers, the French government intervened strongly to reshape the national industrial structure and to increase the country's international competitiveness.

The major exception among the continental European nations to a centralized approach to industrial policy has been West Germany. In part, at least, this reflects the fact that Germany is a federal political unit with power divided between the federal government and the provinces (*Länder*). Although often described as 'light', the federal government's role is far from insubstantial:

> [It] pursues active industrial intervention to achieve its objectives. Subsidies to industry (including tax concessions, grants, loans and interest remission), initially viewed as temporary exceptions to the rule of market competition in the determination of resource allocation, came to be viewed as legitimate industrial policy instruments ... Within the West German framework of the social market economy, the state has a responsibility to regulate and intervene to improve the working of markets (as well as combat cartels and promote social justice).
>
> (Hesselman, 1983, p. 203)

The West German economy is characterized both by a considerable degree of competition between domestic firms and also by a high level of consensus between various interest groups, including labour unions, the major banks and industry. The provincial governments have played an important role in the economy including, in recent times, the aiding of large manufacturing companies in distress. Like the French, the West German government has also intervened to stimulate technological development in key sectors. During the 1970s, in particular, government financial support for research and development became focused on the development of advanced technologies – especially computer technologies – with wide application.

Industry policy in the United Kingdom contrasts in a number of ways with that of the continental European countries. 'The UK industrial system seems to possess little of the "cement" provided by government links with industry in France and by the banks, interfirm cooperation and harmonious labour relations in Germany' (Shepherd, Duchene and Saunders, 1983, p. 18). Perhaps the most consistent feature of UK industry policy in the postwar period has been its *in*consistency. This only partly reflects changes associated with change of government; even the same government has often adopted variable policies. In the early 1950s government involvement in the economy was primarily macroeconomic: the management of domestic demand and the creation of an amenable 'business climate'.

This policy stance altered markedly with the change to a Labour government in the mid-1960s when, for the one and only time, an attempt was made to operate a national plan. The attempt was short-lived, largely because of external pressures on the country's balance of payments and a currency crisis. Successive changes of government in 1970 and 1974 brought, first, a return to government disengagement from economic intervention and then a renewal of government involvement in the form of an explicit industrial strategy. After 1979 the policy emphasis changed again as the Thatcher government's strongly anti-interventionist policy took hold. In concert with these large-scale policy fluctuations the degree of direct state ownership of productive enterprises also waxed and waned.

Even during the more active phases of government involvement in UK industry the precise focus also varied a good deal. In the late 1960s a key element was the encouragement of mergers in sectors such as motor vehicles, electrical engineering – through the Industrial Reorganization Corporation – and steel. In the late 1970s 'small is beautiful' became a dominant theme as government sought to stimulate indigenous investment. The Labour government's strategy of the 1970s was a sector-based approach to industry policy, with selective assistance becoming increasingly important. Subsequently, the focus shifted to one of stimulating advanced technology, including microelectronics, fibre-optics and robotics.

One policy thread, common to all the European states, including the United Kingdom, has been that of *regional policy*: of designating specific geographical areas for special assistance. All the major industrial nations of Western Europe operate a regional industrial policy to varying degrees. In general, the aim is to try to solve problems of economic development – especially unemployment – by providing financial and other incentives for firms to locate in specially designated areas. In France the major thrust of regional policy has been to offset the economic dominance of Paris by developing countermagnets based on major provincial centres, and to stimulate the economies of depressed agricultural regions such as Brittany and of declining industrial regions such as Lorraine. In Italy – where regional differences in economic health are especially acute – policy has focused on the south (the Mezzogiorno). In the United Kingdom broad regional policies have subsequently been replaced by more selective policies, including a specific emphasis on the problems of the inner cities. In addition to national policies towards their own disadvantaged regions the EC operates a Regional Development Fund. Although far from being a common regional policy the Fund does grant aid to areas throughout the Community whose gross domestic product is below the Community average.

A major thrust of regional policies in Europe and, increasingly, of national industry policies has been the enticement of mobile investment. The shortage of internationally mobile investment in recent years has greatly intensified the degree of competitive bidding both between individual nations and also between parts of the same nation. TNCs, especially the very large ones, are assiduously courted, often at the highest government levels. National and regional agencies set out their stalls in lavish advertising and in overseas offices. It follows from such efforts by European countries to encourage foreign investment that their general attitude towards TNCs is one of openness. Even France, which has been most suspicious of American and Japanese investment in particular, has adopted a more favourable attitude, especially towards investments which bring new technology, increase exports or create employment in depressed areas. Several European governments, however, together with the European Commission itself, impose various kinds of *performance requirement* on some foreign investors. Japanese companies in particular have been required to satisfy local content requirements (the chapters on automobiles and electronics in Part Three deal with this issue).

During the 1980s the policy emphasis of virtually all European national governments began to shift away from a direct interventionist position to one of greater disengagement.[9] In both the United Kingdom and France, for example, previously state-owned enterprises – including those nationalized by the Mitterrand government of France in 1982 – were progressively privatized. The former emphasis on nurturing 'national champions' was dropped. The rescuing of 'lame duck' firms and industries was largely abandoned. Some key sectors – such as finance and telecommunications in the United Kingdom – were deregulated and opened up to competition. These policy shifts at the national scale reflect a wave of change which spread through Europe during the 1980s. Even more significant, however, from our particular perspective in this book, is the movement which gathered momentum during the second half of the 1980s towards a Single European Market.

The European Community: towards a single market

Initiatives at the EC level in the fields of industry policy and investment have so far remained very limited. Only in matters of trade have Community-wide policies been extensively implemented and even here considerable internal strains and tensions have persisted. In the case of trade between the EC and other parts of the world the general lowering of tariff barriers negotiated through GATT has been offset, to some extent, by an increased use of non-tariff barriers and the extension of orderly marketing arrangements. Such arrangements have become especially important in the textiles, automobiles and consumer electronics industries, as Chapters 8 to 10 demonstrate. The EC also operates preferential trading arrangements with certain groups of developing countries.[10] For example, since 1972 the Mediterranean countries have been granted free access to the EC market for manufactured goods, though with certain restrictions on 'sensitive' manufactured products. In 1975 the EC signed the Lomé Convention with countries in Africa, the Caribbean and the Pacific – the so-called ACP nations – all of which were former colonies of the European nations. The Lomé Convention is a broad-ranging scheme embracing not only trade but also development aid. It allows for the free entry, at least in theory, of manufactured goods from the ACP countries. However, manufacturing industry is a small proportion of these countries' trade; the most important element of the Lomé Convention relates to commodities and agricultural products.

Whatever its relationships with the outside world the major *raison d'être* of a common market is the removal of trade barriers between member states. As far as tariffs are concerned there is no doubt that this has occurred within the EC. In the early 1980s, however, it was being claimed that

> the Common Market is becoming less common by the day because of the plethora of administrative and other non-tariff barriers to cross-frontier trade within the EEC. The majority of member states feel that in the present recession their best course is to renationalize their

Table 6.2 Removal of barriers as a result of the completion of the Single European Market

1. *Physical barrier controls*
   – customs and excise requirements
   – movement of people

2. *Technical barriers*
   – differences in national product standards and specifications
   – differences in national business laws
   – differences in degree of protection of public procurement markets
   – differential controls on movement of capital

3. *Fiscal barriers*
   – differences in indirect taxation (e.g. value added tax)
   – differences in excise duties

economies, a course of action which brings with it the growing danger of the economic balkanization of the community. Efforts by the EEC Commission or the European Court of Justice to stem the protectionist tide have been as successful as those of King Canute against the rising waters of the North Sea.

(Economist Intelligence Unit, 1983, p. 32)

Individual EC countries were increasingly resorting to tactics which prevented or delayed the import of certain products from other member nations: for example, the insistence on particularly stringent health and safety checks on products, the heavy bureaucracy of customs and frontier procedures, even the awarding of government contracts to domestic producers and the contravention of EC regulations on the sub-sidizing of domestic industries. Each individual member of the EC was guilty of some such practices although some were more active in this respect than others. France and Italy appear to head the league table of cases brought before the European Commission.

It was a growing concern with such internal 'balkanization' and its negative effect on the Community's globally competitive position which led to the Single European Act of 1985 and its proposals to complete the Single European Market by the end of 1992.[11] It is important to emphasize the word 'complete'. The Single European Act is not proposing a radically new European Community; it is merely attempting to *complete* the process of integration begun with the Treaty of Rome in 1957. Its aim is to remove all the remaining internal barriers: physical, technical and fiscal (Table 6.2). Removal of these remaining barriers (several hundred of them) will, according to the Cecchini Report, create a virtuous circle of growth for the European Community as a whole, its member states and for those business firms which successfully take advantage of the changes. Figure 6.6 shows the major components of beneficial change envisaged by the completion of the internal market.

After 1992, therefore, there will be a single *unified* market of approximately 330 million people (including the unified Germany). Potentially this offers the prospect of major restructuring and rationalization of businesses serving that market although the extent to which this occurs will probably vary from one industry to another. Some of the more extreme predictions – for example that half of Europe's factories will disappear as fewer huge plants serve the single market – are almost certainly gross overstatements. So too is the notion that removing internal barriers will lead to a *uniform*, homogeneous market for goods and services. National, regional and local differences in tastes and preferences will certainly not disappear overnight, if ever.

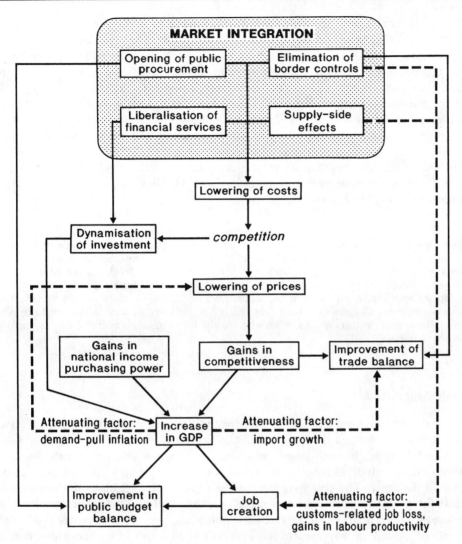

Figure 6.6 Macroeconomic changes associated with the completion of the Single European Market (*Source*: Cecchini (1988), Chart 10.1)

For the world outside the European Community there is a broader concern. Removal of the internal barriers will be set within a single external Community boundary. For firms not physically established in Europe, but anxious to maintain or achieve market access, the overriding concern is the possible emergence of a 'fortress Europe'. Certainly, the existing bilateral trade agreements which individual member states have with non-Community countries will ultimately be replaced by single Community arrangements. As we shall see in Part Three, there are very considerable tensions in such industries as automobiles and electronics. Certainly the imminent completion of the single market is stimulating new waves of foreign direct investment into the Community and leading other European countries to seek membership of the Community.

Thus, whether or not the single market is complete by 31 December 1992 is almost irrelevant. As the American writer Michael Callingaert (1988, pp. 37–8) has perceptively observed: '1992 is a process not an event . . . The process will continue beyond 1992. Indeed, it is unlikely that a specific point will be identifiable at which the internal market becomes "complete".' Whether or not immediate integration occurs is less significant than the fact that the Single European Act has set in motion a whole chain of developments, not least the urgent re-evaluation of trade and investment strategies both by nation states throughout the world economy and by business firms in all industries.

For the individual member states the completion of the single market alters the national policy environment very considerably. As Sharp and Holmes observe in relation to the United Kingdom and France:

> the last 20 years . . . have witnessed the shift of policy from nationally based systems of direct protection and support towards a more complex position in which there are still strands of these older policies, but where the locus of activity is now rapidly shifting towards the Community, and in which, at both national and community level, the future is seen to lie not in direct intervention, but in equipping industry with the capabilities it needs to face up to the competitive pressures of the modern world. That said, there remain deep-seated divisions between those who believe that this is best achieved at least in the short-run, via some element of protection, and those who see the fostering of competitiveness as the key to policy thinking.
>
> (Sharp and Holmes, 1989, p. 232)

North America

Despite the geographical and functional proximity of their economies, national policies towards industry, foreign investment and trade differ substantially between the United States and Canada. In both cases, however, the political structure creates particular tensions between federal government on the one hand and state or provincial governments on the other. The policy stance of the United States reflects both the sheer scale and wealth of the domestic economy and also a basic philosophy of non-intervention by the federal government in the private economic sector. As far as industry is concerned the role of the federal government has generally been a *regulatory* one whose aim is to ensure the continuation of competition. Thus, 'the United States in the postwar period has not had anything that most Americans would think of as an "industrial policy"' (Diebold, 1982, p. 158). The United States remains the clearest example of a regulatory or market-rational state although, as we shall see, strong internal pressures currently exist for a more strategic policy approach.

Apart from the system of investment tax credits and accelerated depreciation schemes, action at the federal level has been based on macroeconomic policies of a fiscal and monetary kind. The aim has been to create an appropriate investment climate in which private sector institutions could flourish. This has not, however, prevented the federal government from rescuing specific firms – especially very large ones – from disaster. Prominent examples include Lockheed in 1971 and Chrysler in 1979. At the other end of the size spectrum, the Small Business Administration has provided aid to stimulate new and small firms. Federal procurement policies are generally non-discriminatory but the sheer size of federal government purchases, particularly in the defence and aerospace industries, has exerted an enormous influence on US industry.

Entire communities and regions are heavily dependent on the work created by federal defence contracts.

Direct federal involvement in regional or area economic development is relatively limited, however. Most of the direct efforts to stimulate industrial development are carried on at the state and local level. All states employ a battery of investment incentives to entice new investment, and most spend vast sums on promotional activities. The result is frenzied and cut-throat competitive bidding between states. The most common form of state aid to industry is the industrial development bond, which is a device to raise funds for investment, although states also operate loan and loan guarantee schemes. States also differ considerably in the tax burden they impose on business. Quite apart from the fact that the level of tax varies from one state to another, states may also offer specific tax concessions to business. States also differ greatly in their attitude towards, and legislation of, labour unions. Some twenty states, particularly in the South, operate 'right to work' legislation.

As we saw in Chapter 3, there has been a very rapid increase in foreign investment in the United States in the past few years. The federal attitude towards foreign investment is extremely liberal but it does not itself attempt to attract such investment to the United States. However, states and local governments certainly do; indeed, such efforts have greatly intensified in what Glickman and Woodward (1989) graphically term 'the mad scramble for the crumbs'. Almost all states have foreign promotional budgets, some have opened offices abroad and many advertise for potential investors in the international business press. However, the late 1980s saw an upsurge of concern within the United States over what some see as the 'buying up of America'.[12] So far, such concerns have not produced any new national policy although foreign-owned firms have come under increasing scrutiny by federal bodies over such issues as tax avoidance through transfer pricing.

One particularly interesting aspect of American trade policy which has had important implications for global shifts in manufacturing industry is the operation of *offshore assembly provisions* (OAPs). Under certain tariff provisions (specifically items 807.00 and 806.30 of the US tariff schedule) an American firm may export domestic materials or components for processing overseas and then reimport the processed product on payment of duty only on the value of the foreign processing. OAPs are especially important in certain types of manufacturing, for example electronics and metal industries. Although applicable to any country, OAP imports to the United States have come to be dominated by a small number of developing countries, notably Mexico and the newly industrializing countries of East and South East Asia. There is little doubt that the existence of OAPs has been one of the forces stimulating US firms to seek out cheap labour production locations in certain manufacturing industries, notably electronics (see Chapter 10).

The 'new protectionism': towards a strategic trade policy

United States policy towards international trade in the postwar period has been one of urging liberalization and the reduction of tariffs. As the strongest economy it has been, like Britain in the nineteenth century, the leading advocate of free trade. Even so, the federal government has intervened with the use of tariff and non-tariff barriers to protect particular interests. It has, for example, negotiated orderly marketing arrangements with various countries, notably Japan, in sectors such as textiles (see Chapter 8), automobiles (Chapter 9) and electronics (Chapter 10). The use of non-tariff barriers on a bilateral basis by the United States was explicitly specified in the Trade Act of 1974. This Act in effect heralded the emergence of a 'new protectionism' in the United States and the beginnings of a movement towards a strategic trade policy.

The United States became increasingly embroiled in a whole series of trade disputes – with Japan, the East and South East Asian NICs and the European Community. Such disputes focused upon US allegations of unfair trading practices by these countries in specific industrial sectors. In that respect, policies were no different from the similar measures being implemented throughout the later 1970s and the 1980s by individual European governments. A major new step in US trade policy came with the introduction of the Omnibus Trade and Competitiveness Act of 1988 (OTCA), stimulated by the persistence of a massive trade deficit and the difficulties being encountered by a number of particular American industries. The OTCA incorporates a strongly *unilateralist* approach to trade negotiations rather than the multilateralist approach enshrined in the GATT and hitherto strongly supported by the United States. The key clause – what some have called the 'crowbar' – is the so-called 'Super 301' clause.[13] The aim is to achieve reciprocal access to what the United States defines as unfairly restricted markets.

The difference between the 1974 and 1988 Acts in this respect is that the Super 301 clause is directed to *entire countries*, not just individual industries. In its first application in May 1989 three countries – Japan, India and Brazil – were named as priority countries. Others were 'warned' that their trade practices were being watched. If, over a defined period of time, the named countries do not abandon what the United States defines as 'unjustifiable' or 'unreasonable' trade practices, the United States may retaliate unilaterally by restricting access to the US market for *all* goods from the countries in question. In a separate, but related, set of measures the United States has negotiated with Japan a 'Structural Impediments Initiative' which aims to remove what are seen to be structural restrictions on access to the Japanese market.

Towards regional integration: the Canada–United States Free Trade Agreement

Another important development in US trade policy with significant implications for the global economy is the signing of the Canada–United States Free Trade Agreement.[14] The main elements of the Agreement are shown in Table 6.3. Hitherto, the United States had not taken advantage of the GATT article allowing for exceptions to the principle of non-discrimination in the case of free trade areas and customs unions. In part, this change of stance reflects the United States' desire to have a 'lever' to achieve more open trade in the world economy by showing that it is prepared to make discriminatory agreements. In other words, it is part of the shift in US trade policy discussed above. Other US motives include the desire to reduce trade barriers with Canada, to improve the treatment of US direct investment in Canada and to achieve specific deals on government procurement policies, trade in services and other areas.

To the Canadian government a free trade agreement offered the prospect of obtaining more secure access to the US market with less risk of being inhibited by various non-tariff barriers, to secure an effective dispute settlement procedure and to protect its cultural industries by agreement. In fact, there is substantial political disagreement in Canada over the costs and benefits of the Free Trade Agreement, reflected not only

Table 6.3 The main elements of the Canada–United States Free Trade Agreement

Agreement reached in January 1988. Implementation began in January 1989.

Major provisions of the Agreement

1. *Tariffs and rules of origin*
   - All bilateral tariffs to be removed over a ten-year period.
   - Goods incorporating offshore materials or components will qualify although in certain cases a 50% local content provision will be applied.

2. *Quantitative restrictions*
   Existing quantitative restrictions will be eliminated.

3. *Technical standards*
   The aim is to make federal standards more compatible to reduce technical barriers to trade, while still protecting health, safety, environmental, national security and consumer interests.

4. *Agriculture*
   All tariffs to be eliminated within ten years. Agreement not to use direct export subsidies on their bilateral agricultural trade.

5. *Energy*
   Agreement to prohibit most restrictions on energy exports and imports.

6. *Automotive trade*
   - The Auto Pact is retained.
   - Automotive duty waivers and remissions to be phased out.
   - Rules of origin provisions encourage the sourcing of more components in North America.

7. *Government procurement*
   Agreement to expand access of each other's suppliers to federal government purchases.

8. *Trade in services*
   - Agreement to extend, in the future, principles of national treatment, right of commercial presence and right of establishment, consistent with the investment provisions to each other's providers of services.
   - In the case of financial services existing market access is preserved and competition in securities underwriting and banking is introduced.

9. *Cultural industries*
   Explicitly excluded from the Agreement.

10. *Investment*
    - National treatment to be provided to each other's investors.
    - Agreement not to impose local content, export or import substitution requirements.
    - Canadian threshold for review of direct acquisitions to be raised to $150 million by 1992. Review of US indirect acquisitions to be phased out.

11. *Emergency measures*
    Agreement to more stringent standards in application of safeguard measures (quotas or restrictions) to bilateral trade.

12. *Dispute settlement*
    Dispute settlement mechanism to be established to ensure the fair application of anti-dumping and countervailing duty laws of both countries.

(*Source*: based on material in Canadian Department of Finance (1988) *The Canada–United States Free Trade Agreement: An Economic Assessment*, Ottawa: Government of Canada)

in the activities of individual pressure groups but also at the national and provincial political level. Such differences of attitude are, of course, made more complex by the continuing constitutional disputes between the Canadian provinces.

In 1990 a further dimension was added to the US espousal of regional integration policies with the announcement of the start of negotiations with Mexico on a possible free trade agreement between the two countries. President Bush even suggested that a longer-term goal could well be the incorporation of Latin America into an Americas free trade area stretching from Anchorage to Tierra del Fuego. Such a scenario is far from inevitable. The prospects of a Mexico–United States agreement are, however, causing apprehension in Canada as well as among the Caribbean countries which are currently covered by a specific US Caribbean Basin Initiative.

In a variety of ways, therefore, the international economic policy position of the United States has been changing. Elements of a *strategic trade policy* have emerged. Its development has been strongly driven by various powerful interest groups and domestic lobbies. Since the boundary between trade policy and industry policy is so blurred in the strategic trade policy area perhaps we are beginning to see a shift in the United States' position, from a market-rational or regulatory state to a plan-rational or developmental state.

Japan[15]

Japan as a 'developmental' state

The basic difference in government's economic role between Japan on the one hand and the Western nations on the other can be summarized most concisely in the two following quotations:

> Historically, the State assumed an active economic role in Western economies in order to correct what were considered to be the private sector's economic and social failures. Japanese historical tradition, on the other hand, grants to government a legitimate role in shaping and helping to carry out industrial policy. Japanese businessmen share with government leaders and officials a sense of the importance of co-ordinated national development and are generally amenable to and, in fact, expect government intervention to advance this goal. Senior businessmen view government guidance of industry as a normal state of affairs. They may not always enjoy the process or approve of the specific actions but they see the process as legitimate and, in the main, useful.
>
> (Magaziner and Hout, 1980, p. 29)

> The Japanese government is extremely intrusive into the privately owned and managed economy, but it does this through market-conforming methods and in co-operation rather than confrontation with the private sector . . . The first priority of the Japanese government in the private sector is not protectionism or neo-mercantilism (as in France), or regulation (as in the United States, with the exception of the defense and agricultural sectors), or welfare (as in Sweden or the Netherlands). Japan's first priority is, above all, developmental – meaning the effort by the government to secure Japan's economic livelihood through public policies based on such criteria as long-term dynamic comparative advantage and international competitive ability. The Japanese government's most important contributions to the economy are think-tank functions and supervision and co-ordination of the structural changes necessary to keep Japan competitive in world markets.
>
> (Johnson, 1985, pp. 61, 62)

There is a high level of consensus between the major interest groups in Japan on the need to create a dynamic national economy. To some extent, this consensus is a cultural characteristic of Japanese society, with its deep roots in familism. But it also reflects the poor physical endowment of Japan and the limited number of options which faced the country when, in the 1860s, it suddenly emerged from its feudal isolation. In other words, consensus is also a pragmatic stance built up over more than a hundred years. Given virtually no natural resources and a poor agricultural base, Japan's only hope of economic growth lay in building a strong manufacturing base both domestically

Table 6.4 The major roles of the Japanese Ministry of International Trade and Industry (MITI)

1. *Constructs medium-term econometric forecasts of the development of, and needed changes in, the Japanese industrial structure*
   - Establishes indicative plans or 'visions' of the desirable goals for the private sector.
   - Makes specific comparisons of cost structures of Japanese and foreign competitors.

2. *Arranges for preferential allocation of capital to selected strategic industries*
   - Involves governmental and semi-governmental banks.
   - Ministry of Finance guides commercial banks to co-ordinate their lending policies with MITI's industrial strategy.
   - Financial support of an industry implies guidance (though not control) by MITI.

3. *Targets key industries for the future and puts together a package of policy measures to promote their development*
   - Pre-early 1980s the major measure was protection against foreign competition in the Japanese domestic market.
   - Protectionism abandoned in early 1980s. Emphasis now on financial assistance, tax breaks, incentives given through administrative guidance, anti-trust relief (to facilitate 'research cartels').

4. *Formulates industrial policies for 'structurally recessed industries'*
   - MITI designates a specific industry as 'structurally recessed'.
   - The ministry responsible for that sector formulates a stabilization plan specifying how the capacity to be scrapped should be shared between enterprises. The plans must be drawn up in consultation with the Industrial Structure Council of MITI.
   - Costs of scrapping production facilities are shared between the government and the private sector.

(*Source*: based on Johnson (1985), pp. 66–7)

and internationally through trade. In this process, the state has played a central role not through direct state ownership but rather by *guiding* the operation of a highly competitive domestic market economy. Indeed, there is relatively little state-owned enterprise in Japan and a generally much smaller public sector than in most Western economies.

The 'guiding hand' of MITI

The key government institution concerned with both industry policy and trade policy – the two are seen to be inextricably related in Japan – is the Ministry of International Trade and Industry (MITI). Since its establishment in 1949, MITI has become the real 'guiding hand' in Japan's economic resurgence, although its role has often been misunderstood and exaggerated in the West. Table 6.4 identifies the major roles played by MITI in the Japanese economy. Until the 1960s Japan operated a strongly protected economy and it was not until 1980 that full internationalization of the Japanese economy was reached. During the 1950s and early 1960s MITI, together with the Ministry of Finance, exerted very stringent controls on all foreign exchange and over the import of technology.

In fact, imported technology played a most significant part in the rebuilding of the Japanese economy. Technology was imported largely through licensing from foreign suppliers and not via the direct investment of foreign firms in Japan itself. The technologies were chosen to meet the needs of particular industries – those regarded by MITI as being the ones necessary to achieve national objectives. The selected industries were further aided by preferential financing and tax concessions and were also protected from foreign competition. Within Japan, however, intense competition was encouraged between rival Japanese firms with the result that domestic production costs were kept down and efficiency increased. Within the selected industries, MITI encouraged mergers to create large-scale enterprises, although such moves were not always successful. For example, MITI failed in its attempt radically to restructure the automobile industry.

Initially, MITI focused its energies on the basic industries of steel, electric power, shipbuilding and chemical fertilizers but then progressively encouraged the development of petrochemicals, synthetic textiles, plastics, automobiles and electronics. The results, in terms of growth in these sectors, were remarkably impressive. The Japanese economy was transformed from one based on low-value, low-skill products such as cheap clothing and textiles to a basis of high-value, capital-intensive products (Figure 6.7). The foundation of this transformation was the clearly targeted, selective nature of Japanese industry policy together with a strongly protected domestic economy. By the early 1970s Japan was the world leader in the production of steel, ships and consumer goods such as motor cycles and cameras. However, it was already becoming evident that the spectacular growth of Japanese industry had created substantial problems within the country itself.

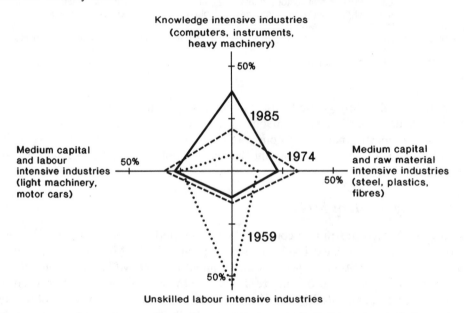

Figure 6.7 Evolution of Japanese industrial structure (*Source*: Magaziner and Hout (1980), Figure II.1)

In 1971 a new industrial policy began to emerge which attempted to meet the problems of environmental pollution, urban congestion and rural depopulation and so on by shifting the focus of Japanese industry towards high-technology, knowledge-intensive industries. Selective government assistance was moved away from the established capital-intensive industries, such as automobiles and steel, and towards the newly emerging high-technology sectors. In 1974 MITI published the first of its *long-term visions* of how the Japanese industrial structure ought to evolve to meet changed circumstances, both domestically and internationally. Appropriately in the Land of the Rising Sun, 'sunset' industries are to be scaled down and 'sunrise' industries encouraged.

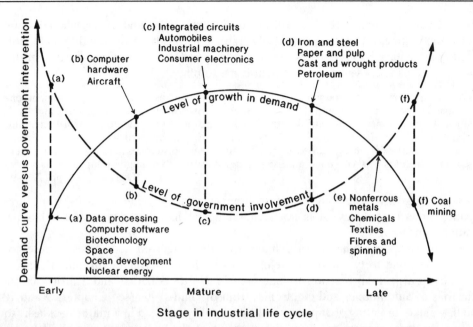

Figure 6.8 The relationship between Japanese government intervention and stage in the industry life-cycle (*Source*: Okimoto (1989), Figure 1.5)

In effect, MITI has used an industrial life-cycle model as the basis for deciding its strategic priorities (Figure 6.8). As Okimoto has shown,

> although the degrees of intervention and selection of policy instruments vary by industry, MITI intervention tends to follow a curvilinear trajectory: that is, extensive involvement during the early stages of an industry's life cycle when market demand is still small, falling off significantly as the industry reaches full maturity and demand reaches its peak, and rising again as the industry loses its comparative advantage and faces the problems of senescence – saturated markets, the loss of market share and excess capacity.
>
> (Okimoto, 1989, p. 50)

Changing attitudes towards overseas investment

Postwar Japanese economic policy has been strongly *mercantilist*. Growth in exports, particularly in the manufacturing sector, has been a major focus along with the building of a strong domestic economy. Among other things, manufactured exports provided the foreign exchange necessary for Japan to import the industrial raw materials which are in such short supply domestically (including, of course, oil). A key element in Japan's 'guiding hand' on the market economy has been the specific treatment of foreign direct investment. For much of the postwar period, both inward and outward investment were extremely closely regulated. The technological rebuilding of the Japanese economy was based on the purchase and licensing of foreign technology and *not* on the entry of foreign branches or subsidiaries. Although the inward investment laws have now been liberalized and foreign firms do indeed operate within Japan, their relative importance remains very small.

As far as overseas investment by Japanese firms is concerned, the situation, as we saw in Chapter 3, changed dramatically during the 1970s and 1980s when Japanese overseas investment grew at a spectacular rate. This was consistent with MITI's policy of internationalizing the Japanese economy and reflected the economically strategic role which Japanese overseas investment came to play:

> For Japan . . . overseas production has suddenly emerged as a national requirement encompassing practically the entire spectrum of her industries and enterprises, small and large alike. The segments of industrial activities that are no longer suitable, environmentally or otherwise, for the Japanese economy need to be transplanted abroad, and overseas resources must now be developed more directly to insure supplies . . . Furthermore, overseas investment is now viewed as an essential device by which to upgrade Japanese industry.
>
> (Ozawa, 1979, pp. 228-9)

Thus, overseas investment by Japanese firms came to be seen as an integral part of Japanese industrial policy. It is not something that has 'just happened': it has been positively encouraged.

Such encouragement has been enhanced by the perceived need for Japanese companies to respond to two major developments in their external environment. The first was the upsurge in protectionist measures in North America and Europe in such industries as automobiles and electronics from the mid-1970s (see Chapters 9 and 10). The 'laser-like' targeting of Japanese industrial policy resulted in a major backlash from various Western economies. As a direct result of the erection of non-tariff barriers Japanese firms began to invest heavily in production facilities overseas. The second external stimulus to increased overseas direct investment was '*endaka*'; the major rise in the value of the Japanese yen which resulted from the 1985 Plaza Agreement among the Group of Five international finance ministers. This political decision stimulated an upsurge in overseas investment by Japanese firms to take advantage of lower production costs, particularly in East and South East Asia.

It should be emphasized that Japan is a *developmental state* but not a centrally planned economy. It is a *guided market economy* in which intense domestic competition is encouraged. The Japanese government role has changed over time: strong and broad-based intervention in the immediate postwar period has now become increasingly selective. Although the government agencies such as MITI are far from the monolithic institutions they are often alleged to be, there can be no doubt of their importance in stimulating Japanese industry and trade.

The newly industrializing countries[16]

The central role of the state

Although they are frequently grouped together, the world's newly industrializing countries are a highly heterogeneous collection of countries. They vary enormously in size (both geographically and in terms of population); in their natural resource endowments; in their cultural, social and political complexions. But they all tend to have one feature in common: the central role of the national state in their economic development. Despite many popular misconceptions, none of today's NICs is a free-wheeling market economy in which market forces have been allowed to run their unfettered course. They are, virtually without exception, *developmental states*; market economies in which the state performs a highly interventionist role. Having said that, the precise role of the state – the degree and nature of its intervention – varies greatly from one NIC to another. In some cases, state ownership of production is very substantial; in others it is insignificant. In some cases, the major policy emphasis is upon attracting foreign direct investment; in others such investment is tightly regulated and the policy emphasis is upon nurturing domestic firms. Thus, although the recurring central theme which runs through the current economic behaviour of all NICs is the role of the state, each individual NIC performs a specific variation on that general theme.

Types of industrialization strategy

Broadly speaking, a developing country may pursue three kinds of industrialization strategy:

1. local processing of indigenous raw materials;
2. import-substituting industrialization – the manufacture of products which would otherwise be imported;
3. export-oriented industrialization.

Which of these strategies can be pursued depends upon a number of factors: the economy's resource endowment (both physical and human); its size (particularly of its domestic market); its international context (especially the rate of growth of world trade and the policies of TNCs); and the attitude of the national government. For example, not all developing countries possess a natural resource endowment which could form the basis of a local processing industry. Even those which have such an asset may experience difficulty in setting up a local industry. Both developed country tariffs and also international freight rates tend to be higher on processed than on unprocessed materials. In addition, where TNCs are involved it may be corporate policy to locate processing operations elsewhere.

In fact, the growth of the major NICs owes little or nothing to local materials processing. Of course, neither Singapore nor Hong Kong, for example, has the material base to support such a strategy anyway. But even in countries such as Brazil, in which primary commodities are an important element in the economy, industrialization has followed a rather different path. The general pattern of industrialization (with few exceptions) has been one of an initial emphasis on import substitution followed eventually by a shift to export-oriented policies.

Import substitution   During both the 1920s and 1930s and the period after 1945 the developing countries faced huge economic problems. In particular, their embryonic manufacturing industries faced the threat of being stillborn by the competition of imports from the more efficient developed country producers. As a result, a number of countries – especially the larger Latin American countries such as Brazil and Mexico – pursued an explicit policy of import substitution. In the postwar period large Asian countries such as India, as well as smaller ones like South Korea, Taiwan and the Philippines, began to follow a similar path. (As we have seen, Japan also pursued a highly protective policy stance.) The aim of import substitution was to protect a nation's infant industries so that the overall industrial structure could be developed and diversified and dependence on foreign technology and capital reduced. To this end, many of the policies listed in Figures 6.1, 6.3 and 6.4 were employed by national governments. In particular, very high tariffs were imposed on those sectors chosen for

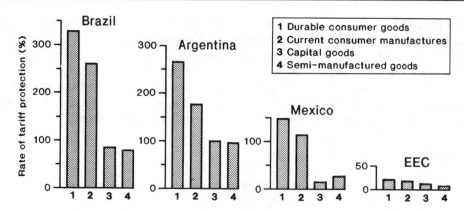

Figure 6.9 Import-substitution policies: tariffs on consumer goods, capital goods and intermediate goods, 1962 (*Source*: United Nations (1964) *Economic Bulletin for Latin America*, Vol. 9, no. 1, Table 5)

protection. Import quotas and other devices, including licences, deposits and multiple exchange rates, were also used, as were incentives to encourage domestic production.

The import-substitution strategy, in theory, is a long-term *sequential* process involving the progressive domestic development of industrial sectors through a combination of protection and incentives:

Stage 1   Domestic production of consumer goods
Stage 2   Domestic production of intermediate goods
Stage 3   Domestic production of capital goods

Invariably, the process began with the heavy protection of domestic consumer goods industries to stimulate local production. As Figure 6.9 shows, tariffs on consumer goods in the major Latin American countries in the early 1960s were many times greater than the tariffs on intermediate and capital goods (and very much higher than EC tariffs). As a result, domestic consumer goods production in at least some of the countries pursuing an import-substitution policy grew considerably although much depended on the size of the domestic market. Although dependence on imported consumer goods certainly declined, however, dependence on the import of intermediate and capital goods – and, therefore, on foreign technology – increased. In most cases progression beyond the production of consumer goods did not occur to the extent anticipated. Hence, various critics described import-substituting industrialization as 'half-way' industrialization or as 'getting stuck' at the consumer goods stage. The hoped-for domestic multiplier effects and the stimulus of a broader industrial structure did not necessarily occur. Where the domestic market was small, local production of consumer goods could not achieve appropriate economies of scale so that domestic prices remained high. The necessarily high level of imports of intermediate and capital goods imposed balance of payments constraints on many developing economies. Yet there were strong pressures from domestic vested interests – especially the protected consumer-goods manufacturers – against reducing the protection afforded to that sector in favour of other manufacturing sectors as originally envisaged.

Export-orientation   The realization that an import-substituting strategy could not, on its own, lead to the desired level of industrialization began to dawn in a growing number

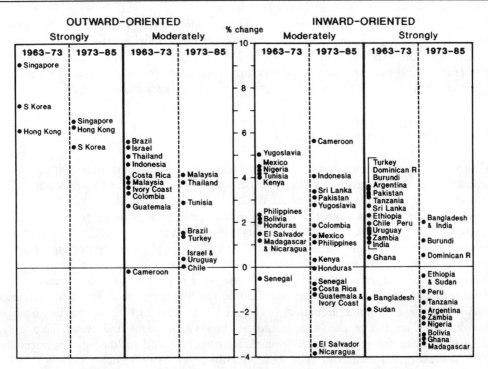

Figure 6.10 The relationship between trade policy orientation and growth in gross national product per capita (*Source*: based on World Bank (1987), Figure 5.3)

of countries: some during the 1950s, rather more during the 1960s. Generally it was the smaller industrializing countries which first began to shift towards a greater emphasis on export-orientation because of the constraints imposed upon such a policy by a small domestic market. Increasingly, an export-oriented, outward-looking industrialization strategy became the conventional wisdom among such international agencies as the Asian Development Bank and the World Bank. Indeed, in its 1987 *World Development Report* the World Bank classified forty-one developing countries by their degree of outward and inward trade orientation and correlated this against growth in GNP per capita for 1963–73 and 1973–85. Figure 6.10 shows that, in general, higher rates of GNP growth were associated with outward-oriented trade strategies. However, to attribute such growth solely to one factor is to oversimplify; other factors may well be involved as well.

A shift towards export-based industrialization was made possible by the kinds of developments we have been discussing in previous chapters: the rapid liberalization and growth of world trade during the 1960s, the 'shrinkage' of geographical distance through the enabling technologies of transport and communications, the global spread of the transnational corporation and its increasing interest in seeking out low-cost production locations for its export platform activities. Such export orientation was based upon a high level of government involvement in the economy. The usual starting point was a major devaluation of the country's currency to make its exports more competitive in world markets. The whole battery of export trade policy measures shown in Figure 6.1 was invariably employed by the newly industrializing countries. In effect, these amounted to a subsidy on exports which greatly increased their price competitiveness. Of course, the major domestic resource on which this export-oriented industrialization rests has been that of the labour supply – not only its relative cheapness (Figure 5.4) but also its adaptability and, very often, its relative docility. Indeed, in many cases, the activities of labour unions have been very closely regulated.

Policies towards foreign investment

The attraction of foreign investment has been an integral part of export-oriented industrialization in many developing countries, although the extent to which particular industrializing countries have pursued this strategy varies considerably.[17] In general, the Latin American NICs have been more restrictive in their attitudes to foreign direct investment than the Asian NICs, although there have been recent shifts towards more liberal investment policies in Latin America. In general, ownership requirements in Latin America have tended to be stricter than in most Asian countries and the number of sectors in which foreign involvement is prohibited rather greater. The most restrictive policies have been operated by the five member states of the Andean Pact – Bolivia, Colombia, Ecuador, Peru and Venezuela. In these countries, foreign ownership of an indigenous firm cannot exceed 49 per cent and there have been stringent regulations governing the remittance of profits and technology contracts. In 1987, however, major relaxation of these measures was introduced.

Mexico has also moved from a strongly restrictive policy towards foreign investment to a more liberal policy. Mexico has generally restricted foreign ownership of Mexican companies to minority shareholding and has excluded foreign investment entirely from certain sectors (although these restrictions do not apply to the *maquiladoras*; see below, p. 184). In the late 1980s, however, Mexico relaxed the rule that there must always be Mexican participation in the ownership of enterprises. Wholly owned foreign subsidiaries are now permitted in certain sectors. For example, Mexico initially refused permission for IBM to establish a wholly owned plant to manufacture personal computers at Guadalajara and then agreed, subject to certain conditions. Other foreign firms, such as Ford, have also been able to establish a wholly owned plant in Mexico.

National attitudes to foreign direct investment among the Asian NICs have been far more varied than is often supposed. Both Singapore and Hong Kong operate open-door policies. Hong Kong's policy has been one of not imposing restrictions but, at the same time, the government has not operated a specific incentives programme. Singapore has industrialized very largely on the basis of a carefully constructed package of incentives to attract foreign companies. In contrast, both Taiwan and South Korea have been more restrictive. Although Taiwan has not generally stipulated a particular level of domestic ownership, it does impose export performance requirements on foreign firms and also local content requirements ranging from 50 to 100 per cent.

The most restrictive of the four leading NICs in Asia, however, has undoubtedly been South Korea. Until 1983 Korea operated strict rules on foreign direct investment which restricted the permitted level of foreign ownership, specified a minimum export performance and local content level. Korean government policy has been to build a very strong domestic sector with particularly strong support for the huge Korean conglomerates (*chaebol*). As a result, in complete contrast to Singapore where foreign firms dominate, changes in Korea's economic structure have been pretty much determined by the Koreans themselves rather than by the activities of foreign-based MNEs. Foreign investors have had some influence in industries like electronics through their exporting activities but their effects on both allocation and sectoral efficiency appear to have been marginal in determining the overall pattern of industrial development in Korea.

(Koo, 1985, p. 306)

Again, however, there are clear signs in Korea of a relaxation of controls on foreign investment. Since 1989 in particular the Korean government has been actively seeking to attract foreign investment in high-technology activities to help the country move more quickly into more advanced industries.

Despite such differences in national attitude, most developing countries engaged in what has been called 'the battle for rapid industrialization' are involved in fiercely competitive rivalry to attract foreign firms. The competition is with both developed and developing countries and few punches are pulled in the comparisons drawn in the national promotional literature. The resource which most developing countries stress in their publicity is undoubtedly that of their *labour force*. Virtually every glossy brochure – and some are very glossy indeed – put out by national economic development agencies eulogizes its quantity, quality and, with some recent exceptions, its relative cheapness. Those industrializing countries located close to major world markets such as North America or the EC, or which have special trading relationships with them, also emphasize this key locational attribute of proximity.

The 'leading edge' NICs of East and South East Asia are, however, making a deliberate effort to move away from the cheap labour image and to upgrade their economic structures to higher skill and more sophisticated technological levels. Hong Kong, Singapore, South Korea and Taiwan are all investing massively in education and training and in information technology. In 1979, for example, Singapore launched its 'Second Industrial Revolution'. In 1986, following the unexpected recession of 1985, the Singapore government announced a further package of measures to move way from being merely a labour-intensive assembly site to a high-technology, knowledge-based economy in which business services and the regional headquarters of TNCs would help to make Singapore a 'total business centre'.

Export-processing zones

Among all the measures used by developing countries to stimulate their export industries and to attract foreign investment one device in particular – the export-processing zone (EPZ) – has received particular attention.[18] EPZs have become the focus of much controversy regarding their economic and social welfare effects. Such issues will be examined in subsequent chapters; at this point the aim is simply to identify the nature of EPZs and to describe their growth and geographical distribution. An EPZ can be defined as:

a relatively small, geographically separated area within a country, the purpose of which is to attract export-oriented industries, by offering them especially favourable investment and trade conditions as compared with the remainder of the host country. In particular, the EPZs provide for the importation of goods to be used in the production of exports on a bonded duty free basis.

(UNIDO, 1980, p. 6)

EPZs are, in effect, *export enclaves* within which special concessions apply, including an extensive package of incentives and, very often, exemption from certain kinds of

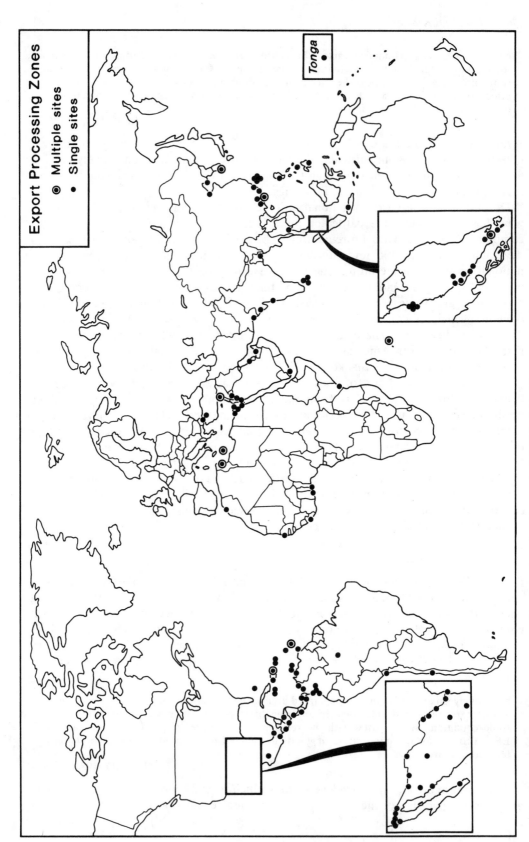

Figure 6.11 Export processing zones in developing countries, 1986 (*Source:* ILO (1988), Table 20; press reports)

Export Processing Zones
- ◉ Multiple sites
- ● Single sites

Tonga

legislation, which do not apply outside the zones. Within an EPZ all the physical infrastructure and services necessary for manufacturing activity are provided: roads, power supplies, transport facilities, low-cost/rent buildings. Special investment incentives and trade concessions apply. In a number of cases the restrictions on foreign ownership which apply in the country as a whole are waived for foreign firms locating in the zone, allowing 100 per cent ownership of export-processing ventures. Within developing countries EPZs have been located in a variety of environments. Some have been incorporated into airports, seaports or commerical free zones or located next to large cities. Others have been set up in relatively undeveloped areas as part of a regional development strategy.

It is often rather difficult to distinguish between EPZs and the many other similar kinds of zone such as freeports or free trade zones. Strictly speaking, freeports and free trade zones are commerical zones only; their functions involve warehousing and transhipment of goods with no change in the nature of the goods themselves. EPZs, on the other hand, are set up for actual manufacturing: the processing and/or assembly of export products from primarily imported materials and components. There are very many more freeports and free trade zones throughout the world than there are export-processing zones. EPZs have developed primarily in recent years in developing countries. Figure 6.11 shows their geographical distribution in 1986. Almost all of the 116 EPZs in operation in 1986 were established after 1971. Before the mid-1960s there were only two EPZs in the developing countries – in India and in Puerto Rico. In 1986, 90 per cent of all EPZs in the developing countries were located in Latin America and the Caribbean (48 per cent) and Asia (42 per cent). By 1986 some seventy-four developing countries either had established EPZs or were planning to do so. The EPZs themselves vary enormously in size, ranging from geographically extensive developments to a few small factories; from employment of more than 30,000 to little more than 100 workers. Total employment in developing country EPZs in 1986 was roughly 1.3 million.

Figure 6.12 shows the relative sizes of the Asian EPZs in terms of employment. Asia contains 60 per cent of all EPZ employment in developing countries. Hong Kong and Singapore are, in effect, entire free zones but with export-processing activities concentrated in a number of industrial estates. In 1986 total employment in such zones was 89,000 and 217,000 respectively. The other major concentrations are in Taiwan (80,469 employed in four EPZs), Malaysia (81,688 in eleven EPZs), South Korea (140,000 in three EPZs) and the Philippines (39,000 in three EPZs). The most interesting recent development in Asia, however, is the establishment of a form of EPZ in the People's Republic of China as the focus of its plan to industrialize with the participation – strictly controlled – of foreign capital.[19]

As part of its modernization programme initiated in the late 1970s, China has established four 'special economic zones' of which the largest and farthest advanced is that of Shenzhen on the southern border adjacent to Hong Kong. This development is not unrelated to the future reversion of Hong Kong to Chinese jurisdiction in 1997. While the People's Republic engages in its first experiment in attracting foreign investment, Taiwan has been busy moving 'up-market' with its construction of the Hsinchu Science-Based Industrial Park, close to Taipei and adjacent to two universities. Unlike the earlier generation of EPZs in Taiwan and most current ones in the developing world, Taiwan's science park is being developed exclusively for high-technology industries. Similar developments are occurring in Singapore.

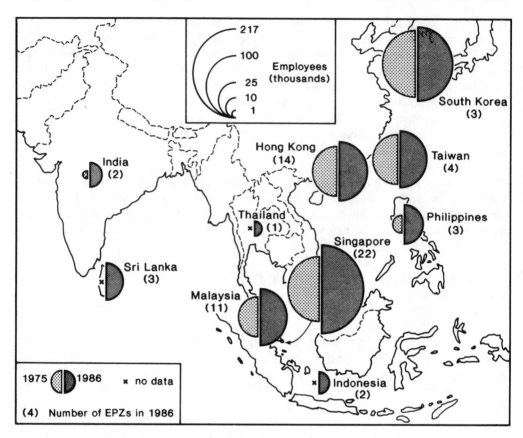

Figure 6.12 Employment in Asian export processing zones, 1975 and 1986 (*Source*: ILO (1988), Table 19)

So far, at least, EPZs have not played such a prominent part in the industrialization programmes of South American countries, with the exception of one or two countries such as Colombia. But the most important Latin American industrial country, Brazil, has only one EPZ and this is located at Manaus in Amazonia, far from the country's economic centre of gravity. The number of EPZs has been increasing quite significantly in the Caribbean but it is in Mexico that the largest programme of EPZ-type activity has occurred. In 1965 the Mexican government introduced its Border Industrialization Programme to help ease the severe economic and social problems of the northern border towns.[20] In effect, the Mexican government made a deliberate effort to 'siphon off' some of the offshore investment by United States TNCs which was flowing to Asian countries in an attempt to reduce production costs. The Mexican programme was very much complementary to the offshore assembly provisions operated by the US government (see above p. 169). Basically, the Mexican government waives its duties on the import of materials and components imported from the United States as long as the end-products are exported. In the US case, duties are paid only on the value added to the components which originated in the United States.

The growth of manufacturing industry in the *maquiladoras* (assembly) plants has been very considerable. In 1965 when the Border Industrialization Programme began

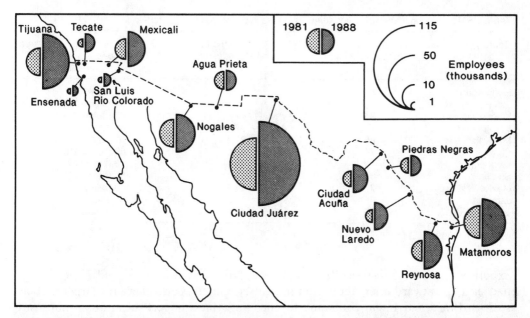

Figure 6.13 Employment in the major centres of the Mexican *maquiladora* programme, 1981 and 1988 (*Source*: Committee for the Promotion of Investment in Mexico)

there were only twelve factories employing a total of less than 4,000 workers. By 1988 the number of plants had increased to 1,383 and employment in them to over 310,000. Figure 6.13 shows the geographical distribution of this employment. The largest concentration was in Ciudad Juárez. The other major centres were Tijuana, Matamoros, Nogales, and Reynosa. The *maquiladora* programme was subsequently broadened geographically to include specific non-border locations. In 1988 a further 77,000 were employed in these locations.

Although EPZs in developing countries come in a great variety of sizes and bear the stamp of their specific national context, they also share many features in common. The overall pattern of incentives to investors is broadly similar, as is the type of industry most commonly found within the zones. The production of textiles and clothing and the assembly of electronics dominates. Almost half the total labour force in the Asian EPZs is engaged in the electronics industry. In the Mexican *maquiladoras* 60 per cent of the workforce is employed in electrical assembly and a further 30 per cent in textiles and clothing. The characteristics of the labour force itself are similarly uniform with a dominance of young female workers (Table 6.5). Some of the implications of these characteristics will be explored in later chapters.

Summary

The pursuit of rapid industrialization in developing countries in the past few decades has followed a fairly regular pattern, although with significant individual variations. In virtually every case the direct role of the state has been extremely prominent. A common initial strategy was one of import substitution. Disenchantment with the results of such a policy, however, led a number of states to shift their emphasis to one of

Table 6.5 The predominance of young female workers in export-processing zones

Country	% of women in EPZ industries	% of women in non-EPZ industries	% of women EPZ workers in specified age group (years)
Hong Kong	60	49.3	85 (20–30)
India	80	9.5	83 (below 26)
Indonesia	90	47.9	83 (below 26)
Korea	75	37.5	85 (20–30)
Malaysia	85	32.9	(Average: 21.7)
Philippines	74	48.1	88 (below 29)
Singapore	60	44.3	78 (below 27)
Sri Lanka	88	17.1	83 (below 26)
Mexico	77	24.5	78 (below 27)
Jamaica	95	19.0	(Average: early 20s)
Tunisia	90	48.1	70 (below 25)

(*Source*: based on ILO (1988), Table 8)

export-orientation. Some smaller countries, particularly Singapore and Hong Kong, had such an outward orientation from the outset. Others, particularly the larger nations such as Brazil and India, continued to encourage their import-substituting industries whilst also increasing their export involvement.

Export-oriented policies encompass a broad range of measures, both investment- and trade-related. However, many developing countries have used the device of the export-processing zone as a primary focus. The proliferation of EPZs seems to be continuing as more and more developing countries attempt to tread the path of export-based industrialization. Even so, EPZs generally represent only a small proportion of a country's employment. They are but a part of the wider national strategy. However, they are especially directed towards the attraction of foreign enterprises in the hope that the policy will create new jobs, raise skills and generally upgrade the technological level of the domestic economy. Some of the costs and benefits of such a reliance on foreign enterprises will be discussed in Chapter 12 after we have examined more closely the global organization of production.

Conclusion

The aim of this chapter has been to demonstrate some of the many ways in which nation states contribute towards the shaping and reshaping of the global economic system. All states engage in a broad variety of trade, investment and industry policies; only the *degree* of involvement varies from one nation to another. A major cause of such variation is, of course, the ideological complexion of the individual state. A significant feature of the last two decades has been the enhanced involvement of governments in the market economies and, conversely, a relaxation of centralized control in the planned economies. The timing of industrialization also appears to be important in determining the degree of state involvement. In general, those nations which have come relatively late to industrialization have pursued the most explicit and strongest national policies. It is no coincidence that almost all of the leading NICs are ones in which the hand of the state is especially firm and often autocratic.

In the pursuit of national goals – whether of inducing or maintaining economic growth or of adjusting to stagnation or decline – states attempt to influence the level, composition and performance of industry within their own boundaries and to regulate their position in the world trading system. A whole range of measures is employed in both cases. One of the most significant developments of recent years has been the intensification of competition between nation states (and between parts of the same nation) to attract international investment. Global competitive bidding is the name of the game, with governments setting out to entice TNCs (in some cases with conditions, in others not) with a whole range of incentives. In the case of developing countries a particularly important development has been the use of export-processing zones. Although some of the 'older' NICs are attempting to change their image, the overriding 'resource for sale' continues to be that of a large, productive and relatively cheap labour supply.

The use of incentives to attract investment or to modify the composition of domestic industry is relatively recent compared with the historical involvement of nation states in regulating their trade. Since 1947, however, such policies have been subject to a set of international rules – the General Agreement on Tariffs and Trade. The major result has been an overall decline in the level of national tariffs imposed on imports and a general liberalization of world trade. But in the case of certain products and more generally since the onset of world recession in the 1970s, there has been a marked increase in the use of non-tariff barriers, particularly import quotas and 'voluntary' trading agreements. Consequently, one of the major sources of international economic tension is that of trade both between the developed economies and also between these nations and the industrializing nations. The spectacular growth of Japan and of a few NICs as major world exporters and the defensive reactions of the older industrialized countries constitute an important ingredient in the reshaping of the global economic system. Currently, however, one of the most significant developments of all is the trend towards an increasing use of 'managed' trade within the context of a strategic trade policy. The further development of major regional trading blocs in both Europe and North America is adding a significant complication to the economic–political relationships between nation states. It also greatly modifies the 'investment potential surface' on which TNCs operate.

Indeed, the actions of nation states form a most important element of the environment within which transnational corporations operate. Policies to encourage or discourage investment, to create a positive or a negative 'business climate', to regulate trade, are all highly significant factors in shaping the surface on which TNCs operate and pursue their profit-based objectives. The growth of international production is not simply a response to geographical differences in economic factors alone; their *political* context is a vital ingredient in the internationalization process. National boundaries, in effect, create major discontinuities in investment opportunities for TNCs. Hence, in today's global economy, the actions of nation states are increasingly bound up with those of transnational corporations. In Chapter 12 we explore the costs and benefits of TNCs to nation states. Before doing this, however, we need to examine how production is organized globally, both within TNCs and between TNCs and other types of firm. In particular, we need to look at these complex processes in the context of specific industries.

Notes for further reading

1. For a discussion of this issue see Radice (1991), Sklair (1991).
2. The concept of 'world system' has been introduced and developed by Wallerstein (1979, 1984) in his historical studies of the evolution of a capitalist world economy. Taylor (1990)

uses Wallerstein's world system famework as the basis for his reformulation of political geography.

3. Mercantilist trade policies formed a cornerstone of international political (and military) relationships. They are based on the notion that a nation's wealth and influence depend upon its ability to regulate and control its external trade at the expense of its rivals.

4. A large literature on strategic trade policy has emerged. For broad reviews of such literature see Stegemann (1989), Cohen (1990), Richardson (1990).

5. The workings of the GATT are described by Macbean and Snowden (1981), Root (1990).

6. Nogués, Olechowski and Winters (1986) examine the extent of NTBs in relation to industrialized country imports; Finger and Laird (1987) compare levels of protectionism between developed and developing economies.

7. The theory of customs unions was established by Jacob Viner in 1950. Viner's ideas are summarized in most basic texts on international economics; see, for example, Root (1990).

8. The most useful general introduction to the economics of the European Community is Swann (1988). Hesselman (1983), Shepherd, Duchene and Saunders (1983) and Jacquemin (1984) discuss industrial policies in European countries. The specific issue of technology policy in Europe is dealt with by Sharp and Shearman (1987), Sharp and Holmes (1989).

9. This development is examined in detail for the United Kingdom and France by Sharp and Holmes (1989).

10. Pomfret (1982) describes the arrangements between the EC and the Mediterranean countries and assesses their effect on foreign investment and exports.

11. The basic case for completing the single market is made by Cecchini (1988). Varying perspectives on the process and effects of complete integration are provided by Callingaert (1988), Pelkmans and Winters (1988), Davis *et al.* (1989).

12. See, for example, Tolchin and Tolchin (1988), Glickman and Woodward (1989), Graham and Krugman (1989).

13. The 1974 Trade Act contained a clause 301 which met the GATT criteria of dealing with unfair trade practices. The 1988 Act adapted this clause to a strongly unilateral measure. For a strongly critical view of the 1988 Omnibus Trade and Competitiveness Act, see Bhagwati (1989). He calls it the 'ominous' trade and competitiveness act.

14. Morici (1989) provides a broad review of the Agreement. Bhagwati (1989) argues that it should be seen as a special case, like Europe 1992, and should not be the basis for further regional arrangements.

15. Reviews of Japanese economic policy are provided by Magaziner and Hout (1980), Johnson (1982, 1985), Dore (1986), Inoguchi and Okimoto (1988), Okimoto (1989). Fransman (1990) discusses Japanese policies in the information technology industries.

16. Particularly useful surveys of government policies in the NICs are provided by OECD (1988), which deals with the four leading Asian NICs plus Brazil and Mexico. The World Bank *World Development Report 1987* contains several chapters on trade and industrialization policies in developing countries. The specific cases of the leading Asian NICs are discussed in van Liemt (1988), Chu (1989).

17. Changing national policies towards foreign direct investment in developing countries are reviewed by UNCTC (1988), chapter XVII. OECD (1988) deals specifically with the situation in Hong Kong, Korea, Singapore, Taiwan, Brazil and Mexico.

18. The most recent comprehensive study of EPZs is by the International Labour Office (1988). Sklair (1986) provides a general critical review while Warr (1987) presents a detailed analysis of EPZs in Malaysia.

19. The Chinese Special Economic Zones are analysed by Wong and Chu (1985), Phillips (1986), Phillips and Yeh (1990).

20. The most comprehensive recent account of the Mexican Border Industrialization, or *maquiladora*, programme is provided by Sklair (1989).

Chapter 7

Global Organization and Reorganization of Economic Activity: Networks of Relationships

Introduction

In Chapter 5 we outlined some of the theories which have been put forward to explain the development of the transnational corporation. We also discussed, in a generalized way, the competitive strategies of TNCs. In Chapter 6 we reviewed the role of states in today's global economy, particularly those governmental strategies which are likely to enhance or inhibit the ability of TNCs to pursue their chosen competitive strategies. Underlying the actions of both of these institutions – TNCs and states – is the pervasive influence of technological change discussed in Chapter 4. The final piece in this explanation of the processes of global shift concerns the *networks of relationships* which exist both within and between firms and the geographical expression of those relationships at global, national and local scales. The focus in this chapter, therefore, is on how economic activities are being organized and reorganized through networks of intra- and inter-firm relationships within the kinds of context outlined in Chapters 4 to 6.

Two sets of relationships are explored. First, we look at the *internal networks of TNCs*; at how the particular strategies pursued are related to particular organizational and geographical structures. Second, we examine the extremely complex *networks of external relationships* which exist between independent and quasi-independent firms, both large and small, transnational and domestic. This allows us to explore the increasingly diversified forms of inter-firm relations, ranging from international strategic alliances, through subcontracting links to more disaggregated network forms. Finally, we take a critical look at some of the more general geographical outcomes of these various processes, ranging from the suggested parallels between the internal organization of TNCs and the broader international division of labour to the re-emergence of new industrial districts.

The basic building block: production chains

In order to understand both the internal network of relationships within TNCs and also the network of relationships between independent firms we need a clear picture of the whole range of activities which are performed within the production system. Here, again, it needs to be emphasized that the term 'production' is used in its broadest sense to include not only the production of goods but also of services. One way of approaching this extremely complex situation is to conceive of the production system in terms of a *chain* of linked functions. In Figure 7.1 we use the term 'production' chain but others use terms such as '*filière*' or, to emphasize the fact that each stage adds value to the sequence, 'value added' or 'value' chain. Whatever the terminology used, each stage in

The basic production chain

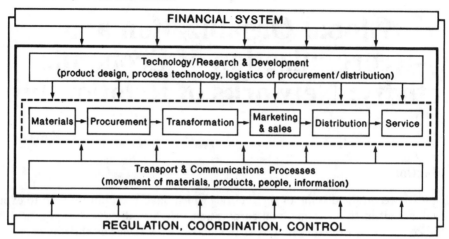

A system of linked production chains

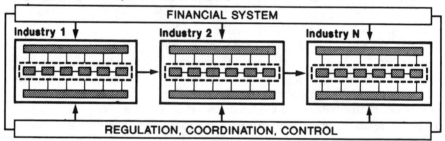

Figure 7.1 The basic production chain

the chain is connected through a set of *transactions*. Just how these transactions are structured organizationally and geographically is the major focus of this chapter.

At one extreme, each function in the chain may be performed by individual, independent firms so that the links in the chain consist of a series of *externalized* transactions between separate firms. In other words, transactions are organized through 'the market'. At the other extreme, the whole chain may be performed within a single firm. In this case, the links in the chain consist of a series of *internalized* transactions within the bounds of a particular firm. Here, transactions are organized 'hierarchically' through the firm's internal organizational structure.[1] As we shall see, these two organizational extremes of markets and hierarchies do not encompass the full range of possibilities. There are other forms of inter-unit co-ordination which are increasingly significant elements of the global economy.

The concept of 'internalization' has already been referred to, in Chapter 5, as one of the suggested explanations of the transnational corporation. In that discussion we noted that a major reason for a firm internalizing some of its transactions, which would otherwise be organized through the market, is the *imperfect* nature of many markets, particularly those for intermediate goods and for information. On the whole, the greater the degree of uncertainty – whether over the availability, price or quality of

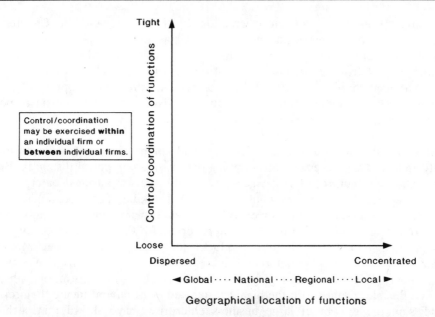

Figure 7.2 The two primary dimensions along which the production chain may be structure (*Source*: based, in part, on material in Porter (1986), chapter 1)

supplies, or of the price obtainable for a firm's output – the greater the incentive for the firm to control these transactions itself.

A basic question, therefore, in considering the global organization of production, is the extent to which the production chain is integrated or disintegrated, both organizationally and geographically. Organizationally, the production chain has to be co-ordinated. This may be achieved either within an individual firm or between individual firms. Geographically, the various components of the chain may be concentrated in particular places or dispersed. The most important development of the past few decades is that the potential geographical scale has become global. Figure 7.2 displays these two primary dimensions and helps to provide the basic framework for what follows. The important point to note, of course, is that the boundary between internalization and externalization is continually shifting.

Networks of internalized relationships: inside the transnational corporation

Coping with complexity: changing organizational structures

One of the basic 'laws' of growth of any organization is that as growth occurs the internal structure of the organization changes. In particular, the *functional role* of its component parts tends to become more *specialized* and the links between the parts become more *complex*. As the size, organizational complexity and geographical spread of TNCs have increased so the internal interrelationships between their geographically separated parts have become a highly significant element in the global economy. The precise manner in which TNCs organize and distribute their production or value-added chains arises from their strategic orientation which, as we noted in Chapter 5, is influenced by a firm's particular history as well as its geographical origins. Internationally competitive strategies, as we have seen, can be regarded as falling along a spectrum ranging from global integration at one end to national responsiveness at the other. Although there has certainly been a trend towards the adoption of globally integrated strategies by an increasing number of TNCs they must remain responsive to national and local differentiation. Both global and local perspectives need to be combined.

Technological innovations in the transport and communications media have enormously enhanced the potential for operating over vast geographical distances. But the co-ordination and control of large, geographically dispersed business enterprises also require an appropriate 'organizational technology'. The kind of organizational structure suitable for the one-person firm or even for the multi-plant firm operating within the bounds of a single country is singularly inappropriate for the transnational corporation operating in a number of different product and geographical markets. Most large organizations still tend to be hierarchical in structure but the sheer size and complexity of the modern TNC demand a highly sophisticated organizational form, one which can cope in a flexible manner with the highly variegated environment facing the firm.[2]

Studies in the growth of large business enterprises have shown that such firms commonly transformed their organizational structures from a *functional* form, in which the firm is subdivided into major functional units (production, marketing, finance, etc.), into a *divisional* form. The divisional structure is one in which organization is by product rather than by function. Each product division is responsible separately for its own functions, particularly of production and marketing, although some functions – especially finance – tend to be performed centrally for the entire corporation. Each product division usually acts as a separate profit centre. The main advantage of the divisional structure is usually seen to be one of a greater ability to cope with product diversity. Thus, as large firms became increasingly diversified during the 1950s and 1960s they also tended to adopt a divisional structure.

Adoption of a divisional structure gave firms greater control over their increasingly diverse product environment. However, operating across national boundaries, rather than within a single nation, poses additional problems of co-ordination and control. Largely through trial and error, TNCs have groped their way towards more appropriate organizational structures. Figure 7.3 shows four commonly used structures. Which one is actually adopted depends upon a number of factors including the age and experience of the enterprise, the nature of its operations and its degree of product and geographical diversity. The form most commonly adopted in the early stages of TNC development – at least when there are several overseas subsidiaries – is simply to add on an international division to the existing divisional structure (Figure 7.3a). This has tended to be a short-lived solution to the organizational problem if the firm continues to expand its international operations.

In such a hybrid structure problems of co-ordination inevitably arise, tensions develop between the parts of the organization operated on product lines (the firm's domestic activities) and those organized on an area basis. The need arises for an organizational form which can integrate both the domestic and international operations of the firm. There are two obvious possibilities. One is to organize the firm on a *global product* basis; in other words, to apply the product-division form throughout the world and to remove the international division (Figure 7.3b). The other possibility is to

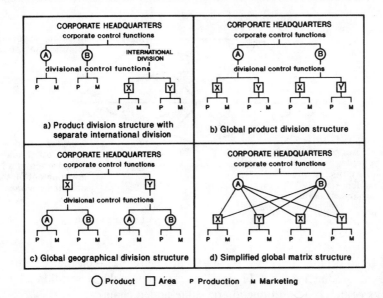

Figure 7.3 Some leading types of organizational structure in transnational corporations

organize the firm's activities on a *worldwide geographical* basis (Figure 7.3c). Davis (1976) suggests that this form of organization is most appropriate to mature businesses with narrow product lines. But he also points out that each of the two global organizational forms solves one set of problems partly at the expense of another. The worldwide geographical structure 'improves the co-ordination of all product lines within each zone, but at the expense of reduced co-ordination between areas for any one product line . . . Global corporations that are organized along product lines have the opposite problem: how to co-ordinate their diverse business activities within any one geographic area' (Davis, 1976, pp. 62, 64). For such reasons some of the largest TNCs have begun to adopt sophisticated *global grid* or *global matrix* structures which contain elements of both product and area structures (Figure 7.3d). Such grid structures are extremely complex and difficult to represent satisfactorily as a simple diagram. In essence, they involve a whole set of dual reporting links between both product and area segments of the firm.

Even matrix structures may be inadequate to enable the global corporation effectively to manage and co-ordinate its multifarious activities. Consequently, some firms are now moving towards a globally integrated network structure (Figure 7.4) in which 'increasingly specialized units worldwide . . . [are] . . . linked into an integrated network of operations that . . . [enables] . . . them to achieve their multidimensional strategic objectives of efficiency, responsiveness, and innovation . . . The strength of this configuration springs from its fundamental characteristics: dispersion, specialization, and interdependence' (Bartlett and Ghoshal, 1989, p. 89).

The growth in organizational sophistication within TNCs enables them to control and co-ordinate their activities more effectively. There is undoubtedly a tendency in some TNCs to move away from a hierarchical, 'top-down' organizational structure in which there is a clear vertical division of control, to a flatter, more complex, networked structure of the kind shown in Figure 7.4. This does not necessarily imply a loosening

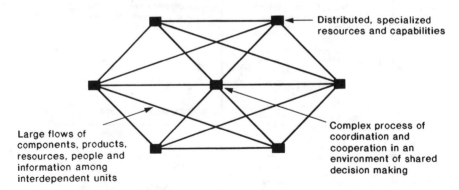

Distributed, specialized
resources and capabilities

Large flows of
components, products,
resources, people and
information among
interdependent units

Complex process of
coordination and
cooperation in an
environment of shared
decision making

Figure 7.4 A globally integrated network structure (*Source*: Bartlett and Ghoshal (1989), Figure 5.1)

of control or co-ordination (the vertical axis in Figure 7.2) – on the contrary. But what about the other dimension of Figure 7.2: the geographical location of the various components of the TNC's production or value-added chain?

The geography of TNC activities: changing locational structures

Writing nearly seventy years ago, the urban economist R. M. Haig (1926, p. 426) observed that 'every business is a package of functions and within limits these functions can be separated out and located at different places'. Over time, these 'limits' have become less and less restrictive. The most obvious reason, of course, has been developments in the enabling technologies, particularly those in transport and communications. But there is more to it than this. The enabling technologies set the outer limits of what is possible; they do not determine what actually occurs. The extent to which a firm separates out its component parts and how it locates them at different places depends upon a number of factors. One of these is the actual strategy being pursued.

The influence of competitive strategy

Figure 7.5 shows four general types of international competitive strategy which a TNC might pursue, in terms of their organizational (co-ordination) and geographical (configuration) dimensions. As the diagram shows, a competitive strategy is defined in terms of its position on both dimensions simultaneously. Porter (1986, pp. 27–8) rightly observes that 'there is no such thing as one global strategy. There are many different kinds . . . depending on a firm's choices about configuration and co-ordination throughout the value chain.' The three sets of arrows in Figure 7.5 suggest possible strategic development paths (the numbered arrows) but it should not be assumed that any one path is inevitable. Each strategic box involves different degrees of geographical concentration or dispersal of value activities. Two of the alternatives – the export-based strategy and the basic global strategy – involve a high degree of geographical concentration of production activities. The other two alternatives – the multidomestic strategy and the complex global strategy – involve a high degree of geographical dispersion.

However, the organizational/co-ordination decision and the geographical location

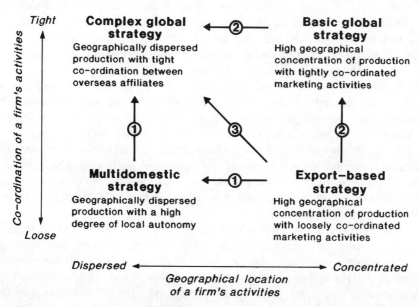

Figure 7.5 A typology of international competitive strategies (*Source*: based, in part, on M. E. Porter (1986) Competition in global industries: a conceptual framework, in Porter (1986), Figure 1.5)

decision have to be made for each element in the firm's production or value-added chain. Some elements may be geographically dispersed; others geographically concentrated. Some elements of the chain may be located in close geographical proximity to one another whereas others may be separately located. 'A firm may standardize (concentrate) some activities and tailor (disperse) others. It may also be able to standardize and tailor at the same time through the co-ordination of dispersed activities, or use local tailoring of some activities (e.g. different product positioning in each country) to allow standardization of others (e.g. production)' (Porter, 1986, p. 35). In fact, combining standardization and local tailoring is becoming increasingly possible with the emergence of flexible production technology (see Chapter 4). The tendency to dichotomize corporate competitive strategies into global versus national, cost versus differentiation, concentration versus dispersal is clearly a gross oversimplification. As pointed out in Chapter 5, TNCs 'must balance pressures for integrating globally with those for responding idiosyncratically to national environments . . . [however] . . . there may not be a single optimal point on the fragmentation–unification continuum but a range of tenable positions' (Kobrin, 1988, pp. 104, 107).

Strategic orientation, therefore, is one reason for a TNC to separate out geographically some or all elements within its production chain. Such separation also depends upon the technological characteristics of the firm's operations. In Chapter 4 we identified three major trends in the evolution of the production process:

1. an increasingly finer degree of *specialization*, which permits greater fragmentation of the process into a number of separate operations;
2. an increasing *standardization* of the individual operations themselves;
3. a move towards more *flexible* production technologies, which alters the relationship between scale of output and cost of production.

But there are big differences between industries in the extent to which these trends have occurred and, therefore, in the degree to which the production sequence can be separated out locationally. Hence, the actual division of labour within a TNC and the extent to which its component parts are interdependent are a function of both organizational – strategic and technological forces.

The particular *spatial* form of this internal division of labour – precisely where the separate parts are located – is the result of the interaction between two sets of factors: organizational and technological factors on the one hand and the relevant location-specific factors on the other. We discussed location-specific factors in general terms in Chapter 5. The importance of the various factors differs, however, according to the particular locational requirements of the individual parts of a TNC's operations. *Different parts of the enterprise have different locational needs*; these needs can be satisfied in various types of geographical location. Each tends, therefore, to develop rather distinctive spatial patterns.

We can illustrate this by looking in turn at three of the most important functions of the TNC: corporate and regional headquarters offices; research and development facilities; production units. Each of these, as we shall see, displays certain geographical regularities: notably a highly uneven pattern of distribution both globally and locally. Other functions in Figure 7.1, such as marketing and sales, distribution and service, tend to be distributed far more widely in accordance with the firm's geographical markets. Indeed, as competition intensifies, firms are increasingly placing an emphasis on the service component of their business. This is as true in industrial products as in consumer products. Such a service-orientation (which involves investment in marketing, local distribution channels and the like) is spreading rapidly through many industries. A local market presence is becoming essential.

Corporate and regional headquarters[3]   The corporate headquarters is the locus of overall control of the entire TNC. Its staff are concerned with making the high-level strategic decisions that shape and direct the whole enterprise – which products and markets to enter or to leave, whether to expand or contract particular parts of the enterprise, whether to acquire other firms or to sell off existing parts. It is concerned, then, with all the major investment and disinvestment decisions. One of its most important roles is financial; it is the corporate headquarters which holds the purse strings and which decides on the level and allocation of the corporate budget between its component units. In the global corporation its horizons are global and its time-span tends to be long rather than short term. Headquarters offices are, above all, handlers, processors and transmitters of information to and from other parts of the enterprise and also between similarly high-level organizations outside. The most important of these are the major business services on which the corporation depends (financial, legal, advertising) and also, very often, major departments of government, both foreign and domestic.

Regional headquarters offices of transnational corporations constitute an intermediate level in the corporate hierarchy. They usually have a broad geographical sphere of influence which covers several countries. The likelihood of a TNC establishing a regional headquarters office will be greater where the following conditions apply:

(1) the larger the number of branch establishments engaging in the main activity of the . . . [TNC] . . . in a particular region; (2) the more internationally (or regionally) integrated a

firm is in its product, process or market strategy; (3) the more the organization set-up of the . . . [TNC] . . . favours a decentralized (but co-ordinated) approach to problem-solving.
(Dunning, 1988a, p. 278)

Regional headquarters perform a distinctive role in the internal affairs of the TNC. Their major responsibility is to co-ordinate and control the activities of the firm's affiliates (manufacturing units, sales offices, etc.) and to act as the intermediary between the corporate headquarters and its affiliates within the region. Branch affiliates generally report to these regional headquarters, which, in turn, report to the corporate head office. Thus, the regional headquarters act as a channel of communication, transmitting instructions from the corporate centre to its affiliates and information from the affiliates back to the centre. Frequently, certain decisions are delegated to the regional headquarters, which may also perform a regional marketing function. Regional headquarters are both co-ordinating mechanisms within the TNC and also an important part of the TNC's 'intelligence-gathering' system.

These characteristic functions of corporate and regional headquarters define their particular locational requirements. Both require a strategic location on the global transport and communications network in order to keep in close contact with other, geographically dispersed, parts of the organization. Both require access to high-quality external services and a particular range of labour market skills, especially people skilled in information processing. Since much corporate headquarters activity involves interaction with the head offices of other organizations, there are, in Alfred Weber's terminology, strong *agglomerative* forces involved. Face-to-face contacts with the top executives of other high-level organizations are facilitated by close geographical proximity. Such high-powered executives invariably prefer a location which is rich in social and cultural amenities.

In Dunning's view, however, regional offices are 'the most footloose of all office activities'. He also argues that

regional offices can adapt functions and locations to changes in the strategy of the parent company or to changes in location factors specific to particular countries including the home country. Hence, questions such as work permits for foreigners, rents, availability of good secretaries, housing conditions and prices, attitudes to foreigners, the taxation of expatriates, trends in the location and function of main activities, the technology of telecommunications, the efficiency and ease of air travel, the pattern of market growth, changes in key executives in the parent company; all of these can and do influence the locations and the changes in the locations of regional offices and their functions.
(Dunning, 1988a, p. 279)

At the global scale only a relatively small number of cities contain a large proportion of both corporate and regional headquarters offices of TNCs. Such *global cities* are the geographical 'control points' of the global economy. Figure 7.6 gives a very approximate picture of these centres (the links shown are diagrammatic only; they are intended solely to give an impression of a connected network of cities). Three global cities – New York, Tokyo and London – stand head and shoulders above all the others. Below them is a tier of other key cities in each of the three major economic regions of the world, Western Europe, North America and Asia, with other representation in Australia and Latin America.

The locational pattern of both corporate and – especially – regional headquarters is far from static. Geographical decentralization of corporate headquarters out of the city centres of New York and London has certainly been occurring, although often such

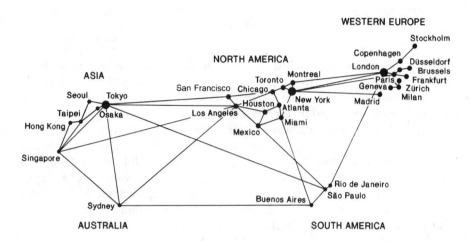

Figure 7.6 Major concentrations of corporate and regional headquarters of TNCs (*Source*: based, in part, on Friedmann (1986), Figure 1; Cohen (1981), pp, 307–8)

shifts are a relatively short distance to the less congested outer reaches of the metropolitan area. But although the geographical relocation of transnational corporate headquarters to other parts of the same country is quite common, it is almost impossible to find any TNC which has relocated its corporate headquarters to a completely different country. With the increasing internal complexity of global corporate organizations and the intensification of global competition, however, some TNCs have begun to relocate some of their divisional headquarters out of their home countries. A recent example is IBM, which has moved its communications systems headquarters from New York to London; the first time the company has located corporate responsibility for a principal line of business outside the United States.

The fact that the locational needs of corporate and regional headquarters of TNCs are satisfied most readily in the very large city – often the capital city – means that they tend not to be spread very widely within any particular country. In the United Kingdom, for example, there are very few regional headquarters of foreign TNCs outside London and the South East; in France few locate outside Paris. In Italy the most important centre is Milan, in the highly industrialized north, which is more important than Rome as a location for foreign TNCs. Such strong geographical polarization is even more apparent in the developing countries.

Research and development facilities[4]   In general, TNCs spend more on R & D than other firms, as part of their drive to remain competitive and profitable on a world scale. Innovation – of new products or new processes – is critically important for such firms in an increasingly competitive global economy. The R & D function is, therefore, highly significant for the TNC. Indeed, it has become even more important with the intensified pace and changing nature of technology. 'The length of time to develop an innovation and the costs of undertaking R & D have risen dramatically during the 1970s and 1980s, as the complexity of the research operation has increased. At the same time, the rate of technological change is speeding up as the life span of new innovations is being steadily eroded' (Howells, 1990, p. 499).

The process of R & D is a complex sequence of operations in which three phases can

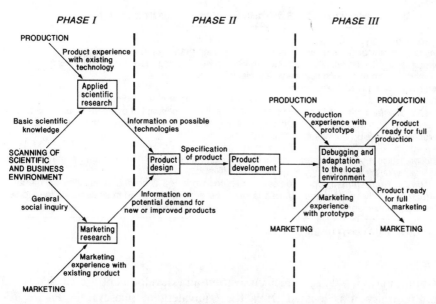

Figure 7.7 Elements of the research and development process (*Source*: Buckley and Casson (1976), Figure 2.7)

be identified (Figure 7.7). In phase I the emphasis is on applied science and marketing research, phase II is concerned with product design and development, while phase III is the 'debugging' of products and their adaptation to local circumstances. Each phase tends to have rather different locational requirements although in each case the TNC has to reconcile several factors. One of these is the advantage of scale economies derived from concentrating R & D against the need to locate R & D closer to other corporate functions or to markets. The primary need in phase I is for access to the basic sources of science and marketing information – universities, research institutes, trade associations, and so on. Phase II tends to require large-scale teamwork, that is, access to a sufficiently large supply of highly qualified scientists, engineers and technicians. Phase III's locational requirements are for quick two-way contact with the users of the innovation: the production or marketing units themselves.

How does the geographical distribution of R & D activity by TNCs reflect these needs? The first point to emphasize is that the bulk of the R & D performed by TNCs still remains concentrated in their home countries. In the mid-1960s only around 5 per cent of total expenditure by US corporations on R & D was located overseas. By 1980 this had risen to approximately 10 per cent. However, as TNCs have developed increasingly large and complex international operations, so, too, their international R & D networks have grown. Much R & D activity, nevertheless, remains heavily concentrated in the OECD group of countries, with only a small proportion in developing countries.

There are important differences in the type of R & D undertaken by TNCs in their overseas locations. Table 7.1 identifies three major categories. The lowest level of R & D activity is the *support laboratory*, whose primary purpose is to adapt parent company technology to the local market and to provide technical back-up. It is the equivalent of phase III in Figure 7.7. The *locally integrated R & D laboratory* is a much more

Table 7.1 Classification of overseas R & D activities of transnational corporations

1. Support laboratory:
   Function: Technical service centre; translator of foreign manufacturing technology.
   Reason for establishment: Response to market growth; differing market conditions; expectation of continuing stream of technical service projects.
2. Locally integrated R & D laboratory:
   Function: Local product innovation and development; transfer of technology.
   Reason for establishment: Improved status of subsidiaries; concept of overseas operations as fully developed business entities; identification of new business opportunities outside home country. Frequently develop out of support laboratories.
3. International interdependent R & D laboratory:
   Function: Basic research centre; close links with international research programme; may or may not interact with the firm's foreign manufacturing affiliates.
   Reason for establishment: Operation of co-ordinated world R & D programmes as part of global product strategies involving the manufacture of a single product line for world markets. Units tend to be created by direct placement.

(*Source*: Hood and Young (1982))

substantial unit, in which product innovation and development are carried out for the market in which it is located. It is the equivalent of phase II in Figure 7.7. The *international interdependent R & D laboratory* is of a quite different order. Its orientation is to the integrated global enterprise as a whole rather than to any individual national or regional market. Indeed, there may be few, if any, direct links with the firm's other affiliates in the same country. Support laboratories are by far the most common form of overseas R & D facility, whilst only a small number of technologically intensive global corporations operate international interdependent laboratories.

According to Behrman and Fischer (1980), the operation of each of these types of R & D activity varies according to the specific market orientation of the TNC. They identify three groups of TNC: home-market firms, host-market firms, world-market firms. Each tends to have rather different R & D patterns. TNCs with a strong home-market orientation tend to carry out little overseas R & D; such R & D as is carried out is usually of the support laboratory type. Home-market firms have tended to regard their foreign sales as not requiring any further R & D beyond that carried out for their domestic market. Host-market TNCs – those oriented towards the national (or regional) market in which their overseas operations are located – operate both support laboratories and also higher-level locally integrated laboratories. The most important locational criteria are proximity to the firm's overseas markets and the fact that the firm's overseas operations are sufficiently substantial to justify separate R & D activities. Such activities tend to be located in the firm's biggest and most important overseas markets. World-market firms are the truly global corporations whose orientation is to world, rather than to national, markets. Their R & D activities may well include both support and locally integrated laboratories but their adoption of a globally integrated production strategy is leading them to establish specially designed international interdependent research laboratories.

The major locational criteria for these world-market R & D activities are the availability of highly skilled scientists and engineers, access to sources of basic scientific and technical developments and an appropriate infrastructure.

Universities were frequently mentioned as an important means of gaining access to the foreign scientific and technical communities that are of such great interest to the foreign exploratory laboratories of world-market companies. Every one of the world-market firms stressed the need for a strong local university system as a prerequisite for choosing an overseas location for R & D.

(Behrman and Fischer, 1980, p. 21)

Such stringent locational demands at present tend to limit these high-level R & D activities to a relatively small number of developed countries.

The R & D activities of transnational corporations have tended to be strongly concentrated in the developed market economies, with a substantial proportion being located in the firm's parent country. The spatial pattern within nations has also tended to be very uneven. The support laboratories are the most widely spread in so far as they generally locate close to the production units, although not every production unit has an associated support laboratory. But the larger-scale R & D activities tend to be confined to particular kinds of location. Their need for a large supply of highly trained scientists, engineers and technicians together with proximity to universities and other research institutions confines them to large urban complexes. These are often the ones which are also the location of the firm's corporate headquarters. A secondary locational influence is that of 'quality of living' for the highly educated and highly paid research staff: an amenity-rich setting including a good climate and potential for leisure activities as well as a stimulating intellectual environment.

Spatial patterns of corporate R & D in both the United States and the United Kingdom illustrate both of these locational influences.[5] In the United States corporate R & D is still predominantly a big-city activity despite recent growth in smaller urban areas. The pull of the amenity-rich environment is illustrated by the considerable concentration of R & D activities in locations such as Los Angeles, San Francisco and San Diego in California, Denver–Boulder in Colorado and the 'Research Triangle' in North Carolina. In the United Kingdom corporate R & D, like corporate headquarters and regional offices, is disproportionately concentrated in South East England. Within this region firms can be both close to London with all its intellectual, social and cultural facilities and also locate in some of the supposedly green and pleasant lands of the Home Counties.

Despite the persistence of such a highly uneven geographical distribution of TNC's R & D activities, however, there is no doubt that its geographical structure is changing. As Howells points out:

> as more companies move from being 'host market' to 'world market' firms, the role of R & D has moved from a direct but secondary role of helping to serve the market via product modification towards a much more integrated mechanism in gaining new markets. Increasingly the sources of new ideas for new products and innovations are coming from the user firms and industries and if firms are to remain competitive and be able to move into new markets they must be able to maintain close relationships with their existing and potential customers.

(Howells, 1990, p. 504)

Howells also points out that another important influence which is stimulating the geographical spread of R & D by TNCs is the growing demand for skilled scientists. This particular labour market is *international* in scale and intensely competitive and is forcing firms to extend their R & D networks in order to capture geographically dispersed scientific workers.

Production units   It is far more difficult to generalize about the locational tendencies of TNCs' production units than about their corporate and regional headquarters or their

a. Globally concentrated production

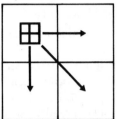

All production occurs at a single location. Products are exported to world markets.

b. Host–market production

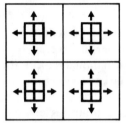

Each production unit produces a range of products and serves the national market in which it is located. No sales across national boundaries. Individual plant size limited by the size of the national market.

c. Product–specialisation for a global or regional market

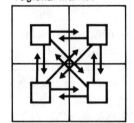

Each production unit produces only one product for sale throughout a regional market of several countries. Individual plant size very large because of scale economies offered by the large regional market.

d. Transnational vertical integration

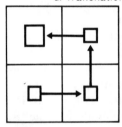

Each production unit performs a separate part of a production sequence. Units linked across national boundaries in a 'chain–like' sequence – the output of one plant is the input of the next plant.

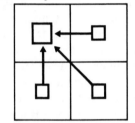

Each production unit performs a separate operation in a production process and ships its output to a final assembly plant in another country.

Figure 7.8 Some major ways of organizing transnational production units

R & D facilities. It is certainly true that, compared with corporate headquarters and R & D facilities, production units of TNCs have become more and more dispersed geographically. But there is no single and simple trend or pattern of dispersal common to all activities, whether at the global scale or within individual nations. The pattern varies greatly from one industry to another. The locational needs of corporate offices and R & D laboratories are broadly similar for all firms regardless of the industries in which they are involved. This is not so for production units. Their locational requirements vary considerably depending upon the specific organizational and technological role they perform within the enterprise and the geographical distribution of the relevant location-specific factors. Figure 7.8 illustrates diagrammatically four types of geographical orientation which a TNC might adopt for its production units.

Globally concentrated production   Figure 7.8a is the simplest case. Here it is assumed that the TNC concentrates all its production at a single geographical location (or, at least, within a single country) and serves its world markets through its marketing and sales networks. This is a procedure consistent with the basic global strategy shown in the upper right cell of Figure 7.5. This is the kind of strategy followed by many Japanese companies until their relatively recent move towards more dispersed global production activity.

Host-market production   Figure 7.8b depicts what has been, so far, the most common TNC production strategy. Here production is located in, and oriented directly to, a specific host market. Where that market is similar in character to the firm's home market the product is likely to be identical to that produced at home. Such production units have been variously termed 'miniature replicas' of home country plants or 'relay affiliates'. In many cases, however, the product may have to be adapted to the circumstances of the local market and involve the establishment of R & D facilities as well. The specific locational criteria for the setting up of host-market plants are, fairly obviously: the size and sophistication of the market as reflected in income levels; the structure of demand and consumer tastes; the cost-related advantages of locating directly in the market. In effect, this kind of production is import substituting. Most of the manufacturing plants established by United States' TNCs in Europe in the post-1945 period were of this kind, in many cases following a product life-cycle sequence (see Chapter 5, pp. 139–42). The more recent surge of European manufacturing investment in the United States is also directly host-market related. Similarly, the large markets of some developing countries, such as Brazil, have attracted considerable numbers of TNC manufacturing affiliates whose role is to serve that market directly.

With the development of the various enabling technologies which have combined to shrink geographical distances the establishment of a production unit in a specific geographical market becomes less necessary in purely *cost* terms. There are, however, two reasons for the continued development of host-market production. One is the need to be close to the market in order to be *sensitive to variations in customer demands, tastes and preferences or to be able to provide a rapid after-sales service.* As we have noted already, sensitivity to local geographical differences continues to be an important issue even where TNCs pursue broadly global strategies for their overall business. The second factor perpetuating the continuation of host market production plants is political: *the existence of tariff and, particularly, non-tariff barriers to trade.* As we saw in some detail in Chapter 6, such barriers are universal, although their scale and type vary.

Tariff barriers have been a significant locational factor from the very early days of transnational investment. In general, of course, tariffs have been falling as a result of the successive GATT negotiations. Today it is the various kinds of non-tariff barrier – especially import quotas and 'voluntary trade agreements' – which have become the major feature of trade policy in many nations. Again, these have acted as a stimulus to TNCs to jump over the barriers and establish direct production units to serve the local market. There is no doubt, for example, that the recent growth of Japanese manufacturing investment in Western Europe and in North America is substantially a response to the actual, or threatened, existence of non-tariff trade restrictions. If such barriers did not exist there is little doubt that, in cost terms at least, Japanese firms in several industries could readily have served European and North American markets by exports either from Japan or from their Asian affiliates.

Product specialization for a global or regional market   Globally concentrated production and host-market production are the longest-established forms of production orientation among TNCs. During the last three decades or so, however, a radically different form of production organization has emerged: production as part of a rationalized product or process strategy. Figure 7.8c shows this kind of product specialization for a global or a large regional market (such as the European Community or North America). Many argue that such large specialized plants will become the norm after the completion of the Single European Market, when all internal barriers to the movement of goods, services and factors of production are removed (see Chapter 6, pp. 165–8). In fact, there is likely to be considerable variation not only from industry to industry but also because the creation of a single market will not necessarily mean the creation of a uniform market in terms of tastes and preferences.

The industries involved in large-scale specialized production units in such regional markets as the EC or continental North America are quite varied, although rationalized production tends to be most common in those sectors in which TNC involvement is especially marked. We identified these in Chapter 3 as being the technologically more advanced sectors; the large-volume medium-technology consumer goods industries and the mass production industries supplying branded goods. The existence of a huge internal market together with differences in factor endowments between member nations facilitates the establishment of very large, specialized units of TNCs which serve the entire regional market rather than single national markets. The key locational consideration, therefore, is the 'trade-off' between the economies of large-scale production at one or a small number of large plants and the additional transport costs involved in assembling the necessary inputs and in shipping the final product to a geographically extensive regional market.

Transnational vertical integration of production   The other kind of rationalized production strategy involves geographical specialization by process or by semi-finished product. As we saw in Chapter 4, technological innovations in the production process permit a number of processes to be fragmented into separate parts and have led to a greater degree of standardization in some manufacturing operations. Parallel developments in the technology of transport, communications and organizational technology have introduced a much enhanced flexibility into the geographical location of the production process. It has become possible for TNCs to locate some of their production units to take advantage of geographical variations in production costs at a global scale, notably in developing countries. Thus, transnational vertical integration, whereby different parts of the firm's production system are located in different parts of the world, becomes feasible. Materials, semi-finished products, components and finished products are transported between the geographically dispersed production units.

In these circumstances there may be no direct link at all between the location of production itself and the national market in which the production unit is located. The traditional market connection is broken. The output of a manufacturing plant in one country may simply be the input for a plant belonging to the same firm located in another country. Alternatively, the finished product may be exported to a third-country market or to the home market of the parent firm. In these circumstances the term 'export platform' is used to describe the role of the country in which production is located. The plants themselves are sometimes termed 'workshop affiliates'; their role is to act as *international sourcing points* for the TNC as a whole. Hence, the process is often called offshore, or international intra-firm, sourcing. Figure 7.8d shows two simplified ways in which such international process specialization might be organized as part of a vertically integrated set of operations across national boundaries.

Offshore sourcing and the development of vertically integrated production networks at a global scale were virtually unknown before the early 1960s. The pioneers were US firms (notably in electronics, as we shall see in Chapter 10), which set up offshore assembly operations in East and South East Asia as well as in Mexico, followed later by some European and Japanese TNCs. The growth of such international production networks was extremely rapid during the late 1960s and throughout the 1970s. If anything, the world recession of the 1970s intensified the search by TNCs for low-cost production locations in order to remain competitive and maintain profitability. But it is important to stress that not all industries or processes are suitable for offshore sourcing.

The activities involved tend to be of two broad groups. First, there are those products at the *mature stage* of the product life-cycle, in which the technology has become standardized, long production runs are needed, and semi-skilled or unskilled labour costs are very important. Second, there are certain parts of the *production process of newer industries* which are also labour-intensive and amenable to the employment of semi-skilled and unskilled labour even though the industry as a whole is highly capital- and technology-intensive. The major factors influencing the location of offshore production in developing countries are:

1. *The labour-intensity of the product or process in developed countries.* The last three words need emphasis because it is those products or processes which intrinsically use a lot of labour which are the obvious candidates for relocation to lower-labour-cost locations. As we have seen, the wages gap between developed and developing countries can be immense (Figure 5.4) whereas differences in labour productivity are often minimal. At the same time, many TNCs have experienced labour problems in their home countries and in some of their operations in other developed countries (e.g. resistance to particular kinds of task, especially the more monotonous or unpleasant operations on assembly lines). Labour militancy and union pressure – on wage levels, working hours, working conditions, fringe benefits – have also contributed towards creating an important 'push' factor.

2. *The degree of standardization of the production process.* As we noted in Chapter 4, industries differ considerably in the extent to which production processes can be standardized. This has important implications for the ability of the firm to utilize unskilled labour and to train it quickly. It is processes possessing a high degree of repetitiveness that can be most easily taught to an unskilled and often uneducated labour force.

3. *The extent to which the production process can be fragmented into individual, self-contained operations and the importance of additional 'distance' costs.* Not only do production processes differ in the extent to which they can be divided into discrete operations but, also, it is worthwhile doing so only if the additional costs of transporting the materials or components to the production site and back again are low enough. A critical consideration is likely to be the weight and bulk involved. Additional 'distance costs' may also be imposed by bureaucratic delays at national borders or at transhipment points. For example, it is usually possible to clear all customs formalities within the same day in Singapore whereas it can take from several days to several weeks in some other developing countries.

4. *Government policies towards offshore processing and export production.* Quite apart

from general government policies which contribute towards the political stability or business climate of a host country, there are two sets of policies which have had a most important influence on offshore production strategies. Both were discussed in detail in Chapter 6 and need only be recalled briefly at this point. A major influence has obviously been the adoption of export-oriented industrial policies by a number of developing countries, including the operation of export-processing zones. In this way, developing countries have undoubtedly attracted the production units of foreign TNCs to act almost entirely as processing or assembly operations serving export markets. Host-country export-oriented policies have been substantially reinforced by the operation of offshore assembly provisions by major developed countries. As explained in Chapter 6, these provisions permit the export of domestic materials from, say, the United States, their processing or assembly overseas, and their reimportation into the United States on payment of import duties only on the value added overseas. There is no doubt that these import concessions have played a very significant role in the development of offshore processing and assembly in developing countries.

International intra-firm sourcing – the setting up by TNCs of production units in developing countries – has become an increasingly important mechanism of global integration of production processes in which

the more mobile factors, such as technology, management and equipment, are moved to the site of the least mobile. Through this method the multinational corporation is able to utilize the labour of the less developed countries in production processes formerly associated only with the more industrialized. It brings together both low-cost labour and advanced techniques.

(Leontiades, 1971, p. 27)

However, the choice of location for a production unit at the global scale is by no means as simple as it is often made out to be. It is not just a matter of looking at differences in labour costs between one country and another or at the incentives offered as part of an export-oriented policy. Despite the enormous shrinkage of geographical distance that has occurred, the relative geographical location of parent company and overseas production unit may still be significant. If it were not for the influence of distance (broadly defined) we might expect the offshore production plants to be drawn to the lowest-labour-cost locations. Yet this is not so. The sheer organizational convenience of geographical proximity may encourage TNCs to locate offshore production in locations close to their home country even when labour costs there are higher than elsewhere. A clear example of this is Mexico in the case of United States TNCs and parts of southern Europe in the case of European TNCs.

Of course, just as geographical proximity may override differentials in labour costs so, too, other locational influences may dominate in any particular case. Not all offshore sourcing arrangements are regional in nature. For the largest TNCs – the global corporations – the world is indeed their oyster. Their production units are spread globally, often as part of a strategy of *dual or multiple sourcing* of components or products. This is one way of avoiding the risk of over-reliance on a single source whose operations may be disrupted for a whole variety of reasons. In a vertically integrated production sequence in which individual production units are tightly interconnected, an interruption in supply can seriously affect the other units, perhaps those located at the other side of the world. In an extreme case, a whole segment of the TNC's operations may be halted.

The location of TNC production units at an *intra-national* scale has been the subject of far less attention than their location at the international scale. Yet their distribution within a country is extremely important from the viewpoint of their economic and social impact, especially in terms of employment (see Chapter 12). Within developing countries TNCs appear to locate their production units in the *core areas* of developing countries: the areas having a relatively high intensity of economic activity and the necessary infrastructure. This means, in most cases, either the major urban centres or the export-processing zones – which are, in effect, foreign production enclaves. There are exceptions, for example where an EPZ or industrial park is deliberately located in a less developed part of the country in order to stimulate local or regional development.

Within developed economies which, we should remind ourselves, contain the bulk of TNC activities in the world economy, the distribution of TNC production units generally follows that of industrial activity in general. Many such units are, of course, the branch plants of domestic TNCs. Both these and the branch plants of foreign TNCs have also shown some tendency to locate in areas of relatively high unemployment in order to tap large pools of labour. In some cases, too, TNCs have responded to government regional development policies in locating their production units. There is some evidence that foreign firms may be more responsive to government persuasion to locate in more peripheral assisted areas. It may simply be that regional incentives merely reinforce prevailing corporate locational trends, however, rather than actually initiating new ones.

It should now be clear that the spatial division of labour within the TNC is extremely complex not only because of the intricacy of its organizational structures but also because of the varying extent to which different parts of the enterprise can be separated both functionally and spatially. Such separation varies according to the nature of the TNC's organizational strategy, the technological characteristics of the industry in which it is involved and the locational requirements of the component parts. Different parts of the TNC have different locational requirements and these can be satisfied in various types of geographical location.

Restructuring and reorganization within the TNC

Transnational corporate networks and their resulting spatial patterns are, by their very nature, in a continuous state of flux. Change is endemic in such enterprises but the precise form change takes may well vary from one part of the TNC to another. At any one time, some parts may be growing rapidly, others may be stagnating, still others may be in steep decline. The functions performed by the component parts and the relationships between them may alter. Change itself may be the result of a planned strategy of adjustment to changing internal and external circumstances or the 'knee-jerk' response to a sudden crisis. Whatever its origin, however, corporate change will have a specific spatial expression. The changes that occur within the TNC itself will be projected into particular kinds of impact on the localities in which the component parts are located, relocated, expanded or contracted.

Forces underlying reorganization and restructuring   The forces which may lead to corporate reorganization and restructuring and, hence, to spatial change, can be divided broadly into two categories: external and internal. In many cases, of course, the two are very closely interrelated and it may be difficult to disentangle one from the other. Individual enterprises may be faced at any time by adverse or deteriorating *external* conditions – declining demand, increased competition in domestic or foreign markets, changes in the cost or availability of production inputs, militancy and resistance of labour forces in particular places, the pressure of national governments to modify their activities or even to cede control. Conversely, changes in external conditions may be positive rather than negative, for example the growth of new geographical markets or the availability of new production opportunities. A good illustration is the formation of regional economic groupings, such as the EC, which undoubtedly alter the pattern of investment opportunities. The creation of a large regional market made up of separate nation states, each with its own resource endowment and production cost attributes but with free movement of materials and products across national boundaries, provides an unprecedented opportunity for TNCs to restructure their production activities to serve the regional market. Investments which had made sense in the context of an individual nation are no longer necessarily rational in the wider context.

Quite apart from external forces there may be *internal* pressures stimulating re-organization and rationalization. Such forces may relate to the enterprise as a whole or to one or other individual parts; sales may be too low in relation to the firm's target, production costs may be too high. In a global corporation the performance of individual plants in widely separate locations can be continuously monitored and compared with one another to assess their efficiency. Studies of corporate change often reveal the key influence of a change in top management – a new chief executive who undertakes a sweeping evaluation of the enterprise's activities and investments and makes changes which stamp his authority: the 'new broom factor'. More generally there may be a perceived need to free capital and managerial resources for more profitable activities or to introduce new technologies, whether in products or processes. The need for reorganization may also be brought about by the firm's use of *acquisition* as a corporate strategy: the acquired firm's operational units have to be integrated into the existing organization. This may lead to the closure of some units, although not inevitably so. Corporate restructuring may occur in a variety of ways (Table 7.2) involving in some cases technological change, changes in work practices, rationalization of corporate activities, changes in the extent to which the value-added chain is internalized, and increased international investment.

The spatial dimension of reorganization and restructuring   Whether corporate re-organization is the result of a consciously planned strategy for 'rational' change or simply a reaction to a crisis (internal or external), its spatial outcome may take several different forms. In Figure 7.9 two broad categories of spatial–organizational change are identified: *in situ* change and locational shift. By far the most common form of adjustment is to the existing network of production units. A major advantage of the multi-plant firm in general and the transnational corporation in particular is that it can make substantial adjustments *in situ* without necessarily engaging in locational shift. The capacity of an existing plant can be increased to achieve economies of scale or reduced (partial disinvestment) to shed surplus capacity; an existing plant's capital stock can be replaced with new technology.

In such ways, the importance and even the actual function of production units can be altered as the TNC reallocates tasks among its existing geographically dispersed

Table 7.2 Major forms of corporate restructuring

Restructuring mode	Forms of restructuring
Intensification	Contractual flexibility Flexible working practices Concession bargaining
Investment and technical change	New technology Automation Flexible manufacturing systems
Rationalization	Divestment Differential expansion (contraction) Changes in product lines Transfers of business
De-integration	Outsourcing Subcontracting Homeworking Intrapreneuring New forms of overseas investment
Collaboration	Multinational ventures
Incremental internationalization	Increased overseas investment

(*Source*: Enderwick (1989), Table 1)

operations. Change at an existing plant may be either a gradual process of incremental adjustment or a more sudden change to its scale or function. The other category of spatial reorganization shown in Figure 7.9 explicitly involves abrupt change because it consists of either an increase or decrease in the number and location of plants operated by the enterprise or even, in rare cases, the physical relocation of an entire plant. The most common locational shifts within a TNC's production network are: acquisition of plants belonging to another firm; disinvestment at an existing plant; greenfield investment at a new location.

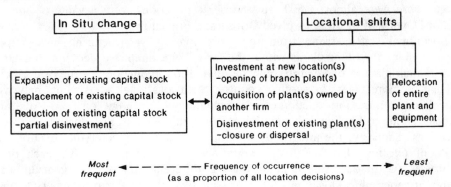

Figure 7.9 Reorganization, restructuring and spatial change: major types of investment-location decision

Acquisition and merger are particularly important mechanisms of corporate adjustment and growth. Until recently most of this activity occurred within national boundaries; it is only within the last two decades or so that international or cross-border acquisitions and mergers have become really significant. However, acquisition

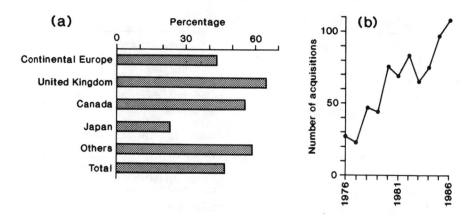

Figure 7.10 The importance of acquisitions and mergers in the entry of foreign firms into the United States (*Source*: US Department of Commerce)

has now become one of the most important ways for firms to expand their activities internationally. It is far more important than new, greenfield investments. For example, in the United States between 1981 and 1988 investment in acquisitions by foreign companies exceeded that in new establishments by between 1.5 and 12 times.

The importance of acquisition as an entry mechanism varied substantially, however, between TNCs from different geographical origins. Figure 7.10a shows that acquisition was almost three times as important for UK firms (64 per cent) investing in the United States as for Japanese firms (23 per cent). Figure 7.10b reinforces this point; it shows the rapid increase in the number of acquisitions made by UK firms in the United States between 1976 and 1986. A similar growth of cross-border acquisitions and mergers has occurred in Europe, partly stimulated by the imminence of the Single European Market but driven more generally by intensified international competition.[6] In 1988–9 alone there were almost 1,500 cross-border acquisitions, with a value of almost $40 billion. Over half of these involved British and French firms.

Locational shifts arising from reorganization and restructuring have important repercussions for other plants within the firm. For example, a decision to establish a new branch plant in one country may be related to a reduction in scale or even closure of one or more plants in another country. Such locational adjustments may well be associated with the introduction of new technology at different locations from those at which existing technology is being replaced or with the shift of production to lower-cost locations. Similarly, the integration of acquired plants may alter the functions or the scale of existing plants. For the larger and more extensive TNCs each of these processes may be occurring simultaneously in different parts of the organization and in different geographical locations. Some existing plants may be expanding or contracting; in some cases this may be associated with functional changes – new plants may be opening and existing ones closed. The whole adjustment process is kaleidoscopic on a global scale.

Thus, reorganization, restructuring and the resulting spatial change are an inevitable aspect of the evolution of transnational corporations. The actual form such change takes depends upon forces both internal and external to the firm. The very large global corporations are developing into what Vernon calls *global scanners*. They use their

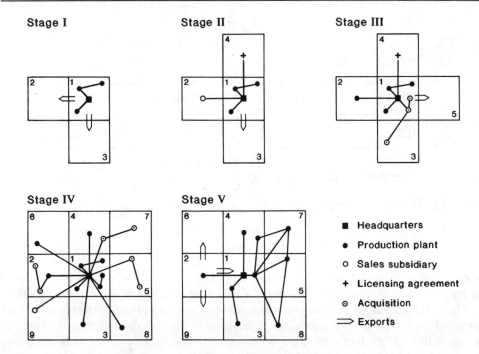

Figure 7.11 Reorganization, restructuring and spatial change in a TNC: an idealized sequence

immense resources to evaluate potential production locations in all parts of the world. The performance of existing corporate units can be monitored and evaluated against the rest of the corporate network and also against potential locations. Those existing plants which fall short of expectations may be disposed of. As plants become obsolete in one location they are closed down; whether or not new investment occurs in the same locality depends upon its suitability for the TNC's prevailing strategy.[7] The chances are, in many cases, that the new investment will be made at a different location, quite possibly in a different country altogether.

Figure 7.11 shows in an idealized manner how such changes may affect a TNC's geographical network. Stages I to IV follow the kind of evolutionary growth sequence discussed in Chapter 5 (see Figure 5.5). Stage V illustrates one possible way in which the complex transnational network of Stage IV might be restructured and rationalized. Such reorganization involves both domestic and overseas operations. The five domestic production plants have been reduced to two. The twelve overseas operations have been reduced to seven but with substantial alterations in their organization and functions. The three plants in country 2 have been rationalized to a single large plant which serves both the overseas markets of countries 6 and 9 and also the firm's home market. In addition a vertically integrated production network has been established in which plants in countries 4, 5 and 7 specialize in specific stages of a production process and final assembly occurs in country 8.

We should, however, beware of over-exaggerating the speed and ease with which TNCs can and do restructure their operations. There are 'barriers to exit' – in many cases production units represent huge capital investments which cannot be written off lightly. Political pressures may also inhibit firms from closing plants, especially in areas

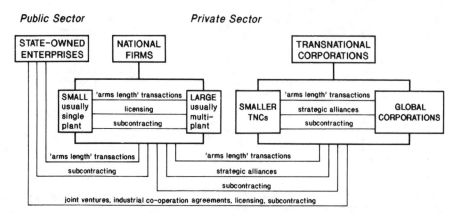

Figure 7.12 Major organizational segments of the global economy and their interrelationships

of economic and social stress. On the other hand, TNCs do have a highly tuned capacity to *switch* and *reswitch* operations within their existing corporate network. They also have the resources to alter the shape of their spatial network through locational shifts. Whatever the precise means adopted, whether through disinvestment or through the alteration of a plant's function within its existing structure, there is no doubt that corporate reorganization and rationalization make a major contribution to global economic change.

Networks of externalized relationships

So far we have concentrated upon the organizational and geographical dimensions of the *internal* relationships between the constituent parts of TNCs. But, although TNCs, by definition, operate complex internal networks of relationships, they are also locked into *external* networks of relationships with a myriad of other firms: transnational and domestic, large and small, public and private. Indeed, all economies are made up of various types of business organization which are interconnected in complex arrangements (Figure 7.12). The firms in each of the major organizational segments operate over widely varying geographical ranges and perform rather different roles in the global economy. But the most important point to be made is that the segments are *interconnected* in complex ways. It is through such interconnections, for example, that a very small firm in one country may be directly linked into a global production network, whereas most small firms serve only a very restricted geographical area.

A major theme of this book is that it is increasingly unrealistic to regard national economies as being watertight containers of economic processes. One reason is that the interrelationships between firms of different sizes and types increasingly span national boundaries to create a set of *geographically nested relationships from local to global scales*. These inter-firm relationships are the threads from which the fabric of the global economy is woven. It is through such links that changes are transmitted between organizations and, therefore, between different parts of the global economy. It is through the transmission of such changes – both positive and negative – that the welfare of nations and communities is modified for better or for worse. We cannot understand either the behaviour or the impact of any enterprise on its immediate locality without an appreciation of where the enterprise 'fits' into this global network of interrelationships.

We cannot hope to capture all the subtleties of these networks of relationships; we have to be selective. In this section, therefore, we concentrate on three major types of global inter-firm relationship: international strategic alliances, international sub-contracting and dynamic networks. Each of these organizational arrangements falls between the two extreme ways of organizing transactions within the production chain discussed earlier in this chapter: the complete internalization of transactions within a firm on the one hand and the complete externalization of transactions through the market on the other. These 'hybrid' organizational forms suggest again the importance of adopting a broad definition of the TNC and emphasize that the most important factor is not the ownership of overseas operations in itself but rather the ability to *co-ordinate* overseas operations even when these may be performed by legally indepen-dent firms.

International strategic alliances[8]

One of the most significant developments in the global economy in recent years has been the growth and spread of strategic alliances – various forms of collaboration – between firms at an international scale. They arise from the kinds of changes which have been occurring in the global economy, notably: intensification of competition, acceleration in technological change, increased costs of developing, producing and marketing new products. The major objective of a strategic alliance is to enable a firm to achieve a specific goal which it believes that it cannot achieve on its own. In particular, an alliance involves the sharing of risks as well as rewards through joint decision-making responsibility for a specific venture. Strategic alliances are not the same as mergers, in which the identities of the merging companies are completely subsumed. In a strategic alliance only *some* of the participants' business activities are involved; in every other respect the firms remain not only separate but also often competitors.

Collaborative ventures between firms across national boundaries are nothing new as Kindleberger (1988) has rightly pointed out. What is new is their current scale, pro-liferation and the fact that they have become *central* to the global strategies of many firms rather than peripheral to them. Most strikingly, the overwhelming majority of strategic alliances are between competitors. Of 839 agreements identified by Morris and Hergert (1987) between 1975 and 1986, 81 per cent were between two firms and no less than 71 per cent were between two companies in the same market. However, it appears that many companies are forming not just single alliances but *networks of alliances*.

According to a Business International study (1987, pp. 113–14), relationships are increasingly polygamous rather than monogamous: 'Few companies have only a single alliance. Instead, they form a series of alliances, each with partners that have their own web of collaborative arrangements. Companies like Toshiba, Philips, AT&T and Olivetti are at the hub of what are often overlapping alliance networks which fre-quently include a number of fierce competitors.' As a result it becomes more and more difficult to establish the precise boundaries between firms.

Although strategic alliances are not confined to particular sizes or types of firm, they are undoubtedly more common between large TNCs with extensive international operations. According to Morris and Hergert, most alliances were formed either

Table 7.3 Types of international strategic alliance

I *Alliance joint ventures*
  Separate legal entities with (occasionally without) equity.
  Co-operation may be limited to a single function or cover a broad range.
  Common for partners to co-operate for one specific product or market segment whilst operating as competitors in other markets.

II *Functional-specific competitive alliances*
  No separate legal entity.
  Co-operation confined to one or a limited range of specific functions, e.g.
  – Research & development ventures
  – Cross-distribution agreements
  – Cross-licensing agreements
  – Joint manufacturing agreements
  – Joint bidding consortia

(*Source*: based on Business International (1987), pp. 21–4)

between firms from EC countries (31 per cent of the total) or between EC and US firms (26 per cent). A further 10 per cent of the alliances were between EC and Japanese firms and 8 per cent between US and Japanese firms. Collaborative ventures are especially prominent in certain economic sectors, although few sectors are totally untouched. Five sectors accounted for 87 per cent of all collaborative agreements in the Morris and Hergert study: motor vehicles (24 per cent) aerospace (19 per cent), telecommunications (17 per cent), computers (14 per cent), other electrical (13 per cent). 'All of these sectors are typified by high entry costs, globalization, scale economies, rapidly changing technologies, and/or substantial operating risks' (Morris and Hergert, 1987, p. 18).

In Table 7.3 international strategic alliances are subdivided into two broad types. *Alliance joint ventures* usually involve the partners in creating a separate entity for a specific purpose in which each holds a share of the equity. Such joint ventures can be either broad or narrow in scope. They may also involve a partnership between private firms on the one hand and public/state-owned firms on the other. In contrast, *functional-specific alliances* usually do not involve the creation of a separate legal entity but rather are very specific types of collaboration.

In the case of R & D ventures, co-operation is limited to research into new products and technologies. Manufacturing and marketing usually remain the responsibility of the individual partners. Cross-distribution agreements offer firms ways of widening their product range by marketing another firm's products in a specific market area. Cross-licensing agreements are rather similar but they also offer the possibility of establishing a global standard for a particular technology, as happened, for example, in the case of compact disc players (Business International, 1987). Joint manufacturing agreements are used both to attain economies of scale and also to cope with excess or deficient production capacity. Joint bidding consortia are especially important in very large-scale projects in industries such as aerospace or telecommunications, where the sheer scale of the venture or, perhaps, the specific requirements of national governments put the projects out of reach of individual companies.

It is claimed that 'by co-operating, companies can combine complementary technologies, R & D capabilities, skills, products, market presence, production capacity etc., in a way that will strengthen each partner' (Business International, 1987, p. 3). But not everybody shares this rosy view. The American writer, Robert Reich, is strongly critical, particularly of alliances between US and Japanese companies, on the grounds that they will severely damage the long-term competitiveness of US firms (Reich and Mankin, 1986). The fear is that entering into such alliances will result in the loss of key technologies by the US partners. More broadly, strategic alliances are clearly more difficult to manage and co-ordinate than single ventures; the potential for misunderstanding and disagreement, particularly between partners from different cultures, is great. Certainly many such alliances have relatively short lives. Nevertheless, the obvious attractions of international strategic alliances in today's volatile and competitive global economy are likely to guarantee their continued growth as a major organizational form.

International subcontracting[9]

A particularly important set of inter-firm relationships is that between customers for, and suppliers of, materials, components, semi-finished products and the broad range of business services. As we saw in our discussion of internal relationships within TNCs, such inputs are frequently procured 'in-house' from within the firm's own branches and subsidiaries. But not even the most highly integrated firm is totally self-sufficient. All firms acquire at least some of their inputs from outside suppliers. Possibly between 50 and 70 per cent of manufacturing costs are spent on purchased inputs (Schroeder, 1989). Some of these purchases will be, as it were, 'off-the-shelf' or 'catalogue' sourcing from independent suppliers at the arm's length market price. However, a significant proportion – perhaps the majority – of such purchases are made through the formal mechanism of *subcontracting*.

Subcontracting is a kind of half-way house between complete internalization of procurement on the one hand and arm's length transactions on the open market. It is

> a situation where the firm offering the subcontract requires another independent enterprise to undertake the production or carry out the processing of a material, component, part or subassembly for it according to the specifications or plans provided by the firm offering the contract. Thus, subcontracting differs from the mere purchase of ready-made parts and components from suppliers in that there is an actual contract between the two participating firms setting out the specifications for the order.
>
> (Holmes, 1986, p. 84)

Types and characteristics of subcontracting relationships   Table 7.4 sets out the basic elements of the subcontracting relationship. It shows that subcontracting occurs in both industrial and commercial spheres; that it can cover not only processes and components but also complete finished products. Generally, the firm placing the order or contract is known as the 'principal firm'; the firm carrying out the order is known as the 'subcontractor'. *Commercial subcontracting* involves the manufacture of a finished product by a subcontractor to the principal's specifications. The subcontractor plays no part in marketing the product, which is generally sold under the principal's brand name and through its distribution channels. The principal firm itself in this case may be either a producer firm, that is, one which is also involved in manufacturing, or a retailing or wholesaling firm whose sole business is distribution. Whereas a producer firm may engage in both industrial and commercial subcontracting, retailers/wholesalers are confined to commercial subcontracting.

Industrial subcontracting can be subdivided into three types according to the motivation of the principal firm. *Speciality* subcontracting involves the carrying out, often on a

Table 7.4 Elements of the subcontracting relationship

1. *Technical aspects of production*
   (a) Subcontracting *processes* ⎫ – *Industrial subcontracting*
   (b) Subcontracting *components* ⎭
   (c) Subcontracting whole *products* – *Commercial subcontracting*

2. *Nature of the principal firm*
   (a) Producer firm (both industrial and commercial subcontracting)
   (b) Retailing/wholesaling firm (commercial subcontracting)

3. *Type of subcontracting (motivation of principal firm)*
   (a) Speciality subcontracting
   (b) Cost-saving subcontracting
   (c) Complementary or intermittent subcontracting

4. *Types of relationship between principal and subcontractor*
   (a) Time period involved: long-term, short-term, single batch
   (b) Principal may provide some or all materials or components
   (c) Principal may provide detailed design or specification
   (d) Principal may provide finance, e.g. loan capital
   (e) Principal may provide machinery and equipment
   (f) Principal may provide technical and/or general assistance and advice
   (g) Principal is invariably responsible for all marketing arrangements

5. *Geographical scale involved*
   (a) Within-border – domestic – subcontracting
   (b) Cross-border – international – subcontracting

(*Source*: based on material in Sharpston (1975); Germidis (1980), pp. 10–36)

long-term or even a permanent basis, of specialized functions which the principal chooses not to perform itself but for which the subcontractor has special skills and equipment. *Cost-saving* subcontracting is self-explanatory: it is based upon differentials in production costs between principal and subcontractor for certain processes or products. *Complementary* or *intermittent* subcontracting is a means adopted by principal firms to cope with occasional surges in demand without expanding their own production capacity. In effect, the subcontractor is used as extra capacity, often for a limited period or for a single operation. The actual relationship between principal and subcontractor can also take a variety of forms, as Table 7.4 indicates. The length of time involved may be long or short. The principal's involvement may vary in terms of finance, technology, design, the provision of materials and equipment. Invariably, however, the principal is solely responsible for marketing the finished products or for arranging further assembly or processing.

Costs and benefits of subcontracting to the participants   The precise advantage of subcontracting to the principal firm depends very much on the type of subcontracting involved (see (3) in Table 7.4). In general, however, it is one way in which the firm may avoid having to invest in new or expanded plant. Subcontracting also offers a degree of flexibility: it is easier to change subcontractors than to close down or reduce the firm's own fixed capacity. At the same time, by entering into a contractual agreement the principal firm gains a certain amount of control over the operation. It is also one way of externalizing some of the risks and costs of certain operations. These are, in effect, passed on to the subcontractor. Small subcontracting firms have been described as performing a 'shock-absorbing' role for large firms. Subcontractors tend to be both expandable and expendable, particularly where they are small firms in an unequal power relationship with large firms. There may be further problems from the sub-contractor's viewpoint when the subcontract work being carried out for a particular customer is a large proportion of the subcontractor's total output.

In effect, the subcontractor becomes part of a vertically integrated operation, but without the full benefits or obligations of such involvement. As such, its freedom to move into new products or new markets may be limited. The problem is greatest for the small subcontractor where the principal firm specifies the product in detail and where the subcontractor depends upon the principal for product and process development. On the other hand, small firms may well gain substantially from their sub-contracting role. Most important is the *access* gained to particular markets via brand names, access which would otherwise be unattainable; continuity of orders (in some cases over a long period of time); injection of capital in the form of equipment; and access to technology. In many respects, therefore, the subcontracting relationship is symbiotic – a division of labour between independent firms – in which each partner contributes to the support of the other.

Subcontracting operations are especially important for small firms. Many firms actually start their lives as subcontractors to larger firms, and it is certainly an import-ant channel through which small entrepreneurial firms can operate. Such observations have led to the view that large and small firms are related in a particular kind of unequal power relationship in which large firms dominate. There is often a great deal of truth in this but it is not the whole story. Subcontracting relationships are not confined solely to those in which large firms dominate the small. In some industries, such as aerospace and automobiles, very large firms act as subcontractors to other large firms.

The spatial dimension of subcontracting As a process, subcontracting is as old as industrialization itself, if not older. The 'putting-out' system was a key element of most industries from their earliest stages. It depended, essentially, on close geographical proximity between firms and their subcontractors. The very fine and intricate net-work of subcontracting relationships based on the externalization of transactions in the production chain (Figure 7.1) often led to the development of highly localized industrial districts. Such tight, functionally and transactionally based, geographical agglomerations of linked economic activities declined in most Western industrial countries with increasing speed between the 1960s and early 1980s, although they persisted in Japan.

Indeed, Japan still has one of the most highly developed domestic subcontracting networks.[10] Each large Japanese firm is surrounded by a constellation of small and medium-sized subcontracting firms which act as suppliers of components or perform specialist processes to the specification, and the timetable, laid down by the controlling large firm. Indeed, the Japanese subcontracting system, with its sharp distinction between the two major segments, has contributed a great deal to the international competitiveness of the Japanese economy. By all accounts, however, conditions in the myriad of small subcontracting firms are very different from those in the major firms. The subcontracting segment has none of the much lauded qualities of lifetime employ-ment and corporate paternalism which exist in the major corporations. Competitiveness within the subcontracting segment is fierce; the small firms are very heavily subservient to the stringent demands of the principal companies.

One of the most significant developments of the last thirty years has been the

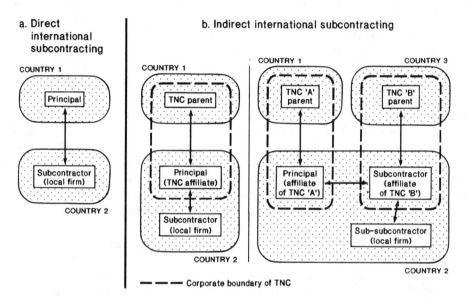

a. Direct international subcontracting

COUNTRY 1

Principal

Subcontractor (local firm)

COUNTRY 2

b. Indirect international subcontracting

COUNTRY 1

TNC parent

Principal (TNC affiliate)

Subcontractor (local firm)

COUNTRY 2

COUNTRY 1

TNC 'A' parent

Principal (affiliate of TNC 'A')

COUNTRY 3

TNC 'B' parent

Subcontractor (affiliate of TNC 'B')

Sub-subcontractor (local firm)

COUNTRY 2

— — — Corporate boundary of TNC

Figure 7.13 Types of international subcontracting relationship (*Source*: based on C. A. Michalet (1980) International subcontracting: a state-of-the-art, in Germidis (1980), pp. 51, 52)

extension of subcontracting across national boundaries: the emergence of *international subcontracting* as an important global activity. The revolution in transport and communications technology, together with developments in the production process itself, have created the potential for firms to establish subcontracting networks over vast geographical distances in the same way as TNCs have established offshore production units of their own (see above, pp. 204–6). Relatively low transportation costs, plus the ability to control and co-ordinate the operation of a long-distance subcontracting system, have allowed firms to take advantage of very low labour costs in developing countries. Various kinds of international subcontracting relationship are possible. Figure 7.13 distinguishes between direct international subcontracting, that is, between independent firms located in separate countries (Figure 7.13a), or indirect, where the principal is an overseas affiliate of a transnational corporation and the subcontractor is either a local firm or perhaps an affiliate of another TNC (Figure 7.13b).

International subcontracting has become common in a wide variety of industries, although it tends to be especially concentrated in those activities which have also seen the development of offshore production units of TNCs. Recall that the key criteria for this have been: the labour intensity of the product or process in developed countries; the degree of standardization of the product process; the extent to which the production process can be fragmented into separate operations; government policies towards offshore processing and export production.

Helleiner provides a concise list of the kinds of activities most commonly involved in international subcontracting in developing countries in the early 1970s. Although the details may well have changed, the picture is still broadly similar:

> Semi-conductors, valves, tuners and other components are manufactured or assembled for a large number of Japanese and American electronics firms in Hong Kong, Singapore, South Korea, Taiwan and Mexico . . . Garments, gloves, leather luggage and baseballs are sewn

together in the West Indies, Southeast Asia and Mexico for American and Japanese firms. (In the case of the baseballs, the leather covers, cotton yarn, thread, latex and cement are all imported from the United States.) Automobile parts are manufactured for British, American and Japanese firms in a wide variety of countries, e.g., radio antennae in Taiwan, piston rings and cylinder linings in South Korea and Taiwan, automobile lamps in Mexico, braking equipment in India, batteries and springs in Thailand, etc. Data are flown to Southeast Asia and the West Indies for punching upon tape by low-wage key punch operators, following which the tapes are flown back again. Swiss watchmakers send jewels to Mauritius for precision drilling . . . Among other industries already engaged in these activities in less-developed countries are those producing electrical appliances (including television and radio, sewing machines, calculators and other office equipment), electrical machinery, power tools, machine tools and parts, motorcycle and bicycle parts, typewriters, cameras, optical equipment, watches, brass valves, aircraft parts, telecommunications equipment, chemicals and synthetic fibres, and musical instruments.

(Helleiner, 1973, p. 29)

Although most of the attention in the literature has been devoted to the involvement of manufacturing TNCs in international subcontracting, the large retail or wholesale buying groups have played an extremely important role. As operators of commercial subcontracting networks, these large non-manufacturing organizations continuously scour the globe for low-cost sources of products. In fact, the author of one of the very few studies of such commercial international subcontracting, Angus Hone (1974, p. 149), claimed that it is the large American, European and Japanese retail, wholesale and general trading companies which have been 'the real key to Asia's expansion in manufactured goods' rather than the establishment of the offshore production units of TNCs.

The changing relationship between customers and suppliers A major incentive for TNCs to engage in international subcontracting has been the availability of low-cost products or processing. In many cases, however, it has led to a *remote relationship* between customers and suppliers – and not just in terms of physical distance. As Sayer observes,

both functionally and geographically the relationship between firms and their suppliers tends to be distant; how the suppliers operate is of no concern to the purchaser, provided the price is right, and this lack of inter-firm contact, together with the infrequent nature of deliveries, allows suppliers to locate at often considerable distances away from the purchaser, if by doing so they can minimize labour costs and other expenses.

(Sayer, 1986, pp. 47–8)

Such a system involves firms in holding large stocks or inventories of materials and components in order to insure against interruptions in supply or faulty components. Schonberger (1982) has called this a 'just-in-case' system.

Recently, however, there has been a move towards a very different kind of procurement system with very different relationships between customers and suppliers. The 'just-in-time' system,[11] like the just-in-case system, is much more than merely a system for the procurement of supplies. It is part of the broader system of the organization of production adopted by the firm. The essence of the just-in-time system is that work is done only when needed and in the necessary quantity at the necessary time. Very small stocks, approaching zero, are held by the firm. Supplies of the needed materials and components are delivered 'just in time' to be used in the production process. These characteristics 'require that orders to the firm's suppliers and subcontractors . . . [be] . . . small and frequent, indeed deliveries may be made several times a day and hence proximity to suppliers is essential' (Sayer, 1986, p. 55). A key question, therefore, is whether a move towards a just-in-time system will lead to a reduction in international subcontracting and the revival of localized agglomerations. The likelihood seems to be a mixture of arrangements involving both long- and short-distance linkages depending upon the particular circumstances in specific industries and firms.

One undoubted development is the tendency for many firms to move towards closer functional relationships with their suppliers. Rather than merely seeking out the lowest-cost supplier and little else, there is a strong move towards the nomination of 'preferred suppliers' with whom very close relationships are developed. Such suppliers are increasingly being given greater responsibility for the quality of their outputs and, indeed, are playing a more direct role in the design of products. More generally, the procuring firm has a variety of options in relation to its suppliers. It can opt for *single sourcing* to gain economies of scale (and lower costs) but with the risk of putting all its procurement eggs in a single supply basket. Alternatively, it can opt for *dual* or *multiple sourcing* and spread its subcontracting network more widely.

Dynamic networks

International strategic alliances and international subcontracting are long-established forms of inter-organizational relationships. We have dealt with them separately although they do, in fact, overlap in intricate ways. It is now being suggested, however, that a new organizational form is emerging which, in effect, contains elements of both strategic alliances and subcontracting networks. These are dynamic networks, or flexibly integrated organizational forms, which involve complex relationships between firms each of which performs a specialist role within a co-ordinated network. Figure 7.14 shows in a highly simplified form the major elements of such a dynamic network, whose

> major components can be assembled and reassembled in order to meet complex and changing competitive conditions . . . Business functions such as product design and development, manufacturing, marketing and distribution, typically conducted within a single organization, are performed by independent firms within a network . . . Because each function is not necessarily part of a single organization, business groups are assembled or located through brokers.
>
> (Miles and Snow, 1986, p. 64)

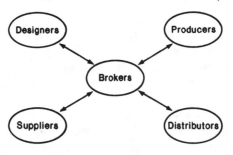

Figure 7.14 A dynamic network (*Source*: Miles and Snow (1986), Figure 1)

Such a dynamic network, if fully developed, goes a long way beyond conventional subcontracting relationships, which usually form only part of a principal firm's activities. Miles and Snow (1986, p. 69) suggest that the network form is becoming especially common in 'labour-intensive industries, where vertical disaggregation is less costly and easier to administer'. They point to developments in the US college textbook publishing industry in which many of the leading firms have now contracted out not just printing and binding functions (often overseas) but also many editing operations, artwork, graphics and design. They predict (op. cit., p. 72) that 'current "merger mania" notwithstanding, it now seems likely that the eighties and nineties will be known as decades of large scale disaggregation and redeployment of resources in the United States and of a reshaping of strategic roles across the world economy'.

Two examples of dynamic network organizations can be given to provide some idea of their characteristics. The first is a US toy manufacturer, Lewis Galoob Toys, Inc., which was, in the mid-1980s, a multimillion-dollar company, yet, in the words of a *Business Week* report, 'is hardly a company at all'.

> A mere 115 employees run the entire operation. Independent inventors and entertainment companies dream up most of Galoob's products, while outside specialists do most of the design and engineering. Galoob farms out manufacturing and packaging to a dozen or so contractors in Hong Kong, and they, in turn, pass on the most labour-intensive work to factories in China. When the toys land in the US, they're distributed by commissioned manufacturers' representatives. Galoob doesn't even collect its accounts. It sells its receivables to Commercial Credit Corp., a factoring company that also sets Galoob's credit policy. In short, says Executive Vice President Robert Galoob, 'our business is one of relationships'. Galoob and his brother, David, the company's president, spend their time making all the pieces of the toy company fit together, with their phones, facsimile machines, and telexes working overtime.
>
> (*Business Week*, 3 March 1986, p. 61)

The second example, less extreme in form, is the Nike athletic footwear company. Donaghu and Barff (1990) present a detailed account of Nike's corporate and geographical development, which displays many of the basic features of a dynamic network. Like other athletic footwear firms, Nike does not wholly own any integrated production facilities but is characterized by 'the large-scale vertical disintegration of functions and a high level of subcontracting activity' (op. cit., p. 539). Its development displays great flexibility in adapting to changing competitive circumstances. As Figure 7.15 shows, Nike consists of a complex tiered network of subcontractors which perform specialist roles.

> Nike's 'in-house' production may be thought of as production from its exclusive partners. Nike develops and produces all high-end products with exclusive partners, while volume producers manufacture more standardized footwear that experience larger fluctuations in demand . . . Nike acts as the production co-ordinator and three categories of primary production alliance form the first tier of subcontractors. A second tier of material and component subcontractors supports production in the first tier . . . [all] . . . production takes place in South East Asia while the headquarters in Beaverton, Oregon, houses Nike's research facilities.
>
> (Donaghu and Barff, 1990, p. 544)

Towards a synthesis of organizational and geographical relationships

The internationalization of business firms involves a great deal more than simply foreign direct investment. It involves, as we have seen in the preceding sections of this chapter, extremely complex internal and external transactions and relationships. Some of this complexity is summarized in Figure 7.16, where the major modes of foreign

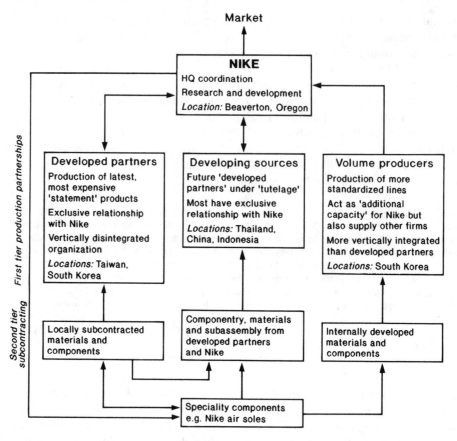

Figure 7.15 The subcontracting network within the Nike company (*Source*: based on Donaghu and Barff (1990), Figure 4 and pp. 542–4)

involvement are related to the more important contributory influences, both economic and political. The situation is complicated enough even if we take a static picture. It is, of course, far more complex than this because all these networks of relationships are in a continuous state of flux. The ways in which the production chain is organized, and the boundary between which functions are internalized within a firm and which are externalized and performed as a division of labour between firms, is extremely fluid. As we have seen, there are signs of *increasingly flexible* forms of organizational relationship in the global economy.

The combination of these various networks of relationships, both within TNCs and between independent and quasi-independent firms, creates a highly complex spatial or geographical structure. The overall scale of this structure is, of course, global but the substructures and the links between them operate at all geographical scales down to the local. Indeed, it is at the local scale, 'on the ground', that the processes within the production chain are actually performed. In fact, the very character of particular places itself exerts a considerable influence on the processes and networks we have been discussing. Specifically spatial relationships within and between territorial complexes of economic activity are, in themselves, an *intrinsic* part of the production system.

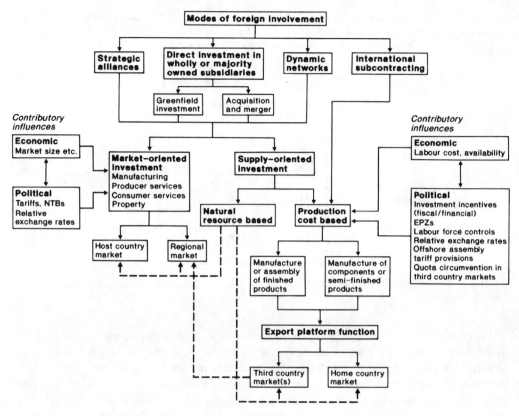

Figure 7.16 Different modes of foreign involvement and their contributory influences

It is a mistake, therefore, to argue that the map of economic activity, at whatever geographical scale, is merely the result of decisions made by business firms or any other organization projected on to the earth's surface. As Walker (1988, p. 385) has rightly observed, 'It is impossible to separate out the organizational from the geographical, much less to treat industrial location as the simple outcome of organizational forms or decisions . . . space is . . . basic to human action . . . all forms or organization are inherently spatial to some degree'. Thus the global economy is made up of a variety of complex *organizational networks* of the kind discussed in this chapter – i.e. the internal networks of TNCs, the networks of strategic alliances, of subcontracting relationships and of other, newer, organizational forms – which intersect with *geographical networks* structured particularly around linked agglomerations or concentrations of activities. But what form do these organizational–geographical relationships take?

Hymer's model of the organizational–spatial hierarchy

One of the best-known attempts to explore this link was made by Stephen Hymer in 1972. According to him,

> the multinational corporation tends to create a world in its own image by creating a division of labour between countries that corresponds to the division of labour between various

levels of the corporate hierarchy. It will tend to centralize high-level decision-making occupations in a few key cities (surrounded by regional sub-capitals) in the advanced countries, thereby confining the rest of the world to lower levels of activity and income.

(Hymer, 1972a, p. 59)

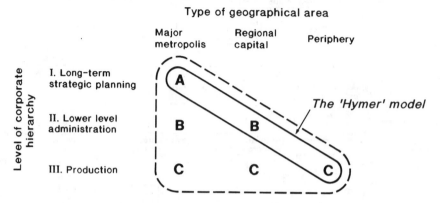

Figure 7.17 Relationships between organizational and geographical hierarchies in a spatial division of labour (*Source*: based on Morgan and Sayer (1983), p. 6)

There is obviously some validity in Hymer's view, as our discussion in this chapter has demonstrated. There *are* recognizable hierarchical spatial tendencies in the location of different parts of TNCs. Corporate headquarters *do* tend to concentrate in a small number of major metropolitan centres; regional offices *do* favour a slightly wider range of cities; production units *are* more extensively spread both within and between nations, in developed and developing countries. Not surprisingly, therefore, Hymer's model of a world being created in the image of the TNC has been widely accepted. But, as some critics have pointed out,[12] it grossly oversimplifies the complexity of the modern global organization of economic activity.

In Figure 7.17 the sequence A, B, C represents Hymer's scheme in a clear and distinct hierarchical arrangement in which the vertical division of labour within the TNC is reflected in an unambiguous geographical division of labour. Interpreted literally this would imply that only those levels of corporate activity would be present at each appropriate level in the geographical hierarchy. The top levels of the metropolitan hierarchy would contain only high-level control functions and associated occupations. But this is clearly not the case. A major metropolis, such as London or New York, does indeed contain the major corporate headquarters of TNCs but it also contains lower-level control functions (B) and, most significant of all in this context, many of the kinds of production unit (C) which are supposed to be present only in peripheral areas of the world. It is more realistic to conceive of different types of geographical location as containing different *mixes* of corporate units in which the actual proportions vary. At the same time, of course, as we have seen, the organizational structure of TNCs is far more complex and differentiated than the simple hierarchy suggested by Hymer.

New flexible production complexes

A very different approach to organizational–geographical relationships has been developed within the flexible specialization school of post-Fordism which was referred to briefly in Chapter 4. According to this view, the replacement of Fordism as the hegemonic form of economic organization by a regime based upon flexible production and flexible specialization is leading to quite new organizational–geographical structures. The growth of vertically disintegrated forms of economic organization (subcontracting, network organizations and the like) with their networks of externalized transactions leads, it is argued, to the spatial agglomeration of economic activities in *new industrial districts*.

These concentrations of interlinked activities – *flexible production complexes* as they have been called – 'form the new growth centres of the world system' (Scott, 1988, p. 178). Such complexes have been identified in most advanced industrial countries and include such areas as Silicon Valley, California, Route 128 around Boston, Massachusetts, Dallas–Fort Worth, Texas, in the United States; Grenoble, Toulouse, Sophia Antipolis near Nice in France; the Cambridge–Reading–Bristol axis in Britain. To these 'high-technology' agglomerations are added new industrial districts based on 'craft, artisanal or design-intensive activities' in areas such as 'the Third Italy', areas in other Western European countries and in the United States. However, as Amin and Robins (1990) point out, this is an extremely heterogeneous group. Such agglomerations are based essentially on the localized linkages of small firms and establishments, using flexible technologies and engaging in flexible specialization. Sabel (1989) also sees flexible specialization arising from the restructuring of large firms and argues that the decentralization of their specialized activities is closely linked to the creation of similar industrial districts.

The debate over the importance of new flexible production complexes or new industrial districts is a very lively one.[13] There is little doubt that such 'new industrial spaces' have emerged. But they have been given both exaggerated importance by their proponents and also explained in excessively uniform terms, given their great diversity. Amin and Robins (1990) provide a particularly powerful critique which emphasizes the interconnections between the local and the global. Far from moving towards a position of greater local economic autonomy,

> we are, in fact, seeing increasing tendencies towards internationalization and the global integration of local and national economies . . . Regional and local economies have to be understood in the context of this global field . . . Discussion of entrepreneurial artisans and of disintegrating and localizing corporations should not blind us to the growing power and influence of global industrial and finance capital. Multinational corporations are (need it really be said?) the real shakers and shapers of the world economy . . . The multinational corporation remains the most powerful agent of the restructuring process . . . What is at work is not corporate fragmentation, but, in fact, more effective corporate integration.
>
> (Amin and Robins, 1990, pp. 26, 27)

The debate continues, however.

Conclusion

Our focus in this chapter has been on the ways in which economic activities in the global economy are being organized and reorganized through dynamic networks of relationships within and between business firms. The most important point to emphasize is the *variety and diversity* of processes and outcomes, both organizational and geographical. The production chain – the basic building block of the economic system – can be articulated in different organizational and geographical ways. One particularly important mode of organization is within the TNC itself, where internal functional specialization and the transactions made between the firm's component parts create a geographical division of labour with certain characteristic features. But TNCs operate within intricate networks of externalized relationships. Such relationships add to the kaleidoscopic complexity of the global economy, in particular because of the tendency of some of these relationships to take on a specifically localized geographical form within the global system. Both internalized and externalized relationships are the threads through with the global economy is integrated, linking together both organizations and geographical areas in complex, interrelated and overlapping divisions of labour.

The TNC is the most important driving force in these networks. The TNC not only directly controls and co-ordinates its own complex internal networks at an international or a global scale but also *indirectly* controls many of the externalized networks in which it is embedded. To repeat the broad definition of the TNC which we have been using: it is an organization which *co-ordinates* production from one centre of strategic decision-making. This is not to argue that all the world's complex production networks are TNC-controlled – they are not – but to emphasize that TNCs play the dominant role. The precise form and articulation of production chains and networks are strongly influenced, although not necessarily determined, by technology. Technologies specific to the particular chain influence both the extent to which the production chain may be fragmented in a technical sense and also the degree of organizational flexibility within and between stages in the chain. The more general space-shrinking technologies of transport and communication determine the potential geographical scale at which the production chain can be distributed.

The precise form and articulation of production networks, however, are also determined by the actions of nation states as they pursue, in particular, policies of trade management and attempt to impose performance requirements on international firms operating within their territories. The shape of the global economy, then, is the outcome of the complex interplay between the strategies of TNCs and nation states set within the context of technological change. The particular mix varies subtantially from one industry to another; it is to such variation that we now turn in Part Three.

Notes for further reading

1. This 'markets and hierarchies' view of the 'governance' of economic transactions was developed by Williamson (1975) and has been further elaborated by others such as Teece (1980). As a concept it derives from the work of Ronald Coase (1937), who addressed the fundamental question of why multi-function firms exist at all.
2. Most standard management textbooks contain a discussion of evolving organizational structures within international firms. See, for example, Czinkota, Rivoli and Ronkainen (1989) chapter 21; Daniels and Radebaugh (1989) chapter 17. Particular attention to network structures is provided by Bartlett (1986).
3. There have been relatively few studies of locational trends in corporate and regional headquarters at the international scale. Examples include Cohen (1981), Heenan (1979), Grosse (1982), Dunning and Norman (1983, 1987). For a summary of trends in headquarters locations in the United States see Dicken and Lloyd (1990), pp. 307–11.
4. Specific research on R & D within TNCs at an international scale remains limited. Important early contributions include Vernon (1974, 1979), Buckley and Casson (1976), Behrman and

Fischer (1980), Hood and Young (1982). Two important recent studies are those by Howells (1990), Pearce (1990).

5. For detailed studies of R & D location in the United States see Malecki (1980). The concentration of R & D activities in South East England is described by Howells (1984).

6. Gray and McDermott (1989), UNCTC (1989b) discuss international merger activity.

7. McDermott (1989) presents a detailed account of the divestment behaviour of TNCs, including a series of case studies.

8. There is a rapidly growing literature on international strategic alliances, including James (1985), Perlmutter and Heenan (1986), Business International (1987), Contractor and Lorange (1987), Morris and Hergert (1987), Wells and Cooke (1991).

9. The earliest comprehensive account of international subcontracting is by Sharpston (1975). A collection of case studies in various parts of the world, together with a discussion of general characteristics, can be found in Germidis (1980). Holmes (1986) discusses the intricacies of industrial subcontracting in general terms. UNCTC (1988, pp. 164–9) describes the role of trading corporations and also original equipment manufacturing arrangements in international subcontracting.

10. Sheard (1983) provides a detailed account of Japanese subcontracting networks in the automobile industry, which will be discussed further in Chapter 9.

11. There is a large literature on just-in-time systems, e.g. Schonberger (1982), Sayer (1986), Schroeder (1989).

12. See, for example, Morgan and Sayer (1983), Ross and Trachte (1983), Walker (1989). Hymer's early death in 1974 removed the possibility of his developing his ideas further.

13. The major features of this debate are captured in Piore and Sabel (1984), Scott (1988), Sabel (1989), Amin and Robins (1990), Lovering (1990).

PART THREE

Global Shift:
The Picture
in
Different Sectors

Prologue

The four chapters of Part Two set out the general ways in which the major processes of change shape and reshape the global economic map. The central theme was that the globalization and internationalization of economic activity arise from the interplay between three sets of processes: the strategies of TNCs, the strategies of national governments and the character and direction of technological change. Particular emphasis was placed upon the interconnections and interdependencies between economic activities and upon the networks of organizational relationships which form the fabric of the global economy.

Part Two, therefore, was concerned with the *general* operation of these complex and dynamic processes. But precisely how they operate and the specific outcomes produced vary substantially between different types of economic activity. TNCs are more directly involved and influential in some industries than in others. Their strategic orientations may vary both within and between industries. The importance of strategic alliances and the nature of materials and components sourcing may differ from one sector to another. Similarly, the involvement of national governments is not uniform across all industries. Some industries are regarded as being more important to governments than others. The precise form of government involvement may differ by sector. The nature and the intensity of the interactions between TNCs and governments may well vary from one industry to another. Technology and technological change are also far from uniform across economic sectors. Indeed, one of the major distinguishing features of an industry is its specific product and process technology. Even the general technologies of transport and communication, which affect all economic activities, may have differential effects.

For all these reasons it is important to look at *specific* cases to see just how the general processes operate and interact to produce particular outcomes. The globalization of economic activity is not a uniform process. In Part Three, therefore, we examine four industries – textiles and clothing, automobiles, electronics, services – each of which has experienced major global shifts in recent decades. It is not suggested that these four industries are in any way 'typical' or 'representative' of all types of industry. Rather, the purpose is to show how the processes of change combine to create particular organizational and geographical forms at the global scale. As we shall see, there are both differences and similarities between the four sectors.

The treatment of each case study follows a broadly similar pattern. In this way it becomes possible to see more clearly the similarities and differences between them. It also makes it easier to compare any one particular characteristic (for example, the importance of strategic alliances or the nature of government policy) across all four sectors.

The structure of each chapter is as follows:

1. General significance of the industry and its general organizational structure.
2. Global shifts in production and trade.
3. The changing pattern of demand.
4. Technology, technological change and the production process.
5. Government policies.
6. Corporate strategies.
7. Employment trends and implications.

Chapter 8

'Fabric-ating Fashion': The Textiles and Clothing Industries

Introduction[1]

The textiles and clothing industries were perhaps the first manufacturing industries to take on a global dimension. They are the most geographically dispersed of all industries across both developed and developing countries. They are organizationally very complex, containing elements of both very new and very old organizational practices. They are changing very rapidly in their geography, their organization and their technology, and these changes are causing intense political friction. Indeed, global shifts in the textiles and clothing industries exemplify many of the intractable issues facing today's world economy, particularly the trade tensions between developed and developing economies.

The textiles industry was the archetypal industry of the industrial revolution of the eighteenth and nineteenth centuries in Britain. In some senses that industrial revolution was a textiles revolution: the cotton textiles industry was the primary engine of growth and its major geographical centre of production – Lancashire – became the exemplar of the nineteenth-century industrial landscape with its oft-described 'dark satanic mills'. Its marketing capital, Manchester, became perhaps the first global industrial city – the 'Cottonopolis' – of an industrial system whose tentacles spread across the globe. All the 'newly industrializing countries' of the nineteenth century – the United States, Germany, France, the Netherlands – also developed large textiles industries employing many hundreds of thousands of workers, often in strongly localized geographical clusters. A similarly concentrated pattern occurred in the rather later development of a factory clothing industry in the second half of the nineteenth century.

The sheer strength – both economic and political – of the British textiles industry in the nineteenth century effectively strangled the development of an indigenous textiles industry in the major colonies, especially India.[2] But such dominance could not last, particularly in an industry which was so ideally suited to the early stages of industrialization. It was possible to produce textiles and clothing using relatively simple technologies and low-skill labour. The traditional craft skills of hand spinning, weaving and sewing were a ready basis for larger-scale industrial application. The capital investment required was relatively modest compared with many other types of industry. Where local supplies of the raw materials were also available there was an even more obvious case for the development of a textiles industry. But materials availability is not a prerequisite; cotton and similar materials are not expensive or difficult to transport. Lack of indigenous supplies has not inhibited the development of highly successful textiles industries in many parts of the world.

Globally the textiles and clothing industries are very large-scale employers of labour. Some 15 million workers are directly employed in the textiles industry worldwide and a further 8 million in clothing manufacture. But these figures grossly understate the actual numbers involved. Countless unregistered workers, employed both in factories and at home, need to be added to reach a true picture, especially in the clothing industry. Despite the changes wrought by new technologies, corporate rationalization and competition from new producers, the textiles and clothing industries remain as important sources of employment in the developed economies. In particular they employ many of the more 'sensitive' segments of the labour force: females and immigrants, often in tightly localized communities. In the developing countries the industries employ predominantly young female workers in conditions often leaving a great deal to be desired and which recall those of the sweatshops and mills of nineteenth-century cities in Europe and North America.

The importance of textiles and clothing as a basis for today's newly industrializing and less industrialized countries, together with their continued, though much diminished, importance in the older industrial economies, have made these industries into an international political football. They are the subject of fierce political controversy between developed and developing countries and, increasingly, between the developed economies themselves. Textiles and clothing are the only industries in the world economy to which special international trade restrictions apply through the *Multi-Fibre Arrangement* (MFA).

The textiles–clothing production system

The textiles and clothing industries form part of a larger production system,[3] as Figure 8.1 shows. Each stage has its own specific technological and organizational characteristics and particular geographical configuration. Each has been changing very substantially in recent decades. The textiles industry itself consists of two major operations: the preparation of yarn and the manufacture of fabric. Both stages are performed by firms of all sizes, from the very small domestic enterprise to the very large subsidiary of the

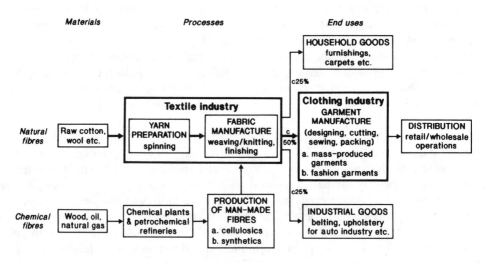

Figure 8.1 The textiles–clothing production system

transnational corporation. The general trend, however, has been for textile manufacturing to become more and more capital-intensive and for large firms to be increasingly important. The output of the textiles industry goes to three types of end-use, of which the clothing industry is by far the most important. Approximately 50 per cent of all textiles production goes into the manufacture of garments.

Despite some recent changes, the clothing industry remains far more fragmented than the textiles industry and is less sophisticated technologically. It is also an industry in which subcontracting is especially prominent. Very often the design and even the cutting processes are performed quite separately from the sewing process, the latter being particularly amenable to international subcontracting. The clothing industry itself produces an enormous variety of often rapidly changing products. A particularly important distinction is between mass-produced staple garments on the one hand and fashion garments on the other. Finally, although not part of the production sequence itself, the role of the distributors of garments – particularly the retailers – is of considerable and growing importance. The increasing dominance of much retail trade by very large firms has enormous implications for the organization and the geography of clothing manufacture.

Global shifts in the textiles and clothing industries

Textiles manufacture

Figure 8.2 maps the world distribution of textiles manufacturing employment in 1987. Although widely spread geographically, some clear concentrations stand out. China and the USSR, followed by India, have the world's largest employment in textiles manufacture. Textiles employment in the United States in 1987 was 831,000, that in Japan around 670,000. Most of the other major textiles activity is in Europe, both West and East, and in Asia. Since the 1960s the pattern of change in textiles production has varied enormously between individual nations. In general, the industry grew most rapidly in the developing market economies and in the economies of Eastern Europe and either stagnated or declined in the developed economies.

These broad trends are shown very clearly in Table 8.1. In the world as a whole textiles *production* was 6 per cent higher in 1987 than in 1980. In the developed market economies textiles production grew very little overall, although production growth in North America was considerable (12 per cent higher in 1987 than in 1980). Conversely, the position of the EC was much weaker. Within the EC, the best overall performance between 1978 and 1987 was achieved by Italy with an annual increase in textiles production of 2.5 per cent. France, Germany and the Netherlands all performed poorly but there was a very marked improvement in textiles production in the United Kingdom. Japan, once a real leader in world textiles production, continued to decline.

The major growth of textiles production in the world economy during the 1980s took place in the developing market economies and, to a lesser extent, in the centrally planned economies of Eastern Europe. In the developing market economies, textiles production was 18 per cent higher in 1987 than in 1980 and had been growing relatively rapidly since 1972. But, with some exceptions, the rate of growth of textiles production in the developing countries has slackened markedly (in this sense, the textiles sector in these countries has been growing more slowly than the clothing sector, as we shall see).

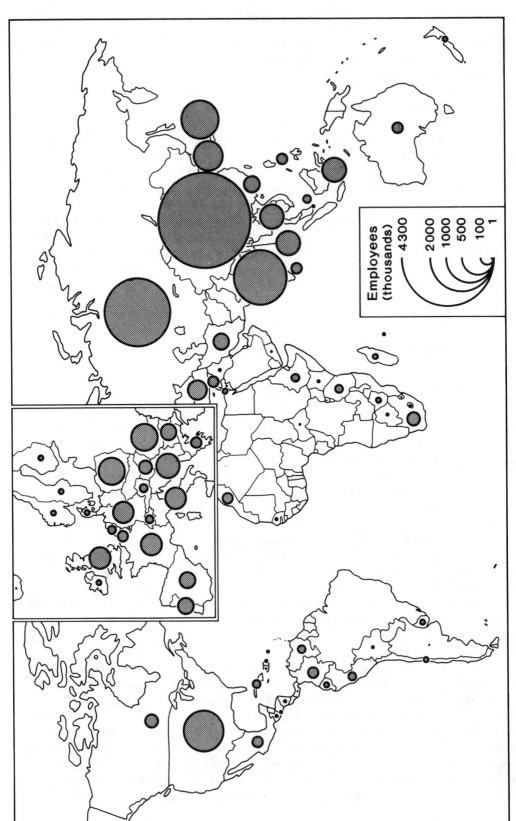

Figure 8.2 The global distribution of employment in textiles manufacture, 1987
(*Source*: United Nations (1989) *Industrial Statistics Yearbook, 1987*, New York: United Nations)

Table 8.1 Changes in production in textiles manufacturing, 1972–87

	1972	(1980 = 100) 1978	1983	1987
Developed market economies	99	98	95	101
North America	95	99	102	112
European Community	105	97	92	100
Japan	—	100	97	92
Developing market economies	79	96	104	118
Asia	75	98	112	123
India	—	100	94	114
South Korea	—	81	119	157
Philippines	—	70	—	256
Indonesia	—	92	—	113
Latin America				
Chile	—	104	—	130
Colombia	—	108	—	125
Peru	—	98	—	140
Venezuela	—	93	—	331
Mediterranean				
Cyprus	—	93	—	102
Morocco	—	113	—	143
Portugal	—	78	—	124
Centrally planned economies	71	97	102	107
World	86	97	99	106

— No data available

(*Source*: United Nations, *Industrial Statistics Yearbook*, various issues)

Among the Eastern European economies, the best performers over the 1978–87 period were Romania, Bulgaria and Czechoslovakia.

Substantial global shifts have occurred in textiles production in the last few decades, with the decline or stagnation of traditionally dominant producers and the emergence of new centres of production. Changes in the pattern of *international trade* in textiles reflect these shifts in production, although not in an identical manner. For example, despite the undoubted growth of textiles manufacturing in the Third World and the relative decline as textiles producers of the older industrialized nations, the two leading textiles-exporting countries are Germany and Italy. As Table 8.2 shows, nine of the fifteen leading textiles-exporting countries are older industrialized countries (including Japan). The fifteen countries as a group generated 78 per cent of world textile exports in 1989. Comparison of the 1980 and 1989 rankings, and of the annual growth rates, indicates substantial volatility in the relative importance of the leading textiles exporters. The major gainers of export shares were the East Asian countries (China, Hong Kong, South Korea, Taiwan) – although not Japan, whose share of world textiles exports declined from 9 per cent in 1980 to 5.5 per cent in 1989.

World textiles exports are dominated by Western European countries on the one hand, which accounted for 46 per cent in 1989, and the five East Asian countries, which generated 32 per cent of world exports. Figure 8.3 shows how the textiles exports from these two major groups of countries were distributed geographically. The most striking feature is that almost four-fifths of Western Europe's textiles exports are intra-regional, the rest being spread fairly evenly. For the five East Asian producers, 51 per cent of

Table 8.2 The world's leading textiles-exporting countries, 1989

Rank	Exporter	Share of world textiles exports (%) 1980	1989	Average annual growth (%) 1980–8	1988	1989
1	West Germany	11.5	11.5	6.5	9.0	5.0
2	Italy	7.5	8.0	7.5	2.5	6.5
3	Hong Kong*	3.0	7.5	17.5	12.5	19.0
4	China	4.5	7.0	12.0	11.5	8.5
5	Japan	9.0	5.5	1.0	−1.5	0.0
6	Taiwan	3.0	5.5	12.5	11.0	19.5
7	South Korea	4.0	5.5	10.5	19.0	11.0
8	Belgium – Luxembourg	6.5	5.5	4.5	8.0	6.5
9	France	6.0	5.0	4.0	9.5	7.5
10	United States	7.0	4.5	0.5	24.5	12.0
11	United Kingdom	5.5	3.5	1.5	14.0	4.5
12	Netherlands	4.0	2.5	4.0	4.0	21.5
13	Switzerland	2.5	2.0	4.5	3.5	−3.5
14	Pakistan	1.5	2.0	9.5	−2.5	13.0
15	India[†]	2.0	2.0	5.5	14.5	—
Above 15 countries		77.5	77.5			

* Includes substantial re-exports
[†] 1978
— No data available

(*Source*: GATT (1990) *International Trade, 1989–1990*, Geneva: GATT, Vol. II, Table IV. 48)

their exports were with one another, 17 per cent were to other Asian countries, 9 per cent went to North America and 9 per cent to Western Europe. In this respect, therefore, the East Asian exporters are more widely involved in world textiles trade than are the other major producers.

Western Europe

$45.5 bn

East Asian suppliers
(China, Hong Kong, Japan, South Korea, Taiwan)
$31.4 bn

Figure 8.3 Geographical distribution of textiles exports from Western Europe and the five leading East Asian suppliers (*Source*: GATT (1990) *International Trade, 1989–1990*, Geneva; GATT, Vol. II, Table IV.47)

Table 8.3 Exports, imports and trade balances in textiles, 1989

	Developed economies				Developing economies		
	Exports	Imports	Balance		Exports	Imports	Balance
	($ million)				($ million)		
Australia	128	1,578	−1,450	China	6,994	2,845	+4,149
Belgium−Luxembourg	5,297	2,921	+2,376	Hong Kong*	7,574	9,218	−1,644
Canada	602	2,358	−1,756	Indonesia	828	511	+317
France	4,967	6,151	−1,184	South Korea	5,371	1,835	+3,536
Germany	11,073	9,017	+2,056	Taiwan	5,442	952	+4,490
Italy	7,966	5,235	+2,731				
Japan	5,534	4,347	+1,187				
United Kingdom	3,605	6,171	−2,566				
United States	4,370	6,417	−2,047				

* Includes substantial re-exports and imports for re-exports

(*Source*: GATT (1990) *International Trade, 1989–1990*, Geneva: GATT, Vol. II, Tables IV. 49, IV. 50)

Thus, world trade in textiles is still dominated by the developed market economies. More than half of all world textiles exports originates from them compared with approximately a third from the developing market economies and 11 per cent from the 'Eastern Trading Area'. In terms of external trade in textiles the EC had a positive balance with each major world group, $2.7 billion in 1988. The largest volume of textiles exports went to EFTA and to the developing countries. In contrast to the EC, in 1989 the United States had a substantial trade deficit in textiles, of $2 billion. Like the EC, Japan also had a trade surplus in textiles but one that was substantially lower, at $1.2 billion. In fact Japan's textiles trade surplus has been falling steadily; in 1982 it stood at $3.5 billion (in 1982 prices). By far the biggest contributor to Japan's textiles surplus is its exports to developing countries. Within Europe, as Table 8.3 shows, there was considerable variation in both the direction and the size of trade balances. Italy, West Germany and Belgium−Luxembourg all had large positive balances; the United Kingdom and France had substantial trade deficits in textiles. Outside Europe, both Australia and Canada were large net importers of textile products.

Among developing countries, the most spectacular export performance has been by China. Although China has long had the largest textiles labour force in the world, its production was, until recently, mostly for its own domestic market. Within a decade, however, China has emerged as a leading textiles exporter in the developing world and fourth in the world as a whole after Germany, Italy and Hong Kong. As Table 8.3 shows, Taiwan has a huge textiles trade surplus (almost as large as the combined surplus of Germany and Italy). China and South Korea are not far behind, both with surpluses exceeding $3.5 billion. Only Hong Kong of the leading Asian textiles producers has a trade deficit.

Clothing manufacture

Like textiles, the manufacture of clothing is very widely spread geographically, as Figure 8.4 shows. The largest concentration of employment is in the USSR followed, a long way behind, by the United States and then Japan. Clothing manufacture remains extremely important in Western Europe – the United Kingdom, West Germany, France and Italy each had between 160,000 and 230,000 employed in the industry in 1987. Poland and Czechoslovakia are the leading clothing producers in the centrally

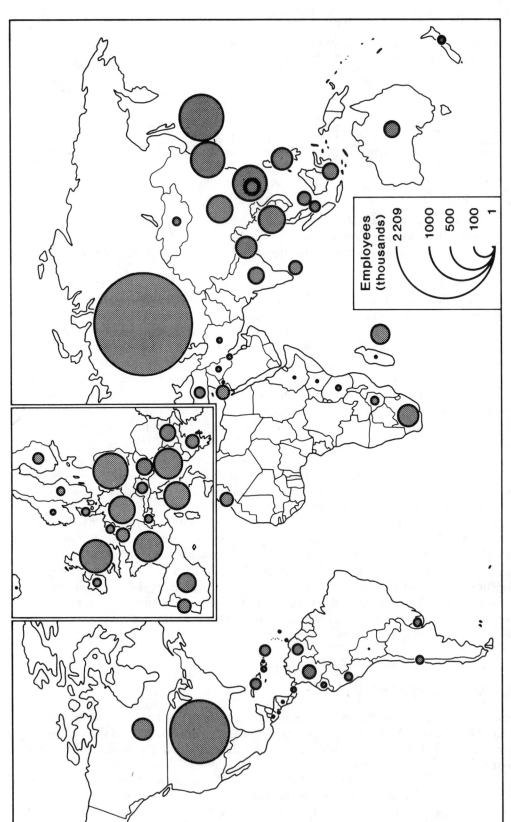

Figure 8.4 The global distribution of employment in clothing manufacture, 1987
(*Source:* United Nations (1989) *Industrial Statistics Yearbook, 1987,* New York: United Nations)

planned economies of Eastern Europe. Among developing countries, Hong Kong is by far the most significant clothing centre. In 1987, 282,000 were employed there in clothing, a very substantial proportion of the colony's total labour force and comparable with the very much larger European economies. Clothing employment in South Korea was around 267,000, with a further 92,000 in the Philippines.

The current geographical distribution of clothing *manufacture* is the result of the substantial global shifts that occurred between the 1960s and the 1970s. The major changes are summarized in Table 8.4. At the world scale clothing production increased at the same rate as textiles manufacture. In the case of the clothing industry, however, the developed economies fared far worse than in textiles. Expressed in relation to a 1980 production level of 100, clothing production in the developed economies in 1987 had fallen to an index of 94. Within the EC, Germany's production index fell to 77, the Netherlands' to 75 and France's to 90. Conversely, the United Kingdom did relatively very well, with its production index rising to 111. In North America, Canada did better than the United States between 1980 and 1987. Australia's clothing production in 1987 was 19 per cent lower than in 1980.

In comparison, clothing production increased very substantially in the developing countries as a whole (a 1987 index of 124). Particularly strong growth of production volumes occurred in South Korea, the Philippines, Singapore, India and Malaysia in Asia; in Venezuela and the Dominican Republic in Latin America, and in Cyprus and Israel in the Mediterranean region. The most rapid growth in clothing production in Eastern Europe occurred in Romania, Poland and the USSR. Overall, the 1987 index of production in the centrally planned economies was 115.

Table 8.4 Changes in production in clothing manufacture, 1972–87

	(1980 = 100)			
	1972	1978	1983	1987
Developed market economies	101	104	94	94
North America	92	105	91	94
European Community	112	105	96	94
Japan	—	102	99	98
Developing market economies	68	92	106	124
Asia	57	96	126	155
India	—	104	126	123
South Korea	—	102	139	206
Singapore	—	77	96	139
Malaysia	—	98	—	170
Philippines	—	54	—	367
Latin America				
Venezuela	—	87	—	214
Dominican Republic	—	70	—	156
Mediterranean				
Cyprus	—	89	—	136
Israel	—	96	—	129
Centrally planned economies	67	92	105	115
World	83	98	99	105

— No data available

(*Source*: United Nations, *Industrial Statistics Yearbook*, various issues)

Table 8.5 The world's leading clothing-exporting countries, 1989

Rank	Exporter	Share of world clothing exports (%)		Average annual growth (%)		
		1980	1989	1980–8	1988	1989
1	Hong Kong*	12.0	14.5	11.5	10.0	18.5
2	Italy	11.0	9.5	9.0	1.0	4.0
3	South Korea	7.0	9.5	14.5	17.5	4.5
4	China	4.0	6.5	14.0	30.0	26.0
5	West Germany	7.0	5.5	8.0	7.5	5.0
6	Taiwan	6.0	5.0	8.5	−5.5	0.5
7	France	5.5	3.5	4.5	8.0	10.0
8	Turkey	0.5	3.0	43.5	7.0	18.0
9	Portugal	1.5	2.5	17.5	11.5	12.5
10	United Kingdom	4.5	2.5	3.5	7.5	−6.0
11	Thailand	0.5	2.5	27.0	21.5	28.0
12	United States	3.0	2.5	3.5	36.0	34.5
13	India	1.5	2.0	13.0	7.5	24.5
14	Netherlands	2.0	1.5	7.0	9.5	4.0
15	Greece	1.0	1.5	15.0	−16.5	22.5
Above 15 countries		67.0	72.0			

* Includes substantial re-exports

(*Source*: GATT (1990) *International Trade, 1989–1990*, Geneva: GATT, Vol. II, Table VI. 55)

Compared with textiles there is a greater dispersion of *clothing-exporting countries*. A comparison of Tables 8.2 and 8.5 shows that the leading fifteen clothing-exporting countries accounted for 72 per cent of the world total, whereas in textiles the top fifteen accounted for 77.5 per cent. Although there are elements in common between the two tables, there are also some very substantial differences. Hong Kong is clearly the world's leading exporter of clothing, followed by Italy and South Korea. Four of the fifteen leading clothing exporters are not on the textiles list at all: Turkey, Portugal, Thailand and Greece are all very important clothing exporters. On the other hand, Japan is no longer one of the world's leading clothing exporters although it is the fifth most important textiles exporter.

In fact, the four East Asian clothing suppliers (China, Hong Kong, South Korea, Taiwan) export almost exactly the same quantity of clothing products as do both Western Europe and North America put together. In 1989 the East Asian four exported $36.9 billion of clothing products compared with Western Europe's $35.2 billion and North America's $2.5 billion. Figure 8.5 shows the geographical distribution of clothing exports from Western Europe and from the four East Asian countries. As in the case of textiles (Figure 8.3), around four-fifths of Western European clothing exports go to other countries within Europe and a mere 7 and 5 per cent go to North America and Asia respectively. Only 15 per cent of the clothing exports of the East Asian four are internal to the group with a further 19 per cent going to other Asian countries. A huge 40 per cent of the four's clothing exports go to North America and a further 19 per cent to Western Europe.

A particularly sensitive issue in the politics of international trade has long been the pattern of trade surpluses and deficits in clothing. Table 8.6 shows the situation in 1989. It is very different from that in textiles (Table 8.3). Apart from Italy every one of the leading developed economies has a big trade deficit in clothing. Most obvious is the

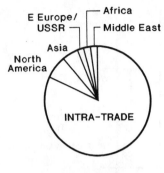

Western Europe
$35.2 bn

East Asian suppliers
(China, Hong Kong,
South Korea, Taiwan)
$36.9 bn

Figure 8.5 Geographical distribution of clothing exports from Western Europe and the four leading East Asian suppliers (*Source*: GATT (1990) *International Trade 1989–1990*, Geneva: GATT, Vol. II, Table IV.54)

enormous US deficit of $23.8 billion, almost half the total clothing deficit of the eleven developed countries listed. West Germany's deficit is also very large, at about 20 per cent of the group total. In contrast, all the developing countries listed had a trade surplus in clothing (recall that Hong Kong has a trade deficit in textiles).

It is interesting to explore the major sources of clothing imports for some of the leading industrialized countries. Figure 8.6 shows some quite significant differences between countries in the relative importance of developing and Western European countries of origin. The United States, Canada and Japan are overwhelmingly dependent on clothing imports from developing countries. This is especially true of the United States, where more than three-quarters of clothing imports are from developing countries. The United Kingdom is almost equally dependent upon developing country

Table 8.6 Exports, imports and trade balances in clothing, 1989

	Developed economies				Developing economies		
	Exports	Imports	Balance		Exports	Imports	Balance
	($ million)				($ million)		
Belgium–Luxembourg	1,467	2,693	−1,226	Greece	1,474	306	+1,168
Canada	318	2,180	−1,862	Hong Kong*	13,994	5,700	+8,294
France	3,626	6,406	−2,780	Malaysia	608	43	+565
Germany	5,632	14,639	−9,007	Portugal	2,585	255	+2,330
Italy	9,449	2,028	+7,421	Singapore	1,393	682	+711
Japan	565	8,972	−8,407	Taiwan	4,735	216	+4,519
Netherlands	1,572	3,700	−2,128				
Sweden	329	2,076	−1,747				
Switzerland	510	2,800	−2,290				
United Kingdom	2,363	5,799	−3,436				
United States	2,211	26,026	−23,815				

*Includes substantial re-exports and imports for re-export

(*Source*: GATT (1990) *International Trade, 1989–1990*, Geneva: GATT, Vol. II, Tables IV. 56, IV. 57)

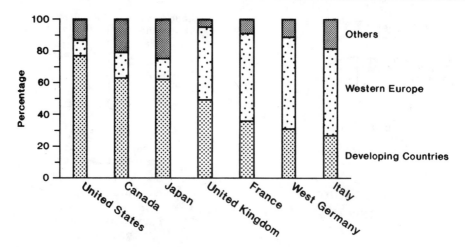

Figure 8.6 Major sources of clothing imports for leading industrialized countries, 1989 (*Source*: GATT (1990) *International Trade 1989–1990*, Geneva: GATT, Vol. II, Tables A6–A13)

and Western European clothing imports while the other three leading European countries derive more than half of their clothing exports from within Western Europe and only between a quarter and a third from the developing countries.

Thus, it is in the clothing sector that the greatest degree of penetration of developed country markets has occurred. But this should not obscure the fact that a large share of the world's clothing trade still occurs between the industrialized countries themselves. There is a very high degree of *interpenetration* of developed country markets, especially within the EC. A particularly interesting aspect of international trade in clothing is the existence of several strong regional patterns. For example, a growing proportion of clothing imports into the United States is from Mexico and from the Caribbean. Between 1982 and 1987 imports into the United States of clothing products from the Caribbean increased by 189 per cent. An increasing share of the EC's clothing imports originates from the Mediterranean rim, including some North African countries; much of Japan's clothing trade is with neighbouring Asian countries. Such propinquitous trade patterns reflect a number of influences, not least geographical distance but also the operation of corporate and political actions, as we shall see below.

Summary

Quite clearly, the global pattern of production and trade in the textiles and clothing industries has undergone dramatic, though very uneven, change. In general, the trend is one of relative decline in the older established centres of production in Europe especially, and a rapid growth in a number of developing countries. So far the developed market economies are succeeding in holding on to a larger proportion of their textiles trade than of their clothing trade but there are substantial differences between one nation and another.

Overall, therefore, recent trends suggest some divergence between the textiles and clothing sectors. Although some developing countries now occupy very significant – even dominant – positions in both sectors they are most significant in clothing production and exports. Among the industrialized countries, only Italy manages to retain a strong positive trade balance in both sectors and is the only industrialized country to buck the general trend. At least some of the industrialized countries have, however, succeeded in maintaining a substantial presence in textiles especially.

A development sequence

To understand the reasons underlying these differential developments we need to examine the major processes involved in these industries. Table 8.7 suggests a sequence of development through which individual producing countries appear to have passed. Six stages are shown, beginning with the embryonic stage typical of the least developed countries through to the maturity and decline of the older industrial countries. The table shows the type of production likely to be characteristic of each stage and the related kinds of trade. It also gives examples of countries which are characteristic of each stage. The sequence is a very useful summary of what has happened so far but, like all such sequential models, it should not be regarded necessarily as being predictive of what will happen in the future – as charting the inevitable course of individual

Table 8.7 A sequence of development in the textiles and clothing industries

	Type of production	Trade characteristics	Examples of countries
Stage 1	Simple fabrics and garments manufactured from natural fibres.	Production oriented to domestic market. Net importers of fibre, fabric and clothing.	Least developed.
Stage 2	Production of clothing for export. Mostly standard items or those requiring elaborate 'craft' techniques.	Export of clothing to developed country markets on basis of low price.	Less advanced Asian, African, Latin American.
Stage 3	Increase in quantity, quality and sophistication of domestic fabric production. Expansion of clothing sector with upgrading of quality. Development of domestic fibre manufacturing.	Much increased international involvement in export of fabric, clothing and even of synthetic fibres.	More advanced ASEAN and Eastern European. China starting to enter this stage.
Stage 4	Further development and sophistication of fibre, fabric and clothing production.	Full-scale participation in international trading system. Substantial trade surpluses.	Taiwan, South Korea, Hong Kong.
Stage 5	Output of textiles and clothing continues to increase but employment declines. Increased capital intensity and specialization.	Facing increased international competition.	Japan, United States, Italy.
Stage 6	Substantial reduction in employment and number of production units. Decline both relative and , in some sectors, absolute.	Severe problems of competition. Substantial trade deficits.	United Kingdom, West Germany, France, Belgium, Netherlands.

(*Source*: based in part on Toyne *et al.* (1984), pp. 20–1)

countries. Although many countries have passed through some or all of these stages, the precise path of development depends upon a number of factors which, together, produce the specific geographical patterns of the textiles and clothing industries.

The changing pattern of demand

Demand is a fundamental influence on the level and location of the textiles and clothing industries. Each sector in the textiles–clothing production sequence shown in Figure 8.1 relates to rather different markets. However, since some 50 per cent of all textiles production goes to the clothing industry, the major influence on the demand for textiles is the demand for clothing.

Demand for clothing is perhaps the easiest to identify since it is final consumer demand. The major general determinant is the level and distribution of personal income. Since personal incomes are so very unevenly distributed (see Figure 5.2) it is the affluent parts of the world which largely determine the level and the nature of the demand for clothing. The generally low incomes in developing countries clearly restrict the size of their domestic clothing markets and this has undoubtedly had an important influence on their adoption of export-oriented policies. But the relationship with income is not total: beyond a certain level of income, demand for clothing tends not to increase at the same rate as incomes increase. In other words, clothing tends to have an elasticity of demand of less than one. As personal incomes rise a relatively smaller proportion is spent on basic clothing.

In the clothing and textiles industries one of the most important ways of offsetting this 'natural' tendency for demand to rise less than proportionally to income is to stimulate more frequent purchasing through *fashion change*. Only the innocent would regard fashion as the spontaneous expression of consumer demand. Fashion – and the resulting demand – are largely induced, not least by the clothing and textiles manufacturers and, especially, the retailers themselves.[4] Fashion change has been especially important to the textiles and clothing industries of the developed market economies: 'Textile activity in high-income countries, and trade between them, has been sustained in recent decades by fashion changes. The opportunities offered by this kind of product differentiation have been crucial to the survival of Western European textiles' (Shepherd, 1983, p. 31). Enormous efforts (and expenditures) have gone into promoting fashion products and creating 'designer' labels. Indeed, one of the most interesting developments in the clothing industry during the past few years has been the incredible proliferation of designer-labelled garments. Such a practice covers a very broad spectrum of consumer income levels from the exceptionally expensive to the relatively cheap. Designer labelling is basically a device to *differentiate* what are often relatively similar products. Nevertheless, in evaluating the influence of demand on the clothing and textiles industries we need to distinguish between fashion goods on the one hand and non-fashion, or staple, goods on the other.

The growing power of the retailing chains

Within the clothing industry, in particular, demand is becoming increasingly dominated by the purchasing policies of the major multiple retailing chains. A smaller and smaller percentage of clothing sales is being channelled through independent retailers. The trend in virtually all industrialized countries has been for retailing in general to become increasingly concentrated into a smaller number of very large retail chains. This is especially the case in clothing sales. In the United States Sears-Roebuck, K-Mart, J. C. Penney and Montgomery Ward accounted for a very large proportion of clothing sales in the late 1980s as did Daiei, Mitsukoshi, Daimaru and Ito Yokado in Japan. In Germany Karstadt, Kaufhof, Schickendanz; in the Netherlands C & A; in France Carrefour; in Britain Marks and Spencer dominated. However, the extent of multiple-retailer dominance over the clothing market varies a good deal from one country to another.

In the early 1980s in Europe such dominance was exceptionally high in the United Kingdom, where almost three-quarters of clothing sales were channelled through the multiple retailers. In 1986 the four leading clothing retailers were responsible for almost 30 per cent of total clothing sales in the United Kingdom (Elson, 1990). The nature of clothing retailing changed substantially during the 1980s as retailers began to focus on segmented markets, such as particular age and income groups. For example,

> one of the dominant areas of retail attention has been the womenswear market for the 25–40 age group. Fashion, design and quality, or at least a quality 'look', have become the source of competition between retailers. The undifferentiated mass market for clothing, dominated by the major retailers, has disintegrated and a 'new wave' of retail chains has grown up associated with this growth of designer-led quality fashion.
>
> (Gibbs, 1988, p. 1223)

These kinds of development, which are occurring to a greater or lesser degree in all developed market economies, have very profound implications for textiles and clothing manufacturers. The highly concentrated purchasing power of the large retail chains gives them enormous leverage over textiles and clothing manufacturers. When the market was dominated largely by the mass-market retailers, the demand was for long production runs of standardized garments at low cost. As the market has become more differentiated and more frequent fashion changes have become the rule, manufacturers are having to respond far more rapidly to retailer demands and specifications. Under such circumstances, the *time* involved in meeting orders becomes as important as the cost. This basic shift in the structure of marketing is having repercussions throughout the textiles and clothing industries and influencing both the adoption of new technologies and also corporate strategies. Although relatively few retail chains are themselves manufacturers of clothing, they are very heavily involved in subcontracting arrangements.

Production costs and technology

The production characteristics of each sector of the textiles–clothing production sequence vary considerably (Figure 8.7), although much depends on *where* production takes place. A process which is relatively capital-intensive in one country may be relatively labour-intensive in another, depending upon relative factor endowments. Similarly, there is considerable variation in the size of production units between one country and another. In clothing manufacture capital intensity is generally low, labour intensity is high, the average plant size is small and the technology relatively unsophisticated.

Production characteristics	Fibres (synthetic)	Textiles	Clothing
Capital intensity	High ←――――――――――――→		Low
Labour intensity	Low ←――――――――――――→		High
Material costs	High ←――――――――――――→		Medium
Average size of production unit	Large ←――――――――――――→		Small
Technology	Sophisticated ←――――――――――――→		Simple

Figure 8.7 Variations in production characteristics between major components of the textiles–clothing production sequence

Variations in labour costs

From the viewpoint of the changing global distribution of these industries the most important consideration is the extent to which the individual production factors are geographically variable in either their cost or their availability. For textiles and clothing manufacture there is no doubt that labour costs are the most significant production factor. Not only are textiles and clothing two of the most labour-intensive industries in modern economies but also labour costs are the most geographically variable of the production costs of the industries. Figure 8.8 shows just how wide the hourly labour cost gap is between different countries. Hourly labour costs in spinning and weaving are shown in US dollar equivalents. The spread is enormous, from $14.60 per hour in Switzerland to less than 25 cents per hour in Indonesia. All the high-cost producers are in Europe, North America and Japan.

In fact, Japanese labour costs have been increasing very rapidly. In 1987 Japan ranked tenth on the hourly labour cost scale; in 1989 fifth. Similarly, some of the leading newly industrializing countries are becoming more expensive producers as their relative labour costs have increased. The difficulty of competing in labour cost terms with other producers was a major reason for the large-scale shift of the US textile industry from New England and of the garment industry from New York to the Southern states. However, in terms of unit labour costs, which allow for levels of productivity, the United States is in a far better position *vis-à-vis* its industrialized competitors. But even allowing for productivity differences the developing countries

Figure 8.8 International variations in labour costs in the textiles industry (*Source*: Werner International, Inc.)

have an enormous labour cost advantage over the developed market economies, particularly in the production of clothing.

The major advantage of low-labour-cost producers lies in the production of *staple* items of clothing, which sell largely on the basis of price, rather than in fashion garments in which style is more important. The difference between the two is one of *rate of product turnover*. Fashion clothing has a rapid rate of turnover which tends to reflect the idiosyncracies of particular markets. Proximity to such markets is important and this helps to explain the survival of many developed country clothing manufacturers:

> The disadvantage of distant producers diminishes, however, for those products with cycles extending beyond a season. Style goods producers thus face a more steeply declining distance function, due to time constraints, despite the fact that their products are less sensitive to cost competition . . . It is the speed and greater reliability of delivery which become more crucial than the cost element of distance, although where wages are particularly low the cost differential may permit volume producers of style wear to distribute finished garments by air . . . [but] . . . the reason why some less developed countries can compete so readily in the distant developed market economies is not only a matter of lower costs but also their specialisation on products with lower rates of turnover or on those attaining or approaching the mature stage of their cycle . . . Low product turnover, or product continuity, eases fabric purchasing, inventory control, work scheduling, and costing. It makes it more feasible to establish longer-term links with wholesale importers as well as retail buyers and to generate larger orders for more distant delivery periods.
>
> (Steed, 1981, pp. 288, 289)

Technological change

Both the cost of production and the speed of response to changes in demand are greatly influenced by the technologies used. Technological innovation may reduce the time involved in the manufacturing process and make possible an increased level of output with the same size – or even smaller – labour force. As international competition has intensified in the textiles and clothing industries the search for new, labour-saving technologies has increased, especially among developed country producers. However, the potential for such innovation varies very considerably between different manufacturing operations within the production sequence.[5] Two kinds of technological change are especially important: those which increase the speed with which a particular process can be carried out and those which replace manual with mechanized and automated operation. On both counts, technological change has been far more extensive in the textiles manufacturing sector than in the manufacture of clothing.

Technological changes in textiles production   One of the most important technological innovations in the spinning of yarn was the introduction of open-ended spinning, which combines what were formerly three separate processes into a single process using rotors instead of spindles. Spinning speeds have been increased at least fourfold and labour requirements reduced by approximately 40 per cent. Textile weaving has been revolutionized by the introduction of the shuttleless loom in which the shuttle is replaced by a variety of alternative devices (for example 'rapier' looms, 'projectile' looms, 'waterjet' and 'air jet' looms). Again, the major result has been a spectacular increase in the speed of the weaving process and this, together with wider loom capacity, has raised productivity by a huge factor. Parallel developments in knitting technology and in finishing – the latter now being especially highly automated – have contributed further to the increased speed of textiles operations and the consequent reduction in the number of workers needed.

Technological changes in clothing production   In contrast, technological developments in the clothing industry have been far less extensive. There was relatively little change in clothing technology between the industry's initial emergence in the late nineteenth century and the early 1970s; the basic sewing machine was not so very different from that in use fifty years earlier. The manufacture of clothing remains a complex of related *manual* operations, especially in those goods in which production runs are short. The problem is that, 'with the exception of the simplest articles . . . a garment is still a relatively complicated product for its price. The materials from which it is made, being limp and relatively delicate, do not lend themselves readily to mechanical handling. Many thicknesses of cloth may be cut out at a time, but each garment has to be put together individually' (Plant, 1981, p. 63).

Most of the recent technological developments in the industry, including those based on micro-electronic technology, have been in the non-sewing operations: grading, laying out and cutting material in the pre-assembly stage and warehouse management and distribution in the post-assembly stage (Mody and Wheeler, 1987). The application of electronically controlled technology to these operations can achieve enormous savings on materials wastage and greatly increase the speed of the process. Hoffman and Rush (1988) point out that the grading process may be reduced from four days to one hour; computer-controlled cutting can reduce the time taken to cut out a suit from one hour to four minutes. But these developments do not reach the core of the problem. The sewing and assembly of garments account for four-fifths of all labour costs in clothing manufacture. So far only very limited success has been achieved in mechanizing and automating the sewing process itself.

Current technological developments in the manufacture of clothing, according to a detailed study by Mody and Wheeler (1987), are focused on three areas. One is to increase the *flexibility* of machines: robots are being designed which can recognize oddly shaped pieces of material, pick the pieces up in a systematic manner and align the pieces on the machine correctly whilst also being able to sense the need to make adjustments during the sewing process itself. A second set of innovations addresses the problem of *sequential operations*, particularly the difficulty of transferring semi-finished garments from one workstation to the next while retaining the shape of the limp material. A third technology being developed is the *unit production system*, which will deliver individual pieces of work to the operator on a conveyor belt system. This will greatly reduce the large amount of (wasted) production time spent by the operator on unbundling and rebundling work pieces. The handling process has been estimated to take up to 60 per cent of the operator's total time.

The drive to introduce new technologies in textiles and clothing has been stimulated mainly by the need for developed country producers to be cost-competitive in the face of the very low labour costs in developing countries. 'In both textile and garment production, the advanced technologies . . . are still so costly that they are optimal only in high labour cost environments' (Mody and Wheeler, 1987, p. 1276). But cost reduction is not the only benefit derived from the new technologies. At least as important, if not more so, are the *time savings* that result from automated manufacture. This has two major benefits: first, speeding up the production cycle reduces the cost of working capital by increasing the velocity of its use; second, it becomes possible for the manufacturer to respond more quickly to customer demand. This is especially important in an environment in which there is strong retail dominance and where the retailer increasingly pushes for frequent deliveries in order to avoid holding too much stock.

In those cases where the manufacturer also operates its own retail operations, either directly or indirectly, the electronic point-of-sale (EPOS) technologies permit a direct, real-time link, between sales, reordering and production. The classic example of this is the Italian clothing manufacturer, Benetton:

> Micro-electronics technology is vital to the Benetton operation . . . At the Italian headquarters is a computer that is linked to an electronic cash register in every Benetton shop; those which are far away, like Tokyo and Washington are linked via satellite. Every outlet transmits detailed information on sales daily, and production is continuously and flexibly adjusted to meet the preferences revealed in the market. Benetton produces entirely to the orders received from the shops.
>
> (Elson, 1989b, p. 103)

The jeans manufacturer Levi Strauss also operates a computerized system both to keep track of the work in progress in each factory (and payments to workers on the basis of completed work) and to control the overall volume of production. According to a company executive, 'at the moment in the US we trigger manufacturing in units of 10,000 . . . The idea is that eventually whenever someone buys a pair of jeans it will automatically trigger the production of another pair' (quoted in the *Financial Times*, 6 June 1990).

Industry structures

The conventional way of measuring the structure of an industry is to calculate its *concentration ratio*, which is simply the percentage of total production of employment in the industry accounted for by a given number of firms (usually five or ten). Unfortunately, there are no really up-to-date statistics which allow us to do this beyond the early 1980s. Since then there have been substantial acquisitions and mergers, particularly in textiles, and the textiles sector is more highly concentrated – that is, more dominated by large firms – than the clothing sector:

> [Although] the bulk of the fabric is produced by a small number of large firms . . . the fabric sector . . . has a dualistic nature: thousands of small- and medium-sized firms each producing a limited range of products and accounting for a comparatively small percentage of total output, and a much smaller number of large firms each producing a wider variety of products and, as a group, accounting for a disproportionate share of total output.
>
> (Toyne *et al.*, 1984, p. 14)

The 'world textile oligopoly'

In the world as a whole, some thirty-five to forty textiles corporations are especially important. These form what Clairmonte and Cavanagh (1983) term the 'world textile oligopoly': the group mainly responsible for reshaping the global textiles industry. Table 8.8 lists the world's major textiles companies in 1986, divided into three size bands (the leading twenty, forty and fifty companies) and arranged by country of

Table 8.8 The world's leading textiles companies by country of origin, 1986

World size band	United States	United Kingdom	Japan	Italy	France	West Germany
	Burlington Industries	Coats Viyella	Kanebo		Prouvost	
	West Point	Courtaulds	Toyobo		DMC	
	Pepperell		Nisshin			
	J. P. Stevens		Gunze			
	Spring Mills		Kurabo			
	Collins &					
	Aikman					
	Fieldcrest Mills					
	Bemis					
	United					
	Merchants					
	Tyco					
	Laboratories					
20						
	Mohasco	Tootal	Nitto Boseki	Benetton	Boussac-St-	Freudenberg
	Industries	Dawson	Nagai Co.	Marzotto	Frères	
	Shaw		Daiwa Spinning			
	Industries		Fuji			
	Guilford Mills		Fukusuke			
	Russell		Kawashina			
			Orimono			
40						
	Salem Carpet		Nippon Keori	Lanerossi		KBC
	Mills		Omikenshi	Gruppo		Textil-
			Atsugi	Tessile		Grüppe
			Shin-ei	Mirogli		Hof
			Shikibo			
50						

(*Source*: based on Anson and Simpson (1988), Table 2.11)

origin. Nine of the leading twenty companies in 1986 originated from the United States, two from the United Kingdom, five from Japan and two from France. Neither Italy nor West Germany had firms in the top twenty. But the degree of large-firm dominance varies enormously between different countries. The United Kingdom has by far the most concentrated textiles industry. In 1980 five firms – Courtaulds, Coats Patons, Tootal, Carrington Viyella and Vantona – accounted for more than half of all textiles employment in the country. Today, as the result of mergers, the industry is dominated by only three large firms.

Within the rest of Europe, France and the Netherlands have roughly a quarter of their textiles activity in their leading five firms. In both the United Kingdom and France concentration increased dramatically between 1965 and the early 1980s, largely through acquisitions and mergers. In comparison, the West German and Italian textiles industries are far less dominated by large firms; in both cases, medium and smaller-sized firms are the major feature. Italy's textiles industry, in particular, has retained a strongly small-firm character. In the United States the leading five firms – of which Burlington Industries, West Point Pepperell and J. P. Stevens were the largest in the mid-1980s – employed roughly 20 per cent of the nation's textiles workforce. However, there were some dramatic changes among the leading US textiles companies in the late 1980s. Burlington Industries, the world's largest textiles corporation, was involved in a leveraged buy-out to escape being taken over. The resulting rationalization and restructuring of the company reduced its size considerably. West Point Pepperell was also taken over. The least concentrated of all the major textiles producers is Japan; even so, large firms are extremely significant, for example Kanebo, Toyobo, Nisshin, Gunze and Kurabo.

The fragmented clothing industry

Of all the major parts of the textiles–clothing production sequence the manufacture of clothing is by far the most fragmented and least dominated by large firms. In part this is explained by the relatively low technological sophistication of the clothing process and the low barriers to entry to the industry. In part it reflects the vagaries of the market for clothing, which restrict long production runs and high-volume production to staple items. Even in this archetypal small-firm industry, however, large firms are becoming increasingly important: only they can afford to invest in the new technologies, to build a worldwide brand image based on mass advertising. Thus, although the clothing industry of most countries is made up of a myriad of very small firms, many of which operate as subcontractors, there is an undoubted trend towards increased concentration.

This is clearly seen in the United States, 'where concentration has been steadily rising since the early 1970s . . . the ten largest clothing firms now account for over 20% of total domestic production. Over the last five years, these firms have grown much faster than the sector as a whole' (Hoffman, 1985, p. 381). In the United States the giant clothing manufacturers include both the specialist manufacturers, such as Levi Strauss and VF Corporation, and also giant industrial conglomerates such as Gulf & Western, General Mills and Interco. In the United Kingdom a similar dualistic structure exists. The dominant firms tend to be part of vertically integrated groups, either textiles firms such as Courtaulds, Tootal and Coats Viyella, or major retailers.

Government policies towards the textiles and clothing industries[6]

In developing economies textiles and clothing manufacture have occupied a key position in national industrialization strategies. Hence, the kinds of import-substitution and export-oriented measures outlined in Chapter 6 have been applied extensively to the encouragement of the industries in most developing countries. But it is in the older-established producing countries of Europe, North America and more recently Japan, faced with increasingly severe competition from low-cost or more efficient producers, that government intervention has been especially marked. For 'no government has been prepared, for strategic if not employment considerations, to allow the industries to disappear completely' (Edwards, 1982, p. 87).

The policies adopted by governments have been of two kinds:

1. those aimed at encouraging the *restructuring and rationalization* of the country's textiles and clothing industries;
2. those aimed at *protecting* the domestic industries from outside competition through the use of *trade restrictions*.

Let us look at each of these in turn.

Policies to encourage restructuring

A major policy pursued by a number of developed country governments – most notably in Europe – has been the encouragement of restructuring in their textiles and clothing industries through the use of subsidies and investment assistance programmes. For example, in the United Kingdom as early as the 1950s the textiles industry once again found itself faced with severe competition from low-cost producers in Asia. In addition to the imposition of 'voluntary' trade restraints, the United Kingdom government implemented its Cotton Industry Act, whose aim was to remove half of the industry's capacity and to stimulate re-equipment and modernization in the surviving firms. Firms were encouraged to engage in such 'voluntary euthanasia' by generous financial compensation for scrapped capacity and grants for re-equipment. Even though the 50 per cent scrapping target was broadly achieved the problem of overcapacity did not disappear and the subsequent decades have seen a continuation of rationalization within the industry.

Large-scale financial assistance to the textiles and clothing industries is common throughout Europe:

> with one exception, all the major EEC governments give subsidies to their textile and clothing industries. Some of these are massive ... One estimate is that in Italy and the Netherlands subsidies totalled $302 and $300 a head respectively in the clothing industries in the mid-1970s, while in Britain the figure was $200 and in France and in Belgium it was about $50. The exception is Germany, which gave subsidies of only about $2 a head, mostly through regional assistance.
>
> (Farrands, 1982, p. 93)

Government involvement in restructuring domestic textiles and clothing industries has not been confined solely to the older producing nations, however. From the late 1960s the Japanese government has actively sought to restructure its textiles industry by reducing capacity, modernizing plant and moving into higher-value products. Even the newer producers such as South Korea and Taiwan have found it necessary to intervene to maintain their competitive position through the encouragement of new investment and the scrapping of old capacity. In both cases, the hand of the government has been firmly involved.

Policies aimed at regulating trade

A rather different kind of national policy which has had important effects on the global geography of the textiles and clothing industries is the use of offshore assembly provisions. Such provisions, part of a country's customs and excise regulations, have been especially common in the United States and West Germany. As explained in Chapter 6, offshore assembly provisions permit a company to export materials, have them made up into garments in another country (invariably one with low labour costs) and then re-import them into the company's domestic market paying duty only on the value added in offshore processing. Some industrialized countries have also negotiated preferential access to their domestic markets with some developing countries. For example, the United States has a special agreement with the Caribbean Basin countries and the European Community has preferential agreements with certain Mediterranean countries.[7]

The international regulatory framework: the Multi-Fibre Arrangement   Individual national policies to stimulate domestic textiles and clothing industries or to facilitate their
rationalization and restructuring have been extremely important in helping to reshape
the industry globally. Of far greater significance, however, is the international regulatory
framework within which the industries have operated for the past two decades. The
textiles and clothing industries are unique in that they are the only industries to which
special international trade regulations apply, under the Multi-Fibre Arrangement
(MFA).[8] Today, a large proportion of all world trade in textiles and clothing is covered
by this agreement. Its provisions and their implementation – and avoidance – have
been a major factor in the changing global pattern of production and trade. Although
protectionism in these industries is not solely a postwar phenomenon, the origins of its
modern variant, the MFA, are to be found in the problems which faced developed
country producers, particularly the United States and the United Kingdom, in the
1950s.

Faced with a massive inflow of low-price imports from Japan, Hong Kong and some
other Asian producers, both the United States and the United Kingdom negotiated
separate 'voluntary' agreements with the Asian exporters to restrict imports for a
limited period. By 1962 such arrangements had become broadened into the Long-Term
Arrangement (LTA), within GATT, which regulated international trade in cotton
textiles. The aim of the LTA was to encourage orderly development of the
international cotton textiles market to allow the developed countries to restructure
their cotton textiles industry. It allowed an importing country to limit shipments from
any source in any cotton textile category which would 'cause or threaten to cause
disruption in the market of the importing country'.

The LTA allowed for a gradual increase in imports from developing country
signatories of 5 per cent a year. It remained in force for eleven years. During that time,
however, the world picture became far more complex. First, there was the massive
growth of man-made fibres which were not covered by the LTA. Second, an increasing
number of developing countries emerged as important exporters of textiles and,
especially, of clothing. The precipitous decline of the industries in the developed
economies continued. In 1973 a much broader trade agreement, which included the
European countries and also covered man-made and other non-cotton fibres, was
negotiated. This was the first Multi-Fibre Arrangement. By Article 1(2), its principal
aim was:

> to achieve the expansion of trade, the reduction of barriers to such a trade and the
> progressive liberalization of world trade in textiles products, while at the same time ensuring
> the orderly and equitable development of this trade and avoidance of disruptive effects in
> individual markets and on individual lines of production in both importing and exporting
> countries.

Like the LTA, the MFA was initially negotiated for a limited period: four years from
January 1974. Like the LTA, too, its aim was to create an 'orderly' development of
trade in textiles and clothing which would benefit *both* developed and developing
countries. Access to developed country markets was to increase at an annual average
rate of 6 per cent, though this was far below the 15 per cent sought by the developing
countries. At the same time, the developed countries were to have safeguards to
protect the 'disruption' of their domestic markets. Within the MFA, individual quotas
were negotiated which set precise limits on the quantity of textiles and clothing

products which could be exported from one country to another. For every single product a quota was specified. When that quota was reached no further imports were permitted.

In practice it has been the disruptive, rather than the liberalizing, aspect which has been at the forefront of trading relationships. Since 1974 the MFA has been renegotiated three times and the current MEA IV runs until the end of 1992. In general, the MFA has become more, rather than less, restrictive. Both the EC and the United States have negotiated much tighter import quotas on a bilateral basis with most of the leading developing country exporters, and in several cases have also invoked anti-dumping procedures.

The effects of the MFA on world trade in textiles and clothing have been immense:

> The MFA has become the key to trade flows of textiles and apparel and has had a particularly significant impact on the growth of textile and clothing exports from the NICs. The data show a marked slowing in the growth of clothing exports from developing countries, principally the NICs, after the MFA came into effect in the mid-1970s. From 1963 to 1976, the real growth in exports of clothing was about 21 percent per annum, whereas real growth from 1976 to 1978 was about 5 percent per annum. The impact of the MFA is unmistakable, since 1976–78 was a period of general economic recovery.
>
> (McMullen, 1982, p. 91)

A major initial beneficiary of this decline in the relative growth of developing country exports was the United States, which greatly increased its penetration of EC textiles and clothing markets during the 1970s. During the early 1980s, however, it was the European producers which greatly increased their penetration of the United States market. Between 1982 and 1984, for example, 'by far the largest increase in the physical volume of apparel imports came from the uncontrolled suppliers – the industrial countries. The top five among them increased their exports to the United States by 191% in these two years' (Cline, 1987, p. 182).

An inevitable consequence of the increased restrictiveness on developing country exports of textiles and clothing has been a parallel increase in efforts to circumvent the restrictions. Such evasive action may take a variety of forms. One method is for a producing country which has reached its quota ceiling in one product to switch to another item. Another – illegal – method is to use false labelling to change the apparent country of origin. Yet another is for firms to relocate some of their production to countries which either are not signatories to the MFA or whose quota is not fully used by indigenous producers. There is no doubt that the Multi-Fibre Arrangement has greatly influenced global trade and investment in textiles and clothing. There have been winners and losers from what is, in many respects, a distortion of the GATT principles. Expressed in rather graphic terms, 'the MFA embodies some clear violations of the principles of the GATT . . . Indeed, that the Textiles Committee is located within the GATT building is as if a brothel were operated inside a cathedral' (Curzon *et al.*, 1981, p. 32).

A major task of the Uruguay Round of the GATT negotiations in the late 1980s was to integrate the MFA into the GATT. As we noted in Chapter 6, the basic principle at the heart of GATT rules on international trade is the most favoured nation principle, which implies non-discrimination between all trading partners. The MFA clearly contravenes this principle. From the standpoint of the industrialized countries, 'the MFA by its structure discriminates among supplier countries, most notably between industrial countries, which enjoy free access, and developing countries (and Japan), which face controls' (Cline, 1987, p. 2321). On the other hand, the developing countries are indicted by the industrialized countries for restricting access to their domestic markets through both very high tariffs and non-tariff barriers.

When the Uruguay Round negotiations broke down in December 1990 (ostensibly because of dispute over agricultural subsidies) the MFA issue had not been finally resolved. Among the plethora of competing proposals two were most widely discussed:

1. The replacement of the MFA by a system of global quotas which would cover all countries, including the industrialized countries. This was the preferred position of the United States which, as we saw earlier, has experienced major import penetration not only from developing countries but also from Europe.
2. The gradual phasing out of MFA quotas over a period of years so that they would eventually disappear. This was the European Community's position but it also contained stringent restrictions which were strongly opposed by developing country producers.

Within virtually all the industialized countries both producer and labour lobbies have been pressuring their governments to continue to protect their domestic textiles and clothing industries, not least because of the fear of continuing job losses in the industries. But whether protection is the best way forward for developed countries is open to dispute. It certainly does not help developing countries for which the textiles and clothing industries are an extremely important element in their economic development strategies.

Corporate strategies in the textiles and clothing industries[9]

Some general observations

As we have seen in this chapter, the manufacture of textiles and clothing is geographically very widespread. They are relatively rare instances of globally significant industries which are extensively present in many developing countries, rather than in just a few. Vast numbers of, mostly small, developing country firms are involved in textiles and clothing production. Nevertheless, the *globalization* of these industries has been driven primarily by developed country firms. Indeed, it is paradoxical that a significant proportion of the textiles, and especially the clothing, imports, which are the focus of such concern in developed countries are, in fact, organized by the international activities of those very countries' *own* firms. But the processes and strategies involved are both complex and dynamic.

In examining the development of these strategies two basic points need to be made. First, the globalization of the textiles and clothing industries cannot be explained simply as a relocation of production from developed to developing countries in search of low labour costs. As Elson (1988) points out, other factors are involved including, in particular, orientation to specific markets. Second, where firms have internationalized their production operations they have used a variety of methods, notably international subcontracting and licensing, which do not necessarily involve equity participation. We discussed such strategies in general terms in Chapter 7. Foreign direct investment has been relatively unimportant in textiles and clothing although it does, of course, occur.

Major factors in how and where the various modes of international involvement have been used are the existence of the Multi-Fibre Arrangement, with its complex system of national quotas, and the volatility of currency exchange rates:

> Exchange rate volatility clearly complicates decisions about where to source. What was a competitive location for production when the pound was worth almost $1 may not be competitive when the pound is worth $1.75 . . . Ironically the very factors that enable firms to contemplate internationalization of their operations also contribute to exchange rate volatility. Firms are able to internationalize production because of ease of communications and deregulation of capital movements. The same factors are at the root of exchange rate volatility.
>
> (Elson, 1990, p. 57)

Technological innovations in production and distribution processes are also part of the strategic equation in so far as they influence both costs of producing textiles and clothing and also the speed of response to changing customer demands. These, in turn, are heavily affected by the dominance of the major retail chains.

Within this multiplicity of influences on the strategic behaviour of textiles and clothing firms it is the need for *flexibility* which is increasingly the dominant consideration. Whether a firm pursues a strategy of cost leadership or product differentiation and whatever its focus (product or geographic) it must be able to respond quickly and flexibly to changing circumstances. Consequently, the internationalization strategies of textiles and clothing companies can best be understood in these terms. In this respect, Oman's (1989) study of 'new forms of investment' (NFI) in these industries is especially useful. The following section draws heavily on this, in looking first at internationalization strategies in the textiles industry and, second, at developments in the clothing industry.

Corporate strategies in the textiles industry

Although the British and European textiles companies have a lengthy history of international involvement in textiles production, the first really major wave of such activity occurred in the early 1960s and was led by the Japanese textiles firms and the general trading companies (the *sogo shosha*). The Japanese textiles firms were already strongly vertically integrated in Japan itself and operated a dualistic-hierarchical network of domestic subcontractors. The *sogo shosha* were responsible for organizing a huge proportion of Japanese imports and exports and already had an intricate international distribution system. When the United States introduced the LTA in 1962 (see above, p. 255) to protect its domestic market from Japanese cotton textiles imports, it triggered the first surge of overseas involvement by Japanese firms. To avoid the problem of quotas both the textiles firms and the *sogo shosha* set up international subcontracting links in other East and South East Asian countries. Very often, as in Japan itself, the principal firms took a small equity share in the subcontractors.

Relatively quickly, therefore, the Japanese textile industry became an international vertically integrated operation which incorporated an extensive network of local Asian producers. According to Oman (1989), by 1980 Japan's nine leading *sogo shosha* were involved in 150 textiles ventures outside Japan. The leading vertically integrated textiles companies (Toray, Asahi, Teijin, Toyo, Mitsubishi) were also involved in a vast array of international operations, primarily in Asia but with some in Latin America and Africa. Some of these took the form of direct investment but most were international subcontracting arrangements.

Compared with the Japanese companies, US and European textiles firms are less internationalized. In the United States the tendency has been to increase the degree of domestic concentration in the industry through acquisition and merger and to upgrade domestic productivity through heavy investment in new technology. In Europe, too, these strategies have been pursued, although some of the very large textiles companies have also become increasingly internationalized. Major examples are the three leading British companies: Courtaulds, Coats Viyella and Tootal. These three companies have been engaged in massive rationalization and restructuring programmes throughout the 1970s and 1980s (Elson, 1989a). Such programmes have involved a varying mixture of: product rationalization and focus; technological innovation to reduce costs and increase flexibility; reduction of production capacity particularly in the United Kingdom but also overseas; and the use of international subcontracting, licensing and other forms of relationship with local firms in developed and developing countries.

As a result, all three companies have reduced their UK labour force by huge amounts: Courtauld's from 124,000 in 1975 to 46,000 in 1986; Tootal's from 20,000 in 1974 to 8,500 in 1984. They also reduced the number of workers they employed directly overseas as they moved to more indirect operations using subcontractors and licensees. Tootal, in particular, is in the process of transforming itself from a textiles and clothing manufacturer to a 'worldwide marketing, distribution and sourcing business'. It remains the second largest producer of thread in the world, however, operating a globally rationalized production and marketing network. 'China has now become the linch-pin of the company's international thread marketing operations ... eventually 80% of Tootal's requirement for the Pacific Basin will be sourced in China' (Elson, 1989a, p. 198). Although Tootal's Chinese joint ventures have certainly been influenced by low labour costs, they also reflect the desire to be close to the major textiles *markets* of East and South East Asia.

Corporate strategies in the clothing industry

International subcontracting, licensing and other forms of non-equity international investment have been even more pervasive and influential in the clothing sector than in textiles. In some cases – such as the Japanese and British firms discussed in the previous section – clothing is part of a firm's vertically integrated activities. But in clothing the influence of the major international retail chains and buying groups is especially pronounced. As we have seen, they exert enormous purchasing power and leverage over clothing manufacturers. Although the production and retailing of clothing may be fragmented in individual markets, international buying operations are highly concentrated:

> Unlike clothing production, where barriers to entry are low, in the international marketing of clothing, buying groups use their capacities as a basis for undertaking non-equity NFI in developing countries' clothing industries. International subcontracting is by far the most important of these NFI, although licensing of trademarks and brand names is of growing importance.
>
> (Oman, 1989, pp. 227–8)

Again, as in the case of textiles, it was Japanese firms that initiated extensive use of international subcontracting in clothing. During the 1960s and 1970s Japanese companies established subcontracting arrangements in Hong Kong, Taiwan, South Korea and Singapore. Their production was mostly exported to the United States and not to Japan's own domestic market. Again, the Japanese *sogo shosha* were at the leading edge of these international subcontracting developments, often using minority investments in local firms. Probably 90 per cent of Japanese overseas clothing operations are still located in East and South East Asia and most were set up in the 1970s. Overall, however, the Japanese clothing industry is far less 'internationalized' than its textiles industry. The opposite is true of the US and European clothing industries, which are far more internationalized than their textiles industries.

The strategies developed by US clothing firms to cope with the intensified competition of the 1970s and 1980s were 'two-pronged'. 'On the one hand, companies concentrated on product and marketing strategies on the leading edge of the fashion market . . . On the other hand, manufacturers focused on investments to cut costs and raise productivity. As a complement, the larger US apparel firms significantly increased offshore processing via subcontracting arrangements with producers in developing countries' (Oman, 1989, p. 228). The geographical pattern of this activity has changed over time; an initial emphasis on Mexico and the Philippines has been superseded by increasing subcontracting activity in the Caribbean region. Some of the larger US clothing companies are also involved in direct investment in overseas operations through wholly or majority-owned subsidiaries. For the most part, such plants are oriented towards local markets rather than third-country exports and are mainly in developed countries.

Some of the most interesting strategic mixes are used by those clothing firms which compete in mass markets but on the basis of brand names supported by extensive advertising. The best examples are probably the jeans manufacturers, Levi Strauss and VF Corporation. Even by the late 1970s Levi Strauss was spending $50 million a year on worldwide advertising. It has made various attempts at product diversification but has recently reverted to a 'core product' strategy. Its current problem is how to adapt to the demographic change which is altering the size of its traditional market segment of fifteen- to twenty-four-year-olds without moving too far from its core product, the denim jean.

Levi Strauss's international production strategy has been to develop its own branch factories in Europe, Latin America and Asia, although it now also has licensing agreements. Levi Strauss is the market leader in this sector; its nearest rival is VF Corporation, which manufactures another heavily branded product, Wrangler Jeans, which it acquired from Blue Bell in 1987. According to Oman, Blue Bell's international strategy was the classic one of the market follower. Instead of developing an international network of directly owned manufacturing plants, Blue Bell opted for the less exposed and less risky strategy of licensing its product to independent manufacturers in more than forty countries.

The adoption of offshore production strategies by European clothing firms has been most pronounced among German and British companies. German clothing firms have been especially heavily involved in international subcontracting arrangements, as the detailed survey by Fröbel, Heinrichs and Kreye (1980) revealed. They calculated that in the 1970s around 70 per cent of all German clothing firms, including some quite small ones, were involved in some kind of offshore production. Roughly 45 per cent of the arrangements involved international subcontracting and a further 40 per cent involved varying degrees of equity involvement by German firms in local partners. German firms tended to concentrate their offshore activities in Eastern Europe but with a growing presence in countries of North Africa and the Mediterranean (notably

Greece, Malta and Tunisia). Overall, Fröbel, Heinrichs and Kreye calculated that more than four-fifths of all West German clothing imports manufactured under subcontracting arrangements came from East Germany, Poland, Hungary, Romania, Bulgaria and Yugoslavia.

Similarly, the three leading British clothing companies – Courtaulds, Coats Viyella and Tootal – have developed a mixture of international strategies. Courtaulds has established clothing factories in Portugal, Morocco and Tunisia. Tootal has developed perhaps the most extensive international network although, as noted earlier, the company has been reducing its own direct production activity:

> The internationalization of the company has not, however, resulted in the relocating of any of its clothing factories abroad . . . Tootal does not own any clothing factories in developing countries. It does not design and cut out clothing in the UK, send it abroad for making up, and re-import it for sale in the UK. Instead of setting up new factories abroad, Tootal has pursued a two-pronged strategy of commercial subcontracting, both internationally and in the UK; and of investment in new technology, often in new factories, for its own in-house production in the UK.
>
> (Elson, 1989b, p. 99)

Of all the European clothing producers the Italians have developed most differently. We noted earlier that Italy is the only major European country whose clothing industry has continued to perform well in the teeth of intensive global competition. We noted, too, the innovative methods used by the Italian company Benetton to link together its production and distribution networks. Benetton has a highly sophisticated and complex global sales and distribution network but it produces almost its entire output in Italy. Its production system is a mixture of directly owned and operated factories and a large network of independent Italian subcontractors. In general, the Italian producers have pursued a strategy of product specialization and fashion orientation with the aim of avoiding dependence upon those types of good most strongly affected by low-cost competition.

In the case of the Italian industry this involves mainly small firms in a decentralized production system. 'Decentralised production has occurred in those areas of textiles where the fashion element (hence the need for greater risk and flexibility) is important. Italy's unique strength in Western Europe is to have created a kind of price-competitive mass-market in fashion, where certain products often enjoy an area "trademark" (Como silk ties, Prato wool fabrics, and so on)' (Shepherd, 1983, p. 42). More recently, however, some Italian firms have established international licensing agreements for production of high-fashion and designer-label products:

> [These] are mostly with producers located in the most advanced developing countries in Asia and Latin America, as well as in the United States and Japan, where demand for Italy's high fashion and branded apparel is growing. The Italian firm usually does not take an active part in the management of the franchise, but demands minimum local advertising expenditures and places restrictions on the diffusion of designs and trademark and on exports.
>
> (Oman, 1989, p. 230)

Strategies of developing country firms

The entire emphasis in this discussion of international corporate strategies in the textiles and clothing industries has been on firms from the industrialized countries. But a growing number of producers from the leading Asian NICs, in particular, are themselves increasingly involved in a variety of internationalization strategies. Firms in Hong Kong, Singapore, South Korea and Taiwan face increasing competitive pressure from the newer wave of Asian producers (notably China, but also Malaysia, Thailand and Indonesia) as well as restrictions on their trade with North America and Europe through the MFA. In fact Hong Kong firms began to shift clothing production to other Asian countries as early as the mid-1960s. With the tightening grip of the MFA in the 1970s, Hong Kong firms set up plants in the Philippines, Thailand, Malaysia and Mauritius and, subsequently, in Indonesia and Sri Lanka to get round quota restrictions. In the past decade, a huge number of investments have been made in China. East Asian firms have also begun to establish plants in Europe and North America (including the Caribbean) to serve developed country markets directly. Significantly, also, a specific emphasis on product specialization and on higher-value and fashion goods is now being adopted by leading manufacturers in the more advanced of the developing country producers, particularly those in Hong Kong, Singapore, South Korea and Taiwan.

Corporate strategies in textiles and clothing: a summary

The strategies adopted by textiles and clothing producers, therefore, are extremely varied and complex. The combinations of technological innovation, different types of internationalization strategy, the relationship with retailers and the constraints of the Multi-Fibre Arrangement combine to produce a more complex global map of production and trade than a simple explanation based on labour-cost differences would suggest. Although firms in these industries do engage in foreign direct investment this is a far less significant practice than other forms of international involvement, especially international subcontracting and licensing arrangements. These are the dominant forms of international production in these industries, particularly in the manufacture of clothing. In this industry, the influence of transnational corporations tends to be more indirect than direct and the involvement of local capital and entrepreneurship greater than in many other industries. The manufacture of clothing is an ideal candidate for international subcontracting. It is highly labour-intensive in the developed countries, it uses low-skill or easily trained labour, the process can be fragmented and geographically separated, with design and often cutting being performed in one location (usually a developed country) and sewing and garment assembly in another location (usually a developing country).

Although international subcontracting in clothing manufacture knows no geographical bounds – with designs and fabrics flowing from the United States and Europe to the far corners of Asia and finished garments flowing in the opposite direction – there are some strong geographical biases in the relationships. For example, most of the clothing manufactured in the Caribbean region and in Central America is organized by and for the US market and operates within the US government's tariff provisions on offshore assembly.

Preferential access to the EC market, together with geographical proximity, have also been important in the development of offshore clothing production in the Mediterranean and North African countries. For example, the clothing industries of countries such as Malta and Tunisia both depend heavily on their links with the EC. Some 40 per cent of Malta's clothing exports go to the European Community. A detailed study of the growth of Malta's clothing industry reveals the importance of the dual factors of preference and proximity:

> Other countries have as much unskilled labour available at as low wages as in Malta, so why should Malta be a favoured location for the clothing industry? From the direction of Malta's exports it is clear that this favouritism applies only to exports in Europe, and not to other high-wage large markets. Distance favours Malta over other low-wage locations. Preferences also favour Malta . . . It is difficult to separate out the impact of distance and preferences, but together they give an edge to producers located in Malta over producers located in countries where wages are lower or labour more used to a factory regime . . . The direction in which the Maltese clothing industry has developed, from gloves as the leading export in the early 1960s to jeans as the dominant product in the 1970s, has been from an easily marketable good to one where brand names enjoy global prestige. Thus the Maltese clothing industry is dominated by foreign-owned firms or by joint ventures in which the Maltese partner handles labour relations and other production matters while the foreign partner takes care of input purchasing and output selling.
>
> (Pomfret, 1982, pp. 245, 246)

Jobs in the textiles and clothing industries

Job losses in the older centres of production

Within the developed market economies the past thirty years have witnessed a massive decline in employment in the textiles and clothing industries (Figure 8.9). Altogether, the five leading EC countries lost almost one and a half million jobs in textiles and around three-quarters of a million jobs in clothing between 1963 and 1987. As Figure 8.9 shows, the early 1970s were a watershed for textiles and clothing employment in most of the leading industrialized countries. The biggest losses were experienced in the United Kingdom, where a total of 780,000 jobs disappeared in textiles and clothing between 1963 and 1987, followed by West Germany (−564,000) and France

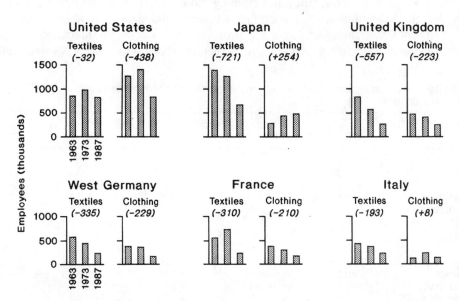

Figure 8.9 Employment trends in the textiles and clothing industries of leading industrialized countries (*Source*: United Nations, *Industrial Statistics Yearbook*, various issues)

(−520,000). Italy lost almost 200,000 jobs in textiles but gained 8,000 in clothing. Almost 500,000 jobs disappeared in the United States textiles and clothing industries.

The popular view – and, indeed, the political view as expressed through such measures as the Multi-Fibre Arrangement – is that these job losses have been caused by the wholesale geographical shift of production to cheap-labour locations in the Third World and to the resulting high levels of import penetration of the domestic market. There is no doubt at all that such imports have adversely affected employment in the textiles and clothing industries of the industrialized economies. But it is misleading to attribute all – or even, in some cases, most – of the blame directly to this single cause.

Employment change within a nation's industry is the result of the operation of several interrelated forces. The most important are (a) changes in domestic demand, (b) changes in productivity (output per worker), (c) changes in exports and imports. Table 8.9 shows the contribution of each component to employment change in the US textiles and clothing industries in the 1962–85 period. Quite clearly the effect of imports on employment change in textiles has been negligible compared with the effect of productivity growth and the growth of domestic demand. The picture in clothing was a little less clear. Particularly in the 1982–5 period the effect of imports was almost as great as that of productivity growth. But, as Cline points out, this was an unusual period of a heavily over-valued dollar which made imports cheaper. Allowing for this, he estimates that 'the broad range of employment erosion attributable to import growth would be in the vicinity of 1 percent annually' (Cline, 1987, p. 96). Earlier studies of the components of employment change in the United Kingdom and West Germany during the 1970s reached broadly similar conclusions: that the biggest source of employment loss in the textiles and clothing industries was productivity growth.

These are, of course, *aggregate* figures and it is quite likely that import penetration has a more significant employment effect in some types of textiles and clothing product than others. But there is a further complication in trying to calculate the employment effects of imports in the textiles and clothing industries. The method of calculation is based upon the assumption that each factor operates independently of the others. This is clearly not the case. At least some of the increase in productivity – through technological change in the process and organization of production – is stimulated by

Table 8.9 Components of employment change in the textiles and clothing industries of the United States, 1962–85

| | Percentage growth rates | | | | |
	1962–7	1967–72	1972–7	1977–82	1982–5
Textiles					
Consumption	9.07	4.12	1.44	−0.56	3.03
Exports	0.17	0.10	0.28	0.09	−0.51
Imports	−0.41	−0.30	0.21	−0.17	−0.65
Productivity	−7.28	−3.86	−3.19	−2.70	−4.27
Total	1.54	0.05	−1.25	−3.34	−2.40
Clothing					
Consumption	4.52	1.61	3.31	1.86	5.99
Exports	0.04	0.03	0.20	0.15	−0.44
Imports	−0.31	−0.58	−0.95	−0.98	−3.25
Productivity	−2.10	−1.04	−3.05	−2.22	−3.79
Total	2.15	0.02	−0.50	−1.20	−1.49

(*Source*: Cline (1987), Table 4.3)

the pressures of competition from overseas, particularly from lower-cost producers. Much of the competition has come from other developed countries, however, and only in certain types of textiles and clothing is the direct impact of developing country imports the major influence on employment loss.

Characteristics of the labour force and conditions of work

Whatever the precise cause of employment loss in the textiles and clothing industries of the developed economies its actual impact is extremely uneven, not only geographically but also in terms of the kinds of worker affected. Textiles and clothing manufacture tends to be strongly concentrated in particular locations within individual countries. In the United Kingdom four regions – the North West, Yorkshire and Humberside, the East Midlands and Scotland – contain most of the country's textiles and clothing employment. In the United States the major concentration is in the Southern states (the industry having migrated on a large scale from the North East in the past few decades). In Canada the province of Quebec is the major location.

In Western Europe, Lorraine in France, the Wallonian region of Belgium and the Mezzogiorno of Italy have a large proportion of their employment in the industries. The manufacture of clothing has also tended to be an inner-city industry, clustering into congested areas in old buildings: the garment districts of Manhattan, London, Manchester and Montreal typify such tight geographical concentration. Hence, the large-scale employment losses in the industries have hit specific areas very hard indeed. In many cases, these are areas with very limited alternative employment opportunities.

The problem is made more acute by the nature of the labour force itself. Some four-fifths of the workers in the clothing industry and more than half of those in textiles manufacture are female. A substantial proportion of the labour force is relatively unskilled or semi-skilled, with no easily transferable skills. The specific socio-cultural role of women, in particular their family and domestic responsibilities, also makes them relatively immobile geographically. A further characteristic of the textiles and clothing workforce in the older-industrialized countries is that a large number tend to be immigrants or members of ethnic minority groups. This is a continuation of a very long tradition, especially in the clothing industry. The early industry of New York, London, Manchester and Leeds was a major focus for Jewish immigrants. Subsequent migrants have also seen the industry as a key point of entry into the labour market. The participation of Italians and Eastern Europeans in both the United States and the United Kingdom has been followed more recently by the large-scale employment of blacks and Hispanics in the United States and by non-white Commonwealth immigrants in the United Kingdom. All of these 'sensitive' segments of the labour market experience real problems of alternative employment when their jobs in textiles and clothing disappear.

In the rapidly growing textiles and clothing industries of the developing countries the labour force is similarly distinctive. Employment in the industries tends to be spatially concentrated in the large, burgeoning cities and in the export-processing zones. The labour force is overwhelmingly female and predominantly young.[10] Many workers are first-generation factory workers employed on extremely low wages and for very long hours: a seven-day week and a twelve- to fourteen-hour day are not uncommon. Employment in the clothing industry in particular tends to fluctuate very markedly in response to variations in demand. Hence, a high proportion of outworkers are employed; women working as machinists or hand-sewers at home on low, piecework, rates of pay. Such workers are easily hired and fired and have no protection over their working conditions. Many are employed in contravention of government employment regulations. Yet there is no shortage of candidates for jobs in the fast-growing clothing industries of the developing countries. Factory employment is often regarded as preferable to un- or underemployment in a poverty-stricken rural environment. A factory job does provide otherwise unattainable income and some degree of individual freedom. Often the wages earned are a crucial part of the family's income and there is much family pressure on young daughters to seek work in the city factories or in the EPZs.

The conditions of employment in many (not necessarily all) textiles and clothing factories in the developing countries hark back to those found in the nineteenth century in Europe and North America. The clothing industry, in particular, has a long history as an industry of 'sweated' labour. This tradition continues in many developing countries today:

> Workers hesitate to stay away from their work even when sick, due to the fear of losing their employment. Illness is quite widespread, however, not least due to insufficient ventilation, high workshop temperatures, overcrowded workrooms, and inadequate safety provisions. Complaints about health hazards are reported, such as fatigue and backstrain caused by working long hours at machines in the clothing industry. Poor lighting and demanding work often cause eyestrain, while certain materials cause skin allergies. Although the risk of industrial accidents in the clothing industry is lower than in some others, health risks due to overcrowding in poorly ventilated rooms are the main causes of widespread premature invalidity
>
> (Robert, 1983, pp. 31–2)

At least in the textiles industries of the developed economies such conditions are now relatively rare; factory and employment legislation have seen to this. But the sweatshop has certainly not disappeared from the clothing industry of the big cities of North America and Europe. The highly fragmented and often transitory nature of much of the industry makes the regulation of such establishments extremely difficult. Some argue that there has been a major resurgence of clothing sweatshops in some big Western cities. In New York City, for example,

> where ten years ago there were fewer than 200 garment factory sweatshops, there are now between 3,000 and 4,500 sweatshops in New York ... They employ between 50,000 and 70,000 persons and a large portion of their employees are illegal migrants from the Caribbean, Latin America, and Oriental countries ... There exists within New York, the global city, a substantial growing segment of the labour force whose conditions of production resemble those of the labour force in the Third World. This segment cannot be considered a mere aberration ... sweatshop labour is a necessary condition of global competition. Sweatshops in New York are the logical consequence of the globalisation of production in the garment industry and the consequent competition for jobs between segments of the global reserve of labour.
>
> (Ross and Trachte, 1983, p. 416)

Thus, despite the enormous global shifts which have occurred in the textiles and clothing industries and their rapid and widespread growth in Third World countries, these industries continue to be significant and sensitive sectors in the older-industrialized countries.

Notes for further reading

1. Comprehensive studies of these industries are provided by Clairmonte and Cavanagh (1981), Toyne *et al.* (1984), Cline (1987), Anson and Simpson (1988).
2. The irony of this is that 'the cotton revolution . . . .began by imitating Indian industry, went on to take revenge by catching up with it, and finally outstripped it. The aim was to produce fabrics of comparable quality at cheaper prices. The only way to do so was to introduce machines – which alone could effectively compete with Indian textile workers' (Braudel, 1984, p. 572).
3. For a detailed discussion of each stage in the sequence, see Clairmonte and Cavanagh (1981), Toyne *at al.* (1984).
4. The increasingly significant role of the retailers is discussed by Clairmonte and Cavanagh (1981), Gibbs (1988), Elson (1990), Peck (1990).
5. Detailed discussions of technological change in the textile and clothing industries can be found in Clairmonte and Cavanagh (1981), Toyne *et al.* (1984), Hoffman (1985), Mody and Wheeler (1987), Gibbs (1988), Hoffman and Rush (1988).
6. National policies in these industries are discussed by Renshaw (1981), Shepherd (1983), Toyne *et al.* (1984), Cline (1987).
7. The Caribbean Basin case is discussed by Steele (1988) and the EC's trade relationship with the Mediterranean countries is discussed by Ashoff (1983), Hamill (1989).
8. Details of the MFA and of the development and effects of international trade restrictions in the textiles and clothing industries can be found in Ashoff (1983), Cline (1987).
9. Excellent discussions of corporate strategies in the textiles and clothing industries are provided by Elson (1989a,b, 1990) and by Oman (1989), who provides a very detailed analysis of new forms of international investment in the industries.
10. Studies of female employment in the textiles and clothing industries of both developed and developing countries include Fuentes and Ehrenreich (1983), Robert (1983), Elson and Pearson (1989), Fernandez-Kelly (1989).

Chapter 9

'Wheels of Change': The Automobile Industry

Introduction[1]

The automobile industry was the key manufacturing industry for most of the middle decades of the twentieth century. Peter Drucker (1946, p. 149) captured its significance at that time: 'the automobile industry stands for modern industry all over the globe. It is to the twentieth century what the Lancashire cotton mills were to the early nineteenth century: the industry of industries.' The internal combustion engine was, quite literally, the major engine of growth for most of the developed market economies until the middle 1970s. The significance of the industry lay not only in its sheer scale but also in its immense spin-off effects through its linkages with numerous other industries. The motor vehicle industry came to be regarded as a vital ingredient in national economic development strategies.

For a while it seemed that the industry was maturing and ageing gracefully and that it had lost its propulsive influence on the industrialized economies. Not so. In the opinion of the researchers on the MIT International Motor Vehicle Program, 'the auto industry is even more important to us than it appears. Twice in this century it has changed our most fundamental ideas of how we make things. And how we make things dictates not only how we work but what we buy, how we think, and the way we live' (Womack, Jones and Roos, 1990, p. 11). Some 4 million workers are employed directly in the manufacture of automobiles throughout the world and a further 10 million in the manufacture of materials and components. If we add the numbers involved in selling and servicing the vehicles, we reach a total of 20 million – not far short of the entire population of Canada.

Organizationally the automobile industry is one of the most global of all manufacturing industries. It is an industry of giant corporations, many of which are increasingly organizing their activities on internationally integrated lines. In contrast to the textiles and clothing industries, for example, the world automobile industry is predominantly an industry of transnational corporations: 'There are few industries in the world where a small number of corporations account for so much of world production and trade as in the international auto industry' (UNCTC, 1983a, p. xiii). The ten leading automobile producers account for no less than 76 per cent of world production. Fully 96 per cent of the world total is produced by a mere twenty companies. All of these either have fully fledged manufacturing operations in different countries or, at the very least, foreign assembly operations.

However, the global pattern of automobile production is not solely the reflection of the locational whims of these transnational firms. Throughout its history, the international location of the industry has been strongly influenced by the trade policies of national governments. Tariff, and especially non-tariff, barriers continue to exert an extremely important influence in both developing and developed economies. In recent years trade in automobiles has become an increasingly sensitive issue. At the same time, national governments have struggled to outbid one another in their efforts to secure the large manufacturing plants of the major automobile manufacturers. Indeed, the automobile industry is a particularly good example of the competitive bidding process for internationally mobile investment which we discussed in Chapter 6. Not surprisingly, the giant global corporations of the industry have developed consummate skills in playing one government off against another to secure the maximum advantage from the situation.

Most national government involvement in the industry is motivated by the desire either to build a new automobile industry or – as in the traditional producing countries – to retain their existing industry. The large scale of automobile production creates very large, strongly localized concentrations of employment. In addition, the high linkage-intensity of the automobile industry means that the closure or contraction of a large assembly plant will almost inevitably have very serious 'knock-on' effects on employment in the components industries.

The automobile production system

The automobile industry is essentially an *assembly* industry. It brings together an immense number and variety of components, many of which are manufactured by independent firms in other industries. As Figure 9.1 shows, there are three major processes prior to final assembly: the manufacture of bodies, of components and of engines and transmissions. The nature of the industry offers the possibility of organizational and geographical separation of the individual processes. How far vehicle manufacturers carry out the separate parts of the production sequence themselves varies considerably. In fact, the production of automobile components comprises a specialized set of industries in which the leading firms have themselves become increasingly transnational as the global character of the automobile industry itself has

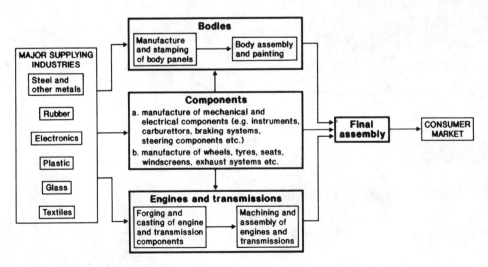

Figure 9.1 The automobile production system

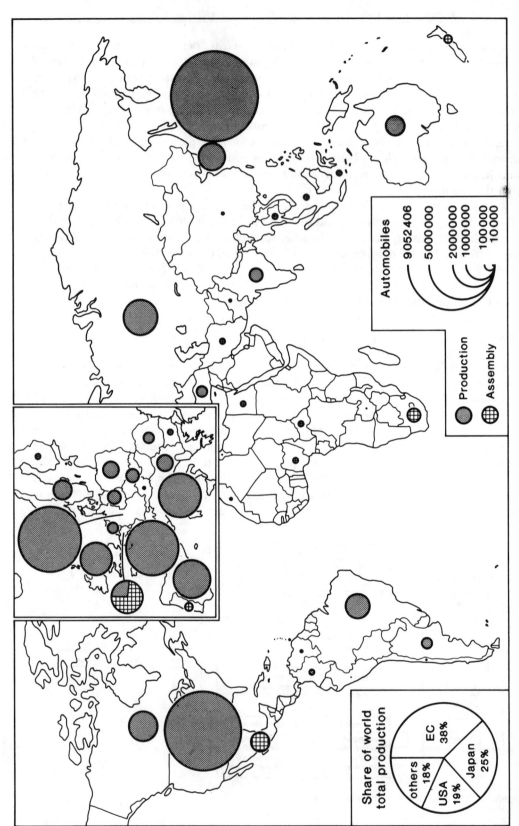

Figure 9.2 World production and assembly of automobiles, 1989 (*Source*: SMMT (1990) *World Automotive Statistics*, London: SMMT, Table 9; United Nations (1989) *Industrial Statistics Yearbook*, pp. 776, 777)

evolved. In this chapter we are concerned primarily with automobile manufacture rather than with component manufacture and with the production of passenger cars. As we shall see, it is a global industry in the throes of a major revolution stimulated, in particular, by the organizational and technological innovations of the leading Japanese companies.

Global shifts in the automobile industry

Production of automobiles

Between 1960 and 1989 world production of passenger cars increased by 173 per cent: from 13 million to 35.5 million vehicles. In fact, more cars were produced in 1989 than in any previous year. During those three decades major changes occurred in the global distribution of the industry. Figure 9.2 is the world automobile map for 1989 while Table 9.1 shows the major changes which occurred in the output of the major manufacturing nations. A distinction is made between those countries which are fully fledged producers of automobiles and those which are assemblers of vehicles using imported components. (In practice, the distinction can be rather blurred but it is, nevertheless, important in terms of its implications for a national economy.)

Table 9.1 Growth of automobile output by major producing countries, 1960–89

Country	1960 Production (000 units)	1960 World share (%)	1989 Production (000 units)	1989 World share (%)	Average annual % change 1960–89
Europe					
Belgium	194	1.5	1,144	3.2	+16.9
France	1,175	9.0	3,409	9.6	+6.6
West Germany	1,817	14.0	4,564	12.9	+5.2
Italy	596	4.6	1,972	5.6	+8.0
United Kingdom	1,353	10.4	1,299	3.7	−0.1
Spain	43	0.3	1,639	4.6	+128.0
Sweden	108	0.8	384	1.1	+8.8
North America					
United States	6,675	51.4	6,823	19.2	+0.1
Canada	323	2.5	984	2.8	+7.1
Asia					
Japan	165	1.3	9,052	25.5	+185.7
South Korea	–	–	872	2.5	–
Latin America					
Argentina	30	1.0	112	0.3	+9.4
Brazil	38	0.3	731	2.1	+62.9
Mexico	28	0.2	439	1.2	+50.6
Australia	–		357	1.0	–
USSR	139	1.1	1,300	3.7	+28.8
Czechoslovakia	–		184	0.5	–
Poland	–		289	0.8	–
World	12,999	100.0	35,455	100.0	+6.0

– Data unavailable

(*Source*: SMMT (1990) *World Automotive Statistics*, London: SMMT; OECD (1983) *Long-Term Outlook for the World Automobile Industry*, Paris: OECD, Table 1)

Automobile production is very strongly concentrated in the developed market economies, particularly Western Europe, Japan and the United States. In 1989 four-fifths of world automobile output was produced in this global triad. With only a small number of important exceptions, most automobile manufacture in developing countries is assembly rather than production.

By far the most dramatic development of the last two decades has been the spectacular growth of Japan as an automobile producer. In 1960 Japan produced a mere 165,000 cars, 1.3 per cent of the world total. In 1989 Japan produced 9 million autos, 26 per cent of the world total. By 1980, in fact, Japan had already overtaken the United States as the world's leading producer of passenger cars. By 1982 one car in every four produced in the world was made in Japan. Particularly heavy relative decline has been experienced by both the United States and the United Kingdom.

In 1960 the United States produced more than half of total automobile output in the world; by 1989 its share had fallen to 19 per cent. Less dramatic, though nevertheless very significant, was the decline of the United Kingdom's automobile industry. In 1960 the United Kingdom produced 10.4 per cent of the world total, more than eight times more cars than Japan; in 1989 the United Kingdom's share was a mere 3.7 per cent. Within the EC, West Germany and France remain the dominant producers of passenger cars; the most impressive growth in automobile production in Europe occurred in Spain. In 1960 Spain produced a mere 43,000 cars; by 1989 its output was 1.6 million.

Outside the core areas of Japan, the United States and Western Europe there are only three important concentrations of automobile production. The first of these is in the Soviet Union and Eastern Europe (notably Czechoslavakia, Poland and East Germany). The second focus of production outside the core regions is Latin America, notably Brazil, Mexico and Argentina. But as Table 9.1 shows, Argentina has declined in relative importance. A third, more recent, centre of automobile production is in East and South East Asia where South Korea, in particular, has very suddenly emerged as an important producer. As recently as the early 1980s, Korea was producing only 20,000 automobiles; in 1989 Korean output was almost 900,000. Malaysia is also developing a domestic automobile industry but, so far, it is still in the assembly stage. Indeed, virtually all automobile activity in the developing countries consists of the assembly of imported kits and components.

International trade in automobiles

The high level of geographical concentration of automobile production in the global economy is reflected in the structure of international trade. As Figure 9.3 shows, this trade is essentially *triangular* between (and within) the three major producing regions: Western Europe, North America and Asia (overwhelmingly Japan). Ninety-six per cent of world automobile trade is accounted for by these three regions. Western Europe is responsible for 50 per cent of the world total. But, as Figure 9.3 shows, almost three-quarters of Western European automobile trade is intra-regional. The biggest inter-regional flows are between Asia (i.e. Japan) and North America and between Western Europe and North America.

Table 9.2 summarizes the changing position for the leading automobile-producing countries and also shows the extent to which each country has a surplus or deficit in its automobile trade. As in the case of production, it has been the growth of Japan which

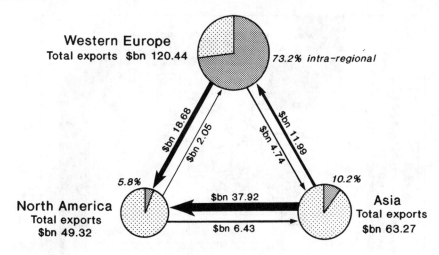

Figure 9.3 The triangular pattern of world trade in automobiles, 1987 (*Source*: GATT (1989) *International Trade 1988–1989*, Geneva: GATT, Vol. II, Table A6)

has been the most dynamic and transforming element in world trade in automobiles. Japan has not only overtaken the United States as the world's largest producer of automobiles but it has also displaced West Germany as the world's leading exporter of cars. In 1963 only 7.6 per cent of Japan's car production was exported; in 1987 more

Table 9.2 Trade in automobiles: leading exporters, 1963 and 1987

Country	Year	Production	Exports	Imports	Balance	Exports as % of production
		(thousand units)				
Japan	1963	408	31	9	+22	7.6
	1987	7,891	4,508	108	+4,400	57.1
West	1963	2,414	1,217	133	+1,084	50.4
Germany	1987	4,374	2,451	1,364	+1,087	56.0
France	1963	1,521	530	147	+383	34.9
	1987	3,052	1,908	1,119	+789	62.5
Canada	1963	532	16	60	−44	3.0
	1987	809	956	1,003	−47	118.0
Spain	1963	79	−	15	−15	−
	1987	1,402	837	290	+547	59.7
Italy	1963	1,105	292	191	+101	26.4
	1987	1,713	639	978	−339	37.4
United	1963	7,644	144	409	−265	1.9
States	1987	7,099	633	4,589	−3,956	8.9
South	1963	−	−	−	−	−
Korea	1987	793	546	−	−	68.9
United	1963	1,608	616	48	+568	38.3
Kingdom	1987	1,142	245	1,047	−802	21.5
Sweden	1963	147	68	153	−85	46.3
	1987	432	217	282	−65	50.2

− Data not available

(*Source*: GATT (1983) *International Trade, 1982–1983*, Geneva: GATT, Table A11; MVMA (1989) *World Motor Vehicle Data, 1989*, various tables)

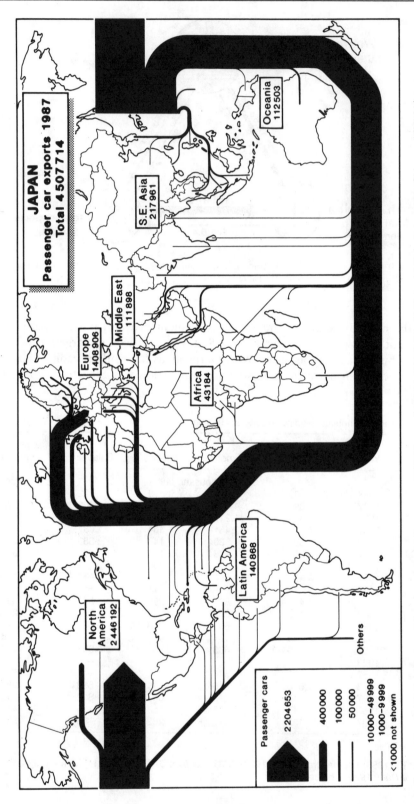

Figure 9.4 The global distribution of Japanese automobile exports, 1987 (*Source:* MVMA (1989) *World Motor Vehicle Data, 1989,* Chicago: MVMA)

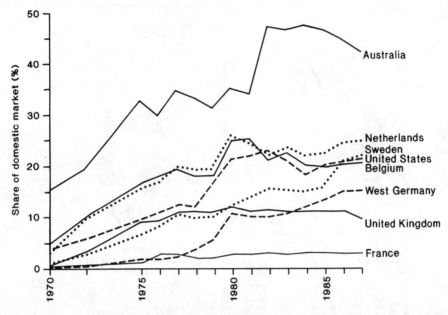

Figure 9.5 Japanese import penetration of major automobile-producing countries, 1970–87 (*Source*: MVMA, *World Motor Vehicle Data*, various issues)

than one car in every two produced in Japan was exported. Japan has by far the largest automobile trade surplus. At the global scale, Japan truly dominates world automobile trade, as Figure 9.4 suggests.

It was during the 1970s, in particular, that Japanese car exports began to penetrate most world markets, including the domestic markets of the major automobile producing nations, as Figure 9.5 indicates. Import penetration increased rapidly in all cases throughout the 1970s although in the United States and some EC countries this penetration was restricted through the operation of 'voluntary export restraints'. Even so, in 1987 almost one quarter of domestic car sales in the United States were of Japanese cars. A similar degree of Japanese penetration was evident in Belgium and the Netherlands. In the United Kingdom and West Germany the level of penetration was 9.7 and 15.1 per cent respectively, but it was much lower in France (2.9 per cent) and negligible in Italy. By far the highest degree of Japanese import penetration into countries with their own domestic vehicle industry was in Australia, where virtually one car in every two sold in 1987 was Japanese-made. This very much reflects the marked reorientation of Australia's economy towards Japan and the Pacific Basin.

Within the EC, West Germany and France were net exporters in 1987 whilst Italy and the United Kingdom were substantial net importers (Table 9.2). The very high export figure for Spain reflects its position within the integrated operations of trans-national corporations. As mentioned earlier, the vast majority of automobile exports from the major European producing countries is intra-regional: trade with other European countries. Import penetration for Western Europe as a whole is relatively low. There is a flow of imports from Japan and a flow of exports to North America. But the geography of automobile exports from the major individual European producing

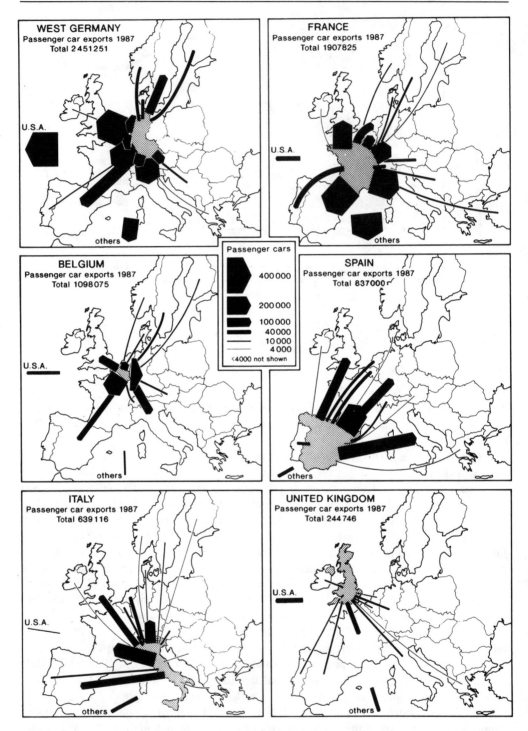

Figure 9.6 Exports of automobiles from major European producing countries, 1987 (*Source*: MVMA (1989) *World Motor Vehicle Data, 1989*)

countries is strongly differentiated, as Figure 9.6 reveals. As we shall see later, this pattern of intra-regional automobile trade is strongly influenced by the strategies of the leading transnational automobile producers.

In North America the very low export share of the United States is a reflection of the sheer size of the domestic market and also the fact that American-built cars have not generally been suited to most other markets. What is striking, however, is the massive increase in import penetration of the US market both by Japanese and European producers. The increase in exports from Canada is explained by the introduction of the Auto Pact in 1965 (see below, p. 288).

Outside the dominant producers of the developed market economies there was considerable growth in exports from some of the developing countries and Eastern European economies although, together, these accounted for little more than 5 per cent of world trade in automobiles. Nevertheless, their emergence as vehicle (and especially component) exporters is a significant development. For the most part, however, there is a strong geographical bias to these exports although there are differences between the export of cars and the export of components. In general, the export of cars from developing countries is more strongly oriented towards other developing countries than is the export of components. Virtually all car exports from Brazil and Argentina, for example, go to other parts of Latin America. Mexico's position is very different, however, being strongly tied in to the US market. Yugoslavia's car and component exports are overwhelmingly to Europe as are those of the USSR and Eastern Europe. Thus, although both production and trade in the automobile industry remain strongly dominated by the core producers of Japan, North America and Western Europe, it is clear that considerable changes have occurred and that some new producers are emerging, although on a much smaller scale than is often suggested.

Stages in the development of a country's automobile industry

Bloomfield (1978) has suggested that four stages can be recognized in the development of a country's automobile industry. Although it is by no means inevitable that all countries will actually pass through this sequence it does provide a useful classification of the world's industry:

Stage 1   *Import of completely built-up (CBU) vehicles by local retailers.* Tends to be limited in scale because of the high transport costs involved and possibly by government trade restrictions.

Stage 2   *Assembly of completely knocked-down (CKD) vehicles imported from the home plants of world manufacturers.* Permits the saving in transport costs and the opportunity to make minor modifications for the local market.

Stage 3   *Assembly of CKD vehicles with increasing locally made content.* Both depends upon and encourages the development of a local components industry. Strongly favoured by national governments.

Stage 4   *Full-scale manufacture of motor vehicles.* Tends to be restricted to a smaller number of countries than stages 2 and 3. It is by no means inevitable that countries which have reached stage 3 will reach the stage of full-scale manufacture. Conversely, it is possible that a country might 'regress' from the status of full-scale manufacturer to that of assembler.

The changing pattern of demand for automobiles

Changes in the level, the composition and the geography of demand have all played a key role in the evolving global map of the automobile industry and in the problems facing many of the traditional producing nations and firms. Automobile production is strongly *market-oriented*. Thus, historically, production has developed within large, affluent consumer markets in which high levels of demand have permitted the achievement of economies of scale. The world's three major vehicle-producing regions – North America, Western Europe and Japan – are also the world's most developed consumer markets. The changing demand for automobiles has three major characteristics: it is highly cyclical; there are long-term (secular) changes in demand; there are signs of increasing market segmentation and fragmentation.

Demand for automobiles is very sensitive to changes in the level of economic activity as a whole. Periods of high and increasing demand are separated by periods of stagnant or declining demand. Thus, there was a very rapid growth in demand for cars during the second half of the 1980s, associated with global recovery. But that demand fell back in 1990 and 1991 very substantially. Industry forecasters predict slow demand growth in the early 1990s and then recovery. However, demand trends vary substantially between different parts of the world. Demand for automobiles is growing most slowly in Western Europe and North America and fastest in some of the East and South East Asian countries. In South Korea, for example, automobile demand increased by 40 per cent a year between 1987 and 1990. In the Asia–Pacific region as a whole, demand increased by 28 per cent in 1989 and is forecast to grow at between 6 and 10 per cent a year through to the mid-1990s. Conversely, the projected growth in demand in Western Europe ranges between −0.5 to +3.6 per cent; in North America the estimated range is from −1 to +2.9 per cent. There is an expectation of a major surge in demand in Eastern Europe in the 1990s but this will clearly depend upon the rate of economic reconstruction there.

The slow growth in demand for automobiles in the Western European and North American markets reflects more than just cyclical forces. There are deeper *secular* or *structural* characteristics in these markets which limit future growth in car sales. Demand for automobiles is broadly of two kinds: new demand and replacement demand. Rapid growth in demand is associated with new demand. But as a market 'matures' and vehicle ownership levels approach 'saturation', the balance between new and replacement demand alters. More and more car purchases become replacement purchases. In the OECD countries today some 85 per cent of total demand for automobiles is replacement demand. Replacement demand is generally slower growing and also more variable because it can be postponed. In such mature markets, manufacturers have long adopted strategies similar to those of clothing manufacturers: they regularly introduce 'new' models to entice existing owners to replace their vehicles. Often, the changes are little more than cosmetic although promoted by massive advertising campaigns as being both significant and highly desirable. New production technologies are, however, making possible a far greater variety of vehicle types.

Until relatively recently there were considerable differences between the major automobile markets, each tending to favour particular types of vehicle. The greatest difference was between North America on the one hand and Western Europe and Japan on the other. Demand in the highly affluent, highly mobile, cheap-energy-based North American market was overwhelmingly for very large cars. In contrast, in

Western Europe and Japan generally lower incomes and higher energy costs plus more congested driving conditions were expressed in a demand for smaller, more fuel-efficient cars. But even within Europe, individual national markets tended to be served separately by domestically based producers. During the past twenty years, and especially since the oil crises of 1973 and 1979, which reduced the attractiveness of large cars, these circumstances have changed dramatically.

In the case of the automobile industry, of course, the impact of oil price rises was especially severe. Virtually at a stroke, the cost to the consumer of owning and running a car accelerated to unprecedented levels. The impact was especially severe in North America: demand for large gas-guzzling cars plummeted (though it later recovered somewhat). The large US manufacturers moved to 'downsize' their range, a trend reinforced by US government regulations to reduce the average gasoline consumption of automobiles. A major reason for the increased import penetration of the US market in the 1970s, therefore, was the inability of the domestic manufacturers to produce small, fuel-efficient cars to compete with European and, especially, Japanese vehicles.

One result of all these changes in the structure of demand has been to reduce some of the geographical differences between individual national markets. Demands have begun to *converge*, at least in the mass market sector. Regional and even global markets have become apparent. Within such markets, however, there are signs of greater market *segmentation* and *fragmentation*: demand for particular types of car for particular uses or for a customized version of a general model. At least in the affluent consumer markets this is leading to more consumer-driven choice, which can, in turn, be satisfied because of the dramatic changes taking place in the way automobiles are made.

Production costs, technological change and the changing organization of production[2]

Mass production

The basic method of producing automobiles changed very little between 1913, when Henry Ford first introduced the moving assembly line, and the 1970s, when a radically new system of production began to emerge in Japan. The automobile industry was *the* mass production industry. It produced a limited range of standardized products for mass market customers. It produced in very large volumes in massive assembly plants using very rigid methods in which each assembly worker performed a highly specialized and narrow task very quickly and with endless repetition. The automobile industry appeared to have abolished craft production for ever, apart from the small number of firms manufacturing for the luxury car market.

As developed by Ford and General Motors, and adopted by all the other Western manufacturers, this 'Fordist' method of production certainly brought the automobile as a commodity into the reach of millions of consumers. But, although very efficient in many respects, it contained one major limitation: its *rigidity*. In order to reduce costs a particular model had to be produced in huge volumes. Assembly lines were 'dedicated' to a specific model and to change models took a great deal of time and money. Individual assembly workers were mere small cogs in the wheel of production, with no responsibilities beyond their very narrow single task.

The automobile industry is an assembly industry requiring hundreds of thousands of components – ranging from entire engines and bodies down to small pieces of cosmetic trim. The big American and European manufacturers developed a particular kind of relationship with their suppliers, based on short-term, cost-minimizing contracts. The supplier–customer relationship was distant not just in functional terms but also, increasingly, in geographical terms. The major producers scoured the world for low-cost component suppliers. The close geographical proximity of customer and supplier, which had been a feature of the early years of the industry, began to break down as technological developments in transport and communication made long-distance transactions feasible. The increased geographical distance between the assemblers and their suppliers made it necessary for the assemblers to hold huge inventories of components at their assembly plants. In this way, the possibility of the assembly line being disrupted by a temporary shortage of components (or by faulty batches) was reduced. This was the 'just-in-case' system, described in Chapter 7.

Production of an automobile, from its initial design stage through to its appearance in the sales rooms, is an immensely complex and expensive process. A five-year development period has been the norm; costs of up to $3 billion have been common for the very large-scale and complex projects. Such enormous costs of development forced the major producers to seek very large production volumes to achieve economies of scale in production. It was generally agreed that a vehicle manufacturer needed to produce approximately 2 million vehicles per year to reap the maximum benefits from scale economies, although the minimum efficient scale of production varied according to the particular manufacturing process involved (Table 9.3). In locational terms,

> Body assembly, paint, trim and final assembly on a particular model should be carried out together, because of the high costs involved in transporting assembled bodies. Transport of unassembled bodies is less expensive and the location of the pressing plant is therefore less crucial. Engine and power train (engine and gearbox) components are easier and cheaper to transport, and the production technology involved is also very different, so there is no particular need to have these plants on the assembly sites. The same would apply to gearbox and axle production.
>
> (Bhaskar, 1980, p. 55)

For the automobile industry as a whole, labour costs may account for between a quarter and a third of total production costs. The importance of labour costs in motor

Table 9.3 Minimum efficient scale in different parts of the automobile industry

Activity	Volume of production required to achieve minimum production costs
Casting of engine blocks	1 million
Casting of other parts	100,000–750,000
Power train machining and assembly	600,000
Axle making and assembly	500,000
Pressing of various panels	1–2 million
Painting	250,000
Final assembly	250,000
Advertising	1 million
Finance	2–5 million
Research and development	5 million

(*Source*: Rhys (1989), Table 3)

vehicle manufacture, like scale economies, varies between different processes. Final assembly is the most labour-intensive stage. For example, the labour involved in body-making and final assembly may account for about 23 per cent of total vehicle production costs, whereas the labour costs involved in engine and transmission production are only 8 per cent of the total. Thus, certain parts of the automobile production process are more sensitive to labour cost differentials than others. As always, of course, it is not simply the cost of labour which is important but, rather, the overall quality and reliability of the labour force. Such considerations have played an important part in the investment-location decisions of most major transnational vehicle manufacturers in recent years. Certainly they have sought out areas of surplus labour and, especially, those in which labour relations seem less likely to be problematical, both within and between individual countries. The importance of geographical differences in labour costs is, however, declining as the industry becomes more capital-intensive.

A major factor here is, quite obviously, *technological change*. The industry's production cost structure has not been static; production cost relations have been altered by developments in the technology of both products and processes. Until recently, technological change in the industry seemed to be creating two distinct – and opposed – tendencies. One was towards even greater scale of production of *standardized* vehicles, the other was towards smaller-scale, more specialized production of vehicles aimed at particular *niches* in the market. Some automobile manufacturers have favoured one path of development while some favoured the other.

Product innovation in the motor vehicle industry has generally been incremental but very rapid and far-reaching changes have occurred since 1973 in the aftermath of the oil crisis. The most important innovations have been those directed towards reducing fuel consumption, for example by making more efficient engines, by reducing the weight of materials used (such as substituting plastics and non-ferrous metals for steel), by reducing the size of cars (especially in North America) and by increased use of electronics to control engine performance. In addition, government safety and anti-pollution regulations in many countries have created pressures for change in car design. As yet, however, the development of cars powered by alternative fuels has been very limited.

'Lean production'

Many of the characteristics of the automobile industry just described remain very important. Since the 1970s, however, and with accelerating speed, major – even revolutionary – changes in the ways in which automobiles are developed and manufactured have been occurring. The source of such changes, which are dramatically reshaping the industry, is the leading Japanese producers, led, initially, by Toyota. Womack, Jones and Roos (1990) use the term 'lean production' to contrast with the mass production techniques which still pervade much of the industry. In their view, 'lean production combines the best features of both craft production and mass production – the ability to reduce costs per unit and dramatically improve quality while, at the same time, providing an even wider range of products and even more challenging work' (Womack, Jones and Roos, 1990, p. 277).

Table 9.4 summarizes the major differences between craft production, mass

Table 9.4 The major characteristics of craft production, mass production and lean production in the automobile industry

Characteristic	Craft production	Mass production	Lean production
Technology	Simple, but flexible tools and equipment using unstandardized components	Complex, but rigid, single-purpose machinery using standardized components. Heavy time and cost penalty involved in switching to new products	Highly flexible methods of production using modular component systems. Relatively easy to switch to new products
Labour force	Highly skilled workers in most aspects of production	Very narrowly skilled professional workers design products but production itself performed by unskilled/semi-skilled 'interchangeable' workers. Each performs a very simple task repetitively and in a predefined time and sequence	Multiskilled, polyvalent workers operate in teams. Responsibilities include several manufacturing operations plus responsibility for simple maintenance and repair
Supplier relationships	Very close contact between customer and supplier. Most suppliers located within a single city	Distant relationships with suppliers, both functionally and geographically. Large inventories held at assembly plant 'just in case' of disruption to supply	Very close relationships with a functionally tiered system of suppliers. Use of 'just in time' delivery systems encourages geographical proximity between customers and suppliers
Production volume	Extremely low	Extremely high	Extremely high
Product variety	Extremely wide – each product customized to specific requirements	A narrow range of standardized designs with only minor product differentiation	Increasingly wide range of differentiated products

(*Source*: based in part on material in Womack, Jones and Roos (1990))

production and lean production in the automobile industry. The contrasts between mass and lean production are striking. In particular,

> lean production . . . is 'lean' because it uses less of everything compared with mass production – half the human effort in the factory, half the manufacturing space, half the investment in tools, half the engineering hours to develop a new product in half the time. Also, it requires keeping far less than half of the needed inventory on site, results in many fewer defects, and produces a greater and ever growing variety of products.
>
> (Womack, Jones and Roos, 1990, p. 13)

However, such flexible system do not, as is so often stated, mean the reduced importance of economies of scale:

> What the new lean production techniques allow is the production of a variety of cars but within a large annual volume. Hence, an assembly plant is still optimum at around 250,000 units a year, although the model specific optimum can be lower . . . What the new techniques do is to make it easier for large companies to make a variety of products, but they do not make it easier for small companies to survive.
>
> (Rhys, 1990)[3]

What they also appear to do is make possible shorter design periods per model. The average design time per model for Japanese manufacturers in 1989 was 47 months compared with 60 months for US manufacturers.

A major requirement of the flexible production of automobiles is ease of assembly. The problem is that

> before you can make things flexibly, you must first make them simple. Simplifying is difficult because each new car is more complex and more costly to develop than the last. So the trick is to get somebody else to develop parts of the car for you – and ideally to build them as well . . . Modular manufacturing involves designing and assembling the entire car as a series of sub-assemblies or modules. New modules can be developed directly to replace an existing one, allowing cars to be changed easily.
>
> (*The Economist*, 29 July 1989)

Such modules may be made by the automobile assembler but, in many cases, they are made by outside suppliers.

The relationship between customer and supplier in such a flexible production system must, however, be very different from that in the mass production system. The relationship has to be extremely close in functional terms, with design and production of components being carried out in very close consultation. Long-term relationships are preferable to short-term relationships. The system adopted by companies like Toyota is one of tiered suppliers, each tier having different tasks. The first-tier suppliers relate directly to the customer, second-tier suppliers to first-tier suppliers, and so on (Figure 9.7). The use of first-tier (or preferred) suppliers tends to involve a smaller number of suppliers than is common in the more remote, cost-driven, procurement systems. Not only are there close functional relationships between customers and suppliers in this system but also there are close geographical relationships. The use of *just-in-time* methods, which we discussed in Chapter 7, encourages geographical proximity.

Thus, a series of profound technological and organizational changes is sweeping through the automobile industry. From being an apparently 'mature' industry in which technologies were relatively stable, the industry has 'de-matured'. As we shall see in a

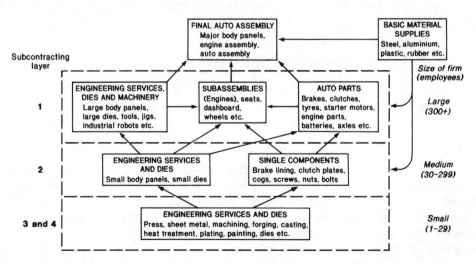

Figure 9.7 Functional tiers in the Japanese automobile supplier system (*Source*: after Sheard (1983), Figure 2)

later section of this chapter, these developments are having profound implications for the corporate strategies of the automobile manufacturers. Before reaching that point, however, we need to consider the influence of government policies on the industry.

The key role of government[4]

The world automobile industry is dominated by a small number of global corporations, yet today, as in the past, their investment-location decisions are greatly affected by the policies of national governments towards the industry. Before looking at the strategies pursued by the major manufacturers themselves in their efforts to remain afloat in the turbulent seas of international competition we should look, therefore, at the general ways in which national governments affect the automobile industry. As Reich (1989) has shown, in an industry like automobiles, which is dominated by transnational producers, there are two key aspects of state policy. The first is the degree of access to its domestic market which the state allows to foreign firms to establish production plants there. The second important aspect of state policy is the kind of support provided by the state to its domestic firms and the extent to which the state discriminates against foreign firms. Reich suggests that each of the major automobile-producing countries has pursued a different mix of these two policies:

1. *France*: limited access granted to foreign firms and discriminatory intervention in favour of domestic producers.
2. *Britain*: unlimited access granted to foreign firms and equal treatment of both foreign and domestic producers.
3. *West Germany*: unlimited access granted to foreign firms and discriminatory intervention in favour of domestic producers.
4. *United States*: unlimited access to foreign firms and equal treatment of both foreign and domestic producers.

To this list we should add:

5. *Japan*: limited access granted to foreign firms and discriminatory intervention in favour of domestic producers.

The shift from tariff to non-tariff barriers

Historically, the existence of national tariff barriers around sizeable consumer markets explains much of the early geographical spread of the automobile industry. Most governments in Europe and elsewhere (including Japan, Canada and Australia, for example) in the period before the Second World War had erected tariffs against automobile imports. These provided an obvious stimulus both to the setting up of foreign branch plants by the major vehicle producers and also to the growth of domestic manufacturers. Even in 1960 both France and the United Kingdom levied tariffs of 30 per cent on motor vehicle imports while Japan operated a 40 per cent tariff. Today, few of the developed market economies levy particularly high import tariffs against automobile imports although they remain high in some cases: in Australia, for example, they are more than 50 per cent, although they are now being reduced progressively to reach 15 per cent by the year 2000. The common EC tariff is 11 per cent, the US automobile tariff is only 3 per cent. Japan now has no import tariffs on motor vehicles although it has been much criticized for its slowness in removing certain non-tariff barriers.

Although tariffs in the major developed country markets are now generally low, there has been a strong recent upsurge in other forms of protection, primarily against Japanese imports. We saw earlier (Figure 9.5) just how rapidly and deeply Japanese import penetration has developed since the 1970s. In response to the growing domestic outcry from major industry pressure groups both the United States and several Western European countries negotiated 'voluntary' export restraints with Japan. In 1981 the US government established such a bilateral agreement as an allegedly temporary measure to last for three years. Although the import ceiling has been raised since then (from 1.68 million Japanese vehicle imports in 1981 to 2.3 million in 1989) the agreement still exists. The overall level was negotiated between the US and Japanese governments. The actual share of the cake between the Japanese firms themselves has been determined by the Japanese government.

In Europe Japanese imports are currently restricted to specific levels: to 11 per cent of the domestic market in the United Kingdom, 3 per cent in France and less than 3,000 vehicles per year in Italy. The completion of the Single European Market in 1992 adds a major complication, however: such individual bilateral agreements will no longer be valid. In the single market it would be quite legal for Japanese automobiles to enter an EC member state which does not have restrictions and then be distributed in other states which do have a voluntary export restraint agreement with Japan.

What is needed, therefore, is a Community-wide policy towards Japanese automobile imports. But this is a hugely controversial political issue,[5] with the French, Italians and Spanish pushing especially hard for a lengthy transitional period before voluntary export restraints are phased out. There is particularly bad feeling between the French and Italians on the one hand and the British on the other over the treatment of cars assembled in Japanese plants in Britain. Are they sufficiently 'European' or are they really Japanese? In the continuing political argument, Britain has been variously described as a 'Trojan Horse' infiltrating Japanese cars into the European market or as an 'aircraft carrier' moored offshore and launching Japanese cars into Europe.

Local content and export requirements

The increased fear of Japanese imports into both North America and Western Europe has stimulated additional policy measures. One is to encourage Japanese vehicle manufacturers to build *overseas production plants* in their major markets to displace imports. The other is to insist on specific levels of *local content* in vehicle manufacture by foreign producers. Again, it is in Western Europe that the greatest controversy has developed over this issue, both over precisely what local content means and also how it should be measured. For a specific EC member state 'local' refers to all Community countries and not just its own domestic territory. But how much local (i.e. EC) content must an automobile contain in order to count as a domestic product? The Treaty of Rome does not lay down any quantitative threshold of local content; it simply states that a product is regarded as originating in the European Community 'if the last substantive manufacturing process takes place in a Community country'. As far as Japanese firms in Europe are concerned, an eventual level of 80 per cent local content seems to be regarded as acceptable (although the French would prefer 100 per cent).

Such measures, though relatively new in the developed country context, have long been important elements in the national policies of developing countries towards the automobile industry.[6] Virtually all developing economies with any kind of motor vehicle industry operate both local content requirements and various types of tariff and non-tariff import restrictions. The use of both local content and other import restrictions together with very high tariffs on vehicle imports formed the basis of the strong import-substitution policies followed by a number of developing countries, particularly during the 1950s and 1960s. The aim was to build a domestic automobile industry to serve the domestic market. Each of the major industrial countries in the developing world – India, Brazil, Argentina, Mexico – as well as countries such as Spain adopted this kind of strategy. In most cases local content requirements were set at between 50 and 90 per cent, usually on a progressively rising scale over a period of several years.

Although domestic market protection remains a prominent feature of most developing countries' automobile industry the general shift of policy towards export promotion (see Chapter 6) has also been reflected in the automobile industry. Apart from the usual battery of financial and tax incentives and concessions, each of the major Latin American countries has export requirements under which manufacturers must export a specified proportion of their output. The increased involvement of Brazil, Mexico and Spain, in particular, in the international automobile trade is a reflection of these efforts and of the response of transnational automobile manufacturers. In the late 1980s Mexico made explicit moves to increase its involvement in the internationally integrated operations of the major transnational firms by substantially modifying its local content requirements, reducing import restrictions on vehicles but still insisting on a proportionate level of exports by the TNCs operating within Mexico.

In most other developing countries, protected domestic assembly is the dominant characteristic with, in some cases, limited export trade of components. An emerging exception to this pattern is South Korea, which has rapidly built a substantial indigenous automobile industry through a vigorous interventionist policy by the government. Until the late 1970s the South Korean automobile industry was totally insignificant, operating as it did within a heavily protected domestic market A major turning point, in the aftermath of the 1979 oil crisis, was the decision of the Korean government to enforce the rationalization of the Korean automobile manufacturers and to develop a strategic direction for the industry. Initially extremely successful, the policy has subsequently been hit by problems of the competitiveness of the leading Korean firms.

General government involvement in the automobile industry

Apart from questions of trade regulation the major developed country governments have also intervened in their automobile industries in a number of ways which vary considerably from one country to another. In general, the Western European governments, especially France, the United Kingdom and Italy, have been more extensively involved in their domestic vehicle industries than the governments of the United States and Japan, and have also gone furthest in attempting to restructure their industries. In the past, of course, Japanese government 'guidance' of the country's growing automobile industry was indeed considerable.

Japanese government involvement was especially marked in the 1950s and 1960s and took a number of forms. First, very tight protective barriers were placed around the domestic industry: strict import quotas and extremely high tariffs were imposed, which prevailed until relatively recently. Second, the direct involvement of foreign manufacturers in the Japanese industry was prohibited for a long period. The much-needed automobile technology was acquired through licensing agreements with foreign manufacturers. Third, MITI attempted to encourage rationalization among the major Japanese vehicle producers although with little success. Fourth, the Japanese government was heavily involved in assisting overseas marketing and exports, through financial and other assistance. These measures are no longer in force; instead the Japanese government's involvement in the automobile industry is now primarily in negotiating voluntary export restraint agreements with its trading partners.

Until recently, the governments of France, the United Kingdom and Italy have been directly involved both in direct ownership of vehicle producers and in large-scale financial support. Renault, Rover and Alfa-Romeo were all state-owned enterprises, although they have all now been privatized. Each of the state-owned enterprises had needed enormous financial subsidies, as did Chrysler before it pulled out of European manufacture in the late 1970s. Even the US government had to assist Chrysler with huge loan guarantees to save it from bankruptcy. The United Kingdom, France and Italy have also used their regional development policies to influence the location of automobile manufacturers and to persuade both domestic and foreign firms to build plants in depressed regions of the country. Massive financial incentives and restrictions on development elsewhere have been used to establish huge assembly plants in areas of high unemployment. Although the US government has not pursued such a spatially discriminatory policy, huge incentive packages have been used by individual states to attract Japanese automobile assembly and component plants to specific locations. The Toyota plant in Kentucky, for example, involved a total incentive package of $325 million. Although other incentive totals were much lower than this, the total for six Japanese projects (including two joint ventures) was almost $700 million.[7]

The implications of regional integration

A particularly signficant aspect of state policy towards the automobile industry is the emergence, or strengthening, of regionally integrated trade areas in both North America and Western Europe.

The Single European Market The establishment of the European Economic Community in 1957 and its subsequent enlargement in 1973, 1981 and 1986 to the present twelve member states have had a dramatic influence on the shape of the automobile industry within Europe.[8] Indeed, it is in Europe that the greatest progress has been made by the automobile TNCs, especially Ford and General Motors, in creating international integrated production networks. The completion of the single market in 1992 will merely add the finishing touches by removing the remaining technical and physical barriers to the flows of vehicles and components. Technical barriers to the single market are especially significant for the achievement of economies of scale in the automobile industry. The practice whereby an automobile manufacturer has to obtain a different type-approval certification for every member state, involving compliance with often minute variations in vehicle specification, will be abolished. This will greatly reduce the costs of detailed design and manufacturing variations. In general, it seems likely that the major beneficiaries of the single market will be those automobile producers which already have the most highly integrated operations across national boundaries within Europe.

It is certain that the European automobile industry will not be the same after 1992. But this will be only partly the result of the Single European Market. More significant in the long run will be the opening up of the Eastern European region. Suddenly, through the political decisions of the Eastern European states to liberalize their economies, a huge contiguous region is being created with the potential of being both a large potential market and also a low-cost production location for sourcing both components and finished vehicles. The political developments of the late 1980s and early 1990s are presenting major strategic opportunities for automobile manufacturers operating in Europe. At the same time, the political transformation of the entire European automobile production space may pose serious problems for countries such as Spain and Portugal. If there should develop a real functional integration of Eastern Europe into the European Community then the EC states on the western periphery will be even further removed from the centre of gravity of the European automobile industry.

Regional integration in North America In North America, too, political agreements on the integration of national markets have had profound repercussions on the structure of the automobile industry.[9] The *1965 Automobile Pact* between the United States and Canada reshaped the industry, producing an integrated structure in which production by the major manufacturers was rationalized and reorganized on a continental, rather than a national, basis. By the early 1970s, in fact, the automobile industry in North America was fully integrated; Canadian plants performed specific functions within the larger continental production and marketing system.

The recently negotiated Canada–United States Free Trade Agreement (FTA) also contains important provisions for the automobile industry. In particular, the FTA redefines the level of 'North American content' necessary for a firm to be able to claim duty-free movement within the North American market. The FTA, by making distinctions between three types of automobile manufacturer based on their current status, will create a 'two-tier' automobile industry in Canada:

> On the one hand, General Motors, Ford, Chrysler . . . so long as they comply with the provisions of the Auto Pact, will be able to continue to bring parts and vehicles into Canada duty-free from any country (including such low cost countries as Brazil, Korea, Mexico, Taiwan and Thailand). On the other hand, Honda, Hyundai, Toyota and any future new producer, 'operating in Canada in the same way as their commercial rivals, will have to pay duty on anything they import from countries other than the United States'.
>
> (Holmes, 1990, pp. 173–4)

Of course, the proposals to integrate Mexico into a North American Free Trade Area will be of the greatest significance in the automobile industry because of the lower production cost characteristics of the Mexican auto industry and the fact that it is increasingly becoming integrated into the North American market anyway through the strategies of the American, Japanese and European producers and the changing attitude of the Mexican government.

Environmental and safety legislation A final aspect of state policy to be mentioned is that of environmental, fuel-efficiency and safety legislation. All governments have some legislation covering these areas, each of which has profound implications for the design, technology and materials used in cars and, therefore, in their cost. Complying with changes in legislation can be especially problematical where it involves fundamental design changes. For US producers, one of the major effects of the 1973 oil crisis on the industry was the introduction of fuel economy standards whose aim was to lower the average fuel consumption level. Legislation to control noxious emissions from automobile engines has been particularly stringent for many years in states such as California. Recent heightened concern with the environment has led universally to intensified measures to clean up engine emissions through the introduction of catalytic converters and the like. The 1990 US Clean Air Act's fifty new requirements will cost the industry between $8 billion and $10 billion a year to implement.

Corporate strategies in the automobile industry

Increasing global concentration

In the early, pioneer days of the automobile industry in North America and Western Europe there were scores of manufacturers each producing a limited range of vehicles for individual national markets. In 1920, for example, there were more than 80 automobile manufacturers operating in the United States, more than 150 in France, 40 in the United Kingdom and more than 30 in Italy. Today, after decades of acquisitions and mergers each major national market is dominated by a very small number – literally a handful in most cases – of massive corporations. The top four firms account for between 88 and 100 per cent of national automobile production in each of the major producing countries. In France and Sweden the entire national output of cars is produced by only two companies; in Italy one firm, Fiat, is totally dominant. But such high levels of concentration are evident not only at the national scale but also internationally: the global automobile industry is in the hands of a small number of very large firms.

Table 9.5 lists the world's leading automobile manufacturers and shows both their individual shares of total world production and also the extent to which their production is transnationally based. The seventeen firms produced nine-tenths of world automobile output in 1989. The two leading firms – General Motors and Ford – produced almost 30 per cent between them; the top five produced 51 per cent and the top ten 76 per cent of the world total. GM and Ford have remained the world's leading automobile producers throughout the postwar period although their share of the world total has declined. Chrysler, for long the world number three, plummeted to eleventh position, shedding its European operations as part of its efforts to stay in business. Other changes in relative position occurred through acquisitions and mergers.

By far the most dramatic shifts in relative position – not surprisingly – involved the leading Japanese firms, particularly Toyota and Nissan (Figure 9.8). Before 1960 no Japanese manufacturer ranked among the world's top fifteen vehicle producers. In 1960 Toyota appeared in the top league for the first time – in fourteenth position. By 1965 Toyota had moved up to ninth position and had been joined by Nissan in eleventh place and Toyo Kogyo (Mazda) in thirteenth place. Five years later, Mitsubishi and Honda had also entered the first division while Toyota and Nissan had moved into sixth and seventh position respectively. By 1989 Toyota was challenging Ford for second place in the world league, Nissan ranked sixth, Honda ninth and Mazda tenth.

Table 9.5 also indicates the extent to which the leading automobile manufacturers

Table 9.5 The world league table of automobile manufacturers, 1989

Rank	Company	Country of origin	Passenger car production*	Share of world total (%)	Percentage produced outside home country
1	General Motors	USA	5,523,134	15.6	41.8
2	Ford	USA	4,060,586	11.5	58.7
3	Toyota	Japan	3,330,380	9.4	8.3
4	Volkswagen	Germany	2,713,671	7.7	30.5
5	Peugeot–Citroën	France	2,320,266	6.5	15.4
6	Nissan	Japan	2,289,123	6.5	13.8
7	Fiat	Italy	2,108,622	6.0	7.1
8	Renault	France	1,755,510	5.0	17.6
9	Honda	Japan	1,604,430	4.5	28.0
10	Mazda	Japan	1,184,166	3.3	18.3
11	Chrysler	USA	1,052,537	3.0	13.0
12	VAZ	USSR	724,740[†]	2.0	–
13	Mitsubishi	Japan	708,418	2.0	–
14	Daimler–Benz	Germany	536,993	1.5	–
15	BMW	Germany	489,742	1.4	–
16	Rover	UK	466,619	1.3	–
17	Volvo	Sweden	423,385	1.2	33.8

*Excludes commercial vehicles
[†]1987

(*Source*: calculated from SMMT (1990) *World Automotive Statistics*, London: SMMT, pp. 28–33)

are *transnational* in their operations. Quite clearly a good deal of variety exists. By far the most transnational producer of all is Ford: in 1989 almost 60 per cent of its passenger car production was located outside the United States. GM's degree of transnationality, however, though substantially lower – 42 per cent in 1989 – has been increasing rapidly in recent years. Of the other very large automobile manufacturers (those producing more than 1 million cars) Volkswagen has the largest share of its operations located overseas (30.5 per cent). But if the big US producers are the most transnational in their activities the big Japanese producers are clearly the least trans-national of the major producers. Apart from some local assembly operations which

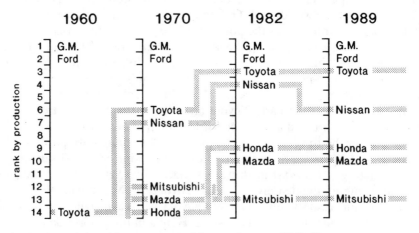

Figure 9.8 The rise and rise of Japanese automobile manufacturers, 1960–89

build cars from imported knock-down kits for local markets, Toyota had no overseas production facilities for passenger cars in 1982, while less than 3 per cent of Nissan's total production was located outside Japan. Among the smaller Japanese companies only Honda had overseas production facilities (in the United States).

However, the transnationality of the Japanese producers changed dramatically during the 1980s. By 1989 almost 30 per cent of Honda's automobile production was located outside Japan, 18 per cent of Mazda's, 14 per cent of Nissan's and 8 per cent of Toyota's. In fact, the figures in Table 9.5 understate the real extent of the leading firms' transnational involvement in at least two ways. First, they refer to production only and exclude simple assembly. Second, they do not reflect one of the most important developments of recent years: the fact that most of the world's automobile manufacturers are increasingly involved in complex transnational sourcing arrangements for components and in collaborative agreements with other manufacturers.

Strategic alliances in the automobile industry

The very high level of concentration in the world automobile industry is largely the outcome of factors discussed earlier in this chapter, in particular the drive to achieve efficiency in design, production and marketing in an increasingly competitive global market. A major problem, however, is that the development of new models, which themselves have a shrinking life, requires massive investment not only in machinery and other equipment but also in research and development (see Table 9.3). Consequently, even the very largest firms are involved in collaborative ventures with other manufacturers, while the very survival of smaller firms seems to depend increasingly on interfirm agreements to supply parts, to produce jointly under licence and to engage in joint R & D.

As a result, a veritable transnational spider's web of strategic alliances has developed, a web which stretches across the globe (Figure 9.9). Most of Ford's and GM's collaborative links are with Japanese and Korean companies. For example, GM has an arrangement with Isuzu (of which GM owns 39 per cent) for the Japanese firm to supply transmissions and axles to GM plants and to manufacture a small car. GM has a joint venture with Toyota to build a small car at Fremont, California (the NUMMI plant) and one with Suzuki in Canada. Ford owns 24 per cent of Toyo Kogyo (Mazda), which supplies transmissions for Ford cars and kits for assembly of vehicles in Asia. The Ford Escort in North America was jointly developed with Mazda, which also sells a version under its own badge in other markets. Ford has recently signed an agreement with Yamaha for the Japanese company to develop a new engine for some Ford cars. Chrysler also has Japanese connections: it owns 24 per cent of Mitsubishi, sells some Mitsubishi cars in the United States and purchases engines from the Japanese company. Chrysler has a joint venture with Mitsubishi to manufacture a small car in the United States based on Japanese design (Diamond Star Motors Corp.).

Within Japan itself there is an exceptionally complex network of relationships between the major automobile manufacturers. Japanese firms also have close links with Korean automobile manufacturers. These, in turn, have close links with GM, Ford and Chrysler. For example, the Korean firms Daewoo and Kia manufacture cars which are sold in the United States under GM and Ford brand names respectively.

Honda and Rover have a complex arrangement to produce cars in the United Kingdom. Within Europe there is a complex collaborative network. Virtually all the

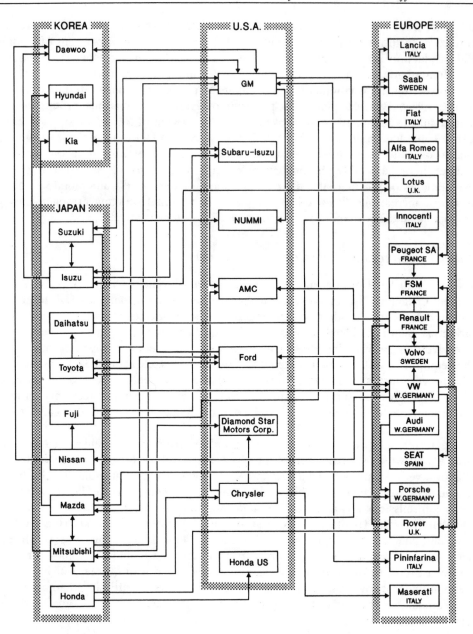

Figure 9.9 Some of the major collaborative relationships in the automobile industry (*Source*: based on the *Financial Times*, 20 October 1988)

major European manufacturers are involved in collaborative deals. For example, Fiat and Peugeot collaborate on the production of engines and steering components; Renault and Volvo have completed a complex share exchange. Six European firms – Fiat, Peugeot, Renault, Rover, VW and Volvo – have a co-operative research programme. Volkswagen and Ford operate a joint venture (Autolatina) for production in Argentina and Brazil.

Transnational collaborative arrangements also involve the economies of the USSR and Eastern Europe. Indeed, automobile manufacturers have been especially prominent in such ventures. Fiat has been particularly active in co-production and licensing agreements in the USSR, Poland and Yugoslavia, as have Renault in Romania and Citroën in Romania, East Germany and Yugoslavia. In fact, much of the Soviet Union's and Poland's postwar automobile industry was built in conjunction with Fiat, including the massive assembly plant at Toglliattigrad. The political changes in Eastern Europe are producing an upsurge in such arrangements between Western automobile manufacturers and local firms in the production of both vehicles and components.

Types of competitive strategy

Although the automobile industry is undoubtedly a global industry, there is a good deal of *variety* in the specific corporate and competitive strategies adopted by the leading companies. In terms of the strategic categories identified in Figure 7.5, we can position automobile companies in each of the four cells, although several companies have been pursuing different strategic trajectories in recent years, as we shall see. During the late 1970s and early 1980s two clearly defined strategic options appeared to be available for the world's major automobile producers.

One option was the so-called 'world car' strategy: the manufacture of mass-produced cars for a world market. The world car strategy was based explicitly upon transnationally integrated production between the overseas affiliates of the parent company. Component manufacture was to be located in the most favourable – least-cost – locations at a global scale. These highly specialized plants would then supply a strategically located network of assembly plants which served specific, large-scale geographical markets. The emphasis was on standardization with minor, often cosmetic, modifications to suit particular markets. Hence, economies of scale were a key consideration both for component manufacture and for the assembly of finished vehicles. The world car, therefore, was not merely to be sold to a world market: more significantly, it was to be *manufactured* on a world scale. The alternative strategy was seen to be one confined to the *luxury car* manufacturers, producing in relatively small volumes for a small market in which premium prices could be charged. Such a strategy was regarded as being less likely to involve transnational production.

For a while it certainly seemed to be the case that the world automobile industry would segment according to these two strategic models. Certainly the two leading US manufacturers, GM and Ford, moved a considerable way along the world car path while firms like BMW, Daimler–Benz and Volvo continued to operate as specialist manufacturers in the luxury car niche. In fact, by the late 1980s the picture had changed dramatically. The stark distinction between mass-produced and lower-volume cars began to disappear in the face of the Japanese development of alternative methods of production (what was referred to earlier as 'lean' production). In examining the evolving competitive strategies of the major automobile manufacturers, therefore, it makes sense to begin with the Japanese companies. Not only have they been drastically changing their global strategies but also they are the main force driving changes in the strategic behaviour of the US and European automobile producers. Without doubt, the Japanese are currently in the driving seat.

The strategies of the Japanese automobile manufacturers

The dramatic emergence of Japan as the world's leading automobile producer and exporter (Tables 9.1, 9.2) and the spectacular rise of the leading Japanese companies up the world league table (Figure 9.8) were achieved almost entirely without any actual overseas production. Japanese firms occupied the top right-hand cell in the strategic matrix of Figure 7.5. Japanese firms had established local assembly operations in a number of countries, such as Australia and a number of developing countries, to put together finished vehicles from imported kits. But before 1982 there was not a single Japanese automobile *production* plant outside Japan. Certainly, as far as the developed market economies of North America and Western Europe are concerned, Japanese vehicle manufacturers have shown themselves perfectly capable of serving such markets by exports from Japan.

The price competitiveness of Japanese vehicle exports was based upon the extremely large-scale, flexibly organized, and highly automated production plants in Japan. Japan's massive technological investment in motor vehicle production during the 1950s and 1960s, aided by government measures, resulted in extremely efficient vehicle production. The very high degree of vertical integration with Japanese component suppliers added to this remarkably efficient system. These production cost efficiencies more than offset any transport cost penalties which arose from Japan's geographical distance from major markets.

Since the early 1980s, however, the Japanese companies have begun to change their global strategy; to move from the upper right-hand cell of Figure 7.5 to the upper left-hand cell. Undoubtedly, the major stimulus for this change in strategy was the increasingly powerful political opposition in both North America and Western Europe towards the growth of Japanese imports. The establishment of 'voluntary' restraint agreements between Japan and a number of countries was the visible sign of the level of trade friction being engendered by Japanese import penetration. Certainly in the

Table 9.6 Japanese automobile plants in North America

Company	Plant location	Date opened	Current planned capacity	Projected plant employment
United States				
Honda	Marysville, Ohio	1982	360,000	2,600
	East Liberty, Ohio	1989	150,000	1,800
Nissan	Smyrna, Tennessee	1983	480,000	3,300
	Decherd, Tennessee	1996	300,000 engines	1,000
Toyota/GM				
(NUMMI)	Fremont, California	1984	340,000	2,500
Toyota	Georgetown, Kentucky	1988	400,000	3,000
Mazda	Flat Rock, Michigan	1987	240,000	3,500
Mitsubishi/Chrysler				
(Diamond Star)	Normal, Illinois	1988	240,000	2,900
Subaru/Isuzu	Lafayette, Indiana	1989	120,000	1,700
Canada				
Honda	Alliston, Ontario	1987	100,000	700
Toyota	Cambridge, Ontario	1988	50,000	1,000
Suzuki/GM				
(CAMI)	Ingersoll, Ontario	1989	200,000	2,000

(*Source*: Press and company reports)

United States, one of the objectives of the trade protection policy was to persuade Japanese automobile manufacturers to locate production facilities there. In so doing, it was believed the Japanese companies would have to operate under the same conditions as the US companies: the playing field would be levelled. But trade protectionist measures, although the most important, were not the only stimulus to Japanese automobile investment in the United States. Given the increasing possibilities of tailoring vehicles to variations in customer demand, it was becoming more important for Japanese firms to locate close to their major markets. The same story is being played out in Western Europe, although, so far, the plot is rather less well developed.[10]

Having been extremely cautious overseas investors, the Japanese automobile manufacturers have all now established full production facilities in North America. As Table 9.6 shows, by 1991 there were more than a dozen Japanese automobile 'transplants' in place in the United States and Canada. In fact, the pioneer was neither of the two largest Japanese producers, Toyota or Nissan, but Honda, which established a manufacturing plant at Marysville in Ohio in 1982. This was soon followed by the Nissan plant at Smyrna, Tennessee in 1983. The leading Japanese manufacturer, Toyota, entered North America as a producer in a very cautious manner: by establishing a 50/50 joint venture with GM to produce cars at GM's Fremont, California, plant. This NUMMI plant began production in 1984; four years later Toyota began production at wholly owned plants in Georgetown, Kentucky, and Cambridge, Ontario. Honda continued to develop its American operations by opening a further, plant in Ohio and one in Ontario. Mazda began production in Michigan in 1987 and Mitsubishi entered a 50/50 joint venture with Chrysler – named Diamond Star – in Illinois.

Each of the major Japanese firms has continued to increase its planned capacity and, in several cases, to make major investments in engine, transmission and components plants. By the early 1990s the Japanese transplant factories in North America had a planned production capacity of 2.7 million vehicles and some 25,000 employees. In other words, during the period of less than a decade an entirely new Japanese-controlled automobile industry has been created in North America in fierce, direct competition with domestic manufacturers. By the end of 1990 Japanese cars produced in the United States had captured 21 per cent of the domestic market. The leading Japanese firms are now exporting some of their output not only to third countries like Europe or Taiwan but also to Japan itself.

As the Japanese plants in the United States have progressively increased their North American content they have been followed by a continuing wave of Japanese component manufacturers. There are now approaching 300 such plants in operation. The locational pattern of both assemblers and suppliers in North America is quite distinctive; perhaps not surprising since, as newcomers, the Japanese have no existing plants or allegiance to specific areas. Mair, Florida and Kenney (1988, p. 361) identify a 'transplant corridor', a 'region stretching from Southern Ontario south through Michigan, Illinois, Indiana, Ohio, and Kentucky to Tennessee. This "transplant corridor" is organized principally along several interstate highways... The single exception to the pattern of regional concentration is the NUMMI joint venture between Toyota and GM, which reopened a previously closed GM plant in Fremont, California.'

Suppliers have generally followed the assemblers because of the use of just-in-time delivery systems. At the finer geographical scale the strong preference of Japanese automobile firms has been for greenfield sites near to small towns in rural areas. The old-established automobile industry centres are not favoured. The major exceptions to this rural orientation are the Toyota joint venture in California and the Mazda plant in Michigan. Choice of greenfield, rural locations has been determined mainly by the desire of the Japanese producers to minimize the influence of the labour unions. Recruitment has been primarily of young workers with little factory experience but with the 'right attitudes' towards the kinds of flexible labour practices employed in the plants. The advantages of being able to start from scratch, on greenfield sites, with newly designed plants and with a hand-picked workforce prepared to accept new working conditions and practices are enormous.

Compared with their entry into North America, Japanese automobile manufacturers have been slower to establish production facilities in Europe. However, the momentum increased markedly at the end of the 1980s (Table 9.7). As far as passenger cars are concerned – as in the United States – Honda was the pioneer. In the European case, however, Honda merely signed a joint venture agreement in 1979 with BL, the predecessor of Rover and then state-owned. Nissan was the first to announce plans for an actual manufacturing plant in the United Kingdom, in 1981, but it was not until 1986 that production actually began. In 1989 both Toyota and Honda announced their plans for both car and engine plants, also to be located in the United Kingdom. Although there has not been the same following inflow of Japanese components suppliers as yet, some have indeed arrived. All three Japanese automobile firms have adopted similar locational criteria as in North America. They have avoided traditional automobile manufacturing areas, they have opted for greenfield sites and they have adopted very specific recruitment policies. All three have insisted on single-union agreements.

The Japanese plants in the United Kingdom are specifically oriented towards the European market and this, as we have seen, has led to political friction within the European Community. Again, both market-access/political factors and market-proximity factors have been influential in the Japanese decisions to locate production in Europe. No doubt the imminent completion of the Single European Market and the fear of a 'fortress Europe' were instrumental in accelerating both Toyota's and Honda's

Table 9.7 Japanese automobile plants in Europe*

Company	Plant location	Date opened	Current planned capacity	Projected employment
Nissan	Sunderland, North East England	1986	200,000 cars	2,750
Toyota	Burnaston, East Midlands	1992	200,000 cars	3,000
	Shotton, North Wales	1992	200,000 engines	300
Honda	Swindon, Wiltshire	1992	100,000 cars 200,000 engines	1,300
Honda/Rover[†]	Longbridge, West Midlands	1992	40,000 cars	n.a.

* Excludes commercial vehicle plants operated by Nissan in Spain, Isuzu in the UK; other non-passenger car plants operated by Toyota in Germany and Portugal, Suzuki in Spain.
[†] Honda and Rover's predecessor company have had a joint venture arrangement since 1979. In 1989, Honda acquired 20% of Rover and Rover acquired 20% of Honda's UK operation.

(*Source*: Press and company reports)

investment decisions. But the perceived need to be closer to the European consumer to assess local market conditions and consumer tastes is also a key factor.

Each of the leading Japanese automobile manufacturers is moving towards the greater transnationalization of their production. Nissan has plans to manufacture 25 per cent of its production outside Japan by the early 1990s. Toyota's objective is to increase its worldwide production capacity by almost a third by the late 1990s in order to become the world's number one automobile manufacturer. But as Figure 9.10 shows, there is a long way still to go. Only Honda has a really large proportion of its output outside Japan and all of that is in North America. Nissan has a major production facility in Mexico whose capacity is about to be doubled. It is clear that, as yet, the degree of transnationalization among Japanese automobile producers is very limited, particularly when compared with the leading US firms, GM and Ford. But it is equally clear that the primary objective of the Japanese producers is to develop a major direct presence in each of the world's major automobile markets – the global triad – of Asia, North America and Western Europe.

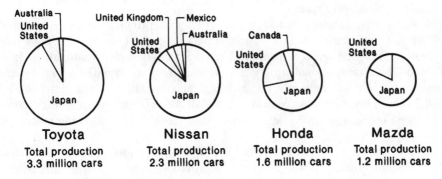

Figure 9.10 International automobile production by the leading Japanese companies, 1989 (*Source*: SMMT (1990) *World Automotive Statistics*, London: SMMT)

Strategic responses to Japanese competition

For a decade and a half the Japanese automobile producers have been perceived as the major competitive threat by the American, and most of the European, producers. In facing up to the intensification of global competition American and European firms have pursued rather different competitive strategies and created different geographical configurations in their production networks. In so far as both US and European producers face a shared competitive threat from the Japanese, however, they have all adopted a varying mixture of the following measures.

1. *Introduction of new technology*, notably flexible manufacturing systems.
2. *Introduction of new work practices* with, again, a particular emphasis on flexibility, multi-skilling and the reduction of the multifarious task divisions which have characterized all American and European automobile plants.
3. *Modification of component sourcing procedures* to move towards a just-in-time system.
4. *Development of strategic alliances* with other firms.
5. *Continued reorganization and rationalization* of their production activities both

domestically and internationally. It is in this latter respect that the strategies of the non-Japanese companies differ most of all.

The strategies of the United States automobile manufacturers

Ford and GM are the longest-established transnational vehicle producers in the world. In general, Ford has expanded internationally mainly by opening new plants; GM primarily by acquiring existing foreign manufacturers. Ford established a manufacturing plant in Canada in 1904, just across the Detroit River in Windsor, Ontario; GM acquired the Canadian vehicle company, McLaughlin, in 1918. Ford opened its first European car assembly plant at Trafford Park, Manchester in 1911 (subsequently replaced by a massive integrated operation at Dagenham in 1931), followed by a plant at Bordeaux in 1913. Ford began to assemble cars in Berlin in 1926 and in Cologne in 1931. GM's entry into Europe began in 1925 with the acquisition of Vauxhall Motors in England, followed, in 1929, by the purchase of the Adam Opel Company in Germany. During the 1920s both firms also established assembly operations in Latin America as well as in Australia and Japan.

These early transnational ventures were triggered primarily by the existence of protective barriers around major national markets as well as by the high cost of transporting assembled vehicles from the United States. In recent years, however, both Ford's and GM's transnational strategies have been concerned initially with expanding and integrating and subsequently rationalizing their operations on a global scale. As Table 9.5 shows, Ford is the more transnational of the two leading companies, although GM has been investing heavily overseas in recent years as part of its intensified global strategy.

Figure 9.11 shows the broad pattern of Ford's and GM's international production. It can be seen immediately how extensively transnational the two firms are. Although there are slight differences of detail, their geographical similarities are very considerable. Both companies have heavily rationalized and restructured their domestic operations with extensive closure and contraction of plants. GM, in particular, has undertaken a massive rationalization programme in the United States, including the

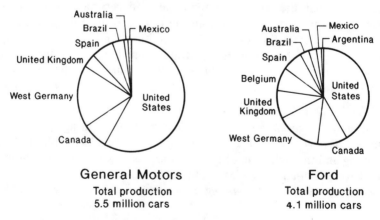

Figure 9.11 International automobile production by Ford and General Motors, 1989 (*Source*: SMMT (1990) *World Automotive Statistics*, London: SMMT)

closure of seven plants between 1987 and 1990 with a combined capacity of 1.3 million vehicles. On the other hand, GM has gambled heavily on a totally new car project, Saturn, located at Spring Hill, Tennessee, whose objective is to produce a totally new range of small (sub-compact) cars to compete with the Japanese. It is the first fully integrated automobile production facility to be built in the United States since 1927, incorporating engine, gearbox and finished car production on the same site. The project, whose first cars rolled off the assembly line in October 1990, is an attempt by GM to change completely the way it has made cars and to develop a high level of collaboration with the workforce, including the Union of Auto Workers (UAW). In addition, as we have seen, GM operates a highly successful joint venture (NUMMI) with Toyota at Fremont in California.

Figure 9.11 breaks down the world distribution of Ford's and GM's automobile production. More than two-thirds of Ford's overseas car production is located in Europe. West Germany is the major production focus, with 27 per cent of the overseas total, followed by the United Kingdom with 16 per cent, Spain (12 per cent) and Belgium (13 per cent). Outside Europe, Ford's foreign production operations are primarily in Latin America (Brazil, Argentina), Mexico (where Ford is doubling its plant capacity) and Australia. It is within Europe, however, that Ford's strategy of transnational integration is furthest developed.

Ford was the first automobile producer to take full advantage of the development of the European Community. In 1967 Ford reorganized its entire European operations – previously separately focused on the United Kingdom, West Germany and Belgium – into a single organization, Ford Europe. Since 1967 the separate national operations of Ford plants in Europe have been transformed into a transnationally integrated operation, with each plant performing a specified, often specialist, role within the corporation to achieve economies of scale. In the early stages of the process Ford simply reallocated the production of certain models to specific European plants. However, the first real product of Ford's transnationally integrated strategy in Europe was the Fiesta in the 1970s, followed later by the Escort. Both of these projects illustrate very clearly how the strategy of transnational integration in the automobile industry works.

Key elements in Ford's European strategy from the early 1970s were the desire to gain a foothold in the Spanish market and produce a small car. The Spanish car market in the early 1970s was heavily protected: not only were tariffs on imports very high (81 per cent on cars, 30 per cent on components) but also cars built in Spain had to have a 95 per cent local content. In addition, no foreign company could own more than 50 per cent of a company operating in Spain. This latter point was very much contrary to Ford's policy of 100 per cent ownership of its overseas subsidiaries. On the other hand, the Spanish government was anxious to build up its motor vehicle industry and, especially, to increase vehicle exports. Two years of negotiations at the highest level led to an agreement whereby the local content requirement was reduced to 50 per cent provided that two-thirds of production was exported; tariffs on imported components were reduced to 5 per cent and Ford was allowed full ownership of its new subsidiary company in Spain. In return, Spain gained a massive boost to its export trade and a major injection of new jobs.

The Spanish operation became, from the beginning, a key node in the transnational network of production for the Ford Fiesta (Figure 9.12). The Spanish plant is one of three final assembly locations for the Fiesta, the others being at Dagenham and Saarlouis. By locating assembly at three separate locations Ford can not only serve

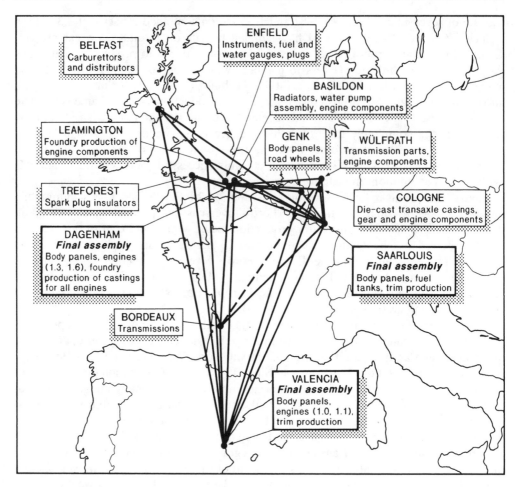

Figure 9.12 The Ford Fiesta production network in Europe

specific geographical markets but also make up for shortfalls in production at one location by shipping cars from one of the others. The three assembly plants are, in turn, locked into a highly complex network of component suppliers, both inside and outside Ford.

Figure 9.12 shows the geographical pattern of this network within Ford itself (external components suppliers are not shown). Approximately 45 per cent of the value of a Fiesta is produced in Ford's own plants. Some components are single-sourced to take advantage of economies of scale: for example, all carburettors are supplied by the Belfast plant; all transmissions are built at Bordeaux in a plant specially built to manufacture front-wheel-drive units for the Fiesta; Basildon supplies radiator assemblies; Treforest in South Wales makes the spark plugs. The result is a highly complex network of cross-border flows of finished vehicles and components, a network made even more complex by the involvement of hundreds of outside suppliers. Figure 9.13 shows the more extensive procurement network developed by Ford for its Escort model. Clearly, it is in Europe that Ford has gone furthest in integrating its

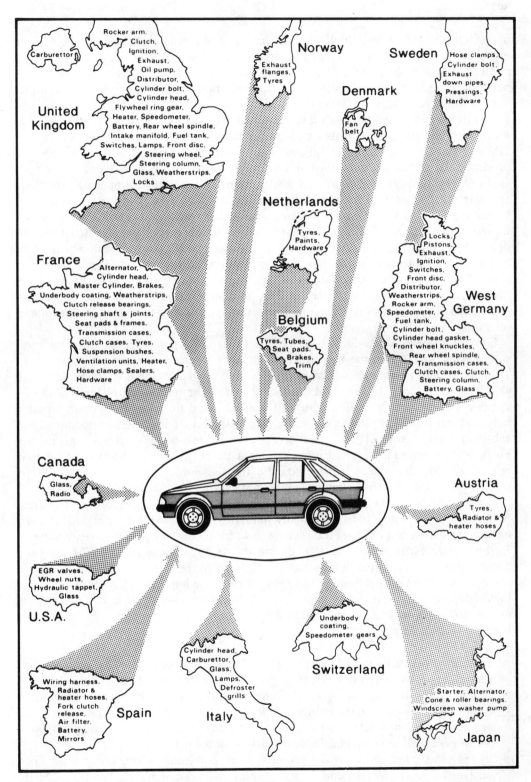

Figure 9.13 The component sourcing network for the Ford Escort in Europe during the late 1970s (*Source*: based on US Department of Transportation (1981) *The US Automobile Industry, 1980*, Washington, DC: USGPO, p. 57)

manufacturing operations. In effect, Ford has created a production system which is, in many ways, a microcosm of a globally integrated system on a European scale. Ford's strategic emphasis has now shifted towards reorganization and rationalization of the network, however, rather than further expansion. Before looking at this it is useful to turn to the development of General Motors' international strategy.

GM now has a global production system much like Ford's in its general structure but the nature and the timing of its development have been different. Figure 9.11 shows GM's global manufacturing operations. Like Ford, two-thirds of GM's overseas car production is in Europe but with a much greater emphasis on West Germany, where GM's Opel subsidiary is responsible for 63 per cent of the firm's European production. In contrast, the United Kingdom is far less important to GM than it is to Ford. For both companies, however, Spain occupies an especially significant position as a location for the production of small cars. Outside Europe, Canada is the most significant area of operations, followed by Brazil, Mexico and Australia. Mexico, as with Ford, is an increasingly important sourcing point for the North American market. Again, however, it is in Europe that the greatest degree of transnational integration has occurred. Until the early 1980s GM was far weaker in Europe than Ford; now it produces more cars there than its rival and has very radically reshaped and rationalized its European operations.

GM's European operations had long been based upon two separate national subsidiaries: Vauxhall in the United Kingdom and Opel in West Germany. During the 1970s Vauxhall's performance became progressively weaker as the investment emphasis shifted towards Opel. In 1979 GM drastically redrew the production responsibilities of the two European subsidiaries. Vauxhall was relegated to being a domestic supplier only and much of this production was simple assembly of imported components and sub-assemblies from Opel. This situation has only recently begun to change again. Like Ford, GM set up a major new European engine plant (in Vienna) to supply its global operations, together with a new automatic transmission plant at Strasbourg. More significantly, GM also built a major new manufacturing plant in Spain (at Zaragossa).

Having arrived at their current transnational structures by rather different routes, both companies are now pursuing broadly similar strategies. In terms of Figure 7.5 both have moved from a multidomestic strategy towards a complex global strategy. Within Europe, after expanding production during the 1970s, both Ford and GM are now concentrating on increasing the productive efficiency of their *existing* assembly plants in order to match not only the competitiveness of Japanese imports but, more specifically, that of Japanese producers within Europe; the transplants which are coming on stream in the United Kingdom. Both United States TNCs, then, are currently pursuing a four-pronged strategy:

1. Enhancing productivity and decreasing costs at their existing plants by implementing radically new working practices, including twenty-four-hour, three-shift working.
2. Building new strategic component plants, particularly for the new generation of engines which have to meet stringent emission requirements. Ford is building a dual engine source in South Wales and in Germany; GM's new engine plant is in Merseyside, adjacent to its Ellesmere Port assembly plant.
3. Moving into prestige/luxury car production by acquiring existing companies. Ford now owns Aston Martin and Jaguar; GM owns Lotus and has control of Saab.

Indeed, in early 1991, Saab's Malmö plant was closed and Saab's operations became more tightly integrated into GM's European network.
4. Developing new production and distribution networks in Eastern Europe: GM, in particular, is building major new production plants in eastern Germany and in Hungary while Ford is establishing a components factory in Hungary.

The strategies of the European automobile manufacturers

The major European automobile producers are significantly more parochial locationally than either GM or Ford. Figure 9.14 summarizes their production geography. Of the four leading European companies, Volkswagen has pursued the most extensive and systematic international strategy. Outside Europe, VW is the majority partner in the Autolatina joint venture with Ford, which organizes their Latin American operations. VW's production facilities in Mexico have effectively replaced much of the production formerly located at VW's Pennsylvania plant, which was closed in 1988. Mexico is now the sole production location for the VW Beetle. Within Europe, prior to the opening up of Eastern Europe, VW concentrated its production in two countries in a clear strategy of spatial segmentation. High-value, technologically advanced cars are produced in West Germany; low-cost, small cars are produced in Spain where VW is undertaking a massive investment programme. During 1990, however, VW moved very rapidly to establish production of small cars in eastern Germany and to take a major stake in the Czech firm, Skoda.

While VW has been expanding its European production base to incorporate Spain, the Italian automobile firm, Fiat, has been moving in the opposite direction and reconcentrating production in its home market. Fiat is now by far the most dependent on its domestic market of all the major European producers. Fiat currently manufactures its entire EC output in Italy; 70 per cent of its EC sales are also in its home market. Such geographical reconcentration is the outcome of a major strategic shift which began in the early 1980s. But such parochialism may not last. A major element in Fiat's current strategy is to create an extensive production network in the Soviet Union and Eastern Europe where it has the longest-established links of any automobile producer in the world. Fiat's unique history of collaborative ventures in those regions is now the basis for new developments in the changed political environment. Fiat's 'grand

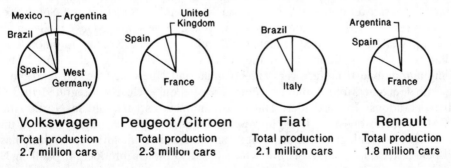

Figure 9.14 International automobile production by the major European manufacturers, 1989 (*Source*: SMMT (1990) *World Automotive Statistics*, London: SMMT)

European design' is to build a manufacturing network extending from the Mediterranean to the Urals.

Between 1982 and 1989 VW increased the proportion of its output produced overseas from 25 to 30 per cent. During the same period, Fiat reduced its overseas production from 20 to 7 per cent. The two French automobile companies, Peugeot–Citroën and Renault, fall between these two positions. Both are strongly home-country oriented in their production. Peugeot–Citroën was formed by a state-induced merger of the separate Peugeot and Citroën companies in 1975, followed by the purchase of Chrysler's European operations in 1978. It has faced major problems in digesting the results of these mergers and has become the leading lobbyist in favour of a strong political stand against Japanese automobile companies in Europe. Eighty-five per cent of Peugeot–Citroën's production is located in France with a further 11 per cent in Spain and 5 per cent in the United Kingdom. It has no production plants outside Europe.

Renault has been, for more than forty years, the French government's national champion, supported by massive state aid, which served to constrain its activities. State control has now been removed and Renault, like Peugeot–Citroën, has been involved in major restructuring. An important strategic development is its new agreement with Volvo. Like Peugeot, Renault has a substantial presence in Spain which is increasingly integrated with its French operations. It also has some production in Argentina. But Renault has sold its major non-European operations – its involvement with the American Motors Corporation (AMC) in North America – to Chrysler and closed its Mexico plant.

Even allowing for the case of VW, the most internationally oriented European automobile producer, the major European companies are far narrower in global terms than the US and, increasingly, the Japanese firms. Apart from some limited involvement in Latin America (and excluding simply local assembly plants in various countries), they are entirely European in their production networks. It is especially notable that, while the Japanese have been busy building large manufacturing plants in North America during the 1980s, the Europeans have not been doing this. Indeed, both VW and Renault have pulled out of North America.

The emergence of NIC automobile producers

The global automobile companies based in Japan, the United States and Europe exert such market dominance that there have been virtually no new entrants in the industry during the past twenty-five years. The single exception to this is the South Korean manufacturers who, in the space of a few years, have emerged as significant international players. As we noted earlier, the development of a Korean automobile industry has been strongly influenced by the policies of the Korean government.[11] The two leading automobile manufacturers are Hyundai and Daewoo. Daewoo is a joint venture with General Motors and its operations are quite closely integrated into GM's global operations. In particular, Daewoo manufactures cars sold in North America by GM's Pontiac division. It is, however, Hyundai which has emerged as a unique case in the developing world: an indigenous automobile manufacturer which has not only penetrated global markets but which has also built a production plant in North America.

Hyundai developed on the basis of acquiring technological know-how but without the direct involvement of foreign automobile producers. Mitsubishi has a minority stake in

Hyundai but control rests firmly with the Korean management. Its first car, the Pony, produced in the 1970s, drew on technology and design from a bewildering variety of foreign sources, including Italy, the United Kingdom, Japan and the United States. The Pony was succeeded by a new car, the Excel, based on Mitsubishi designs, and built in a huge new plant at Ulsan. Both the Pony and the Excel were exported in large volumes, especially to North America. Hyundai's strategy was to compete with the Japanese in a very narrow product range and entirely on price. Such price competitiveness rested on low Korean labour costs, which were a fraction of those in developed countries. Initially, Hyundai's export success was remarkable. In 1986, 300,000 cars were exported to North America, a level more or less maintained for the following two years. On the strength of this success, Hyundai built a second new plant in Korea. More ambitiously still it built a plant in Canada, at Bromont near Montreal, with the capacity to produce 120,000 cars and to employ 1,200 workers directly.

By the end of the 1980s, however, Hyundai was experiencing major problems. Wage costs rose substantially in Korea as part of the country's transition to democracy. North American demand declined dramatically because of the perceived unreliability of the Korean vehicles. The Canadian plant in 1990 was operating at only 40 per cent of its planned capacity and was hit by Chrysler's decision not to purchase 30,000 cars to sell under its own brand name in the United States. Thus, Hyundai, the only world-scale automobile company to emerge from a developing country, faces the difficult process of continuing to compete at a world scale in the face of the radical changes currently sweeping through the industry. It faces the particular challenge of catching up technologically and of moving from being an old-fashioned, low-wage producer to a firm capable of competing with the Japanese. Nevertheless, Hyundai's success in such a short time has been truly remarkable and should not be underestimated.

Apart from Korea, and those countries like India which have resisted the large-scale incursion of foreign TNCs, the role of the developing countries in the TNC's global strategies is quite specific. In the majority of cases they are little more than mere local assemblers of imported CKD kits for local, protected markets. Only a few NICs – none of them in Asia – have developed into more than local assemblers, although in 1985 Malaysia introduced its 'national' car, the Proton, built in close co-operation with Mitsubishi. By far the most important are Brazil, and Mexico.[12] In these few cases a dual role tends to be performed. First, their subsidiary plants produce cars for large local and regional markets. In some cases production is of models which are no longer produced in the TNC's home country. In others, plants have been opened specifically to manufacture totally new models for more extensive geographical markets. Second, this small number of developing country producers acts as sourcing points for specific components which are exported to other parts of the TNC's production network.

This sourcing role is apparent in Brazil but more especially in Mexico, which has the added advantage of proximity to the US market. Since the late 1970s the three leading US producers together with Nissan and Volkswagen have established major operations in Mexico. Much of the installed capacity is for the production of engines which are then exported to other parts of the TNCs' international production networks. Given favourable political developments, Mexico could well come to play a much larger role in an integrated North American automobile production system. As Womack, Jones and Roos (1990, p. 266) observe, 'a new configuration of production for the entire North American region could emerge. GM, Ford, Chrysler, Nissan and VW might assemble in Mexico – for sale to the entire North American market – cheap, entry-level cars and trucks that use parts produced by production complexes in northern Mexico, near the assembly plants.'

Summary

In summary, the strategies of the major automobile producers are more diverse than is often realized. They are also in a condition of flux as major technological and organizational changes sweep through the industry, as 'mass' production is replaced by 'lean' production. There is, undoubtedly, a global battle raging in the automobile industry – 'car wars' – in which there will certainly be casualties. It seems that the next battleground will be Europe as Japanese penetration increases the pressure on local manufacturers. This battle has both an economic and a political dimension as governments, as well as firms, strive to protect and enhance their positions. The long-term strategy of the leading Japanese companies is clearly to establish a major integrated production system in each of the three global regions: Asia, North America, Western Europe. GM and Ford are moving in the same direction and have the advantage of already operating a sophisticated integrated network within Europe. In comparison, the European manufacturers remain far more limited geographically. Only Volkswagen has much of an internationally integrated system. Its withdrawal from production in the United States is being replaced by its growing involvement in Mexico, which could give VW increasing access to the North American market.

Jobs in the automobile industry of the older-industrialized countries

Changes in demand for automobiles in the developed market economies, technological and organizational changes in the production process itself and the increased tendency for the larger TNCs to expand overseas, to integrate their production operations transnationally and to engage in international sourcing of vehicles and components have all combined to produce major changes in employment in the automobile industries of the older-industrialized countries. High levels of import penetration, particularly by Japanese manufacturers, have also adversely affected the domestic motor industry of individual countries, reducing the size of their labour forces.

All the major American and European vehicle manufacturers have been restructuring their operations and, in so doing, have greatly reduced the size of their labour force. Between 1978 and 1989 employment in the US automobile industry fell by 24 per cent, from 470,000 to 355,000. Plant closures and contractions continue and the effects are felt not only by blue collar workers but also by white collar workers. GM is to cut its white collar staff in the United States by 15,000 by 1993. Plant closures have been under way in the United States since the late 1970s. During 1979 and 1980

> domestic automobile manufacturers closed or announced the imminent shutdown of twenty facilities employing over 50,000 workers. As a consequence of these closings and of the output reductions in other auto plants, suppliers of materials, parts, and components to the automotive industry closed nearly 100 plants, eliminating the jobs of about 80,000 additional workers . . . Among the major permanent closings in the last years of the decade were thirteen Chrysler plants employing nearly 31,000 workers, five Ford plants including the huge facility at Mahwah, New Jersey, and seven plants in the General Motors system. Of these twenty-five shutdown, eleven were located in Michigan and six more were in other midwestern states. But even the Sunbelt lost some of its automobile capacity. Ford shut

down its large Los Angeles assembly plant in 1980, while Chrysler closed a small facility in Florida.

(Bluestone and Harrison, 1982, p. 36)

Between 1987 and 1990 GM and Chrysler closed a further ten assembly plants in the United States. Set against these closures, of course, are the newly opened Japanese transplants and the new GM Saturn plant. Even so, there has certainly been a major net job loss in the US automobile industry.

In Europe Ford cut its blue collar labour force by 26 per cent (104,000 to 77,500) and its white collar labour force by 34 per cent (18,300 to 12,000) between 1979 and 1988. A further 3,000 jobs will disappear at Ford's Merseyside plant between 1990 and 1995. Renault's French workforce fell during the 1980s from 89,000 to 77,000. Peugeot–Citroën is cutting its Spanish workforce by 22 per cent. Saab (now part of GM) is reducing its labour force from 17,000 to 11,500 and closing its Malmö plant – which opened only in 1989. In Brazil the Autolatina joint venture between Ford and VW is reducing its labour force in 1991 by 8,000 (16 per cent). Ford's Australian workforce is being reduced by 13 per cent (16,000 jobs) as part of a major restructuring programme.

The large size of automobile plants and the historical tendency for the industry to be geographically concentrated into specific localities pose severe problems. The employment ramifications of the closure of an automobile assembly plant are far more extensive than those caused by the closure of, say, a steel plant or a textile mill. The very nature of the automobile industry – its use of a myriad of materials and components from many different industries – means that the knock-on effects are far greater than in almost any other industry. For example, Bluestone and Harrison (1982) estimated that a loss of 5,000 automobile assembly jobs in Michigan would lead to a further 3,000 job losses in the US automobile industry as a whole and an additional 12,000 job losses in other related industries.

In Europe additional problems have arisen because of the large numbers of immigrant workers employed in the motor vehicle industry. In West Germany and France, in particular, thousands of migrant workers from Turkey, North Africa and southern Europe flocked to the automobile assembly plants during the boom years of the 1960s. For example, more than half of the 17,000 workers employed at the Peugeot/Talbot plant at Roissy in France in the early 1980s were African immigrants. With the drastic down-turn in the fortunes of the European automobile industry in the 1980s and the severe rationalization programmes being pursued by VW, Peugeot and Renault the problems of these so-called 'guestworkers' have become acute. Minority groups in the United States have also been seriously affected by contraction in the motor vehicle industry. For example, 'In August 1979 virtually 30 percent of Chrysler's national employment was made up of black, Hispanic, and other minorities while over half of its Detroit work force was nonwhite' (Bluestone and Harrison, 1982, p. 54).

Each of the traditional motor vehicle-producing nations is, therefore, experiencing major problems of employment change and employment adjustment. In each case, the strong geographical localization of the industry and its tentacular reach into many other industries makes such problems especially serious in particular places. Hardly any of these job losses can be attributed to a relative shift of automobile production to Third World countries for, as we have seen, such geographical relocation has been very limited. They are the result, primarily, of a combination of circumstances: a profound change in the structure of demand for automobiles in the developed market economies, the adoption of new technologies in the production process and the international production and sourcing strategies of the major TNCs as they seek to survive and flourish in an increasingly competitive global market. If the predictions of the replacement of mass production by lean production turn out to be accurate, then there are many more job losses to come. The problem is that

> the current Western automobile workforce is in precisely the opposite position of craft workers in 1913. The introduction of mass production created new jobs for craft workers – these workers made the production tools needed by the new system. By contrast, lean production displaces armies of mass-production workers who by the nature of this system have no skills and no place to go.
>
> (Womack, Jones and Roos, 1990, pp. 235–6)

Notes for further reading

1. There is a huge literature on the automobile industry. The most comprehensive reviews are those arising from the International Automobile Program at MIT: Altshuler *et al.* (1984), Womack, Jones and Roos (1990). See also UNCTC (1983a).
2. Womack, Jones and Roos (1990) provide a highly readable account of the history of technological change in the automobile industry and, especially, the basic characteristics of mass production on the one hand and 'lean' production on the other. Kaplinsky (1988) interprets technological and organizational change within the industry in terms of the restructuring of the labour process.
3. This quotation comes from a letter to the *Financial Times* (18 October 1990) by Garel Rhys.
4. Reich (1989) presents a detailed historical analysis of the evolution of government policy towards automobile producers in France, Britain, West Germany and the United States.
5. See Dicken (1991).
6. Discussion of state policies towards the automobile industry in developing countries can be found in UNCTC (1983a), OECD (1988), Nuñez (1990), Womack, Jones and Roos (1990).
7. These figures are taken from Glickman and Woodward (1989), Table 8.1. See also Mair, Florida and Kenney (1988).
8. The effects of the Single European Market on the strategies of the international automobile producers are discussed in detail in Dicken (1991).
9. Holmes (1990, 1991) provides detailed analyses of the 1965 Auto Pact and also of the effects of the 1988 Canada–United States Free Trade Agreement on the North American automobile industry.
10. The growth and location of Japanese automobile investment in North America is analysed by Mair, Florida and Kenney (1988), Reid (1990), Womack, Jones and Roos (1990) and in Europe by Dicken (1987, 1991).
11. Development of the indigenous Korean automobile industry is discussed by Mukerjee (1986), Amsden (1989), Oman (1989), Womack, Jones and Roos (1990).
12. The automobile industry in Brazil, Mexico and in other Latin American countries is discussed by Jenkins (1984b), Oman (1989), Nuñez (1990), Womack, Jones and Roos (1990).

Chapter 10

'Chips and Screens': The Electronics Industries

Introduction[1]

Of all the manufacturing industries discussed in this book the electronics industry is both the youngest and also the one with the most far-reaching implications for the future economic evolution of developed and developing economies alike. Micro-electronic technology has emerged as the dominant technology of the last three decades of the twentieth century, extending its transformative influence into all branches of the economy and into many aspects of society at large. Although the early stirrings of an electronics industry became apparent in 1901 with the introduction of radio, its really spectacular development is entirely a product of the last four decades. The first step in creating the modern electronics industry was the development of the transistor in the Bell Telephone Laboratories in the United States in 1948. The transistor replaced the thermionic valve or vacuum tube and made possible the development of a *micro*-electronics industry. Unlike the valve or tube, the transistor is a solid-state device which uses materials such as silicon which, with the addition of chemical impurities, can be made to act as a semiconductor of electric current. Within a few years of the initial development of the transistor, semiconductors were being manufactured commercially in the United States. The end of the 1950s saw the emergence of a second major innovation – the *integrated circuit* – which consists of a number of transistors connected together on a single piece or 'chip' of silicon. By the early 1970s it had become possible to incorporate a number of very sophisticated solid-state circuits on to a single chip the size of a finger nail to create a *microprocessor*. This was, in effect, a computer on a chip. It was able to perform the functions which only two decades earlier had taken a whole roomful of valve computers.

The progressive refinement of these basic innovations over a very short period has dramatically increased the power of electronic components and also spectacularly decreased their size. It is now possible to pack millions of individual circuits on to a single chip of silicon less than one square centimetre in size. Increased *miniaturization* has been a most important development for it permits the incorporation of electronic components into a vast range of products, from pocket calculators and pocket tele-visions to highly complex computers, industrial robots and aircraft guidance systems. It is the *increasingly pervasive application* of electronics, rather than its absolute size, which makes the industry of such very great significance. Indeed, the micro-electronics industry is more aptly described as today's 'industry of industries' than the automobile industry.

The extensive ramifications of the electronics industry not only for other sectors of the economy but also for telecommunications and national defence have made all governments increasingly anxious to avoid being left out or left behind in what is a rapidly moving technological scene. The electronics industry, like textiles, steel and automobiles before it, has come to be regarded as the touchstone of industrial success. Hence, all governments in the developed market economies, as well as those in the more industrialized developing countries, operate substantial support programmes for the electronics industry, particularly microprocessors and computers. At the same time, shifts in the global pattern of production of consumer electronics – particularly television and audio products – have led to considerable trading frictions and the implementation of various protectionist measures against imports, especially from Asia to North America and Western Europe.

The electronics production system

In this chapter we are concerned with just two branches of the electronics industry: semiconductors and consumer electronics (primarily television). Figure 10.1 shows where these two subsectors fit into the electronics industry as a whole. The electronics industry proper 'consists of those products or systems that use electronic circuits handling small currents which incorporate "active" components capable of modifying the flow of electricity' (Cable and Clarke, 1981, p. 4).

The core of the electronics industry is the *components* sector. Its most important elements are the active components, based on the semiconductor, which control the flow of electrical current. Semiconductors themselves can be divided into two major categories: *memory chips*, which contain pre-programmed information, and *microprocessors*, which are, in effect, 'computers on a chip'. Semiconductors and related components are employed in a bewildering variety of applications which, for convenience, can be classified into two categories: electronic equipment and consumer electronics. Within the electronic equipment sector there is obviously much overlap, particularly because the computer – the biggest single user of semiconductors – is universally involved in the other equipment sectors. The recent spectacular growth of

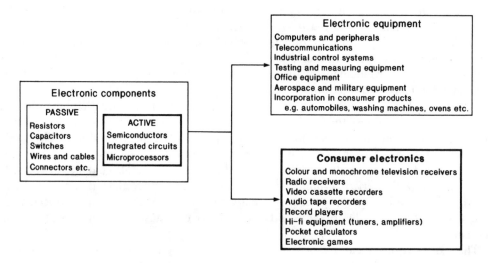

Figure 10.1 The electronics industry

the personal computer has put what was formerly a product used only by industry, commerce or the public sector into individual households. In this sense, the computer has begun to invade the consumer electronics sector as well. Of course, other types of electronic equipment have also been incorporated into various personal and household products, such as automobiles, washing machines, ovens and similar items. But consumer electronics as a sector is usually defined in terms of 'complete' electronic products such as radios, televisions, video and audio recorders, hi-fi equipment, and so on.

Both of the sectors examined in this chapter are highly capital-intensive industries in which very large transnational firms tend to dominate. However, certain activities have also been open to new, dynamic and initially small, electronics firms, particularly in the United States. As we shall see, the semiconductor and consumer electronics industries differ somewhat in their global distribution and, therefore, in the degree to which developing countries are involved. The key to such variation is the extent to which parts of the production process – those which are highly labour-intensive – can be geographically separated from the other stages in the sequence. Finally, the nature of employment in the electronics industry in both developed and developing countries poses a number of significant problems.

Global shifts in the electronics industries

The semiconductor industry

The growth of *production* of semiconductors has been truly phenomenal since their commercial introduction in the 1950s. In terms of volume of production, output virtually doubled every year throughout the 1970s. Geographically, the commercial production of semiconductors began in the United States during the 1950s and for nearly two decades the United States dominated world production. During the 1980s, however, Japan overtook the United States to become the world's leading producer. Figure 10.2 shows the geographical distribution of the production of 'active' electronic components, which include semiconductors, integrated circuits and microprocessors. In 1989 Japan accounted for 42 per cent of the world total, the United States for 26 per cent and the whole of Europe for only 12 per cent. Within Europe the leading producers are West Germany (31 per cent of the European total), France (19 per cent) and the United Kingdom (16 per cent). Outside these three core areas of production the major centres are in East and South East Asia, notably South Korea, Malaysia, Taiwan, the Philippines, Thailand and Hong Kong.

Japanese dominance of DRAM (dynamic random access memories) production, the most important segment of the active electronic components industry, is even more pronounced. Figure 10.3 shows how the United States' dominance of this segment of the industry has been dramatically eroded. In 1974 the United States produced the entire world output of DRAMs. By 1988 its share had fallen to below 20 per cent. Japan now produces four-fifths of the world total. European production remains extremely small and has recently been overtaken by others, notably the group of East and South East Asian producing countries.

The global pattern of *trade* in semiconductors has been described as a 'war of supremacy' between the United States and Japan, with Western Europe striving to

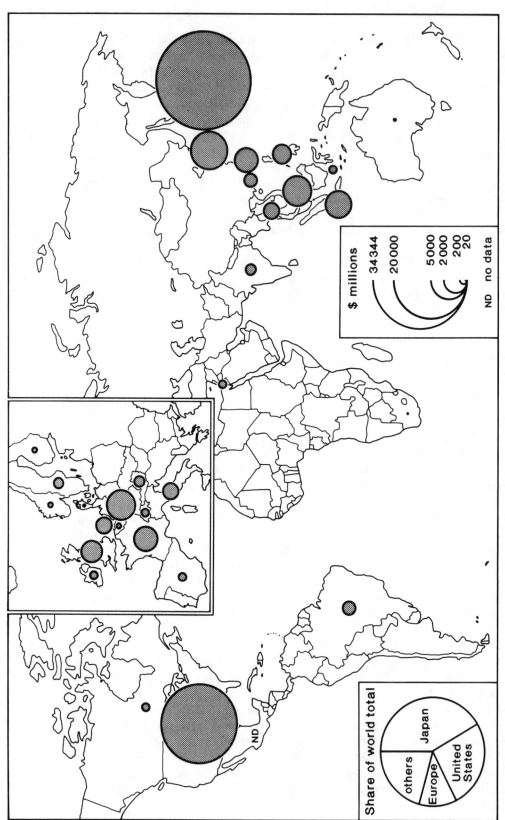

Figure 10.2 World production of active electronic components, 1989 (*Source: Yearbook of World Electronics Data, 1990,* Oxford: Elsevier Advanced Technology)

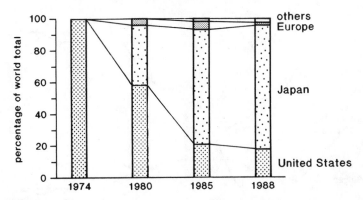

Figure 10.3 The changing location of the production of dynamic random access memories, 1974–88 (*Source*: press reports)

maintain a presence. Figure 10.4 depicts this trading situation. It shows that Japan had a massive trade surplus in electronic components of $19 billion in 1987 while both the United States and Europe were running deficits of more than $5 billion each. The group of newly industrializing countries had a substantial trade surplus in electronic components of $1.3 billion. Figure 10.5 shows the geographical composition of such exports to the United States under the offshore assembly provisions of the US tariff code. It shows some significant changes since the late 1960s. In 1969, 97 per cent of all the imports of semiconductors into the United States under the special tariff codes originated from just four Asian countries. Hong Kong was by far the most important with almost 50 per cent, followed by South Korea (23 per cent), Taiwan (15 per cent) and Singapore (10 per cent). By 1983, however, the picture was very different. The share of the original four had fallen to just over a third of the total. Countries which had no such activity until well into the 1970s had become very important, notably Malaysia (37 per cent of the total) and the Philippines (21 per cent). But important as these exports of semiconductors are, they need to be kept in perspective. The majority

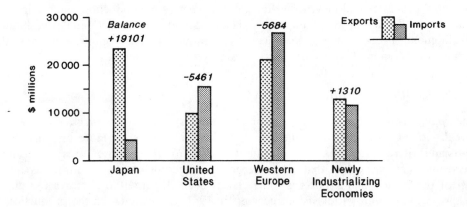

Figure 10.4 Trade balances in active electronic components, 1987 (*Source*: Electronic Industries Association of Japan (1989) *Facts and Figures on the Japanese Electronics Industry*, Tokyo: EIAJ)

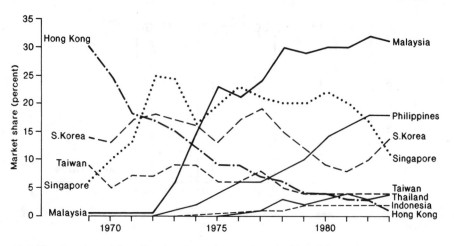

Figure 10.5 The changing origins of semiconductor imports into the United States from Asian countries (*Source*: Grunwald and Flamm (1985), Table 3.7)

of world trade in semiconductors occurs within the *triangular* framework of the United States, Japan and Western Europe.

A number of significant changes have been occurring in the global geography of semiconductor production and trade in recent years, however, changes affecting both the established production areas of the United States and Europe and also the newer centres within East and South East Asia. Within the United States, the locational core of the semiconductor industry remains the Santa Clara valley of California – 'Silicon Valley'.[2] But Silicon Valley's dominance is less marked than it was as new centres of production have emerged in such states as Colorado, Oregon and Utah. Within Western Europe there has been considerable development of semiconductor production in certain more peripheral regions, most notably in Ireland, Wales and Scotland.[3] In the latter case, the Central Valley of Scotland has acquired a substantial semiconductor industry, giving rise to the label 'Silicon Glen' and its description as 'the largest "chipshop" in Europe'. Within East and South East Asia, too, locational change is becoming evident as the 'gang of four' move into more sophisticated products and processes and new, lower-technology, centres of production emerge in such countries as Malaysia, the Philippines, Thailand and Indonesia.[4] Similarly, limited development of certain stages of semiconductor production has begun to emerge in the Caribbean region in close geographical proximity to the US market.

The consumer electronics industries

The *manufacture* of consumer electronics products is generally much more widely spread globally than that of semiconductors. It also displays far more obvious global shifts and, therefore, a greater involvement of Third World countries. It is in consumer electronics, in particular, that the emergence of certain Asian countries as major foci of production and exports is most clearly seen. Figure 10.6 portrays the global pattern of production of television receivers in 1987. Compared with the production of semiconductors, television manufacture is far less concentrated geographically. Whereas well over 90 per cent of semiconductor production is located in the United States,

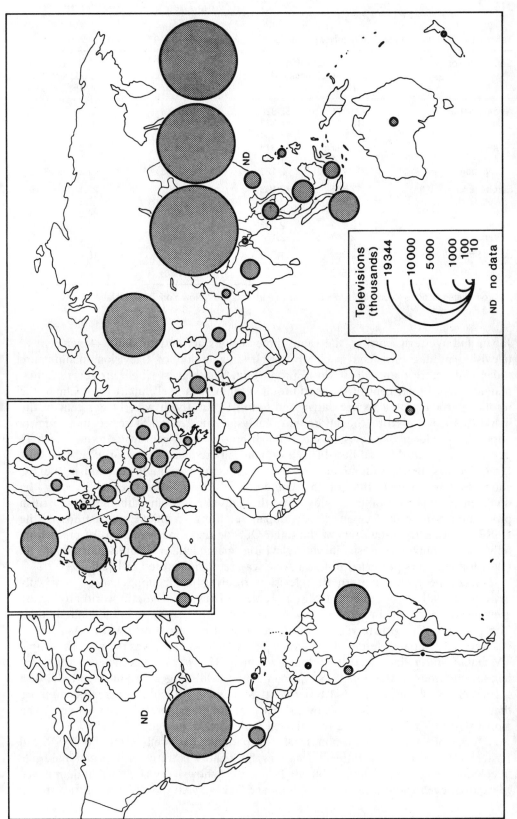

Figure 10.6 World production of television receivers, 1987 (*Source*: United Nations (1989) *Industrial Statistics Yearbook, 1987*, New York: United Nations)

Table 10.1 Changes in television receiver production by major region, 1978–87

Country/region	Production (000 units)		% Change 1978–87	Share of world production (%)	
	1978	1987		1978	1987
Asia (excl. USSR)	21,175	56,826	+168.4	35.0	55.7
Japan	13,116	14,777	+12.7	21.7	14.5
South Korea	4,826	14,664	+203.9	8.0	14.4
China	517	19,344	+3,641.6	0.9	19.0
Malaysia	150	1,240	+726.7	0.3	1.2
Singapore	725	2,123	+192.8	1.2	2.1
Europe (excl. USSR)	17,656	19,193	+8.7	29.1	18.8
West Germany	4,391	3,537	−19.5	7.3	3.5
United Kingdom	2,417	3,022	+25.0	4.0	3.0
Italy	2,172	2,233	+2.8	3.6	2.2
France	2,101	2,184	+4.0	3.5	2.1
United States	9,309	11,310	+21.5	15.4	11.1
World total	60,592	101,985	+68.3	100.0	100.0

(*Source*: United Nations (1989) *Industrial Statistics Yearbook, 1987*, New York: United Nations)

Japan and Western Europe, the same three regions account for only 44 per cent of television production. Asia is clearly the leading region of television manufacture today, with more than half of world production. The biggest single producer, China, manufactures 19 per cent of the world total. Within Asia itself, apart from China, the most important producers are Japan and South Korea (both with 14 per cent of the world total), followed by Singapore and Malaysia. The United States remains a fairly important producer of television sets (though not by American-owned firms), with 11 per cent of the world total but this is a dramatically reduced share for what was, only three decades ago, the dominant television-producing nation. Production in the EC countries now exceeds that of the United States, with West Germany being the dominant European producer followed by the United Kingdom, although most British production is now by foreign, mostly Japanese, firms. Within Eastern Europe, the USSR produces three-quarters of the total. Outside these major regions of production only Brazil ranks as a significant television manufacturer. Three-quarters of all television receivers produced in Latin America are manufactured in Brazil.

In fact, the global pattern of television receiver production changed markedly between the late 1970s and late 1980s, as Table 10.1 reveals. Overall, world production grew by 68.3 per cent. In Europe growth was only 8.7 per cent and in the United States 21.5 per cent. The growth region was Asia (+168.4 per cent) but, within Asia, there was very substantial differential growth. Japan, which, in 1978, was the world's leading TV manufacturer, increased its output by a mere 12.7 per cent. China, on the other hand, came from nowhere in 1978 to become the world's largest producer in 1987, with a 1978–87 growth rate of 3,641.6 per cent! Malaysia's TV output increased by more than 725 per cent, South Korea's by 204 per cent and Singapore's by 193 per cent. The global shift of TV manufacture to Asia was very marked indeed.

Absence of data for China and Taiwan in particular, as well as for the USSR and Eastern Europe, makes it difficult accurately to identify patterns of international *trade* in colour television receivers. Table 10.2, however, shows very clear Asian dominance of exports, even without figures for China and Taiwan. All the Asian countries shown

Table 10.2 International trade in colour television receivers, 1987*

Country/region	Exports ($ million)	% of world total	Imports ($ million)	% of world total	Trade balance ($ million)
Asia	3,651,794	51.6	1,077,171	15.0	+2,574,623
Japan	1,550,399	21.9	53,842	0.8	+1,496,557
South Korea	880,351	12.4	6,375	0.1	+873,976
Singapore	604,804	8.6	235,466	3.3	+369,338
Hong Kong	310,565	4.4	301,550	4.2	+9,015
Malaysia	286,198	4.0	30,269	0.4	+255,929
Europe	3,224,191	45.6	3,935,626	54.9	−711,435
EC	2,642,509	37.3	3,275,054	45.7	−632,545
West Germany	948,705	13.4	816,197	11.4	+132,508
Belgium−Luxembourg	522,544	7.4	181,120	2.5	+341,424
United Kingdom	358,766	5.1	407,202	5.7	−48,436
Italy	312,536	4.4	651,160	9.1	−338,624
Finland	244,937	3.5	48,309	0.7	+196,628
Sweden	87,588	1.2	173,998	2.4	−86,410
Austria	229,615	3.2	136,735	1.9	+92,880
Netherlands	124,454	1.8	406,974	5.7	−282,520
France	211,478	3.0	532,654	7.4	−321,176
Denmark	74,744	1.1	78,602	1.1	−3,858
Portugal	36,430	0.5	78,845	1.1	−42,415
Spain	46,533	0.7	71,866	1.0	−25,333
United States	—	—	1,372,948	19.2	
Canada	—	—	250,838	3.5	

* Market economies only; excludes China, Taiwan, USSR, Eastern Europe
— Data unavailable

(*Source*: United Nations (1989) *Yearbook of International Trade Statistics, 1987*, New York: United Nations, p. 669)

in the table have a large trade surplus in colour television receivers. Conversely, the United States alone accounted for almost 20 per cent of world imports of colour TVs and had negligible exports. Europe's trade performance is more respectable than might have been expected. Several countries have a trade surplus in colour TVs even though the region as a whole is in deficit.

The very rapid increase in Asian exports of television sets in the last fifteen years has resulted in deep *import penetration* of European and, especially, North American markets:

> From 1948 till 1962, the USA was totally self-sufficient in television manufacture. In 1962 the first Japanese imports developed in the black-and-white sector. The first imported colour television sets were introduced into the US market in 1967 and within three years, imports were making up 17 percent of the US market for colour sets and about 50 percent for black-and white . . . Beginning in late 1975, there was a marked acceleration of this import penetration.
>
> (Turner, 1982, p. 55)

Although Asian import penetration has been greatest in North America, it has also been very considerable in Western Europe. Production and export figures tell us only *where* the sets originate, however; they do not tell us *who* actually produces them. In fact, a substantial proportion of European production is in foreign-owned, primarily

Japanese and other Asian, plants located in Europe. This is even more true of the United States, where only one indigenous manufacturer, Zenith, remains in business.

These shifting global patterns of production and trade in semiconductors and television sets (as well as in other consumer electronics products) reflect a complexity of causal factors. Perhaps more than in any other industry the global geography of the electronics industry is highly volatile in response to both technological and organizational developments within the corporate strategies of the major producers and also the actions of national governments. It is to these factors that we now turn.

The changing pattern of demand

Semiconductors

Demand for semiconductors depends upon the growth of demand for products and processes in which they are incorporated: it is a *derived* demand. In the United States, at least initially, most of the stimulus came from the defence–aerospace sector and this explains much of the initial growth of the semiconductor industry in the United States. In Europe, and especially in Japan, defence-related demand has been far less important than industrial and, particularly, consumer electronics applications. Nevertheless, governments have often been dominant customers, either directly or indirectly, throughout the semiconductor industry's short history. Overall, the biggest end-user is the computer industry.

Quite apart from defence and military applications, however, demand for semiconductors has grown spectacularly. A major reason has been the phenomenal decline in their selling price: the fall has been vertiginous. Partly because of technological developments (discussed in the next section) and partly because of fierce price competition between producers the price of semiconductors is a tiny fraction of that prevailing only a short time ago. Between the mid-1950s and the late 1970s, for example, 'the cost per electronic function . . . [fell] . . . by a factor of 100,000 in less than two decades. The cost per memory bit, $50 in the mid-1960s, dropped to less than $0.0005 in the 16,000-bit random access memories (16k RAMs) of the later 1970s' (Siegel, 1980, p. 3). However, the memory capacity of semiconductors has grown phenomenally since the 1970s. 16k RAMs now seem ludicrously limited. The industry standard by the late 1980s was the 1-megabyte DRAM, whose price in early 1991 was a mere $4.50. Higher-capacity chips are appearing so quickly that each affects the price of the others. 'For the first time in its brief history, the semiconductor industry is going to be making, and trying to sell, three generations of chips in large volumes at the same time. Prices should tumble as each cannibalises sales of the others' (*The Economist*, 23 February 1991).

Such dramatic price reductions have led to the application of semiconductors in an increasingly wide range of other industries. As we saw in Chapter 4, the microelectronics industry, and especially the semiconductor, is at the heart of the information technology revolution which is transforming both products and processes throughout the global economy. Such wider use extends the market for semiconductors and makes possible further price reductions, and so on.

But growth in demand for, and therefore the price of, semiconductors has been far from continuous and uninterrupted. The industry is subject to spectacular *fluctuations in demand*; gluts and shortages are endemic. Supply gluts further intensify price competition; at other times, severe shortages of chips have created major problems and raised prices for end-users. For example, at the end of 1986 a standard memory device, bought in large quantities, sold for less than $2. By early 1987 its price had risen to $2.70; by summer 1988 the same item was costing between $8 and $10. Then the price began to tumble again as the chip cycle pursued its erratic course. The situation is made more complex by the strongly *segmented* character of demand for semiconductors. There are six major types of semiconductor, each of which has rather different patterns of demand:

1. *Standard devices*, which are technically indifferent to their final use.
2. *Exclusive devices*, which, like standard devices, are technically indifferent to final use but which are produced by one or a few producers with a technological monopoly.
3. *Specific devices*, which are mass produced but which can be used only in specific applications.
4. *Custom devices*, which are manufactured to very particular requirements of a specific user.
5. *Microprocessors*, which can be mass produced and used for multiple purposes but which can be programmed for particular applications.
6. *Semicustom devices*, in which certain parts are mass produced but which can then be tailored to a particular use by leaving the final connections to be decided in the light of the user's requirements. Such devices are also known as *application-specific integrated circuits* (ASICs). Demand for them has been growing exceptionally rapidly.

In general, the geography of demand for semiconductors correlates closely with the geography of production. It is overwhelmingly concentrated in the three major regions of the United States, Japan and Western Europe. But as suggested above, the nature of demand varies between these three areas with the defence and aerospace demands being especially significant in the United States.

Consumer electronics

The pattern of demand for consumer electronics products such as television receivers differs considerably from that for semiconductors. First, of course, they are *final-demand products*. Second, they are products for which demand is income-elastic, whereas this is less so for semiconductors in which industrial and government demand is especially important. In many ways, in fact, demand for television receivers is similar to that for automobiles. Demand for sets grew especially rapidly throughout the 1960s and well into the 1970s, with a strong geographical concentration of demand in the higher-income countries. But by the end of the 1970s, in addition to the effects of economic recession on consumer purchasing, demand for television sets was approaching saturation in the developed market economies. Replacement demand had become the major characteristic of the industry, creating severe problems for producers because, as we saw in the previous chapter, the growth of demand in a saturated, replacement market is inevitably very much slower and more volatile than in a less developed market.

Hence, the most likely increases in demand for television sets in the coming decades will be in the NICs and in Eastern Europe, assuming that incomes in these countries continue to grow. In the traditional markets of North America, Europe and Japan, however, television manufacturers face real difficulties in adjusting to prevailing market conditions. It is for such reasons that the consumer electronics companies are constantly seeking ways of developing new, related products. During the 1980s the major development was the video cassette recorder (VCR). Demand for VCRs is still relatively immature. There is still a strong 'first-time buyers' market in most countries in Europe. Recent estimates suggest that in Europe as a whole 49 per cent of households own a VCR compared with 81 per cent in the United States and 116 per cent in Japan (*Financial Times*, 11 March 1991).

Production costs and technological change

In the fiercely competitive environment of the global electronics industry the drive to push down production costs is the paramount consideration. This is certainly true in the semiconductor and television sectors. Hence, both have become increasingly captial-intensive whilst retaining a considerable degree of labour intensity in certain parts of their production sequence. The urge to minimize production costs is reflected both in the nature and the rapid rate of technological change and in the changing geography of production at various spatial scales.

Semiconductors

The pace of technological change has been especially rapid and far-reaching in semi-conductor production, where it is now necessary for production equipment to be replaced or updated every two or three years. The cost of establishing a semiconductor plant has escalated accordingly. In the 1960s such a plant could be set up for roughly $2 million; by the early 1970s between $15 million and $20 million was needed; ten years later, in the early 1980s, an investment of between $50 million and $75 million was necessary. By the late 1980s a new chip production facility was costing $150 million. As newer generations of chips are developed, the costs escalate even further. In 1990 alone, Japanese semiconductor companies spent around $7 billion on production facilities for 4 Mb chips. An experimental plant for 16 Mb chips cost $1 billion. Even allowing for inflation it has obviously become a much more costly operation in which the *capital barriers to entry* have increased very substantially. Not surprisingly, therefore, the industry is one of the most capital- and research-intensive of all manufacturing industries.

Two considerations are especially important in the manufacture of semiconductors. One is the reliability of the product itself. A great deal of effort and expenditure is devoted to improving the 'yield' of the manufacturing process and reducing the number of rejected circuits. The second is the need to pack as many circuits as possible on to a single chip. One of the most remarkable features of the semiconductor industry is the fact that the number of components per chip has increased phenomenally (Figure 10.7). In 1965, each chip could store 1 k (1,000) 'bits' of information, by 1970 the capacity had increased to 10 k, by 1977 to 16 k. Since the 1970s a new generation of memory chips has appeared approximately every three years; each one has had four times the memory capacity of the previous generation.

It is the combination of these two needs – to increase both the yield and the number of circuits per chip – which accounts for the steep increase in the cost of setting up a

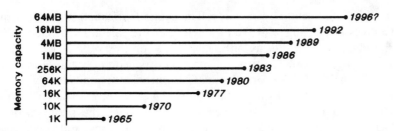

Figure 10.7 The increasing capacity of memory chips (*Source*: based, in part, on the *Financial Times*, 7 February 1990)

new semiconductor plant. Every major increase in the number of components per chip demands far more sophisticated and expensive manufacturing equipment which, in turn, becomes obsolete more quickly. At the same time, the trend is towards larger, more capital-intensive and research-intensive production. But to understand the changing global structure of the semiconductor industry we need to look more closely at the production process itself. We also need to take into account the different types of semiconductor being produced, because the potential for mass production varies greatly between the standard device and the custom device. Production of standard devices demands large-scale mass production; manufacture of custom chips demands much smaller runs. However, the development of semicustom and application-specific (ASIC) devices has extended the use of mass production techniques into new areas.

The general sequence of semiconductor production is shown in Table 10.3. The differing characteristics of each stage have very important implications for the spatial organization of the industry at a global scale. The process begins with the *design* of a new circuit (1). Its precise form will obviously depend upon the function it is to perform. Production of complex circuits involves the superimposition of a series of separate layers, each one being produced initially as a pattern or *mask* (2), from which

Table 10.3 Stages in integrated circuit production

Stage
1. *Design of a new circuit* The precise location of each element in the circuit and the connections between them.
2. *Production of masks* Each mask represents an individual circuit layer and contains hundreds of identical images next to one another.
3. *Fabrication of wafers* (a) Production of pure silicon crystal. (b) Slicing into wafers roughly 0.5 mm thick. (c) Etching of the patterns contained on the masks on to the silicon wafer using photolithography techniques and a variety of chemical 'dopants'. The complete device is built up in layers. (d) Separation into individual chips and mounting into separate package.
4. *Assembly of integrated circuit* (a) Bonding or wiring of circuits to external electrodes using extremely fine wires. (b) Sealing of packages.
5. *Final testing and shipping*

(*Source*: UNIDO (1981), pp. 69–71)

the actual circuits will eventually be made. The actual *fabrication* stage (3) consists of a number of processes starting with the production of the silicon itself. Generally, this is done by specialist chemical firms although a few of the very large semiconductor manufacturers produce their own silicon. The silicon crystal, formed into a cylindrical rod roughly 100 mm in diameter, is sliced into individual wafers 0.5 mm thick. The circuits are etched on to the wafers, layer by layer, using the masks in a photolithographic-chemical process. Each wafer contains large numbers of identical circuits which are separated out and individually packed. These individual circuits are *assembled* (4) into the final integrated circuit or microprocessor using a bonding/wiring process. They are then *tested* (5) and *shipped* to the customer for use in the final product.

A particularly important distinction exists between the design and wafer fabrication stage on the one hand and the assembly stage on the other. Each has different production characteristics and neither needs to be located in close geographical proximity to the other. Design and fabrication require high-level scientific, technical and engineering personnel while the fabrication stage itself requires an extremely pure production environment and the availability of suitable utilities (pure water supplies, waste disposal facilities for noxious chemical wastes). In contrast, the assembly of semiconductors is carried out using low-skill labour, notably female, and although there is still a need for a 'clean' production environment this is less critical than for wafer fabrication. The low-weight/high-value characteristics of semiconductors permit their transportation over virtually any geographical distance. Hence, it is the assembly stage of the production sequence which has been most susceptible to relocation to low-labour-cost areas of the world.

Developments in the technology of semiconductor production – new lithographic techniques, new methods of wafer fabrication and the introduction of automation at all stages of the process – have created considerable changes in the industry. In particular, the relative importance of labour costs has been reduced substantially. In addition, labour costs are relatively far less important in the production of complicated and high-value integrated circuits than in simple devices. At the same time, there have been increased pressures to enlarge the scale of production plants: 'a wafer fabrication plant, in order to be profitable, needs to produce at least 10 to 30,000 wafers per month, i.e. several million complex circuits belonging to the same family per annum. Increasing minimum plant size in turn requires an expanding turnover and thus is bound to lead to a further intensification of worldwide competition for market access' (UNIDO, 1981, p. 169).

Overall, therefore, the manufacture of semiconductors is a highly capital- and research-intensive industry but one in which there are distinct 'breaks' in the production sequence. Such breaks can be used by producers, as we shall see, to create a complex global geography of production.

Consumer electronics

In general, the manufacture of television receivers, like that of most other consumer electronic products, is less advanced technologically than the production of semiconductors. In product cycle terms, television sets have come to be regarded as mature products in which the technology has become standardized and in which the emphasis is on reducing production costs through economies of scale and the use of less-skilled labour. There is a good deal of validity in this view. But colour television production in recent years has shown signs of technological rejuvenation; a 'de-maturing' similar to that evident in the automobile industry. This is reflected both in new process technology and also in the development of new functions for television receivers and of new, related products such as video recorders and teletext systems as well as the imminent introduction of high definition television (HDTV). These are part of broader developments in communications systems in which the domestic television receiver is seen, potentially, as the centre of a sophisticated entertainment, educational, commercial and general information system. At the same time, developments in HDTV are closely related to the increasingly sophisticated visual display requirements of industrial and commercial users, particularly in the information technology industries.

The production process itself consists of three related stages. As in semiconductor manufacture each stage has rather different characteristics which have important organizational and locational implications. The design stage is highly research-intensive, particularly as efforts are made to develop new functions for television sets and to improve the efficiency of the production process. The manufacture of components, particularly that of television screen tubes, is heavily capital-intensive. It is a stage in which economies of scale are especially significant: the optimal scale of production for television tubes is approximately 1 million per year compared with roughly 400,000 for complete sets. Again, it is the assembly stage which is the most highly labour-intensive, employing large numbers of low-skilled, mainly female, workers. The potential advantage of low-labour-cost areas is clear.

There have been two particularly important technological developments in television manufacture in recent years which have affected the production process very considerably. One is the progressive simplification of set engineering, as reflected in the number of components used. During the 1970s alone, the number of components in the average television set declined from 1,400 to 400. The other major development is increased automation of assembly, particularly the use of automatic insertion techniques, which were pioneered by Japanese firms but which are now widely spread throughout the industry. Reductions in the number of components used and increased automation of the production process have together altered the needs for different types of labour. In general, fewer skilled workers are now needed to manufacture a television set.

The role of governments in the electronics industries

Semiconductors

National governments throughout the world involve themselves in the electronics industry to a greater or lesser degree. Such intervention has important implications for the kinds of strategies pursued by producing firms, which are examined in the following section. Government involvement is most extensive and fundamental in the semiconductor sector as well as in computers and telecommunications. In these sectors, government interest dates from the industry's earliest days.[5] It is easy to appreciate why governments have attempted to intervene in the development of the semiconductor industry. Semiconductor production is recognized as a key technology with enormous ramifications throughout the economy. If a country is to benefit fully from it

– or if it is to avoid being left behind – it must have access to what is an expensive and rapidly changing technology.

This access may be achieved in several ways. One is to build an indigenous production capacity based upon domestically owned firms. Another is to attract foreign semiconductor firms to establish production units. A third is to purchase semiconductors on the open market and concentrate on developing the end-uses. Problems are inherent in all three options. Setting up a viable domestic industry may be beyond the means of many countries. On the other hand, relying on foreign investment or the open market may lead to problems of dependency of supplies on foreign sources. Such potential vulnerability may be important not only for industrial applications but especially for defence. Semiconductor technology is at the heart of all modern defence systems, hence the sensitivity of most national governments to developments in this industry.

The particular policies pursued by governments reflect their specific national circumstances, including the country's relative position in the global semiconductor industry. In the United States, the country in which the semiconductor industry originated, the dominant influences have been the federal defence and aerospace sectors. In the early days of the semiconductor industry these set the direction and nature of the industry's development because they were its dominant customers. Although defence-related forces are now less important to the American semiconductor industry their influence persists. In recent years, as the United States' lead in semiconductor production has been eroded, there has been much broader criticism of the direction (even the existence) of US policy towards the micro-electronics industry. The sharp deterioration in the country's global position as a semiconductor producer has led to strong political and industry lobbying for targeted policies. The relative decline of the industry is seen as a threat to national security, both economic and military.

Apart from some Defense Department support for Sematech, a consortium of US semiconductor firms, however, no specific government policy towards the industry has yet emerged except in the area of trade. In 1986 the US government signed a 'semiconductor pact' with Japan. This arose out of allegations by the United States that Japanese producers were 'dumping' chips at excessively low prices in the United States (and, therefore, undercutting US producers) and also that the Japanese were unfairly restricting access to their domestic market. The pact expires in 1991 and is being renegotiated with a particular US emphasis on access for American semiconductor manufactures to the Japanese market. The dumping issue has receded.

Advocates of an explicit US government strategy towards the semiconductor industry point, in particular, to the case of Japan. In Japan policy has been directed to the industrial and consumer applications of semiconductors. We have already discussed the general nature of Japanese industrial policy (Chapter 6) but we should recall that micro-electronics and the related computer and information technologies form the central focus of the drive to develop knowledge-intensive industries. Japanese policy in this sector, as in others, has been one of general guidance through MITI and the establishment of a suitable framework in which the private sector firms can operate competitively. The initial aim was to avert technological dominance by the United States by discouraging direct foreign participation in the Japanese semiconductor industry and acquiring foreign technology through other means. Protection of the industry was extremely tight until the end of the 1970s, both through import controls and restrictions on inward investment.

Since that time Japanese policy has become more liberal. But the Japanese determination to get ahead of the competition in specific types of micro-electronics remains very strong. A clear example is the very large scale integration (VLSI) project begun in 1976 on the initiative of MITI. This involves research collaboration between the five major Japanese companies to develop highly sophisticated integrated circuits for the next generation of computers. The Japanese government funded some 40 per cent of the total cost. Although MITI's involvement has certainly been important, Fransman (1990) argues that its actual contribution has been less significant than is often believed.

Turning to Western Europe, the third major focus of semiconductor production, we find a very uneven policy picture. Not only does the European semiconductor industry lag a good way behind those of the United States and Japan but also much of its production capacity is in American-owned and, more recently, Japanese-owned plants. In general, European governments have not only welcomed such investments but also, in some cases, assiduously courted the foreign semiconductor firms. According to Dosi (1983), Western European policies towards the semiconductor industry have evolved in three stages. First, the period to the mid-1960s was largely one of non-intervention (apart from defence-related R & D and some bias towards national producers in government purchases). Second, between the mid-1960s and mid-1970s government focus on the computer industry gave some stimulus to semiconductor research. But in neither period was government involvement particularly influential. The third period, from the mid-1970s onwards, marked a major intensification in government involvement, this time focused on the information technologies including micro-electronics.

Since the early 1980s European governments have been involved in supporting major collaborative ventures in information technologies in general (the ESPRIT programme) and specifically in semiconductors. The $4 billion (£2.4 million) JESSI programme – the Joint European Submicron Silicon Initiative – is concerned with developing advanced microchip technology. There is also increasing pressure on non-European semiconductor manufacturers to locate more than merely assembly operations in Europe. As 1992 approaches, the objective is to persuade US and Japanese companies to locate more of their design and fabrication plants in Europe. The aims of these European initiatives are threefold: to protect European capacity in the core technologies of micro-electronics; to accelerate innovation; and to encourage cross-border links between national electronics firms within Europe. A major problem, however, is whether or not to allow non-European firms with major European operations to participate in the collaborative programmes.

The European Commission has also been increasingly active in protecting the semiconductor industry against what are seen to be unfair trading practices. In 1990 it secured a voluntary agreement with the Japanese on the minimum prices at which standard memory chips would be sold in the EC. Subsequently, the Commission initiated anti-dumping measures against Korean semiconductor manufacturers who were alleged to be dumping chips at below acceptable prices. In effect, this was the price agreed with the Japanese, who were regarded as the lowest-cost producers.

Consumer electronics

In general, government involvement in the consumer electronics industry has been far less extensive than its involvement in semiconductors. The involvement has also been primarily defensive.[6] Until the 1970s, of the major industrial nations only Japan and, to a lesser extent, France had adopted specific policies towards consumer electronics. In

Japan consumer electronics, and especially television receivers, were regarded as an important sector for national development not only in itself but also as a significant complement to the semiconductor and electronic components sectors. Thus, consumer electronics in Japan were given the 'usual treatment' reserved for targeted growth sectors. Emphasis was placed on creating a highly efficient and productive sector able to achieve world leadership in products such as colour television sets. The policy was both protective and stimulative. Of the European countries, only France had adopted a long-term strategy explicitly for consumer electronics, mainly by encouraging Thomson as the major television manufacturer and by protective measures.

In Western Europe as a whole, in fact, protection of the television market has operated since the early 1960s through the existence of exclusive transmission systems (notably the PAL and SECAM systems), which, until they began to expire in the 1980s, had been used to keep out Japanese competition. Before 1970 licences were not granted at all to Japanese manufacturers; after 1970 some limited access was allowed. But protection was mainly against the larger sets and did little to prevent the inflow of smaller sets, for which demand was growing most rapidly.

The rapid increase in import penetration led to most Western countries adopting quantitative trade restrictions against television imports. In 1977 a 'voluntary export restraint' agreement was reached between Japan and the United States which substantially reduced the volume of Japanese imports. Restrictions were subsequently extended to South Korea and Taiwan producers. At the same time some European governments, notably the British, set out to attract Japanese consumer electronics firms to set up production plants. The French were far more reluctant although they have subsequently softened their attitude. France, however, continued to follow a 'national champion' policy in its support of Thomson. Voluntary export restraints also apply, at the European Community scale, to VCRs.

Currently, the major policy issue facing both US and European governments in this industry relates to the new technology of high definition television (HDTV). Again, the fear is that the Japanese will gain the initial advantage in a product which, it is believed, will be a huge income generator. The US government, the European Commission and the Japanese are locked in battle over agreement on a global HDTV standard. Each is trying to get its own system accepted and thus give a major advantage to its own domestic industries.

Among developing countries it is the three Asian NICs – South Korea, Taiwan and Singapore – which have adopted the most positive policies towards the consumer electronics sector as part of their export-oriented industrialization strategy.[7] Singapore and Taiwan have been especially open to investment by TNCs, particularly from Japan. In contrast, South Korea's strategy has been one of developing an indigenous industry and providing far less scope for the direct involvement of foreign firms. For example, the South Korean government has had a major influence on the development of the country's electronics industry using a mix of policies, including:

1. protectionist trade barriers, both tariffs and a ban on the import of foreign-made electronics;
2. provision of low-interest capital for companies in targeted sectors, which has given the state an influence on industry decision-making;
3. heavy government investment in electronics R & D, particularly in the design of semiconductors, the results of which are made available to Korean producers.

'The result of these co-ordinated initiatives by capital and the state is that ... South Korea is the only one of Asia's newly industrializing countries (NICs) "that has developed a manufacturing base to produce high value-added goods sufficient to sustain high growth rates over the next decade"' (Henderson, 1989, pp. 66–7). A broadly similar state electronics strategy has been followed in Taiwan.

Faced with increasing competition in the global semiconductor industry, the South Korean government has been urging Korean electronics companies to adopt a multi-pronged strategy, including:

1. diversifying their geographical markets to reduce dependence upon the United States;
2. establishing overseas production facilities either to achieve lower production costs or to circumvent trade barriers;
3. developing their own brand names to reduce their dependence on original equipment manufacturers;
4. raising expenditure on R & D in order to lessen dependence on foreign technology;
5. moving their product emphasis towards higher value-added products.

Corporate strategies in the electronics industries

In such a volatile technological and competitive industry as electronics, firms inevitably employ a whole variety of strategies to ensure their survival and in pursuit of growth. However, corporate strategies – whether offensive or defensive in nature – increasingly are being implemented at a *global*, rather than a purely national, scale. Indeed, in the face of fierce global competition, firms have been systematically *rationalizing* and *reorganizing* both their domestic and overseas operations. It is the inexorable pursuit of such strategies – set within the context of rapidly changing market, technological and political forces – which does most of all to shape and reshape the global map of production and trade in the electronics industry. Of course, many factors influence the particular mix of strategies employed. One important variable is size of firm: large firms tend to operate in rather different ways from smaller firms. Another influence is, undoubtedly, a firm's geographical origins: the domestic context in which it has developed. There are substantial differences in behaviour between American, Japanese and European electronics firms.

A strategy common among some firms has been to specialize in specific market segments, to pursue a 'niche' strategy such as the manufacture of high-value micro-circuits for very specific applications. Among other firms, however, the preferred strategy has been to increase the degree of vertical integration. Among television manufacturers, particularly American and European firms, a common strategy has been that of product diversification to reduce their dependence upon a product in which competition from Japanese and other Asian producers is especially fierce. Cutting across these strategies are those of increasing internationalization of production, of automation and, overall, of rationalization and reorganization on a global scale. One characteristic common to each major region of production and to both semiconductor and television manufacture is the very high degree of firm concentration. Increasingly, a relatively small number of very large transnational corporations dominate production, a reflection of the technological and production characteristics of these sectors (the accelerating necessity of very large capital investments brought about by the rapid and highly expensive nature of technological change). In such circumstances, small-scale operations become less and less viable.

Strategies in the semiconductor industry

The transformation from a fragmented industry structure, in which small and medium-sized firms were the norm, to one of large-firm dominance and to a more restricted role for the small firm, has been especially dramatic in the semiconductor industry of the United States. During the late 1950s and through much of the 1960s, entry into the semiconductor industry was relatively easy as the proliferation of new small firms – particularly in Silicon Valley – demonstrated. New firms sprang up overnight, often spinning off directly from established firms (many of which were, themselves, relatively new).

The starting point was the departure of William Shockley, one of the pioneers of transistor technology, from Bell Telephones to set up on his own in Palo Alto, California. But this was only the beginning:

> In 1957 eight of his young assistants . . . quit the company to form, with the backing of Fairchild Camera and Instruments, Fairchild Semiconductor. Although Fairchild soon became one of the top merchant semiconductor houses and introduced a number of key technological advances, its Eastern management was unable to satisfy the innovators. In clusters of twos and threes, a large number of Fairchild's researchers and executives, including the eight founders, quit to found or re-organise other semiconductor firms in the area. The first group formed Rheem Semiconductor (Raytheon) in 1959, and two other groups . . . formed Signetics and Amelco (now Teledyne Semiconductor) in 1961. In 1967 Fairchild Semiconductor's general manager . . . left to rebuild a merger of Molectro (another Fairchild spinoff) into National Semiconductor as a major competitor. In 1968 . . . [two other Fairchild founders] . . . quit to form Intel. And the following year, market manager . . . took seven other Fairchild employees with him to form Advanced Micro Devices. In 1971, 21 of the 23 semiconductor firms in the Bay Area could trace their ancestry to Fairchild . . . The Santa Clara Valley proved a fertile breeding ground for semiconductor companies, as well as other high technology ventures. Stanford University, by carrying out millions of dollars in electronics research for the Pentagon, creating the Stanford Research Institute, and leasing Stanford land to high technology companies, had consciously created a 'community of technical scholars'.
>
> (Siegel, 1980, p. 4)

The result was the emergence of a remarkable geographical cluster of semiconductor and high-technology industries.

The proliferation of new small firms in the US semiconductor industry slowed down as the barriers to entry increased during the 1970s. The dominance of large firms became especially marked in the manufacture of standard semiconductor devices which depend on mass production technology. Outside the United States, the small entrepreneurial firm, starting its manufacturing life in the owner's garage, was never as important in the semiconductor industries of Japan and Western Europe. In these areas production tended to be within large, diversified electronics companies from the beginning.

Table 10.4 lists the world's ten leading semiconductor firms. Together this small group produced almost 60 per cent of the world output in 1989. The leading four firms alone accounted for almost a third of world production. Comparison with Table 9.5 shows that semiconductor production is far less highly concentrated than automobile production. Six of the top ten semiconductor firms are Japanese, including the top

Table 10.4 The world's leading semiconductor producers, 1989

Rank	Company	Country of ownership	Share of market 1989 (%)	Rank 1978	Change in rank 1978–89
1	NEC	Japan	8.9	4	+3
2	Toshiba	Japan	8.8	8	+6
3	Hitachi	Japan	7.0	6	+3
4	Motorola	United States	5.9	2	−2
5	Fujitsu	Japan	5.3	20	+15
6	Texas Instruments	United States	5.0	1	−5
7	Mitsubishi	Japan	4.7	14	+7
8	Intel	United States	4.4	11	+3
9	Matsushita	Japan	3.4	12	+3
10	Philips	Netherlands	3.0	5	−5

(*Source*: Industry sources; press reports)

three; a further three are US firms and only one (in tenth place) is European. In fact, if IBM were included, it would rank first but, because all IBM's semiconductor production is consumed in-house, it tends not to appear in industry league tables. As the final column of Table 10.4 reveals, there were major changes in rank between 1978 and 1989. The biggest jump in rank was achieved by Fujitsu, which moved from being the twentieth world semiconductor producer in 1978 to fifth in 1989. Conversely, both Texas Instruments and Philips fell five places.

Such a simple listing of the leading semiconductor firms masks some important differences between them. Semiconductor producers can be classified into three broad types:

1. *Vertically integrated captive producers*, which manufacture semiconductors entirely for their own in-house use.
2. *Merchant producers*, specialist firms which manufacture semiconductors for sale to other firms.
3. *Vertically integrated captive-merchant producers*, which manufacture semiconductors partly for their own use and partly for sale to others.

Most US firms have tended to fall into categories (1) and (2). IBM is the biggest vertically integrated captive producer but there are others such as AT&T, General Motors and Hewlett-Packard. At the other extreme are the merchant producers which manufacture entirely for the open market. The leading US merchant firms are Texas Instruments, Motorola and Intel. In contrast, the leading Japanese and European semiconductor firms are primarily vertically integrated captive-merchant producers. They are, in general, parts of complex, diversified electronics companies operating across a very broad spectrum.

A general trend among the large semiconductor producers, especially in the United States, has been towards increased vertical integration. On the one hand, many large-scale users of semiconductors, particularly computer manufacturers but also automobile firms and others, have been integrating backwards into semiconductor production. On the other hand, the large specialist merchant producers began to integrate forwards into various user industries. Among the already highly integrated Japanese and European firms the strategy has been to strengthen such functional connections and to develop new applications in both industrial and consumer electronics. As in the United States, many of the major users of semiconductors have become increasingly involved in the manufacture of custom devices for their own applications.

Acquisitions, mergers and strategic alliances

In striving to compete in global markets semiconductor firms from the United States, Japan and Europe have followed the two routes of acquisition and merger and of strategic alliances. During the 1970s and 1980s there was a substantial wave of mergers and acquisitions, both domestic and international. Within the United States, for example, Mostek was acquired by United Technologies, Synertek by Honeywell, Intersil by General Electric, American Microsystems by Gould, Signetics by Philips. Particularly vulnerable to takeover were the fast-growing smaller firms; by 1980 only seven of thirty-six post-1966 start-up companies were still independent. One of the biggest acquisitons of the late 1980s was the takeover of the British company ICL by the Japanese firm Fujitsu. Within Europe itself the biggest merger was that between the Italian firm SGS and the French firm Thomson to form SGS–Thomson, which is now the second largest semiconductor producer in Europe after Philips. In 1989 SGS–Thomson bought the sophisticated British producer INMOS, from Thorn-EMI.

International strategic alliances have become increasingly common in the semiconductor industry for all the reasons discussed in Chapter 7. In particular, the massive costs of R & D, the incredibly rapid pace of technological change and the escalating costs of installing new capacity all contribute towards the attractiveness of forming strategic alliances:

> Peace and love, wherever you are. Intel has done it with NMB, Siemens has done it with IBM. Just as, a little while back, SGS–Thomson did it with OKI, Texas Instruments with Hitachi, and Motorola with Toshiba. Everywhere, big chipmakers have joined up with foreign rivals to design, make or sell the chips called dynamic random-access memories (DRAMs).
>
> (*The Economist*, 3 February 1990)

In addition to such international alliances, both US and European companies have been forming *local* alliances in an attempt to compete more effectively with the Japanese. In the United States a consortium of leading semiconductor firms formed Sematech to undertake joint research. Early in 1990 six firms (including IBM) joined together to purchase a leading US manufacturer of semiconductor equipment to prevent it being acquired by Japanese firms. In Europe an alliance of producers from different European companies formed European Silicon Structures (ES2) to manufacture custom chips.

Global organization of production

The most significant characteristic of the semiconductor industry is not simply that its markets are global but, rather, that its *production is organized globally*. The electronics industry was the first to which the label 'global factory' was applied because of its early use of offshore assembly. It is in the semiconductor industry that a *spatial hierarchy of production* at the global scale is most apparent, with clear geographical separation between different stages of the production process. It is overwhelmingly the *assembly* stages that have been partly relocated to certain developing countries. In general, the higher-level design, R & D and more complex stages of production have tended either to remain in the firm's home country or to be established in other developed countries where the necessary labour skills and physical infrastructure are more readily available. This pattern is changing, however, as we shall see. In the case of semiconductors the first two stages in the production sequence shown in Table 10.3 have generally not been relocated overseas. Wafer fabrication, however, is carried out by American and Japanese firms in Europe and a great deal of the final assembly stage has been transferred to developing countries, especially in South East Asia. Overall, the degree of offshore production varies somewhat between American, Japanese and European firms.

United States producers   Offshore production in the semiconductor industry first occurred in the early 1960s when a number of American firms began to seek out low-labour-cost locations for their more routine assembly operations. In the semiconductor industry the initial stimulus was the intensifying competition within the United States itself as new firms entered the industry and the need to reduce production costs accelerated. In the 1960s the differential between US labour costs and those in developing countries was especially great, as Figure 10.8 clearly shows. Hence,

> during the 1960s, prompted by the particular needs of production, the semiconductor industry established a unique form of international production, the integrated global assembly line. Midway in the process of manufacturing transistors or integrated circuits, the producers shipped the unfinished components abroad for assembly – bonding – and they then shipped the assembled chips back to the US for testing. This global scheme, though much more complex today, still guides the industry.
>
> (Siegel, 1980, p. 7)

More specifically, 'integrated circuits may have been fabricated in Japan, shipped to Malaysia for packaging, sent to Singapore for testing and passed on to Hong Kong for

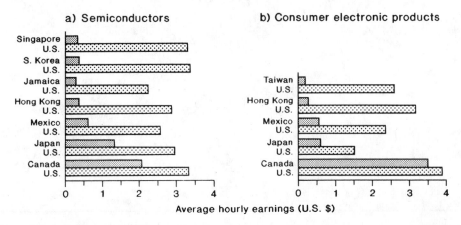

Figure 10.8 Contrasts between the United States and developing countries in labour costs in the semiconductor and consumer electronics industries, late 1960s (*Source*: United States Tariff Commission)

insertion in a PCB which is then exported to a US consumer electronics company'
(Tanzer, quoted in Ernst, 1985, p. 342).

The first offshore assembly plant in the semiconductor industry was set up by
Fairchild in Hong Kong in 1962. In 1964 General Instruments transferred some of its
micro-electronics assembly to Taiwan. In 1966 Fairchild opened a plant in South
Korea. Around the same time, several US manufacturers set up semiconductor
assembly plants in the Mexican Border Zone. In the later 1960s US firms moved into
Singapore and subsequently into Malaysia. During the following decade semiconductor
assembly grew very rapidly in these Asian locations and spread to Indonesia and the
Philippines. Despite some developments elsewhere (for example in Central America
and the Caribbean) most of the growth of semiconductor assembly remained in East
and South East Asia.

By the early 1970s, therefore, every major American semiconductor producer had
established offshore assembly facilities, a tendency greatly encouraged by the offshore
assembly provisions operated by the US government. In 1978, 82 per cent of all imports
under these provisions were of semiconductors from offshore locations. Figure 10.9
shows the global distribution of US semiconductor assembly plants in the late 1980s.
There were clusters in Mexico and the Caribbean as well as in Europe but by far the
largest concentration was in East and South East Asia.

Some especially significant developments have occurred in the organization of US
semiconductor production in that region:[8]

> A definite system of regional specialization is now emerging . . . especially in South East
> Asia . . . First, the activities of US-owned semiconductor branch plants in Hong Kong and
> Singapore have been steadily upgraded in recent years, and much test, burn-in, and board-
> assembly work has been substituted for the more routine semiconductor assembly functions
> which were formerly performed in these two jurisdictions. Second, assembly as such is being
> steadily pushed out to the extensive margins of the region in an incessant search for lower-
> cost production sites . . . Devices are shipped to Hong Kong and Singapore from all over

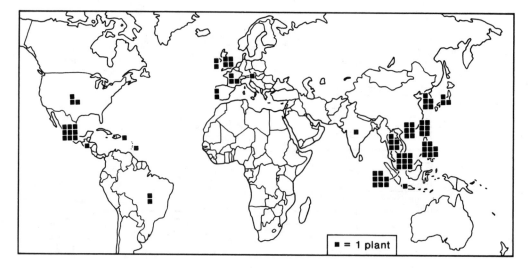

Figure 10.9 The global distribution of United States-owned semiconductor assembly plants (*Source*: based
on Scott and Angel (1988), Figure 5)

South East Asia (especially from Indonesia, Malaysia, the Philippines, and Thailand) for final test. Thus, a localized spatial and international division of labour seems to have made its definite historical appearance in South East Asia as a secondary articulation within the wider global system of production.

(Scott and Angel, 1988, pp. 1063–4)

Two other significant developments are occurring in the organization of the US offshore semiconductor industry in East and South East Asia. One is the emergence of substantial numbers of locally owned subcontractors and suppliers of services and inputs to the US plants (Scott, 1987). The other development is the establishment of regional headquarters in the region by US semiconductor firms: for example, National Semiconductor in Singapore, Motorola, Sprague, Zilog in Hong Kong (Henderson, 1989). These latter developments reflect the growing importance of the region as a *market* for semiconductors and not merely as a low-cost assembly location.

Japanese producers   Japanese semiconductor companies have also developed extensive offshore assembly operations with an even greater emphasis on East and South East Asia. Of course, proximity to the Japanese domestic production base facilitated such developments and helped to create a complex intra-regional division of labour. In 1988 almost 70 per cent of Japan's overseas production facilities in electronic components and devices were located in East and South East Asia. As Figure 10.10 shows, the major locations within the region were Taiwan and South Korea, followed by Singapore and Malaysia. Offshore production in the region by Japanese companies accelerated in the second half of the 1980s, after the major revaluation of the Japanese yen against the US dollar in 1985. The search for lower-cost production locations led to greatly increased offshore production in such Asian counties as Malaysia and Thailand. But whereas much American offshore production of semiconductors has been for eventual

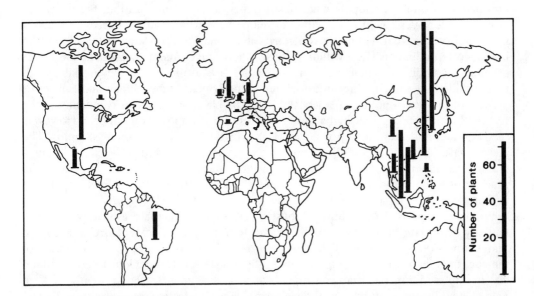

Figure 10.10 The global distribution of Japanese electronic component plants, 1988 (*Source*: Electronics Industries Association of Japan (1989) *Facts and Figures on the Japanese Electronics Industry*, Tokyo: EIAJ, Table 5)

reimport to the United States, most Japanese offshore assembly plants in Asia supply Japanese plants outside Japan, particularly the growing consumer electronics operations set up by Japanese firms in both the developing and, more recently, the developed, countries of the United States and Europe.

European producers   Compared with US and Japanese semiconductor producers, the European producers have not engaged in offshore production to anything like the same extent, although Philips, SGS–Thomson and Siemens all have assembly activities in Asia. Philips has substantial joint venture operations in China manufacturing integrated circuits for the consumer electronics industry. It also has been increasing its sourcing of semiconductors and integrated circuits from the Far East (de Smidt, 1990). But none of the leading European firms operates the kinds of global production network which have been developed by the Japanese and, especially, the US producers which are so heavily dependent on East and South East Asia.

United States and Japanese companies in Europe   Yet, despite the undoubted growth of offshore production by semiconductor companies in some Third World countries, the fact remains that much production by American, Japanese and European firms is located within developed countries, where there is a great deal of *cross-investment*. American electronics firms have long been established in Europe; more recently, European electronics firms have been increasing their number of production operations in the United States. In addition, since the 1970s Japanese firms have begun to locate increasing numbers of production plants in both the United States and Europe. The type of international production located by electronics firms in these developed countries is substantially different from that located in the Third World. Whereas offshore production in most developing countries is aimed at cost-minimization, the dominant motivation for overseas location within developed countries is market access, though it is true, of course, that within a given geographical market firms may well seek out the lowest-cost production site, including those where government investment incentives operate. In addition, semiconductor companies are attracted by the availability of highly skilled scientific and technical labour in the United States, Japan and Europe.

Most of the semiconductor plants established by US semiconductor companies in the EC can be explained primarily by the 17 per cent tariff imposed on imports. However, the completion of the Single European Market in 1992 has added a new stimulus. The increasingly stringent attitude by European governments towards full-scale local production (including the more advanced stages of design and manufacture) and the practice of using anti-dumping measures are also stimulating further direct investment in Europe by US, Japanese and now South Korean firms.

Within Europe there is a long-established US semiconductor presence with a particularly heavy concentration in Scotland and, to a much lesser extent, Ireland. United States firms dominate the Scottish semiconductor sector, with more than 90 per cent of the country's semiconductor employment in the mid-1980s (Henderson, 1989). Scotland has a long tradition of American manufacturing investment, a good supply of both high-skilled labour from the well-developed higher education sector and of female assembly workers. In addition, the Scottish Development Agency has been a strenuous, and successful, seeker of electronics companies to an area with a substantial package of investment incentives. Ireland, too, has strong links with the United States (one has only to think of the number of American presidents with Irish ancestry) and it, too, has adopted an aggressive strategy to attract foreign electronics firms on the basis of generous financial and tax incentives and a good labour supply. The need to operate behind the EC tariff wall has, therefore, attracted large numbers of US semiconductor firms into Europe. In 1988 five of the ten leading semiconductor companies operating in Europe were American: Texas Instruments, Motorola, Intel, National Semiconductor and AMD. These were in addition to the largest US company of all, IBM, which has an extensive European production network.

The establishment of semiconductor production facilities by Japanese firms in both Europe and North America is very much more recent than that of American firms in Europe. As we have already seen, Japanese firms have tended to concentrate their overseas plants in developing countries, particularly in Asia, mainly to reduce production costs. Until the 1970s there was virtually no direct Japanese investment in electronics in the developed countries. As Figure 10.10 shows, however, there are now substantial numbers of Japanese electronic component firms operating plants in both the United States and Europe. In the latter case more than two-thirds of the total are in the United Kingdom and Germany. Major Japanese semiconductor plants in the United Kingdom include the large-scale NEC fabrication plant in Scotland and the Shin-Etsu Handotai silicon manufacturing plant, also in Scotland, both of which arrived in the mid-1980s.

The combination of the imminent Single European Market and the revalued yen – but especially the former – have stimulated a new, and highly significant, wave of Japanese semiconductor investment in Europe. This new wave was led by Fujitsu's decision to build a $100 million chip fabrication plant in north-east England in 1989. The other leading Japanese companies are following this lead. Hitachi, for example, is building an integrated semiconductor manufacturing plant in western Germany. Japanese semiconductor production in the United States in much more firmly established than in Europe. In fact four out of every ten Japanese electronic component plants outside Asia are located in North America (Figure 10.10). Again, the same basic locational factors apply, with a particular emphasis on market access and proximity and scientific and technical labour.

South Korean producers   The newest entrants into the global semiconductor industry – South Korean firms – are very new indeed. We noted earlier (p. 326) the targeted policies of the Korean government towards electronics. The speed with which the four leading Korean conglomerates – Samsung, Goldstar, Hyundai and Daewoo – have developed as extremely advanced semiconductor producers has been little short of spectacular.

None of the Korean electronics firms was in the semiconductor business until Samsung entered the business in the mid-1970s. Goldstar followed in 1979, Hyundai and Daewoo (which were not then electronics companies) entered in 1983. Initially the Korean firms were totally dependent on technology acquired from the United States and Japan. They focused on simple types of semiconductor of low capacity: the products which were being abandoned by the American and Japanese companies. In the mid-1970s, the technology gap between Samsung and the industry standard was around thirty years; today it has virtually disappeared. Indeed, Samsung has totally caught up technologically with the industry leaders. Production of 16Mb DRAMs will begin in 1992 on the basis of massive R & D investment. Samsung's aim is to be in the world's top ten semiconductor producers during the 1990s.

As noted above, each of the Korean firms has established production facilities in the United States and will undoubtedly move into Europe. The objective of the US plants was to be close to the technological heart of the semiconductor industry – to absorb state-of-the-art technology. Increasingly, however, plants will be needed in Europe to circumvent protectionist barriers. The Korean firms are rapidly developing their own technologies; Samsung now claims that it no longer needs foreign technology. But it is closely linked with IBM under a cross-licensing agreement, with Intel and with another leading US firm, Micron Technology (Mody, 1990).

Rationalization in the face of competition

In view of the intense competition from US, Japanese and now Korean firms, it is not surprising that the European producers feel threatened. This competitive threat, and the general weakness of European semiconductor activity, were the major stimulus for the 1987 merger between SGS, the Italian state-owned company and the French company, Thomson. The merged company has been drastically rationalizing and reorganizing its activities, closing plants and increasing the proportion of its assembly located in the Far East. Philips, the leading European semiconductor producer, and the only one in the world's top ten, is currently in immense difficulties in its entire, highly diversified, operations. In semiconductors Philips recently stopped pilot production of 1Mb static random access memory (S-RAM) chips and withdrew from its leadership of a European semiconductor collaborative project. The company is also reorganizing its activities to create a totally separate semiconductor division.

Substantial rationalization has also been occurring among US merchant semi-conductor firms. Intensification of competition from Japanese and now, Korean, producers, together with the collapse of the market for memory chips in the mid-1980s,

Table 10.5 The world's leading consumer electronics manufacturers, 1988

Rank	Company	Country of ownership	Rank 1978	Change in rank
1	Matsushita	Japan	1	—
2	Philips*	Netherlands	2	—
3	Thomson[†]	France	11	+8
4	Sony	Japan	6	+2
5	Hitachi	Japan	9	+4
6	Toshiba	Japan	7	+1
7	Sanyo	Japan	5	−2
8	JVC	Japan	—	—
9	Sharp	Japan	—	—
10	Mitsubishi	Japan	—	—

* Includes Grundig, GTE–Sylvania
[†] Includes General Electric, RCA, Ferguson, Telefunken

(*Source*: Industry sources; press reports)

'forced all leading US merchant firms, with the sole exception of Texas Instruments (TI), to leave the market altogether' (Ernst, 1987, p. 47). Without the markets available to the more diversified captive producers the US merchant producers have faced increasing difficulties. In general, however, the major problems are faced by the European producers:

> More than in any other industry, competition in the world semiconductor industry is dominated to a considerable degree by two actors, namely US and Japanese companies and the patterns of conflict and co-operation prevailing between them. European companies, even the most powerful ones, have been reduced to playing a secondary role . . . In short, global competition in the semiconductor industry is still close to an exclusive US–Japanese affair.
>
> (Ernst, 1987, pp. 45–55)

Corporate strategies in the consumer electronics industry

There are clearly very close links between semiconductor production and consumer electronics, both in general terms and also within those electronics companies which are in both businesses. As we saw earlier, such vertically integrated operations are especially common among Japanese and European electronics firms and less common among US firms. Some of the developments we have discussed in the case of the semiconductor industry apply also to colour television production and related consumer electronics products like VCRs. At the same time, there are important differences.

Production of colour television receivers is even more highly concentrated than semiconductors. Table 10.5 shows the ten leading consumer electronics manufacturers in 1988. They produced almost 80 per cent of the world colour TV total (the comparable figure for semiconductors was 56 per cent; see Table 10.4). The two leading firms – Matsushita and Philips – alone accounted for more than a quarter of world production and were a long way ahead of their closest rivals. But the most striking aspect of Table 10.5 is the fact that eight of the ten leading colour TV firms are Japanese. In 1978, only five of the top ten firms were Japanese; the 'newcomers' were JVC, Sharp and Mitsubishi. Each of these Japanese producers is part of a highly integrated corporation having both component operations and also highly diversified industrial and consumer electronics activities. The second striking feature of Table 10.5 is that there is not a single US firm – whereas in 1978 there were three. Of those three, only one – Zenith – still survives as an independent producer of colour TVs. On the other hand, the two leading European firms, Philips and Thomson, are very clearly global players. The European presence in consumer electronics is much more substantial than it is in semiconductors.

Acquisitions, mergers and strategic alliances

The colour television industry has undergone major reorganization through acquisitions and mergers. The number of colour TV manufacturers in Europe has declined dramatically since 1980, when there were more than thirty.

> In Britain the last major British-owned manufacturer – Thorn-EMI – sold its Ferguson division to the French Thomson company in 1987, at the same time as the latter acquired the RCA division of US General Electric . . . The German industry – the biggest in Europe

– now has only one major German–owned producer, Bosch, which has a majority share-holding in the medium–sized CTV producer, Blaupunkt. Managerial control of Grundig has passed to Philips, and Thomson owns Telefunken, Saba and Nordmende. Sony acquired Wega, and the ex-ITT subsidiary SEL is now owned by Nokia of Finland. This leaves Loewe-Opta, a small up-market manufacturer, as the only other CTV producer in German ownership.

(Cawson, 1989, p. 61)

Even greater changes have occurred in the ownership of the US colour TV industry. Acquisitions and mergers have been rife in the television sector, as many American firms quit an increasingly difficult market. For example,

Admiral sold out to Rockwell in 1973. In the next year, Motorola sold out to Matsushita, Magnavox to Philips, and Ford Motor sold its Philco consumer electronics business to GTE–Sylvania, which, in turn, sold its TV business to Philips in 1980. In 1976, Sanyo bought out Warwick, which was primarily a supplier to retailers like Sears.

(Turner, 1982, p. 57).

In 1966 there were sixteen US-owned television manufacturers; by 1980 this number had fallen to three. Today there is only one left – Zenith – as General Electric absorbed RCA and then sold out to the French company Thomson. Zenith is an especially unusual case; it recently disposed of its computer business (in which it was a highly successful producer of portable and lap-top machines) to concentrate on television production.

Strategic alliances have been less common in colour television production; the preference has been for outright merger or acquisition. In the light of the battle for the high definition television standard, however, Philips and Thomson have formed an alliance to develop a European technical standard. In related product areas, however, strategic alliances have been more common. Philips and Sony jointly developed the compact disc technology which had been invented by Philips. In VCRs a joint venture was created in Europe between the Japanese firm JVC, the British company Thorn-EMI, the German company Telefunken and the French company Thomson. With Thomson's aggressive acquisition strategy this is now a joint venture between just JVC and Thomson. On the other hand, there have been important failures in strategic alliances; both the GEC–Hitachi and Rank–Toshiba colour television ventures in Britain were abandoned and became wholly Japanese-owned.

Global organization of production

United States producers As in the case of semiconductors, internationalization of production is a fundamental competitive strategy of all the major consumer electronics companies. In both cases, it was US producers which initially moved some of their activities offshore in the early 1960s. But the precise stimulus for such a strategy was rather different from that in semiconductors. In that sector the major push factor towards offshore assembly was intensifying competition within the United States itself as new firms entered the industry and as domestic production costs accelerated. In consumer electronics, specifically television manufacture, the major stimulus for US companies to move offshore was intensifying competition from low-cost overseas competitors, notably the Japanese.

Among American television manufacturers offshore production increased further in the 1970s as Japanese import penetration of the domestic market accelerated. For example:

– Admiral came to make both colour and black-and-white sets in Taiwan.
– General Electric moved the manufacture of circuit-boards and other colour components to Singapore.
– Magnavox (a Philips subsidiary) imported from a Taiwanese plant owned by its parent.
– RCA used Taiwan and, in 1975, began moving colour-chassis operations to Mexico.
– GTE Sylvania opened operations in Mexico in 1973 and bought a plant in Taiwan in 1975 for production of both kinds of televisions.
– Zenith built up subsidiary plants in Taiwan and Mexico, and, in late 1977, announced a major shift of jobs to them.

(Turner, 1982, p. 56)

In fact, all of the television sets imported into the United States from Mexico during the 1970s were produced under contract for American firms.

Japanese producers Japanese consumer electronics manufacturers also developed substantial offshore assembly, notably in the East and South East Asian countries, during the 1970s. The motivation for Japanese offshore plants in the Asian NICs was not only one of seeking low labour costs, however, although this was probably the primary stimulus. An additional consideration was the desire to get around the growing US import restrictions by producing in countries not covered by these restrictions. This helps to explain the establishment of television plants by Matsushita, Mitsubishi, Sony and others in Taiwan, Singapore, Malaysia and South Korea during the early 1970s.

Figure 10.11 shows the global distribution of Japanese consumer electronics plants in 1988. Comparison with the map of Japanese electronic component plants (Figure 10.10) reveals some interesting differences. In both cases, the largest concentration of Japanese plants within Asia is in Taiwan but the degree of geographical localization is much lower in consumer electronics. The number of Japanese consumer electronics plants in South Korea is very low, particularly when compared with the components

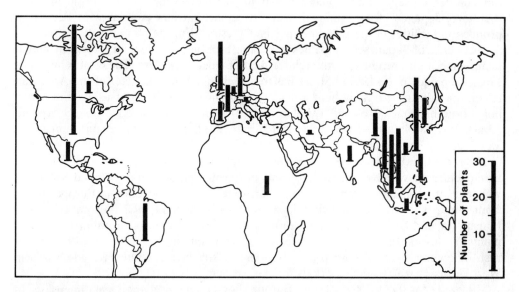

Figure 10.11 The global distribution of Japanese consumer electronics plants, 1988 (*Source*: Electronics Industries Association of Japan (1989) *Facts and Figures on the Japanese Electronics Industry*, Tokyo: EIAJ, Table 5)

sector. This reflects the restrictive policies of the South Korean government towards foreign consumer electronics firms. On the other hand, Japanese consumer electronics plants are spread over a wider range of developing countries than is the case for components. This applies not only within Asia but also in Central and South America, Africa and Oceania.

It is the developed economies of North America and Western Europe, however, that have become an increasingly important focus for offshore production by Japanese consumer electronics firms. Europe and North America contained 36 per cent of Japanese overseas consumer electronics plants in 1988 compared with 22 per cent of Japanese semiconductor and component operations. Within Europe, the United Kingdom has the largest number of Japanese plants, followed by Germany. Clearly, the reasons for such investment are not a search for lower production costs. As in the automobile industry, Japanese firms had shown themselves to be perfectly capable of serving North America and Western Europe from Asia. In television manufacture, for example, the advantages of Japanese producers lay in (Franko, 1983, pp. 79–80):

1. labour costs half or one-third as high as those in Germany or the United Kingdom;
2. designs requiring up to 30 per cent fewer components than Western European or US sets because of a greater use of integrated circuits;
3. automation in the assembly of sets (of between 65 and 80 per cent of total components used as against between 0 and 15 per cent in German and UK plants), meaning that a Japanese company could produce a colour TV set with an average of 1.9 man-hours against 3.9 in West Germany and 6.1 in Britain;
4. large scale of plant operations (output per plant of around 500,000 sets a year) with double the capacity of European plants;
5. superior-quality components.

The introduction of protective measures by American and European governments severely affected these advantages, however, by imposing quantitative restrictions on the number of sets which could be sold in these markets. In the classic manner, therefore, Japanese firms 'jumped over' the trade barriers and invested heavily in production facilities in the United States and Western Europe.

The influx of Japanese companies has been especially evident in television manufacture and, more recently, in the related field of video recorders. The establishment of Japanese plants in the United States has totally changed the character of the American television industry, as 'Sony built a plant at San Diego in 1972; Matsushita and Sanyo bought out existing US-owned TV plants by 1976; and Mitsubishi, Toshiba, Sharp and Hitachi have all launched US factories' (Turner, 1982, p. 58). Only Zenith remains as an indigenous manufacturer.

The same trend has occurred in Western Europe. All the leading Japanese television manufacturers now have production plants in Europe, notably in the United Kingdom, which has the highest concentration of Japanese television production in Europe. Apart from Sony, which set up the first Japanese television plant in Britain in 1974, and Matsushita in 1976, all the other ventures have involved the acquisition of existing plants. Following on from the surge of television plants in the 1970s, the 1980s saw a rush of openings of video recorder factories, mostly in the United Kingdom and in West Germany, the leading markets for this product.

The battle for the VCR market in Europe shows its intensely political dimension. In the early 1980s the French adopted the idiosyncratic device of insisting that all imported VCRs had to enter through the small town of Poitiers. This certainly slowed down the speed of entry but it did not push back the tide. Neither did three years of a voluntary export restraint (VER) agreement protect the European VCR producers who were trying to sell a less acceptable technical system.

> One important effect of the trade war over VCRs was the acceleration of Japanese investment in VCR manufacture within the EC. In 1985–87 there was an extraordinary surge in Japanese VCR capacity, with no fewer than seven new entrants to VCR production . . . By 1987–88 the focus of trade politics had shifted from the import of finished machines to the manufacturing activities of the Japanese plants in Europe . . . After the failure of the VER, the EC is increasingly turning to anti-dumping measures in an attempt to protect the European-owned manufacturing base in consumer electronics . . . In addition there is pressure from European governments for the Japanese to become more fully integrated into Europe by establishing R & D centres and developing new products in Europe.
>
> (Cawson, 1989, pp. 68–9)

A combination of trade barriers, the revalued yen and also a growing realization of the need to have a direct presence in key geographical markets is undoubtedly beginning to change the international nature of Japanese consumer electronics activity. All the leading companies are now developing explicitly *global* strategies though at varying speeds. All have set targets to increase the proportion of their total output to be produced outside Japan; the rate of growth of exports in several products has been declining. Production of low-value products is more and more being relocated in lower-cost Asian countries. Undoubtedly, some of the design and R & D functions will be shifted to the advanced markets of North America and Europe – some have already done so, notably Sony and Matsushita. But the extent of such shifts so far remains very limited:

> Sony keeps more than 90% of its engineers in Japan. Matsushita, by far the world's largest consumer electronics company, has 30 development laboratories and 9,300 engineers in Japan. It has only two laboratories abroad – one in Taiwan and one in New Jersey. They have a total staff of 115 engineers . . . For the past ten years, every interesting new product in consumer electronics . . . has come out of Japan. The manufacturing has tended to stay close to these engineers because production and technology are increasingly linked . . . All these technologies will eventually be shifted abroad . . . [but] . . . in consumer electronics the Japanese investment diaspora starts at the lower-tech end of the business.
>
> (*The Economist*, 12 March 1988, pp. 76, 81)

European producers Compared with American and Japanese firms, European electronics producers have shown rather less of a propensity to establish offshore production facilities in developing countries, but they have by no means been immune to the practice. The bigger firms, in particular, have gone offshore but, apart from Philips and Thomson, less so than the Japanese. Philips has a particularly extensive international production network in developing countries across a whole range of electronics activities. For example, Philips has an especially important manufacturing complex in Singapore as well as plants in Taiwan, Hong Kong, South Korea and Mexico. Indeed, a major part of its current strategy to compete with the Japanese and with the bigger US firms is to shift a greater proportion of its labour-intensive activities to Third World locations. The aggressive French consumer electronics company Thomson also operates extensively in East and South East Asia, where it has plants in Malaysia, Singapore, Thailand, Taiwan and China. Thomson also has operations in

Mediterranean countries such as Morocco, Tunisia and Turkey and in the Mexico–United States border zone. Both Philips and Thomson use their Asian operations to produce lower-cost consumer electronics products. Largely through acquisitions both companies also have important production activities in the United States which manufacture higher-value products for the North American market.

Despite these extensive international operations, both Philips and Thomson have their major bases in Europe.[9] Both have been engaged in large-scale rationalization and reorganization of their European production networks in the context not only of changing circumstances within Europe itself but also in the global context. Philips has been trying to rationalize its television production network in Europe for most of the 1980s. Because it began manufacturing TV sets before the emergence of the European Community Philips developed TV production plants in every one of the major Community countries. More generally, Philips had long operated a strategy of giving a high degree of autonomy to its national subsidiaries. This strategy is in the process of major change. In consumer electronics major rationalization and plant specialization have already occurred:

> The company has now been reorganized into product divisions, with each category of product (except for CTV) concentrated in 'world centres'. VCRs are produced only in Austria and Germany, CD players in a single plant in Belgium, and the Philips centre for tape recorders and radios is in Singapore; the company still had, in 1988, nine CTV plants in as many countries. In September 1988, however, Philips announced plans to concentrate production of CTVs in three plants in Belgium, France and Italy.
>
> (Cawson, 1989, p. 62)

NIC producers   Finally, to an even greater extent than in semiconductors, a number of consumer electronics firms from East Asia (notably from South Korea and Taiwan) have become significant global players. Increasingly, this involves them in establishing production plants overseas, not only in other Asian countries but also in North America and Europe.[10] The major Korean conglomerates – Samsung, Goldstar and Daewoo – all began producing consumer electronics products under subcontracting agreements with major Japanese and US firms. Until relatively recently the Korean firms had no 'brand name' identity outside Korea. Even in 1990 the leading company, Samsung, still manufactured about 40 per cent of its output for firms such as Sony and GE. But the leading Korean firms are now making intensified efforts to establish their products as global brands with considerable success. 'Samsung' and 'Goldstar' are names which now appear in the mass advertising slots at international sporting events and on airport baggage trolleys, along with longer-established Japanese, American and European brands. Both Samsung and Goldstar have built colour TV plants in the United States (in New Jersey and Alabama respectively). Samsung has manufacturing plants in the United Kingdom, Spain and Portugal while Goldstar has two plants in western Germany. A similar pattern is being followed by Tatung, the major television manufacturer in Taiwan. Tatung has plants in Japan and Singapore as well as in the United States and the United Kingdom.

Rationalization in the face of competition

Rationalization in the consumer electronics industry is not only being implemented on a global scale but also goes hand in hand with increased automation. This is especially so among American and European firms as they strive to compete with the already highly automated Japanese firms and with increasing competition from Asian NICs. Such increased levels of automation may well have significant implications for the future development of the consumer electronics industry in the developing countries. It has been suggested that at least some of the present offshore activities could well be 'brought home' by developed country manufacturers as increased automation of the production process comes to offset the developing countries' labour cost advantages. As yet, however, there is little substantial evidence of this. Indeed, as we have seen, some of the East and South East Asian countries have been growing very rapidly as producers of consumer electronics. What may threaten at least some of that activity, as well as the development of electronics production in other developing countries, is the possibility of expansion in Eastern Europe. This, together with the opportunities presented by the completion of the Single European Market, could be a major influence on the future shape of the consumer electronics industry. For the major European producers, and perhaps for others too, Eastern Europe is an obvious future production location.

Summary of global trends in the electronics industries

In a number of ways, therefore, the global pattern of electronics production shifts and changes as a result of the evolving corporate strategies of the major firms. Fierce competition forces firms to increase their degree of functional integration, to diversify into new product lines, to relocate production in more favourable locations in terms of markets or costs, and generally to rationalize their operations on a global basis. There has been considerable geographical shift to some developing countries, mostly in East and South East Asia. In the case of television manufacture, global shifts can be interpreted in terms of the product life-cycle. Television is a technologically mature product in which scale economies and the minimization of production costs are of key importance. But, like motor vehicles, the industry has experienced some technological rejuvenation which, together with government protectionist pressures, has kept a good deal of production within the developed countries. In both the United States and Western Europe, however, there has been a substantial change in the ownership of production as Japanese firms in particular have established production bases there.

Semiconductors, integrated circuits and microprocessors, on the other hand, are by no means mature products. However, one stage in the production process – assembly – is amenable to geographical separation and to location in low-labour-cost areas. It is a clear example of the global combination of highly capital-intensive technology with low-cost, labour-intensive production. But the simplistic global division of labour characteristic of the 1960s and early 1970s no longer applies. Today's global map of semiconductor production is far more complex. Ernst (1985) identifies four basic trends:

1. Locational shifts between the three leading global production regions – the United States, Japan, Western Europe – involving:
   (a) the initial wave of investment by US semiconductor firms in Western Europe;
   (b) subsequent direct investment by Japanese and Western European semiconductor firms in the United States and by Japanese firms in Europe;
   (c) more recent investments by US semiconductor firms in Japan.

2. Locational shifts from the centre of US and Japanese firms in Scotland, Ireland and Wales.
3. An increased sophistication of semiconductor functions being carried out in the 'first wave' of offshore assembly platforms. Hong Kong, South Korea, Taiwan, Singapore (and to some extent Malaysia) now perform fabrication and testing (and even design in some cases).
4. A consequent relocation of simple assembly operations from these original offshore locations to new sites in countries such as Thailand, the Philippines, Indonesia, China and the Caribbean Basin.

To these four trends we should add a fifth: the development of international production by some leading firms from South Korea and Taiwan which have been establishing production bases in the United States and Europe as well as in Asia.

Jobs in the electronics industries

At first sight, the employment situation in the electronics industry would seem to be very different from that in the two industries we have examined in the preceding chapters. In both of those industries the story was one of substantial employment decline in the older-industrialized countries. There is no doubt that electronics is a growth industry. In semiconductors, for example, production has increased at a truly phenomenal rate since the early 1960s. But, for a number of reasons, this growth in output has not been accompanied by a comparable growth in employment. Employment decline after 1974 certainly owed a good deal to the effects of world recession, particularly in the consumer electronics sector. But this does not fully explain these employment changes. In addition, the consumer electronics sector of North America and Western Europe has been hit hard by import penetration from lower-cost producers, particularly in Asia (including, of course, Japan). There can be little doubt that a proportion of the job loss in the consumer electronics industries can be attributed to import penetration. But just how much is very difficult to assess and import penetration is not the whole story. Two related factors are particularly important in any explanation – technological change and corporate rationalization, both of which have reduced the need for labour.

Technological change has been extremely rapid in the electronics industry, as we have seen. A major effect has been to increase greatly the productivity of the labour force. All the major firms have been investing heavily in automated production and assembly and, thus, greatly reducing the number of workers required for a given level of production in both semiconductor and consumer electronics. In virtually all the major electronics companies, therefore, productivity has soared whilst employment has either not kept pace or has actually fallen.

At the same time, the strategies of rationalization and reorganization on a global scale have adversely affected employment, especially in the TNCs' home countries. Whether or not there is a viable alternative to increased international production by firms in today's world is a general issue which we shall examine in Chapter 12. In the electronics industry there is, indeed, an element of truth in the 'runaway plant' charge whereby firms close a plant in one country and transfer its production (and jobs) to another country. But Siegel (1980) argues that the semiconductor industry, in particular, is not a classic runaway industry to the extent that consumer electronics and

Table 10.6 Global employment change within Philips, 1973–6

Region	Balance of employment gains & losses	Region	Balance of employment gains & losses
Europe		*Subcentres*	
(a) Netherlands	−10,000	(a) Australia, Japan, South Africa	0
(b) Other EC countries	−5,000	(b) Developing countries (Argentina, Brazil, Colombia, Chile, India, and others)	−5,000
(c) Periphery (Ireland, Portugal, Spain,Greece, Turkey, Finland)	+3,000		
North America			
USA and Canada	+15,000	(c) 'Runaway' countries (Mexico, Hong Kong, Taiwan, Korea, Singapore, and others)	+12,000
Total for Europe and North America	+3,000	Total for subcentres	+7,000

(*Source*: Teulings (1984), Table 1)

clothing may have been. In fact, geographical shifts in employment do not necessarily follow a simple path of reduction in countries where labour costs are high and growth in countries where they are low.

The geographical complexity of employment change within global electronics corporations is far greater than this. Table 10.6 shows, for example, that between 1973 and 1976 total world employment within Philips increased by 10,000. But this was a *net* figure which was the outcome of a complex pattern of gains and losses in different regions. Within Europe, it was Philips' home country, the Netherlands, which suffered the greatest job loss; indeed the home country's loss was equivalent to the total gain worldwide. In the rest of the EC there was a net loss of 5,000 jobs made up primarily of large employment increases in Philips' operations in West Germany and substantial job losses in the firm's operations in the United Kingdom. It was in the periphery of Europe that gains were mostly made. In total, therefore, Philips' labour force in Europe as a whole fell by 12,000 between 1973 and 1976. In contrast, Philips' employment in North America increased by 15,000, mostly through acquisition.

Table 10.6 is simply used as a general illustration of the complexity of employment change resulting from corporate global reorganization. In fact, Philips continues to face enormous rationalization problems which will greatly reduce its labour force, particularly in Europe.

Over the past three years Philips has closed or merged 75 of its 346 plants spread over 50 countries. It has also shed some 38,000 employees (17,000 by selling businesses) out of a total of 344,000. But that still leaves it with too many factories, over half of them in Europe, where labour costs are often high
(*The Economist*, 7 April 1990).

In general, electronics plants tend to be small or medium-sized in employment terms. Even the largest plants may employ only a few hundred workers. In this respect, therefore, they do not compare in magnitude with automobile factories or even many textile mills in terms of their local employment impact. However, some distinctive geographical concentrations of electronics employment have evolved in both developed and developing countries. In the United States, the Silicon Valley complex is the biggest and best-known but by no means the only important area of production. Of similar vintage is the agglomeration of electronics and high-technology firms along Route 128 around Boston in New England, while newer electronics clusters have been developing rapidly in Colorado, Northern California, Oregon and North Carolina. Within Europe substantial geographical clusters of electronics production have emerged in the Central Valley of Scotland, the M4 corridor in South East England and in the Grenoble area of France, among others. Within the developing countries most of the electronics production is concentrated in the export-processing zones which have proliferated during the past two decades. As we noted in Chapter 6, roughly half of all the employment in the Asian EPZs is in the electronics industry.

But these geographical concentrations of electronics production are not identical. They occupy different positions, and perform different roles, in the spatial hierarchy of production. Most importantly, they offer different kinds of job opportunity. Perhaps more significant than the simple geographical concentration of electronics employment is the pronounced *spatial differentiation of skills and occupations*. This is a direct consequence of the spatial and functional separation of the stages in the production process, especially in the semiconductor industry.

There is a clear *polarization* of skills in the industry between highly trained professional and technical workers on the one hand and low-skilled production workers on the other. Increased automation has led to a much steeper relative decline in the number of production workers employed, particularly the less skilled. But there is also a clear spatial pattern of skills in the electronics industry. Overall, professional and technical occupations are far more important in the developed countries than in the developing countries, a reflection of the fact that it is mainly the more routine assembly tasks which have been located in developing countries while the more sophisticated design, R & D and wafer fabrication stages remain mostly in developed countries. Hence, most, though by no means all, of the semiconductor employment in Silicon Valley and the other production clusters in the United States and Japan is in the higher-skill categories; conversely, virtually all the electronics employment in the EPZs of the Asian countries and in the Mexican border plants is of production workers.

In the European electronics clusters there is also a degree of spatial segmentation of labour skills though in a less extreme form. For example, the Scottish electronics plants include both assembly and fabrication activities and, therefore, have a more even mix of skills than the developing country plants. Even so, the relative absence of higher-level design and of R & D activities in the Scottish branch plants creates a relative shortage of job opportunities for particular segments of the labour force. In contrast, the electronics cluster in South East England is predominantly one of high-level administrative, design and R & D units within large firms together with highly innovative and specialized small electronics firms.

A further fundamental aspect of employment in the electronics industry is the predominance of female workers in the assembly stages.[11] This is apparent throughout the world and is not confined to the developing countries alone. An overwhelming majority of electronics assembly jobs are occupied by young female workers on relatively low wages. In this respect, there are clear parallels with the situation in the textiles and clothing industries (see pp. 265–6). Female workers are the norm in the assembly processes wherever the plants are located:

Most of them – more than 90% – are young women. Most do assembly, the bonding of hair-thin wires to semiconductor chips, and the associated packaging. Though the work requires good eyesight and dexterity, little training is required. Women generally achieve peak productivity within a few months ... In most countries, companies only hire unmarried women, of specific ages. The workforce, therefore, ranges in age from about 16 to 26. By avoiding the employment of married women except during times of severe labour shortages, companies avoid paying maternity benefits and ensure that an employee's primary loyalty will be to the company, not to her family or household. More important, by hiring young women, employers can get away with low wages and poor working conditions.

(Siegel, 1980, p. 14)

However, more recent survey work of the electronics labour force in Malaysia and Singapore shows that changes are occurring (Lin, 1987). For example, the proportion of married women in the workforce has increased; neither marriage nor childbirth now automatically leads to a woman worker leaving her job; education levels of entrants are higher.

Although labour regulations are generally more stringent in developed countries, some undesirable characteristics may still apply. According to a study of the participation of women in the 'global factory',

In Silicon Valley ... 75 percent of the assembly line workers are women. The pattern is repeated along Route 128, outside Boston, and in North Carolina, an anti-union, 'right to work' state now favoured by the electronics industry. As in the US garment industry, immigrant women comprise a significant chunk of those workers: 40 percent. On the west coast, Filipinas, Thais, Samoans, Mexicans and Vietnamese have made the electronics assembly line a microcosm of the global production process. Management exploits their lack of familiarity with English and US labour law. Often, companies divide the assembly line according to race and nationality – one line may be all Vietnamese while another is all Mexican – to encourage competition and discourage cross-nationality alliances.

Among the non-immigrant workers, many are married women, assuming paid employment after years of being homeworkers. Wages for semiconductor assembly in the US, while vastly superior to those overseas, are among the lowest in all of US industry ... Since there is little upward mobility for women, especially Third World women ... only 19 percent of all technicians and less than 9 percent of managers are women.

The illness rate in the electronics industry is one of the highest. Women are regularly exposed to solvents which may cause menstrual and fertility problems, liver and kidney damage, cancer, and chemical hypersensitisation.

(Fuentes and Ehrenreich, 1983, pp. 53–4, 55)

Again, however, there are some signs of change in these conditions. Lin's (1987) study of Malaysian and Singaporean electronics workers found that although occupation-related health problems remain significant, the frequency of accidents and injuries appears to be declining.

The gleaming plate glass and metal plants of the modern electronics industry look, at first sight, to be the complete opposite of the dark, satanic mills of the nineteenth century and the squalid sweatshops of the garment industry. In some senses, of course, the contrast is undoubtedly what it seems. For many of the workers employed in the industry wages are high, the work is stimulating and conditions are superb. But for others the story is rather different. For the female workers of the export-processing zones and for some in developed country plants as well, the contrast is less evident. At worst, conditions are just as bad, with long working hours, an unpleasant and noxious working environment and little or no job security. It is indeed paradoxical that an industry which epitomizes all that is new and up to date at the same time harbours some of the oldest and least desirable attributes of work in manufacturing industry.

Notes for further reading

1. Recent surveys of the electronics industry include Grunwald and Flamm (1985), UNCTC (1986), Ernst (1987), Morgan and Sayer (1988), Henderson (1989).
2. Siegel (1980), Rogers and Larsen (1984), Saxenian (1985), Scott and Angel (1987), provide detailed accounts of the evolution of Silicon Valley.
3. See Morgan and Sayer (1988), Henderson (1989) for discussions of the development of the electronics industry in Scotland and Wales respectively.
4. Grunwald and Flamm (1985), Scott (1987), Salih, Young and Rasiah (1988), Henderson (1989), Mody (1990), discuss the changing structure of semiconductor production in East and South East Asia.
5. Dosi (1983) provides a broad treatment of national government policies in the electronics industries but with a specific focus on Europe (see also Sharp and Shearman (1987), Sharp and Holmes (1989)). Japanese policy is discussed by Okimoto (1989), Fransman (1990), United States policy by Cohen and Zysman (1987).
6. See Turner (1982), Cawson (1989).
7. Henderson (1989) and Mody (1990) discuss government policies towards the electronics industries in the Asian NICs.
8. Scott (1987), Scott and Angel (1988), Henderson (1989), chapter 4, discuss the changing intra-regional division of labour in semiconductor production in East and South East Asia.
9. Cawson (1989) describes the changing strategies of Philips, Thomson and Nokia, the Finnish consumer electronics company. de Smidt (1990) provides a detailed analysis of strategic reorganization within Philips.
10. Developments in the strategies of South Korean and Taiwanese consumer electronics firms are discussed by Mody (1990). Cawson (1989) discusses the activities of Korean companies in Europe.
11. See Lim (1980), Siegel (1980), Fuentes and Ehrenreich (1983), Lin (1987).

Chapter 11

'Making the World Go Round': The Internationalization of Services

Introduction: the growing importance of services in the global economy

One of the most significant developments of the last few decades in the global economy has been the rapid growth of the service industries. As we saw in Chapter 2, services account for the *largest share of gross domestic product* in all but the lowest income countries (Table 2.12). They are increasingly the *major source of employment* in all the developed market economies and in many developing countries as well. Indeed, Riddle (1986) argues that the services sector has been absolutely *central* to the economic growth of developing countries, a contribution which is often overlooked in the popular emphasis on manufacturing activities. During the 1980s, again as we saw in Chapter 2, international trade in commercial services accelerated to become an important driving force in its own right in the global trading system. The services sector has also been attracting an increasing share of world foreign direct investment, as we saw in Chapter 3. Table 3.6 shows that services now account for a larger share of foreign direct investment, for the leading industrialized countries as a whole, than manufacturing industry.

Services are, without doubt, therefore, becoming increasingly internationalized:

> Beginning in the 1970s, the sleeping giant of the world-economy awoke and began to stalk the globe. Until then, service industries had tended to remain domestically bound. But in the 1970s and 1980s they began to internationalize, supported by a wide range of factors including the increased tradability of some services, occasioned by advances in tele-communications and information technology, as well as better packaging ('customization') and government deregulation . . . By the 1980s service industries were growing faster than any other sector of the world-economy.
>
> (Thrift, 1989, p. 32)

As such, services became a *major political football* within the protracted international trade negotiations of the Uruguay Round of the GATT in the second half of the 1980s. In particular, the developed economies, led by the United States, have been pressing for the inclusion of services in a new GATT agreement. Their objectives are to ensure unhindered access to developing country markets for services. Conversely, the developing countries are especially concerned to prevent their own service industries from being swamped either by trade or by direct investment. Some service industries are regarded as being extremely sensitive, either culturally or strategically, and are closely controlled and regulated by national governments.

In fact, the whole debate over the 'liberalization of trade in services' and, indeed, of services in general, is highly confused. Certainly it is often very confusing. The reasons for this state of affairs lie in the rather complicated nature of services themselves. The aims of this final case study chapter, therefore, are twofold. First, we need to disentangle the confusion over exactly what services are: what functions they perform, how they relate to the production of goods, how they operate at an international scale. Second, in line with our approach in the preceding three chapters, we need to look more closely at specific cases. Hence, one particularly important example – the *financial service industries* – will be examined in detail.

The nature of services: disentangling the confusion

Everybody knows what manufactured goods are: they can be seen; they can be handled; they are tangible. Some have at least a degree of permanence, although they may, of course, wear out or become obsolete. Others such as food products are consumed within a short period of time, although all goods can be stored in some way or other for future use long after they have been produced. Goods can also be acquired or consumed far away from their place of production: they can be traded at all geographical scales, including the global scale. But what, exactly, is a service? To many it is virtually the opposite of a manufactured good. Services are commonly regarded as being intangible, perishable, requiring their consumption at the same time and in the same place as their production. Most services, by this definition, are not tradable. In the classifications used by government statistical agencies, services are often seen as a residual category: activities which are neither dug out of the ground nor manufactured.

The diversity of the services sector

The services sector, as Figure 11.1 shows, is extremely diverse; far more so, in many respects, than the manufacturing sector. It ranges from the highly sophisticated, knowledge- and information-intensive activities performed in both private and public sector organizations to the very basic services of cleaning and simple maintenance. It includes retail and wholesale distribution and entertainment as well as health care and education services. It encompasses construction activities, transport activities, financial activities, communications activities, professional services. The list is (almost) endless.

One way of classifying this incredibly heterogeneous group of activities is to see them as part of a sequence of economic activities – primary, secondary, tertiary, quaternary – in which each element is futher removed from direct involvement with the earth's physical resources.[1] Thus, the primary sector is concerned with the direct extraction and manipulation of the earth's physical resources. Its outputs of raw materials and commodities are taken up by the secondary sector, which transforms them into manufactured products. These are then sold to other producers (producer or intermediate goods) or to private individuals or households (consumer or final goods). However, the distribution of consumer goods is rarely undertaken directly by the manufacturing company itself. This is one of the functions of the tertiary sector, which is itself made up of a vast array of services in retailing and other activities. At one time, the whole of the non-production sector was classified as 'tertiary' activity but this was not very helpful. After all, there is a world of difference between, say, the functions of a supermarket check-out assistant or a petrol station attendant and a nuclear physicist or a consultant surgeon. Hence, it became common to separate out tertiary services from a category of 'quaternary' services whose primary functions involve high-level

CONSTRUCTION SERVICES
Site preparation
New constructions
Installation and assembly work
Building completion
Maintenance and repair of fixed structures

TRANSPORTATION SERVICES
Freight services
Passenger transport services
Charter services
Services auxiliary to transport
    (incl. cargo handling, storage)
Travel agency and tour operator services
Vehicle rental

FINANCIAL SERVICES
Banking services (commercial and retail)
Other credit services (incl. credit cards)
Services related to administration of financial
    markets
Services related to securities markets
    (incl. brokerage, portfolio management)
Other financial services
    (incl. foreign exchange, financial consultancy)

BUSINESS SERVICES
Rental/leasing of equipment
Real estate services
Installation and assembly work
Professional services
    (incl. legal services, accountancy,
    management services, advertising, market
    research, design services, computer services)
Other business services
    (incl. cleaning, packaging, waste disposal)

TRADE SERVICES; HOTEL AND RESTAURANT SERVICES
Wholesale trade services
Retail trade
Agents fees/commissions related to distribution
Hotel and similar accommodation services
Food and beverage serving services

COMMUNICATION SERVICES
Postal services
Courier services
Telecommunications services
    (telephone, telegraph, data transmission,
    telemetrics, radio, TV etc.)
Film distribution and related services
Other communications services
    (incl. news and press agency, library and
    archive services)

INSURANCE SERVICES
Insurance on freight
Non-freight insurance
    (incl. life insurance, pensions, property,
    liability)
Services auxiliary to insurance
    (incl. brokerage, consultancy, actuarial)
Reinsurance

EDUCATION SERVICES

HEALTH-RELATED SERVICES
Human health services
    (incl. hospital services, medical and dental
    services)
Veterinary services

RECREATIONAL AND CULTURAL SERVICES

PERSONAL SERVICES
(not included elsewhere)
e.g. house cleaning/maintenance; nursing,
    day-care services

Figure 11.1 Categories of service activities (*Source*: based on GATT (1989) *International Trade 1988–1989*, Geneva: GATT, Appendix II)

transactions, analysis, research, decision-making, and the like. The quaternary sector consisted of those activities which control, administer and co-ordinate economic and non-economic activities and which also provide 'higher-order' services such as finance, research, education, health care.

A major difference between these different sectors was seen to be the particular *nature of the transactions* they performed, especially their material or non-material nature. Thus, the primary and secondary sectors are supposedly characterized by flows of materials and material products through the transport system. Similarly, the tertiary sector also generates material flows through buying and service trips. The quaternary sector, on the other hand, is regarded as transmitting, receiving and processing information rather than materials; its component parts are linked together by flows of information, rather than materials, through the communications media, as well as by direct face-to-face contact. This kind of classification of economic activities encouraged a linear view of change and development: as societies moved up the development ladder the balance of their economic activities shifted further and further towards the quaternary sector. But, the world is not so simple.

The interdependence between goods and services

One of the major weaknesses of this kind of sectoral classification is that it implies a degree of separation between goods and services which does not really exist; it does not capture the interdependencies of the real world.

> In many respects, the distinction between goods and services is a false one . . . First, most goods purchased are intended to provide a service or a function . . . Second, there are few 'pure' goods or services. Not only do virtually all goods require non-factor services in the course of their production, as most services require physical assets and intermediate goods; but, at the point of sale, most are jointly and simultaneously supplied. For example, travel by airline requires the joint use of tangible assets, such as the aircraft and a whole set of communications-related and airport paraphernalia, and advertising needs the printed word or audio-visual equipment to be effective . . . Similarly, many 'hardware' goods need to be combined with 'software' services if they are to be effective; for example, computers require programmers, air traffic needs air traffic controllers, industrial machinery and equipment, domestic boilers and burglar alarms need to be properly maintained and repaired, etc.
>
> (Dunning, 1989, p. 1)

'Producer' services and 'consumer' services

As the services sector has become an increasingly significant – even the dominant – element in many economies, particularly in terms of its contribution to GDP and employment, a rather different approach to classifying services has become popular. Services, it is suggested, can be separately categorized into *consumer* services and *producer* (or *business*) services. This classification is based on what is seen to be the final output or customer of the particular service. It is derived from the distinction commonly made in the manufactured goods sector in which producer/intermediate goods are distinguished from consumer/final-demand goods. This is quite a clear and reasonable distinction to make in the goods-producing sector. There is relatively little overlap between intermediate and final-demand goods. But even a cursory glance at Figure 11.1 shows that this is emphatically not true in the case of services. Although there are some services which are clearly either 'producer' services on the one hand or 'consumer' services on the other, many services apply to both categories at the same time. In other words, most services are 'mixed' in that they are *both* producer *and* consumer services. Obvious examples are transportation and communication services, hotels, many financial services, insurance, legal and accountancy services. Thus, several of the activities listed in Figure 11.1 under 'business' services are also used by private consumers.

Services within the production chain and in the sphere of circulation

This suggests that we need to think about services in a different way either from seeing them as being totally distinct from goods or, within the services sector itself, seeing them as being clearly divisible into producer and consumer services.[2] In Chapter 7 we introduced the concept of the *production chain* (Figure 7.1) to emphasize the interconnected nature of the production system. If we think of services in relation to the production chain then we can begin to see the kinds of role services play in the economy. Figure 11.2 shows that service inputs are needed at each stage of the production process. It also shows an increasingly significant aspect of today's competitive environment: the fact that 'services have become a major source of value-added. Downstream services in particular are both a factor contributing to competitive strength and a source of value-added' (UNCTAD, 1988, p. 178). But the value of services to the competitiveness of products is more general than this and applies throughout the production chain: 'In a number of product markets it is the "services"... design, styling, research, marketing, delivery, packaging, consumer credit – which determine the competitiveness of agricultural and manufacturing investment. As the length of production chains increases, so services are responsible for a greater share of value added to products' (Britton, 1990, p. 538). Many services are deeply embodied in goods.

The provision of such service inputs to production may be carried out within the producing firm itself (i.e. it can be *internalized*) or some or all of them may be put out

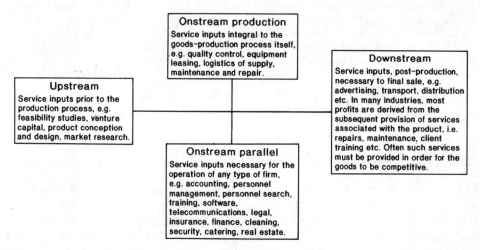

Figure 11.2 The interconnections between services and production in the production chain (*Source*: based on UNCTAD (1988), p. 177)

to independent, specialist service firms (i.e. provision may be *externalized*). As we saw in Chapter 7, the boundary between the internalization and externalization of functions is continuously changing. Functions may be hived off or, alternatively, specialist service firms may be taken over or they may operate on a subcontracting basis. The development of dynamic organizational networks invariably involves the changing of the organizational – and often the geographical – location of the firm's various functions and altering the balance between production and service activities:

> there has . . . been a tendency for corporations to divest themselves of these lesser value-added components in the productive process. Some corporations have been described as 'hollow', in that what were previously considered to be manufacturing firms have ended up not producing any physical goods, but only providing the various high value-added services. To the extent that value-added is derived more from the service component rather than from labour costs or raw material inputs, the question of where a product is produced is becoming of lesser importance than who supplies those services that constitute an increasingly important component of such value added.
>
> (UNCTAD, 1988, p. 178)

In one sense, therefore, services are not readily separable from the production of goods but, rather, are an *integral* part of such production. On the other hand, 'a substantial volume of transactions are generated within and between service industries themselves: the output of many producer service industries go to other services . . . Riddle (1986: 25) argues that one of the fastest-growing segments of the service sector is those services that either distribute the product of, service the input needs of, or act as market intermediaries for, other service industries' (Britton, 1990, p. 534). Indeed, some estimates suggest that more than 50 per cent of the output from service industries is sold to other service industries.

There is no doubt, therefore, that labelling services as either 'producer services' or 'consumer services' can be misleading. It works in some cases but not others:

> While advertising, marketing and R and D are readily identified as producer services . . . the commercial and financial services which mediate and abbreviate the exchange processes within the economy are neither producer nor consumer services. They are circulation services, services produced within the process of circulation and for circulation, and not intermediate services produced primarily for other branches of industry or final services produced for consumers. Circulation services . . . are concerned primarily with the velocity of turnover, whether of commodities, money or money capital . . . Whereas goods production is restricted to the sphere of production, service production occurs within both the sphere of production and the sphere of circulation.
>
> (Allen, 1988, pp. 18, 19)

Hence the title of this chapter: in many respects it is service activities that make the world go round, which lubricate the wheels of production, distribution and exchange.

The internationalization of services

The internationalization of most economic activity generally occurs via two means. One way is through international *trade*; the other is through some kind of *presence* in a particular overseas location or market (e.g. via foreign direct investment, licensing, joint venture, subcontracting and the like, as shown in Figure 7.16). How far do services fit this pattern? They clearly fit the second of the two forms of internationalization: service firms may establish a presence in a foreign market to provide their services to local or locally based customers. But to what extent are services *tradable*? We have already referred to the growth of international trade in services in Chapter 2. However, precisely what is meant by trade in services is a complex issue, bound up with the ways in which balance of payments statistics are calculated by governments. Such technical matters need not concern us here.[3] What is clear, however, is that trade statistics vastly understate the extent to which services have become internationalized. This is because they do not fully capture some of the most important mechanisms involved: the embodiment of services in goods which are themselves traded, the trade in internalized services which occurs within transnational corporations.

In fact, many services are not tradable in the strict sense of the term because they need to be consumed at the point of production: they are not storable. It is claimed that technological innovations, especially in information technology, have made it possible – at least in principle – for some kinds of service transactions to be conducted with geographical separation between the producer and consumer of the service. For example, the development of sophisticated information technologies permits the international – or transborder – flow of data and information. In that respect, they allow 'information-based services' to occur even where the producer and consumer are geographically separated, that is, they can be traded internationally. We shall see later in our examination of financial services that this is a particularly significant feature of current activity. But even in those cases, there is generally a need for the supplying firm to have an actual presence in the foreign market to deliver the service more efficiently and effectively.

Hence, the real issue in the current debate on the liberalization of international trade in services is not trade in itself but, rather, the conditions under which providers of services are permitted to establish an actual direct or indirect presence in a specific national market.[4] In other words, it is really about foreign direct investment and the other modes of international involvement which firms may use (see Figure 7.16). As Gibbs (1985, p. 221) accurately observes, to the service TNC 'the distinction between trade and investment is academic, what they require is a presence in the market, the ability to compete free of regulations which place them at a disadvantage vis-à-vis domestic suppliers, the only trade which is vital is that in information, transborder data flows constituting their "lifeblood"'. However, it is the very capacity of TNCs to control information vital to many modern service industries which places developing countries at a disadvantage. Clearly, therefore, two of the major issues at the heart of both the processes and the politics of the internationalization of services are, first, *information technology* and, second, *government regulation*. Both of these will be examined more closely when we look at the case study of financial service industries.

We will look also at the specific ways in which such internationalization has occurred in those industries. For many service industries, particularly those which are primarily intermediate inputs into the production chain and those which are circulation activities, the initial stimulus to their internationalization was the rapid growth and global spread of TNCs in manufacturing industries. As manufacturing TNCs have proliferated globally so, too, have the major banks, advertising agencies, legal firms, property companies, insurance companies, freight corporations, travel and hotel chains, car rental firms and credit card enterprises. In many respects, the internationalization both of manufacturing and of these business service activities has become *mutually reinforcing*. The major business service corporations have tended to follow their manufacturing clients abroad. Conversely, the existence of an extensive global network of the familiar business service corporations helps to guide the further evolution of transnational manufacturing activities. A truly *symbiotic* relationship has developed. As these services have themselves become internationalized they have also acted as a stimulus for the internationalization of other service activities so that the process no longer is simply one of services following manufacturing.

The emergence of transnational service conglomerates[5]

The global package deal

Many service companies are single-product organizations, specializing in a specific service function and supplying it in many different countries through a network of subsidiaries, branches, joint ventures and other forms of partnership. However, some services are strongly *complementary* to other services, for example there is said to be a 'natural' relationship between airlines and international hotel chains, between different kinds of financial services, between accountancy and management consultancy, between advertising, public relations and the communications media. Hence, there has been a strong development of transnational service conglomerates; firms which operate internationally in a number of related service industries and aim to offer a 'global package deal'.

A good example is the recent development of the global advertising industry.[6] Through a whole series of aggressive acquisitions and mergers, since the early 1980s, the leading advertising agencies have become true global players with a vast network of overseas offices. The UK firm Saatchi and Saatchi, for example, has 79 offices in 51 countries, including 25 offices in the United States. The US agency Young and Rubicam operates 76 offices in around 50 countries. But some of the leading advertising companies have moved beyond simply offering advertising services. They have deliberately adopted the strategy of becoming much broader: *marketing–service conglomerates*. They offer a global package which includes advertising, marketing, promotion, media services, public relations, management consultancy and similar functions.

The rationale underlying this strategy is that it increases the firm's potential to serve global clients by offering an integrated package of services within a single co-ordinated operation. The argument is the old one – now somewhat discredited in manufacturing industries – of *synergy* (the whole is more than the sum of the parts). A client for one of the marketing–service conglomerates, it is argued, can be cross-referred to other parts of the same business. Currently there are four especially large marketing–service conglomerates, each of which is based upon a nucleus of global advertising companies. The four conglomerates are two headquartered in the United Kingdom (WPP Group, Saatchi and Saatchi) and two US firms (Interpublic and Omnicom). Each of the four groups operates a global network of offices in some fifty countries.

WPP and Saatchi and Saatchi are by far the biggest of these groups. Both have grown very rapidly and recently through a frantic series of mergers and acquisitions. Both consist of two very large advertising agencies at their nucleus: WPP owns J. Walter Thompson and Ogilvy and Mather; Saatchi and Saachi's advertising arms are Saatchi and Saatchi Advertising World Wide and Backer, Spielvogel Bates Worldwide. These are surrounded by a galaxy of other marketing, media and consultancy companies. Currently, however, both WPP and Saatchi and Saatchi are experiencing major problems. Neither seems to have been totally successful in actually integrating its diverse activities into a coherent functional unit. The one-stop shop seems to be something of a mirage in these particular industries. We shall look at diversified service companies again in the case study of financial services.

The Japanese sogo shosha

The most highly developed form of the transnational services conglomerate – involved in a vast array of mining, agricultural, manufacturing as well as service activities – is by no means a recent phenomenon. The giant Japanese general trading companies (the *sogo shosha*) are undoubtedly the most remarkable examples of this form of organization, with a long history in the development of the Japanese economy.[7] They have also been especially influential in the development of Japanese overseas direct investment as a whole (see Chapter 3). The common translation of the term *sogo shosha* is 'general trading company'. But they are very much more than this. The leading *sogo shosha* are gargantuan commercial, financial and industrial conglomerates with a massive network of subsidiaries and thousands of related companies across the globe. Nevertheless, their primary function is that of the organization of trade. Their huge size and the extent of their geographical operations give the major *sogo shosha* an enormous importance both in Japan's economic affairs and in the global economy as a whole. Each of the leading *sogo shosha* handles tens of thousands of different products. In the second half of the 1980s, they were responsible for roughly '8% of world trade, in addition to a large share of Japan's domestic trade. The share of the *sogo shosha* in the total trade of Asian and Pacific countries has been estimated at about 17% at the beginning of the 1980s, when they also handled as much as 10% of United States exports' (UNCTC, 1988, p. 384).

Although there are vast numbers of trading companies in Japan, six companies dominate: Mitsubishi, Mitsui, C. Itoh, Marubeni, Sumitomo, Nissho-Iwai. This is the true Japanese general trading oligopoly, each member of which has a major co-ordinating role within one of the Japanese *keiretsu* or bank-centred groups (Yoshino and Lifson, 1986). Each of the leading *sogo shosha* operates an enormous global network of foreign affiliates and offices. They have massive bargaining power and a remarkable internalized information system. In the case of Mitsui, the second-ranking *sogo shosha*,

> the company's communication lines total 500,000 kilometres, greater than the distance between the earth and the moon. In 1986, the daily volume of messages amounted to an estimated 110,000 despatches and receipts containing 10 million words... Mitsui's communication expenditures-to-sales ratio surpasses easily that for R & D-to-sales of most manufacturing firms. For trading companies, the establishment and maintenance of a communication system is an investment, just as R & D is an investment for industrial firms
>
> (UNCTC, 1988, p. 386).

The *sogo shosha* were the first Japanese companies to invest on a large scale outside Japan. Such investments were to set up a global marketing network. Once in place, this network, with all its supporting facilities, enabled a whole range of Japanese firms to engage in overseas investment. Indeed, a good deal of the early overseas investment by Japanese manufacturing firms was organized by the *sogo shosha*. The kind of comprehensive services provided by the *sogo shosha* include, according to Young (1979):

1. *Financial services*: credit, loans, loan guarantees, venture capital.
2. *Information services*: up-to-date information on thousands of product markets across the globe, including details of national regulations, competition in national and international markets, technological developments.
3. *Risk-reduction services*: a 'buffer' role between banks and small producers enabling banks to avoid the high costs associated with small loans; hedging risks in exchange-rate fluctuations.
4. *Organization and co-ordination services*: organization of contracts and links between suppliers and customers.
5. *Auxiliary services*: paperwork, insurance, wholesaling, transportation.

Thus the giant *sogo shosha* are both major transnational firms in their own right and also have been the 'umbrella' under which many individual, often small, Japanese firms have expanded overseas. The vast and complex network of export, sales and marketing investments established over the years by the *sogo shosha* provided a uniquely valuable infrastructure for the transnational expansion of smaller Japanese companies. The *sogo shosha* are a uniquely Japanese institution with deep historical, cultural and social roots. However, similar organizations exist elsewhere in Asia. In South Korea, for example, the trading arms of the giant domestic conglomerates or *chaebol* have similar characteristics to the *sogo shosha* although on a smaller scale.

The internationalization of financial services: the 'fixers' of the global economy[8]

> What is going on now is a revolution: a revolution in the way finance is organized, a revolution in the structure of banks and financial institutions and a revolution in the speed and manner in which money flows around the world.
>
> (Hamilton, 1986, p. 13)

In Chapter 2 we noted the extent to which international financial flows and foreign currency exchanges now dwarf the value of international trade in goods. Massive changes have been, and are, occurring both in the nature and composition of financial services themselves and also in the extent to which they have become internationalized. The global financial system has become extremely complex and extremely volatile. Its ramifications for the entire global economy are so enormous basically for two reasons. First, because financial services are *circulation* services they are totally fundamental to the operation of every aspect of the economic system. Each element in the production chain of Figure 7.1 depends upon necessary levels of finance to keep the chain in operation. This is true not just of manufacturing itself but also of all other intermediate and consumer services in the system.

The second reason for the enormous significance of the global financial system is that many of its activities are *speculative*: financial investments aimed at making short- or long-term profits as ends in themselves. Hence, changes in the level and the geography of the flows of international portfolio investments (investments in stocks and bonds) can greatly affect the financial well-being of entire national economies. Much of the huge US current account and budget deficit is actually financed by the inflow of foreign portfolio investment, particularly from Japan. Any major shift in such investment would have enormous repercussions on the US economy and, therefore, on households and individuals as well as on businesses. Although, in general terms, the speed and geographical complexion of international portfolio flows are determined by geographical differences in interest rates, exchange rates and political risk, the scale of international portfolio flows is now so great that they can, in themselves, help to determine exchange rates.

The changing pattern of demand for financial services

The international financial system is made up of a great variety of different kinds of institution, each having a specific set of *core* functions. Table 11.1 lists the major types of financial institution and provides a brief description of each. In fact, the boundaries between them have become increasingly blurred. There are many reasons for this, as we shall see. An important factor, however, has been the changing pattern of demand for financial services as a whole.[9] 'Ever since the mid- and late-1970s, financial services around the world have been characterized by a sharp intensification of competition and a rapid transformation of markets. These changes are leading to the emergence of a new, more dynamic market environment, quite unlike that which characterized the industry during the previous decades' (Bertrand and Noyelle, 1988, p. 16).

The 'new competition' consists, in Bertrand and Noyelle's view, of four major forces:

1. *Market saturation.* From the late 1970s, in particular, it became apparent that the traditional financial services markets were reaching saturation. There were fewer and fewer new clients to add to the list; most were already being served, particularly in the commercial banking sector but also in the retail sector in the more affluent economies. A similar high level of market saturation was becoming apparent in insurance services.
2. *Disintermediation.* This is the process whereby corporate borrowers in particular make their investments or raise their needed capital without going through the 'intermediary' channels of the traditional financial institutions, particularly the banks.

   The single most important factor that triggered financial disintermediation was the wave of inflation of the mid-1970s. As the administrative costs of financial services rose and as the interest spread between interest charged for financial services and interest paid on deposits widened . . . large customers abandoned commercial banks and insurance companies in droves and went elsewhere to find better-priced services. In banking, for example, commercial customers abandoned traditional bank borrowing and turned to the

Table 11.1 Major types of financial service activity

Type of financial institution	Primary functions
Commercial bank	Administers financial transactions for clients (e.g. making payments, clearing cheques). Takes in deposits and makes commercial loans, acting as intermediary between lender and borrower.
Investment bank; securities house	Buys and sells securities (i.e. stocks, bonds) on behalf of corporate or individual investors. Arranges flotation of new securities issues.
Credit card company	Operates international network of credit card facilities in conjunction with banks and other financial institutions.
Insurance company	Indemnifies a whole range of risks, on payment of a premium, in association with other insurers/reinsurers.
Accountancy firm	Certifies the accuracy of financial accounts, particularly via the corporate audit.

commercial paper market for short-term funds, later to the bond market for long-term financing.

(Bertrand and Noyelle, 1988, p. 18)

A similar trend has been occurring in the retail markets, where private investors have switched funds from traditional savings deposit accounts to higher-yielding funds such as investment trusts and mutual funds.

3. *Deregulation of financial markets.* Financial services markets have traditionally been extremely closely regulated by national governments. One of the most important trends of the last few years, therefore, has been the increasing deregulation or liberalization of financial markets. We shall examine the regulatory environment of financial services, and the trend towards deregulation, in some detail in a later section. At this point we should note that deregulation has been aimed primarily at three areas: the opening of new geographical markets, the provision of new financial products, and changes in the way in which prices of financial services are set.

4. *Internationalization of financial markets.* Demand for financial services is no longer restricted to the domestic context: financial markets have become international and, in some cases, *global*. Three types of demand are especially significant in this context. First, the massive growth in international trade described in Chapter 2 has created a much increased demand for commercial financial services at an international scale. Second, the global spread of transnational corporations has created a demand for financial services way beyond those corporations' home countries. Third, the vastly increased institutionalization of savings – the channelling of savings into the pension funds and other institutions – has created an enormous pool of professionally administered investible capital seeking the best return, wherever that might be achieved.

Each of these forces for change in the demand for financial services is interrelated. Taken together they create a *new competitive environment* in which financial services companies now operate. Before looking specifically at the strategic responses of the major institutions, however, we need to consider two issues in more detail: technological change and government regulation (and deregulation) of financial services.

Financial services and changing technologies

In Chapter 4 we discussed in some detail technological developments in communications technologies and in the broader sphere of information technology. We have seen in the last three chapters that these technologies have been extremely important in the global evolution of the textiles and clothing, automobile and electronics industries. But they are even more fundamental to the service industries in general and to financial services in particular.[10] This is because, to a far greater extent than in our manufacturing case studies, *information is both the process and the product of financial services.* Their raw materials are information: about markets, risks, exchange rates, returns on investment, creditworthiness. Their products are also information: the result of adding value to these informational inputs. A particularly significant piece of added value is embodied in the *speed* with which financial service firms can perform transactions and the global extent over which such transactions can be made.

Thus, 'the world of finance has been fundamentally changed by technology. More than in any other area of activity, the growth of international communications, the development of the data-processing capability of the big computer and the personal desk-top facility and the arrival of the day of the wired society have revolutionized the way in which finance is transacted' (Hamilton, 1986, p. 33). As a result, 'Since 1964, the real cost of recording, transmitting and processing information (including financial information) has fallen more than 95%. This tremendous cost reduction makes it cheaper to assess the risks and rewards of financial assets, cheaper to record and process trades, cheaper to manage portfolios and cheaper to match the users with the suppliers of capital' (Huertas, 1990, p. 263).

Development of *satellite communications systems* has been especially important to the development of international financial markets. Warf (1989) reports that in the mid-1980s, 50 per cent of all international financial transactions were made by satellite-linked telephone calls. Indeed, as he points out, financial services firms are probably the heaviest users of telecommunications systems. Warf summarizes the major effects of the information technologies on financial services as follows:

1. They have vastly increased productivity in financial services.
2. They have altered the patterns of relationships or linkages both within financial firms and also between financial firms and their clients.
3. They have greatly increased the velocity, or turnover, of investment capital. For example, the ability to transfer funds electronically – and, therefore, instantaneously – has saved billions of dollars in interest payments which were formerly incurred by the delay in making transfers.
4. At the international scale, they have enabled financial institutions both to increase their loan activities and also to respond immediately to fluctuations in exchange rates in international currency markets.

Not surprisingly, therefore, all the major financial services firms have invested extremely heavily in information transmission facilities:

> The 1000 largest US banks between 1972 and 1985 increased the proportion of their operating expenses dedicated to telecommunications from 5 to 13% . . . many . . . installed private networks as they expanded into the global marketplace. Citicorp's Global Tele-communications Network (GTN), for example, is the largest private telecommunications network in the world. It links offices in 94 nations, transmits over 800,000 calls per month, allows Citicorp to trade $200 billion daily in foreign exchange markets round the world . . . Securities firms have also invested heavily in new telecommunications technologies. Integrated work stations on the trading floors . . . give brokers rapid access to multiple clients and to the computerized trading programs with which they can sell large blocks of stocks . . . Merrill Lynch, the largest American securities firm, spends $400 million annually on its telecommunications system.
>
> (Warf, 1989, pp. 261–2)

From a technological viewpoint, therefore, it is now possible for financial services firms to engage in global twenty-four-hour-a-day trading, whether this be in securities, foreign exchange, financial and commodities futures or any other financial service. The ability to transmit data electronically over vast geographical distances creates the potential for continuous financial transactions worldwide, whatever the time of the day or the night. Figure 11.3 shows the way in which the trading hours of the world's major financial centres overlap. True twenty-four-hour trading is currently limited to certain kinds of transaction partly because, although the technology is available, either the

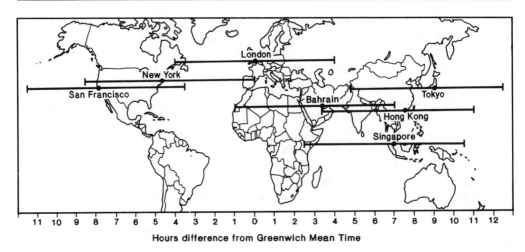

Figure 11.3 The trading hours of major world financial centres (*Source*: Warf (1989), Figure 5)

organizational structure or the national regulatory environment remains an obstacle. But there is no doubt at all that global twenty-four-hour trading will become standard in virtually all kinds of financial service.

To the extent that such electronic transactions do not require the direct physical proximity between seller and buyer they are a form of 'invisible' international trade. In that sense, therefore, financial services are one form of service activity which is *tradable*. Electronic communications have also contributed greatly to the bypassing of the commercial banks and the trend towards the greater *securitization* of financial transactions. Securitization, in the broad sense, is simply the conversion of all kinds of loans and borrowings into 'paper' securities which can be bought and sold on the market. Such transactions may be performed directly by buyers and sellers without necessarily going through the intermediary channels of the commercial banks.

The *global integration of financial markets* brings many benefits to its participants: in speed and accuracy of information flows and rapidity and directness of transactions, even though the participants may be separated by many thousands of miles and by several time zones. But such global integration and instantaneous financial trading also have their costs. 'Shocks' which occur in one geographical market now spread instantaneously around the globe. The collapse of prices on the notorious Black Monday of 19 October 1987 surged immediately round all the world's stock markets. Billions of dollars were wiped off company stocks and, even though the world's markets recovered, the incident brought home to everybody just how sensitive and volatile the global financial system can be as a consequence of the telecommunications revolution.

The influence of telecommunications technologies on the speed and geographical extent of financial transactions between sellers and buyers of financial services is one expression of the effect of technological change on the financial services industries. Another is the effect of information technology on the *internal operations* of financial services firms (Bertrand and Noyelle, 1988). For example, banks and insurance companies were among the earliest adopters of computer technology to automate the internal processing of financial transactions; the so-called 'back-office' functions which are otherwise highly labour-intensive. Subsequently, computerization and related technologies were applied to the firms' 'front offices': to the direct interface between firms and their customers. In retail banking, for example, counter clerks or tellers now invariably operate on-line computer teminals; automated teller machines and cash dispensers have become the norm, giving customers access to certain services outside normal banking hours; home banking by computer terminal, however, has been far slower to catch on.

The equivalent of such services for the large corporate customers is the development of electronic cash management systems in which the corporate customer's computers are linked directly to the bank:

> In their most sophisticated version, these electronic cash management systems provide corporate treasurers with up-to-date account-by-account balance information and give them control over the transfer of funds among the company's accounts. This allows... [corporate] ... treasurers to invest their unused cash flow into money market instruments and to optimize their use of funds in a world economy that has become characterized by greater volatility in interest, inflation and ... exchange rates.
>
> (Bertrand and Noyelle, 1988, p. 22)

Innovations in telecommunications and in process technologies, therefore, have helped to transform the operations of financial services firms. But there has also been a variety of product innovations. In this sense,

> the number of financial markets has proliferated in the 1970s and 1980s as 'financial engineering' (the invention of new financial instruments) has become an important 'art form' of the late twentieth century. Markets like those in options and futures (which trade in forward contracts in commodities, money and shares) and equities (which trade in stocks and shares) have taken their place alongside the Eurodollar and Eurobond markets as important global markets.
>
> (Thrift, 1989, p. 38)

A whole new array of financial instruments has appeared on the scene which can be categorized into two broad types: those which provide *new methods of lending and borrowing* and those which facilitate *greater spreading of risk*.[11] Table 11.2 gives some

Table 11.2 Product innovations in financial markets

Type of financial instrument	Basic characteristics
Floating-rate notes (FRNs)	Medium- to long-term securities with interest rates adjusted from time to time in accordance with an agreed reference rate, e.g. the London Inter-Bank Offered Rate (LIBOR) or the New York bank's prime rate.
Note issuance facilities (NIFs)	Short- to medium-term issues of paper which allow borrowers to raise loans on a revolving basis directly on the securities market or with a group of underwriting banks.
Eurocommercial paper	Non-underwritten notes sold in London for same-day settlement in US dollars in New York. More flexible than longer-term Euronotes of 1, 3 or 6 months' duration.
Loan sales	The sale of a loan to a third party with or without the knowledge of the original borrower.
Interest-rate swaps	A contract between two borrowers to exchange interest rate liabilities on a particular loan, e.g. the exchange of fixed-rate and floating-rate interest liabilities.
Currency swaps	Financial transactions in which the principal denominations are in different currencies.

(*Source*: based on UNCTC (1988), Box VII.2; Lewis and Davis (1987), pp. 415–31)

examples of those financial innovations which have greatly increased the product diversity of financial markets.

Without question, therefore, technological developments, particularly in tele-communications and information technology, as well as in product innovations, have transformed the financial services industries. The global integration of financial markets has become possible, collapsing space and time and creating the potential for virtually instantaneous financial transactions in loans, securities and a whole variety of financial instruments. However, completely borderless financial trading does not actually exist, for the simple reason that most financial services remain very heavily supervised and regulated by individual national governments. Let us now see how the regulatory system operates and how it is changing.

The influence of governments; regulation and deregulation in financial services[12]

*A **tightly regulated system***   Before the 1960s there was really no such thing as a world financial market. The IMF, together with the leading industrialized nations, acted to ensure a broadly efficient global mechanism for monetary management based, initially, on the postwar Bretton Woods agreement. At the national level, financial markets and institutions were very closely supervised and regulated for a whole variety of reasons:

> One is the apparent vulnerability of banking and financial systems to crisis and panic. Second, the financial system is seen by many as playing an important role in business cycles and the determination of macroeconomic activity. Third, an aversion to undue concentration of economic and financial power often finds specific expression in preventing 'money trusts' by means of branching restrictions and limitations upon bank shareholdings. At the same time, a desire to avoid conflicts of interest sees constraints placed upon the range of activities of financial firms. Fourth, governments have sought to implement various social objectives through the financial system in the form of, for instance, finance for housing and agriculture. Finally, the central role of the payments system in facilitating production and exchange has given governments a particular interest in ensuring its smooth and efficient operation.
>
> (Lewis and Davis, 1987, pp. 189–90)

Financial services, therefore, have been probably the most tightly regulated of all economic activities, certainly more so than manufacturing industries. The forms of such regulation can be divided into two major types: those which govern the *relationships* between different financial activities and those which govern the *entry* of firms (whether domestic or foreign) into such activities. In the first case, it has been common in most countries to restrict, or even prohibit, firms from participating in a range of financial service activities. In other words, each national financial services market has been *segmented* by regulation: banks operated in specified activities, securities houses in other areas of activity; the two were not allowed to mix across the boundaries. Neither was allowed to perform the functions of the other. In the United States, for example, commercial banks are separated by law from investment banks under the Glass–Steagall Act of 1933.

Restricted entry into the different financial services markets has also been virtually universal in all countries, although the precise details differ from country to country. Most countries, however, have restrictions governing the entry of foreign firms into financial services activities. At the very least, national governments retain the discretionary power to restrict inward investments and all countries use their anti-trust/ competition laws to regulate foreign acquisitions of domestic financial activities. Governments are especially wary of a too-ready expansion of the branches of foreign banks and insurance companies. Unlike subsidiary companies, which have to be separately incorporated, branches are far more difficult to supervise because they form an integral part of a foreign company's activities. In almost every case, there are limits on the degree of foreign ownership permitted. There are regulations, too, which restrict the activities of finance-related services such as accountancy. Such restrictions largely explain the particular organizational forms adopted by international accountancy firms (see below, p. 371).

'The crumbling of the walls'[13]   Many of these restrictions on the operation of financial services are still operative. The regulatory walls have been crumbling, however, slowly at first but with increasing speed in the past decade. As yet, however, there remain huge differences in the extent and nature of financial services regulation in different countries: the international regulatory environment is highly asymmetrical. Such differences, of course, have a powerful influence on the locational strategies of international financial services firms. The pressure towards deregulation have come from several sources, most notably the increasing abilities of international firms to take advantage of 'gaps' in the regulatory system and to operate outside national regulatory boundaries.

One of the most important developments in the creation of an increasingly global financial market was the emergence of the Eurodollar markets in the 1960s:

> The Eurodollar market was perhaps the most important international financial development of the 1960s, growing, within a decade, from a market worth a few hundred million dollars and handling issues of between $2 million and $5 million each in the mid-1960s to a market worth several hundred billion dollars a year and handling issues of between $200 million and $500 million a time. Now it is worth $300 billion a year, and it is a market whose development can be charted very much in terms of reaction to regulation within the domestic financial systems of the USA, Europe and Japan.
>
> (Hamilton, 1986, p. 21)

In origin, Eurodollars were simply dollars held outside the US banking system, largely by countries like the USSR and China which did not want their dollar holdings to be subject to US political control. 'Once it was appreciated that Eurodollars . . . were free of US political control, it did not take bankers long to recognise that the dollar balances were also free of US banking laws governing the holding of required reserves and controls upon the payment of interest' (Lewis and Davis, 1987, pp. 225, 228).

The rapid growth of this currency market outside national regulatory control was certainly one of the major stimuli towards an international financial system. It was reinforced by the revolutions in telecommunications and information technologies, discussed earlier, which made possible the internationalization of financial transactions. Pressures built up, too, from the desires of banks and other financial services institutions to operate in a less constrained and segmented manner, both domestically and internationally. The internationalization of financial services and the deregulation of national financial services markets are virtually two sides of the same coin. Forces of internationalization were one of the pressures stimulating deregulation; deregulation is a necessary process to facilitate further internationalization.

Major deregulation has occurred, or is occurring, in all the major economies.[14] A series of changes in the United States since the 1970s has both eased the entry of foreign banks into the domestic market and facilitated the expansion of US banks overseas. In 1981 the United States allowed the establishment of International Banking Facilities (IBFs), which created 'onshore offshore' centres able to offer specific facilities to foreign customers. Earlier, in 1975, the New York Stock Exchange had abolished fixed commissions on securities transactions. Currently, the federal government is trying to introduce a major reform of the US financial system aimed at allowing banks to become involved in a whole variety of financial services and to operate nationwide branching networks. One of the paradoxes of the US system is that the 'non-bank banks', such as General Motors' Acceptance Corporation, can operate without many of the restrictions which currently apply to the 'proper' banks.

In the United Kingdom the so-called 'Big Bang' of October 1986 removed the barriers which previously existed between banks and securities houses and allowed the entry of foreign firms into the Stock Exchange. In Japan the restrictions on the entry of foreign securities houses have been relaxed (though not removed) and Japanese banks are now allowed to open international banking facilities. In France the 'Little Bang' of 1987 is gradually opening up the French Stock Exchange to outsiders and to foreign and domestic banks. In Germany foreign-owned banks are now allowed to lead-manage foreign DM issues, subject to reciprocity agreements.

The progressive deregulation of financial services is the most important current development in the internationalization of the financial system. However, 'though deregulation is consistent with the market-oriented approach of present governments in most developed market economies, it would be a mistake to imagine that it has been effected solely, or even mainly, as an act of principle. In fact, it has often been more a matter of abandoning regulations which, because of changing circumstances, have become a source of distortion in financial markets or have been merely rendered ineffective' (UNCTC, 1988, p. 106). Deregulation in financial services is also being reinforced by deregulation in the *telecommunications* industries which form the essential infrastructure without which the international financial system could not operate. As Warf (1989) points out, the deregulation of the telephone industry, in particular, is crucial to the operation of financial services firms because the telephone is their 'workhorse'.

Increasing deregulation of financial services is a fundamental part of the changes associated with the strengthening of regional trading blocs. As Table 6.3 shows, the Canada–United States Free Trade Agreement contains clauses extending the principles of national treatment, right of commercial presence and right of establishment to each other's providers of services. In the particular case of financial services, existing market access is preserved and competition in securities underwriting and banking is introduced. In the case of the European Community's proposals to complete the single market after 1992 the position is more complex, not least because of the existence of twelve independent national jurisdictions.[15]

The European Commission's proposals for integrating financial services markets are threefold. The first is to ensure the *free flow of capital* between all member states, with the abolition of national exchange controls where they still exist. The second proposal relates to the establishment of *free trade* within the Community in all financial services together with the right of financial services firms to establish a *presence* anywhere within the single market. This requires that each state recognize any financial institution licensed in another Community country to operate locally, as long as it does so under local rules . . . The effect of these regulations should be to make European financial markets more contestable. Banks will no longer face the bureaucratic obstacles of obtaining licences in every country in which they wish to operate. More substantively, they will not face many of the implicitly protective regulations that have limited their incentives to enter profitable markets – regulations such as those that impose capital requirements on their individual branches, rather than on their operations as a whole.

(Davis and Smales, 1989, p. 95)

The third proposal is designed to *standardize banking technology* throughout the Community to facilitate, for example, Community-wide use of plastic cards for automatic teller machines regardless of which bank issued the card. Achieving all of those objectives will not be easy; as in other aspects of the Single European Market, expectations and promises may well exceed the reality. Nevertheless, the mere existence of proposed change is forcing both governments and financial services companies to rethink their strategies.

Corporate strategies in financial services

Shifting patterns in the demands for financial services, technological innovations which affect how these services can be delivered and the changing regulatory framework interact together to form the environment in which financial services companies must operate. But, of course, such firms are not simply responding to environmental changes over which they have no influence. The processes are dynamic and interactive. Indeed, the strategies and actions of the major financial services firms – the banks, securities houses, insurance companies and the like – have themselves been influential in changing that environment. There is no simple cause–effect relationship but, rather, a complex interplay between actors and processes. In that respect, financial services are no different from any of the other industries we have been examining.

In this section we focus on the specialist financial services companies and exclude the in-house financial activities of TNCs in other industries. All TNCs, of course, operate large-scale and sophisticated financial operations. The corporate treasury departments of major companies are significant entities in their own right and are larger than many specialist financial services companies. However, the primary functions of these in-house operations are to contribute towards the efficiency of their own corporate systems by, for example, optimizing internal flows of funds, investing surplus capital effectively on the financial markets, and hedging against the foreign currency fluctuations which are endemic in the operations of any TNC, whatever its line of business. As far as the specialist financial services firms are concerned, two major strategic trends will be examined: the *internationalization* of their operations and the *diversification* of companies into new product markets. The two sets of strategies are very closely related.

Internationalization[16]

Banks have always engaged in international business. They have dealt in foreign exchange, extended credit in connection with foreign trade, traded and held foreign assets and provided travellers with letters of credit. Historically, the banks have carried out all this and some other types of business from their domestic locations. There was no need for a

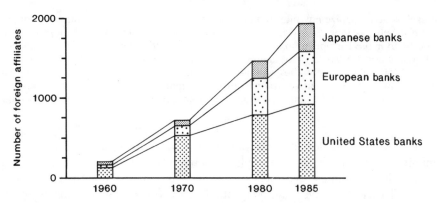

Figure 11.4 Growth in the number of foreign affiliates of banks from OECD countries (*Source*: based on *OECD Observer*, 1989, Vol. 160, p. 36)

physical presence abroad. Business that could not be carried out by mail or telecommunications was handled by correspondent banks abroad.

(Grubel, 1989, p. 61)

A small number of banks certainly set up a few overseas operations towards the end of the nineteenth century. Even in the early part of the twentieth century, however, the international network was very limited. Almost all international banking operations were 'colonial' – part of the imperial spread of British, Dutch, French and German business activities. In 1913 the four major US banks had only six overseas branches between them. By 1920 the number of branches had grown to roughly a hundred but there was little further change until the 1960s. During that period, Citibank had by far the most extensive international network of any US bank.

As with TNCs in manufacturing, the most spectacular expansion of transnational banking occurred in the 1960s and 1970s. Again, the initial surge was dominated by US firms, a reflection of both the focal role of the United States in the postwar international financial and trading system and also the rapid proliferation of United States TNCs. But European, and later Japanese, banks increasingly internationalized their operations. Figure 11.4 shows that the number of foreign affiliates of banks headquartered in the OECD countries increased from 202 in 1960 to 1,928 in 1985. Growth in this international network was especially rapid in the 1960s and 1970s. At the same time the *geographical composition* of the international banking network has changed. United States banks are now less dominant than in the 1960s and 1970s, European and Japanese banks have increased the size of their international branch network. Although Figure 11.4 does not take us beyond 1985, there is reason to believe that the overall size of the foreign network of transnational banks has not increased greatly since then. What *has* certainly been happening is a change in the composition of the overall network, with particular international expansion by Japanese and some European banks and some network consolidation by US, UK and German banks.

As in other areas of economic activity, it has been the rise of the Japanese banks which has been especially dramatic. Table 11.3, which is the league table of the world's twenty-five leading banks, is based on asset values – although these can be misleading, particularly as a measure of transnationality. Nevertheless, the table does indicate major changes in ranking since 1975. In that year, there were only two Japanese banks

Table 11.3 The world league table of commercial banks, 1989

Rank 1975	Rank 1989	Bank	Home country	Total assets 1989 ($ billion)	Total employment	Foreign network 1985 (number)
7	1	Dai-Ichi Kangyo Bank	Japan	388.9	18,411	44
11	2	Sumitomo Bank	Japan	377.2	16,038	48
10	3	Fuji Bank Group	Japan	365.6	15,032	20
12	4	Mitsubishi Bank	Japan	352.9	14,271	53
13	5	Sanwa Bank	Japan	350.2	13,946	41
16	6	Industrial Bank of Japan	Japan	268.5	5,253	38
—	7	Norinchukin Bank	Japan	243.6	3,130	2
—	8	Crédit Agricole	France	242.0	73,714	2
4	9	Banque Nationale de Paris	France	231.5	60,333	128
2	10	Citicorp	US	230.6	92,000	240
20	11	Tokai Bank	Japan	225.8	11,865	37
—	12	Mitsubishi Trust & Banking	Japan	212.0	6,288	8
5	13	Crédit Lyonnais	France	210.7	61,508	102
21	14	Mitsui Bank	Japan	206.9	10,256	49
17	15	Barclays Bank	UK	204.9	116,500	187
3	16	Deutsche Bank	Germany	202.3	56,580	67
—	17	Bank of Tokyo	Japan	198.3	17,040	101
—	18	Sumitomo Trust & Banking	Japan	197.9	6,759	16
19	19	National Westminister Bank	UK	186.5	113,000	30
—	20	Long Term Credit Bank of Japan	Japan	183.3	3,488	28
—	21	Mitsui Trust & Banking	Japan	180.0	6,173	9
22	22	Taiyo Kobe Bank	Japan	174.5	12,931	26
—	23	Yasuda Trust & Banking	Japan	171.7	5,627	14
6	24	Société Générale	France	164.8	45,950	89
—	25	Daiwa Bank	Japan	156.5	8,688	27

— Not in top 25 in 1975

(*Source*: based on *Euromoney*, June 1990, p. 119; *Fortune*, 30 July 1990, pp. 108–9; UNCTC (1988), Annex Table B.2)

in the top ten (ranked seventh and tenth). The two leading banks were both American (Bank America and Citicorp); the others in the top ten were either French or German. In 1975 nine of the leading twenty-five banks were Japanese and four were American. In 1989 seventeen of the top twenty-five banks were Japanese and only one was American. Seven of the top ten – indeed the *first* seven – were Japanese. The spectacular rise to international dominance of the Japanese banks is readily apparent.

In terms of both employment size and their international spread, however, Japanese banks are rather less significant than their asset values suggest. One reason for the far smaller employment size of the leading Japanese banks is that their roles are primarily to serve the major Japanese business corporations and *sogo shosha*. Their retail business is far less extensive than that of the American and European banks. Table 11.4 shows that the average number of foreign operations per bank was far lower among the leading Japanese banks than in banks from the other leading source countries, although the German figure was also low. The Japanese average of 26.0 was less than half that of American and British banks and only one-third that of French banks. Table 11.4 also lists all the banks in the top fifty which had more than fifty overseas operations; only two were Japanese. Thus, although the Japanese banks have undoubtedly internationalized very rapidly in the last decade, the really extensive global networks are still those operated by some of the American and European banks.

Table 11.4 Foreign networks of the leading banks

Source country	Average number of foreign operations by country's banks in top 100	Banks with more than 50 foreign operations
United States	58.9	Citicorp 240; Bank America Corp 184; Chase Manhattan 107; J. P. Morgan 85; Manufacturers Hanover 67
United Kingdom	50.8	Barclays 187; Midland 69
France	76.4	Crédit Lyonnais 102; Banque Nationale de Paris 128; Société Générale 89
Germany	26.3	Deutsche Bank 67; Dresdner Bank 83; Commerzbank 58
Japan	26.0	Bank of Tokyo 101; Mitusbishi Bank 53
All 100 banks	46.6	

(*Source*: based on UNCTC (1988), Table VII.2, Annex Table B.2)

A similarly strong tendency towards the internationalization of their operations is displayed by other financial services sectors. Table 11.5 gives an estimate of the extent of the internationalization of some of the leading securities and financial firms and accountancy firms in the mid-1980s. Both of these sectors have been changing so rapidly since then that the table is only a very rough guide. It does show, however, the

Table 11.5. International spread of some leading securities and accountancy firms

Rank	Company	Home country	Employment	No. of foreign affiliates*
Securities and financial firms				
1	American Express	United States	78,747	122
16	Trilon Financial Corp	Canada	n.a.	106
2	Salomon Inc.	United States	7,800	54
3	Merrill Lynch Co.	United States	45,100	48
14	Nomura Securities Co.	Japan	10,471	21
19	Orient Leasing Co.	Japan	1,273	17
20	Yamaichi Securities Co.	Japan	7,410	16
13	Daiwa Securities Co.	Japan	7,419	15
18	Nikko Securities Co.	Japan	8,059	14
7	Morgan Stanley Group	United States	5,332	7
Accountancy firms				
9	Klynveld Main Goerdeler[†]	Netherlands	30,894	498
8	Touche Ross Intl.	United States	27,500	414
2	Coopers & Lybrand	United States	38,500	397
11	Grant Thornton	United States	13,666	375
4	Price Waterhouse	United States	32,725	337
4	Ernst Whinney[‡]	United States	28,800	332
7	Deloitte Haskins Sells	United States	26,774	302
3	Peat Marwick[†]	United States	32,183	297
10	Binder Dijker Otte	Netherlands	13,027	294
6	Arthur Young[‡]	United States	29,000	288
1	Arthur Andersen	United States	36,117	141

*Accountancy firms consist of a series of nationally based partnerships
[†] KMG and Peat Marwick now merged as KPMG
[‡] Ernst Whinney and Arthur Young now merged as Ernst & Young

(*Source*: based on UNCTC (1988), Annex Tables B. 3, B. 6)

high degree of internationalization among both securities firms and accountancy firms. The structure of accountancy firms is particularly distinctive. They consist of a variety of international organizational forms but almost all are sets of *linked partnerships*. In some cases, the firm uses the same name (brand) in all countries; in others it emphasizes the identity of the local partner (Daniels, Thrift and Leyshon, 1989).

Although the specific reasons for, and the pace of, internationalization of banks and other financial services may well vary from one case to another, there is an overall general pattern. In the immediate post-1945 period, the major international function of the small number of US transnational banks was the provision of finance for trade. But as the overseas operations of United States TNCs in manufacturing and other activities accelerated in the early 1960s, the functions of the US transnational banks evolved to meet their particular demands. Thus, the staple business of US banks in London became that of servicing the financial needs of their major industrial clients who were rapidly extending their operations outside the United States.

In the second half of the 1960s transnational banking functions took on a significant additional dimension with the growth of the Eurodollar market. This, as we have seen, was a market outside the regulatory control of the US government. As a consequence, US banks could raise money there for re-lending domestically as well as overseas. The attractions of international banking widened progressively as both the global capital market evolved and as local capital markets overseas developed. As banks from the other industrialized economies also internationalized, the process became *self-reinforcing*. All large banks had to operate internationally; they had to have a presence in all the leading markets.

> The internationalization of financial markets received fresh impetus in the 1970s from two sources. First, the oil-exporting countries emerged from the price increases of 1973–1974 with huge surplus savings which they sought to lend abroad, and much was recycled by the TNBs [transnational banks] to the deficit developing countries. Secondly in the mid-1970s, the Federal Republic of Germany and the United States dismantled their exchange controls on capital movements, and they were followed towards the end of the decade by Japan and the United Kingdom. The events of the 1970s catapulted the TNBs into prominence as the institutions dominating the world financial markets; but with the advent of the debt crisis in 1982, their pre-eminence as suppliers of international finance began to wane. This was in part because the flow of private loan capital going through the banks to the developing countries dropped to a trickle. But it was also because the dismantling of exchange controls on capital movements and other financial deregulation led to a large and vigorous growth in the international securities market, mainly among the developed market economies.
>
> (UNCTC, 1988, p. 103)

Figure 11.5 summarizes the major phases of development of international banking. As with all such models it should be seen as a broad general framework which captures the main features of the process although the detail may well vary from case to case. However, it certainly fits the general experience of both US and Japanese transnational banks. These broad developments, intensified by the further deregulation of financial markets in many countries and by technological innovation, have pulled more and more securities firms into international operations. As Scott-Quinn (1990, p. 281) points out, 'Up to 1979/80, the US multinational investment bank had little more than a large office in London and perhaps some much smaller ones in other European countries, perhaps an Arab country and possibly (though less likely) Japan. From 1980 onward, the development of the US investment bank as a multinational changed qualitatively.'

	Phase I National banking	Phase II International banking	Phase III International full– service banking	Phase IV World full– service banking
Internationalization of customer companies	Export–import	Active direct overseas investment	Multinational corporation	
International operations in banking	Mainly foreign exchange operations connected with foreign trade. Capital transactions are mainly short–term ones.	Overseas loans and investments become important, as do medium– and longer–run capital transactions.	Non–banking fringe activities such as merchant banking, leasing, consulting and others are conducted. Retail banking	
Methods of internationalization	Correspondence contracts with foreign bank.	To strengthen own overseas branches and offices.	By strengthening own branches and offices, capital participation, affiliation in business, establishing non–bank fringe business firms, the most profitable ways of fund–raising and lending are sought on a global basis.	
Customers of international operations	Mainly domestic customers	Mainly domestic customers	Customers are of various nationalities.	

Figure 11.5 The major phases in the development of international banking (*Source*: Fujita and Ishigaki (1986), Table 7.6)

The same process applied to the leading Japanese *securities houses*, such as Nomura, Daiwa, Nikko and Yamaichi, although with differences of detail. Finally, as both industrial and service companies spread globally a further boost was given to the increased internationalization of *accountancy companies*. As a result, all of the transnational financial services firms are now basing their strategy on a direct presence in each of the major geographical markets and on providing a local service based on global resources. In the words of Nomura's current advertisement: 'Local Commitment, Global Capacity'. It is, in fact, selling an *international brand image*, with the clear message that a global company can cope most easily and effectively with every possible financial problem which can possibly face any customer wherever they are located.

The drive for diversification   It is a short, and supposedly logical, step from this kind of global strategic orientation to the notion that global financial services should not only operate globally in their own core area of expertise but also that they should supply a *complete package of related services*. The financial conglomerate or the financial supermarket has arrived on the scene. Deregulation has been a major stimulus for change; it is becoming increasingly possible for banks to act as securities houses, for securities houses to act as banks, and for both to offer a bewildering array of financial services way beyond their original operations. At the same time, entirely new non-bank financial services companies have emerged. Among all the leading acountancy firms, for example, there has been a strong move into management consultancy, including information systems. A leading UK bank's current portfolio of offerings includes: clearing banking, corporate finance, insurance broking, commercial lending, life assurance, mortgages, unit trusts, travellers' cheques, treasury services, credit cards, stockbroking, fund management, development capital, personal pensions and merchant banking.

Perhaps the best example of the diversified financial services company is American Express, which offers the following package (UNCTC, 1988, p. 399):

1. *International banking services*, via American Express Bank. Its sixteen US and seventy-one foreign affiliates supply private banking, trade finance, correspondent banking, treasury and foreign exchange, merchant banking.
2. *Financial services*, via IDS Financial Services. Its fifteen US and one foreign affiliate supply insurance, investment certificates, mutual funds, annuities, tax-advantaged investment, brokerage.
3. *Investment services*, via Shearson Lehman Brothers. Its sixty-two US affiliates and eleven foreign affiliates supply private-client services, investment banking, capital market services, asset management and real estate services.
4. *Insurance services*.
5. *Travel-related services*. Its twenty-six US and thirty-two foreign affliates supply American Express cards, travellers' cheques, travel services, data-based services, merchandise, publishing, insurance services.

Although some diversification has taken place by new greenfield investment the majority has occurred through the process of acquisition and merger.[17] This has been particularly the case in the rush of the securities houses to gain a foothold in newly deregulated markets, like the City of London in 1986, and in the creation of totally new financial services conglomerates. The rationale for diversification, both into new product and new geographical markets, is the familiar one of economies of scale and economies of scope:

> In financial services, the arguments are that there are large fixed costs in running a network of branches or agents. These can be spread over a lot of customers in large operations . . . Another supposed economy derives from the benefits of scale in participating in capital markets. Unit transaction costs tend to be lower, the larger you are. Consumer recognition is another advantage to the large company who might expect to attract customers most easily . . . a large diversified company can afford to cross-subsidize price wars in one sector with the profits made in another . . . As well as economies of scale, it is generally supposed that there are economies of scope in financial services. The bank doing business in all major European cities will find it easier to serve clients than a similar size bank that is locally concentrated. The international bank would be better placed to serve international clients, and would gain strategically valuable information about a range of markets from its dealings abroad.
>
> (Davis and Smales, 1989, p. 99)

The arguments for a strategy of internationally diversified financial service operations, then, are that it enables such a company to offer an entire package of services – a 'one-stop shop' – to customers. Their supply by a large, internationally recognized brand name – backed up, of course, by lavish advertising expenditure – is supposed to give a reassurance to potential customers that they will receive the highest quality of service. Whether or not economies of scale and scope really do exist to a significant extent is a matter over which there is considerable disagreement. The large financial companies themselves certainly seem to think so. Although there have been divestments as some financial services firms have shed parts of their diversified portfolio of activities, the diversification and internationalization trend continues. On the other hand, there are those who argue that there are distinct benefits in geographical specialization. Davis and Smales (1989), for example, suggest that:

1. the administrative benefits of size can be exaggerated;
2. local knowledge may be a major advantage in assessing local risks and opportunities;

3. the benefits of recognition through size may be less in national markets where names have already established themselves.

In other words, 'localization might be a marketing advantage'. This, of course, has been recognized by the major financial services companies in their emphasis on providing a locally sensitive service within a global organization.

In fact, Davis and Smales argue that there is a major difference between the retail and the wholesale/commercial financial sectors. In their view,

> no very wide branch network is necessary to serve corporate clients. Geography is simply less important at the corporate level. This means that these services can be exported and imported, and do not have to be provided locally. Commercial banking and commercial insurance and reinsurance are all things that can be provided relatively costlessly across borders, along with the wholesale functions of retail banking services. It should thus turn out that retail services remain local in their delivery, and that wholesale services are international . . . particularly in wholesaling, the European market is unified enough for most economies of scale to already have been taken up . . . In wholesale services, therefore, any exploitation of comparative advantage can be pursued from the security of a domestic base, possibly with a small local office.
>
> (Davis and Smales, 1989, pp. 102, 110)

Geographical structure of financial services activities

At first sight, technological developments in communications systems would appear to release financial services companies from the spatial constraints on the location of their activities. Financial services firms, in particular, would appear to be especially footloose. They are not tied to specific raw material locations; at least some of their transactions can be carried out over vast geographical distances using telecommunications facilities. And yet at both the global and national scales the major financial services activities are extremely *strongly concentrated geographically*. They are, in fact, more highly concentrated than virtually any other kind of economic activity other than those based on highly localized raw materials. However, there are some subtle variations according to the particular function; there is a *division of labour* within financial services firms, parts of which may show a greater degree of geographical decentralization.

Centralization: the hierarchy of international financial centres[18]   Figure 11.6 maps the hierarchy of international financial centres identified by Reed (1989). They are classified on the basis of sixteen statistical variables, of which the most important are: the volume of international currency clearings, the size of the Eurocurrency market, foreign financial assets, and headquarters of the large international banks. Reed suggests that the hierarchy of international financial centres has three major levels. Interestingly, he places Tokyo in the second tier although the structure of the global financial system is increasingly becoming articulated along the *tri-polar axis* of New York–London–Tokyo. Nevertheless, New York and London still stand apart from Tokyo as truly global financial centres. The international significance of Tokyo rests primarily on the strength of the Japanese economy itself:

> Whereas London and New York have a history of international financing, Tokyo has had a mainly domestic orientation. It is only since the 1970s that Japanese banks have made a concerted move into international banking. Yet this internationalization of banking was not

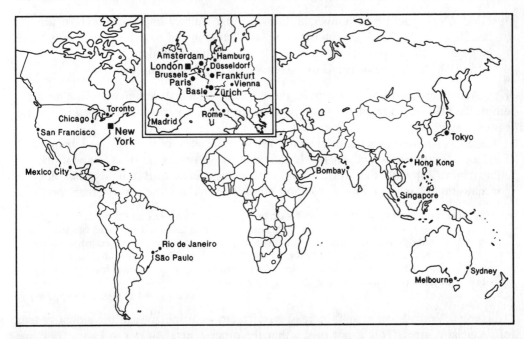

Figure 11.6 The hierarchy of international financial centres (*Source*: based on Reed (1989), Figure 1)

accompanied by the internationalization of the yen nor of Japanese money and capital markets . . . Tokyo is now the world's second largest stock market . . . but at the end of 1985 only 20 listed companies, out of a total of 1,497 listings, were foreign . . . Japan stands out . . . in comparison with . . . especially the UK and USA, in terms of the relatively small number of foreign banking institutions operating in Japan.

(Lewis and Davis, 1987, pp. 243–4)

In 1985, in fact, there were only 76 foreign banking institutions in Tokyo compared with well over 400 in London and around 350 in New York.

Thus, despite the undoubted increase in Tokyo's international significance, on those criteria which measure both the breadth and depth of global financial activity New York and London occupy the apex of the international financial centre hierarchy. London is the more international of these two leading centres in terms of both foreign exchange and Eurocurrency transactions. In the mid-1980s the daily turnover on the London foreign exchange market was almost as large as the turnover in New York and Tokyo put together. Lewis and Davis (1987) attribute London's current great significance as a global financial centre to the following:

1. The historical evolution of the City as a world centre has created both a large pool of relevant skills and also an almost unparalleled concentration of linked institutions within a very small geographical area. 'Unlike the position in some other countries, insurance, commodities trading, futures markets, stock broking, bond trading and legal services in the UK are all concentrated around the City' (op. cit., p. 236).
2. Its geographical position in a time zone between New York and Tokyo.
3. The regulatory environment, which has encouraged the growth of international banking and is favourably disposed towards international bankers. Foreign banks in

London can operate as 'universal' banks, a particularly important feature for US and Japanese banks. They can combine both banking and securities businesses there in ways which are prohibited so far in their domestic operations.
4. London is the centre of the Eurocurrency and Eurobond markets.

However, there is some concern in the London financial community that its pre-eminence may be threatened by competition from rapidly growing European financial centres, especially Paris and Frankfurt.

London's strength as a financial centre rests on the scale of its foreign exchange and Eurocurrency business and its newly deregulated securities markets. New York, in comparison, is by far the world's largest securities market. But it also has a huge concentration of international banks and other financial activities. More than this,

> New York is the nerve centre of worldwide operations of those US banks with the largest presence in the Eurocurrency markets . . . Second, US banks use their overseas branches to switch US dollar funds between the Euro and domestic deposit markets according to the relative cost of funding. These operations help to tie Eurodollar interest rates to domestic money markets which revolve around New York. Finally, New York is the location for the majority of IBFs (International Banking Facilities).
>
> (Lewis and Davis, 1987, p. 239)

Thus, New York, London and, increasingly, Tokyo sit at the apex of the global system of financial centres. This is not to say that the other centres shown in Figure 11.6 are not important. They certainly are, particularly those in the second level, some of which clearly have aspirations to gain promotion to the super-league. Centres in the third level serve mainly as regional financial centres which, again, is not to belittle their international significance but merely to put it in perspective.

The twenty-five cities shown in Figure 11.6 effectively control almost all the world's financial transactions. It is a remarkable level of geographical concentration. But they are more than just financial centres. There is clearly a close relationship with the distribution of the corporate and regional headquarters of transnational corporations which we examined in Chapter 7 (see Figure 7.6). These global cities are, indeed, the *control points of the global economic system*. As far as financial services firms are concerned, the international financial centres – especially those at or near the top of the hierarchy – are their 'natural habitat' (Thrift, 1987). Financial companies agglomerate together in these centres for the following reasons:

> The first reason is to be near clients, especially the headquarters offices of banking and industrial corporations, but also major state departments and other commercial capital firms. (Commercial capital is a social as well as an economic network.) The second reason is to be in close proximity to relevant markets, many of which operate from fixed exchanges and all of which operate under quite severe spatial constraints. The third reason is to tap into information on markets and the operations of banking and industrial corporations and the state rapidly and efficiently. It is no surprise, then, to see most concentrations of commercial capital in selected cities.
>
> (Thrift, 1987, p. 208)

Thrift goes on to list the key functions which are invariably present in these 'trading posts serving the international market':

1. *International commodity exchanges.* These cover trading in every conceivable commodity from the traditional kind to newer linked commodities like international property. In recent years there has been a rapid growth in futures exchanges in

which customers can buy almost anything for delivery at a given price on a specified date. In reality, the commodities themselves are rarely bought; it is the 'future' that is traded. Futures exchanges are now becoming integrated into twenty-four-hour international markets.

2. *International monetary exchanges.* The clearing of international currency transactions is one of the most important functions of international financial centres and, therefore, 'access to clearing is an equally important determinant of location in international financial centres by offices of banking and industrial corportions' (Thrift, 1987, p. 210).

3. *International securities dealing.* Until very recently, securities trading was predominantly a domestic activity. But, as we have seen, pressures from the big institutional investors to buy and sell stocks in different foreign markets have been growing rapidly. This pressure, together with the increasing deregulation of many national securities markets has led to an enormous growth in international securities dealing with 'books' being passed on from one international centre to another across time zones. Again, it seems probable that true twenty-four-hour trading between international centres will ultimately occur.

4. *International services.* The sheer concentration of financial and other business activities in international financial centres inevitably acts as a nucleus around which a whole galaxy of related services gathers. Hence, there are associated concentrations of insurance, accountancy, legal, advertising firms in each major international financial centre, which further increases the centres' scale and dominance.

In fact, Reed claims that the leading international financial centres have a global significance in themselves; that the whole is, in fact, greater than the sum of the parts. In his view, 'the reach, scope, and influence of the centre is far greater than the total reach, scope and influence of its various institutions when measured independently' (Reed, 1989, p. 254).

Decentralization? The geographical rearrangement of 'back office' functions   All of the above would seem to suggest that the potential for other cities, outside the favoured few shown in Figure 11.6, to develop as significant centres of finance and related activities is likely to be very limited. In the United Kingdom, for example, the sheer overwhelming dominance of London makes it extremely difficult for provincial cities to develop more than a very restricted financial function. London, in that sense, is akin to the notorious upas tree, 'a fabulous Javanese tree so poisonous as to destroy all life for many miles around'. Thus, in 1989, some 328 foreign banks had offices in London but only thirty-four had offices elsewhere in the United Kingdom outside London, eleven of which were ethnic banks serving localized populations. The largest concentration of foreign banks outside London was in Manchester – with a mere twelve! (Tickell, 1992).

It is, of course, the 'higher order' financial and service functions which are especially heavily concentrated in the major international financial centres. So-called 'front-office' functions, by definition, must be close to the customer – hence the huge branch networks of the retail banks and other financial services supplying final demand. The essence of all of the financial services, indeed of all the services we have been discussing in this chapter, however, is the transformation of massive volumes of information. Much of that activity is routine data processing performed by clerical workers. Such 'back-office' activity can be separated from the front-office functions and performed in different locations.[19] The early adoption of large-scale computing by banks, insurance companies and the like from the late 1950s initially led many of them to set up huge *centralized* data processing units. To escape the high costs (both land and labour) in the major financial centres such units were often relocated in less expensive centres or in the suburbs. Access to large pools of appropriate (often female) labour was a key requirement.

The introduction of microcomputers and networked computer terminals made such centralized processing units unnecessary and the tendency in recent years has been to *decentralize* back-office functions more widely in much smaller units. In effect,

> telecommunications accelerate a spatial bifurcation within many large finance firms by enhancing the attractiveness of downtown areas for skilled managerial activities while simultaneously facilitating the exodus of low wage, back office sectors. This process mimics the separation between headquarters and branch plant functions widely noted in manufacturing. In both cases, a vertical disintegration of production takes place, accompanied by the dispersal of standardized, capital-intensive functions and the concomitant reorganization of skilled labour-intensive functions around large, densely populated urban areas.
>
> (Warf, 1989, p. 267)

At the same time, however, the distinction between back-office and front-office functions is becoming less clear as distributed computer technology has been developed further. In fact, it is not only routine back-office activities that have been decentralized. It has become increasingly common for some of the higher-skilled functions to be relocated away from head office into dispersed locations. For example, Crédit Suisse, headquartered in Zürich, has recently set up a series of 'satellite offices' in other Swiss centres for teams of computer specialists.

> The principal reason . . . was and still is the shortages in the labour market of the Zürich region . . . In addition to the labour shortage, there is an increasing demand for flexible working hours, shorter distances to work, a generally higher quality of life, and for the improvement of environmental conditions . . . The choice of location for these workcentres was determined by language region (French, German, Italian), the size of the agglomeration and proximity to a major subsidiary.
>
> (Erzberger and Sonderegger, 1989, p. 140)

Jobs in the financial services industries

In all the major industrialized economies, as well as in major financial centres such as Hong Kong and Singapore, the rapid growth and internationalization of financial services have, so far, created impressive employment growth. Figure 11.7 shows that growth in bank employment between 1970 and 1985 was positive in all five countries, apart from Japan between 1980 and 1985. However, over the entire 1970–85 period, employment growth in financial services was substantially higher in the United States and Japan than in the other three countries. Whilst employment in manufacturing industries in many of the industrialized countries has been growing slowly, or even falling in some cases, jobs have been proliferating in financial services (as well as in other service sectors). Of course, it has to be remembered that such job growth is inevitably strongly concentrated within urban areas in general and in the big metropolitan areas in particular. In financial services, the biggest employment increases have been in the international financial centres themselves, as the experiences of both

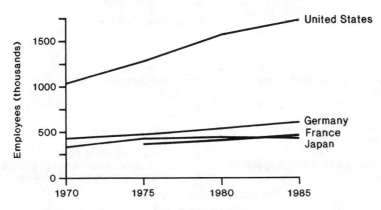

Figure 11.7 Growth in banking employment, 1970–85 (*Source*: based on Bertrand and Noyelle (1988), Table 1.1)

New York and London demonstrate. In both cases, deregulation of financial services, and the boost this gave to internationally oriented transactions, has resulted in very substantial employment increases.

In New York, for example, employment in financial services increased by more than 30 per cent between 1979 and 1987, from 345,600 to 450,900. Growth was particularly spectacular in the securities industries (+106 per cent). There was also parallel employment growth in the related services of advertising, consulting, accounting and the like (Daniels, 1991). Employment growth in financial services was apparent in virtually all major cities. 'Many of Britain's provincial financial centres enjoyed quite phenomenal rates of growth . . . for example, between 1974 and 1981, . . . purely financial service employment expanded by at least 25% in . . . 11 centres' (Leyshon, Thrift and Tommey, 1989, pp. 170, 172). In the United States, producer service employment grew by 60 per cent between 1977 and 1985 in metropolitan areas and by 59 per cent in non-metropolitan areas, although from a much smaller base (Beyers, 1989).

The story is not, however, simply one of progressive employment expansion in the financial services industries. Technological and organizational developments are drastically changing the nature of work at all levels: redefining skills and increasing flexibility. Table 11.6 summarizes some of the major skills changes. In particular, there has been a

> widespread tendency for a dramatic decline in the volume of clerical processing work performed, until recently, manually by lower-tier personnel (with some assistance of mainframe computers for data crunching). This remarkable contraction in old-fashioned clerical processing work is the result not only of automation but also of the transformation of work done by personnel in the middle and upper rungs of a firm's occupational structure . . . Paralleling this transformation in data processing and data handling, increasing competition is generating new demands for both sales and assistance personnel and for specialists able to identify new markets, conceive new products, develop new systems and sell the new, often complex services (swaps, futures, etc.) . . . The outcome of this profound process of skill transformation is the emergence of a new matrix of competencies that may be viewed in terms of new skills that are being substituted progressively for older ones. Some of these new competencies are common to both middle- and upper-level workers; others are specific to various groups within the occupational hierarchy.
>
> (Bertrand and Noyelle, 1988, pp. 40–1)

Table 11.6. The changing nature of skills in banks and insurance companies

Old competencies	New competencies
Common emerging competencies	
1. Ability to operate in well-defined and stable environment.	Ability to operate in ill-defined and ever-changing environment.
2. Capacity to deal with repetitive, straightforward and concrete work process.	Capacity to deal with non-routine and abstract work process.
3. Ability to operate in a supervised work environment.	Ability to handle decisions and responsibilities.
4. Isolated work.	Group work; interactive work.
5. Ability to operate within narrow geographical and time horizons.	System-wide understanding; ability to operate with expanding geographical and time horizons.
Specific emerging competencies	
Among upper-tier workers	
1. *Generalist competencies*. Broad, largely unspecialized knowledge; focus on operating managerial skills.	*The new expertise*. Growing need for high-level specialized knowledge in well-defined areas needed to develop and distribute complex products.
2. *Administrative competencies*. Old leadership skills; routine administration; top-down, carrot-and-stick personnel management approach; ability to carry out orders from senior management.	*The new entrepreneurship*. Capacity not only to manage but also set strategic goals; to share information with subordinates and to listen to them; to motivate individuals to develop new business opportunities.
Among middle-tier workers	
1. *Procedural competencies*. Specialized skills focused on applying established clerical procedural techniques assuming a capacity to receive and execute orders.	*Customer assistance and sales competencies*. Broader and less specialized skills focused on assisting customers and selling, capacity to define and solve problems.
Among lower-tier workers	
1. *Specialized skills* focused on data entry and data processing.	Disappearance of low-skill jobs.

(*Source*: Bertrand and Noyelle (1988) Table 4.1)

Partly as a result of these changes in the demand for particular types of labour, associated with technological and organizational change, and partly because of the effects of recession in the early 1990s, the labour market in financial services is no longer as buoyant as it was. Even in accountancy the claim that 'like funeral parlours and the law business, the majors have discovered that accountancy is immune to economic downswings' (Clairmonte and Cavanagh, 1984, p. 268) no longer seems quite so valid. In banking many of the major firms are currently initiating major rationalization and reorganization programmes which are resulting in substantial job losses and 'voluntary redundancy' measures. In the formerly hot-house labour market of the City of London's securities firms, for example, the 'golden hellos' of the post-Big Bang period are being replaced by less-than-golden goodbyes.

More general employment reductions may well result from continued deregulation of financial markets. The regulatory protection of national markets has almost certainly allowed 'inefficient' banking markets to exist. As the regulatory walls come tumbling down in Europe, the United States and elsewhere, and as the full impact of the internationalization of the industry takes hold, phrases such as 'overcapacity' and 'over-banking' have become commonplace. Thus, in addition to employment losses because of recession, there may well be job losses associated with such *structural* adjustments. It is quite possible, therefore, that a period of 'jobless growth' is in prospect for the financial services industries.

Notes for further reading

1. This particular way of categorizing economic activities is discussed in Dicken and Lloyd (1981), pp. 139–43.
2. Allen (1988), Dunning (1989), Britton (1990) explore some of these conceptual issues.
3. The IMF definition of trade is that it constitutes a transaction between residents and non-residents of a country. Hence, it covers far more than the 'commonsense' definition of international trade as the movement of goods from one country to another. For example, the movement of tourists from, say, the United States to Britain is classified as British exports of tourism, even though the consumers actually move to the point of production of the tourist service.
4. There is a large literature on the question of international trade in services. Gibbs (1985) provides a particularly clear treatment of the issues. See also UNCTAD (1985).
5. For a detailed discussion of this trend across a whole range of service industries see Clairmonte and Cavanagh (1984).
6. Clairmonte and Cavanagh (1984), Harris (1984), UNCTC (1988), Perry (1990) examine the globalization of the advertising industry.
7. The *sogo shosha* are examined in detail by Kojima and Ozawa (1985), Yoshino and Lifson (1986), Young (1979). Briefer treatments are provided by Ozawa (1979), Nakase (1981), UNCTC (1988).
8. The term 'fixer' in this context is borrowed from Thrift (1987). General discussions of developments in the international financial system are provided by Hamilton (1986), Pardee (1988), Thrift and Leyshon (1988), UNCTC (1988). International banking is examined by Yannopoulos (1983), Lewis and Davis (1987), Grubel (1989), Huertas (1990), Scott-Quinn (1990).
9. Bertrand and Noyelle (1988), chapter 2, provide a concise summary of the major developments in the markets for financial services. This section draws extensively on that source.
10. Specific treatment of these technologies in the context of financial services is provided by Hamilton (1986), Bertrand and Noyelle (1988), UNCTC (1988), Warf (1989). Warf, in particular, provides a very comprehensive, but concise, account of the relationships between telecommunications and the globalization of financial services.
11. Further details of these product innovations in the financial markets are provided by Hamilton (1986), Lewis and Davis (1987), UNCTC (1988).
12. Hamilton (1986), Lewis and Davis (1987), UNCTC (1988) present details of the changing regulatory environment for financial services.
13. The term is borrowed from Clairmonte and Cavanagh (1984), p. 240.
14. UNCTC (1988), Box VII.I provides a useful summary of measures to liberalize financial markets in the 1980s in the United States, Europe and Japan.
15. Davis and Smales (1989) provide a detailed discussion of the integration of European financial services in the Single European Market.
16. Yannopoulos (1983), Lewis and Davis (1987), UNCTC (1988), Grubel (1989) discuss trends in international banking as a whole. Huertas (1990) and Scott-Quinn (1990) examine the internationalization of US commercial and investment banks respectively, while Fujita and Ishigaki (1986) provide a treatment of the internationalization of Japanese banking. The internationalization of accountancy firms is described by Thrift (1987), Daniels, Thrift and Leyshon (1989).
17. Clairmonte and Cavanagh (1984) and UNCTC (1988) provide some detailed examples of

acquisitions in the financial services industries. Daniels, Thrift and Leyshon (1989) describe acquisitions and mergers among international accountancy firms.

18. International financial centres are examined in some depth by Lewis and Davis (1987), Thrift (1987), Park (1989), Reed (1989), Warf (1989).

19. Nelson (1986), Bertrand and Noyelle (1988), Erzberger and Sonderegger (1989) discuss the separation of back-office functions.

PART FOUR

Stresses and Strains of Adjustment to Global Shift

Prologue

The focus throughout the preceding eleven chapters has been on the patterns and processes of global shift; on the *forms* being produced by the increasing globalization of economic activities and the *forces* producing those forms. With few exceptions we have not been concerned so far with the *effects* of these profound structural and geographical changes. In the final chapters of this book we turn to address precisely these issues. Part Four consists of two substantial chapters, which deal with specific areas of impact, and a brief final epilogue.

Transnational corporations have been at the centre of all our discussions in the previous chapters, either as entities in themselves or in terms of their interactions with other institutions in the global economy, notably nation states. Ever since they appeared on the world scene, TNCs have been the focus of fierce controversy. Chapter 12 examines the debate over the costs and the benefits of TNCs to host and home countries. Its objectives are to identify in general terms the major ways in which TNC impact may be transmitted to host and home economies. The point is made that a *general* assessment of TNC impact is extremely difficult not only because of the complexity of the processes involved but also because of their highly contingent nature. In Chapter 12 we also examine the nature of the bargaining relationship between TNCs and nation states and briefly outline the attempts which have been made to devise international regulations for the conduct of TNCs.

Chapter 13 takes a broader perspective and examines the problems which face countries at different levels of economic development in their attempts to adjust to global change. A particular emphasis is placed upon the *employment* implications of global shifts because it is largely through the incomes earned from employment (including self-employment) that levels of material well-being are determined. The question 'where will the jobs come from?' is a crucial one throughout the world. Again, we find a complex picture. The major employment changes that have been occurring in both developed and developing economies are the result of a complex interaction of processes. Job losses in the developed market economies, for example, cannot be attributed simply to the relocation of production to developing countries. The industrialized economies face major problems of deindustrialization. However, the problems facing Third World countries in a global economy are infinitely more acute. The spectacular success of a small number of NICs should not blind us to the fact that the majority of developing countries face enormous problems of economic survival in an increasingly globalized economic system.

Lastly, we turn to the future. No attempt is made to construct possible future scenarios or to predict what might happen in the global economy. Rather, a series of questions is posed which arise from some of the major current global trends, notably the emergence of regional trading blocs, the apparent polarization of the global economy and the shift in the centre of gravity of the global economy towards Asia.

Chapter 12

Beauty or the Beast?
The Costs and Benefits of
Transnational Corporations

Introduction: polarization of viewpoints

The title of this chapter reflects the extreme differences of opinion which surround the question of the impact of the transnational corporation. According to viewpoint, TNCs either expand national or local economies or exploit them; they are either a dynamic force in economic development or a distorting influence; they either create jobs or destroy them; they either spread new technology or pre-empt its wider use, and so on. At one extreme the charge is one of political interference in national affairs or of bribery of national officials. At the other extreme are the views of those who regard the TNC as a greater force for international economic well-being than the parochially bounded nation state. The list of contrasting views is almost endless. Indeed, virtually every aspect of the TNC's operations – economic, political, cultural – has been judged in diametrically opposed ways by its proponents and its opponents.

In the face of such dogmatically extreme views, and a debate which has generated more heat than light, the wisdom of Solomon would seem to be required to weigh the evidence. The overriding problem is that it is impossible to make a simple, all-embracing assessment of impact. The problem is a *counterfactual* one: to evaluate fully the impact of TNC activity we need to measure it against a realistic alternative. What *would* have happened *if* the activity did *not* exist? This fundamental dilemma bedevils the entire debate. In other words, we generalize about the impact of TNCs at our peril. A realistic assessment must be based on a careful evaluation of specific cases and against a realistic alternative. What is true in one set of circumstances may not hold in other circumstances. Different types of transnational activity are likely to have different repercussions. Thus, we should avoid 'knee-jerk' reactions, whether favourable or unfavourable.

In this chapter, therefore, no attempt is made to arrive at a *general* evaluation of the transnational corporation. Rather, the major ways in which TNCs *may* affect various aspects of national and local economies are outlined as a framework within which specific cases can be assessed. The aim is to present a reasonably balanced account of conflicting opinions. For the most part our concern will be with the *economic* impact of TNCs although it is often unrealistic to separate these out from the political, social and cultural implications. The chapter is organized into two broad segments: the effects of TNCs on *host countries* and on *home countries* respectively.

By their very nature TNCs not only span national boundaries but they also, in effect, incorporate *parts* of national economies within their own boundaries. Their effects will differ, therefore, according to whether the country involved is host to the operations of a foreign firm or is the home base from which a firm extends its operations overseas.

Overall, of course, the sharp distinction between home and host countries has become blurred as the level of *cross-investment* between the industrialized economies has increased. In particular, the United States is now both the world's largest home and host nation for foreign direct investment (see Chapter 3). However, using a national 'lens' for viewing the impact of TNCs is consistent with the argument developed throughout this book that the TNC and the nation state are the two leading actors in the global economy.

Transnational corporations and host economies[1]

Some general considerations

The establishment of an overseas facility by a TNC incorporates a package of qualities – financial, technological, managerial, marketing – which, together, have far-reaching implications for the host economy. But the precise nature of these implications, and the extent to which they create a *net benefit* or a *net cost* to the economy, are extremely difficult to unravel. This is partly because of the counterfactual nature of the problem but also, as Figure 12.1 shows, a number of variable factors are involved. One set of factors concerns *the nature of the foreign-controlled plant* itself: the method of entry employed, the functions performed by the plant and its operational attributes. *How* the foreign plant is established – by setting up a completely new facility, by acquiring an existing indigenous enterprise or by forming a joint venture with local capital (private or state-owned) – is a most important consideration. The setting up of a completely new 'greenfield' plant is generally regarded with greater favour by host countries than acquisition of existing capacity. A new plant obviously adds initially to the host country's stock of productive capacity whereas acquisition merely transfers ownership of existing capacity to a foreign firm. Of course, the outcome is never as clear cut as this. Much depends on the effect of a new plant on other firms and, in the case of acquisition, on whether the acquired plant's performance is enhanced to the benefit of the domestic economy. The method of entry used depends in part on the opportunities available. In general, foreign acquisitions have been more prevalent in developed host economies, simply because there are many more suitable candidates for takeover, but they are by no means absent in developing countries.

A second influence on the nature of the TNC's impact relates to the *function* performed by the foreign-controlled plant. Figure 12.1 reminds us that foreign branch plants tend to be established for one of three reasons: to exploit a localized material resource, to serve the host market itself by substituting for imports, or to use the host country location as a platform for exporting either finished products or components. Closely related to the primary function of the foreign plant are its *operational attributes*, including industry type, technology employed, scale of operations and the extent to which the plant is integrated into the parent company structure.

Thus, each of the major areas of impact shown in Figure 12.1 – capital and finance, technology, trade and linkages, industrial structure and employment – may be influenced in various ways according to the nature of the foreign plant involved. Plants with different functions, different attributes and different methods of entry are likely to affect the host economy in different ways. But it is not only the nature of the foreign plant which is important. The *nature and characteristics of the host economy* itself also need to be taken into account. Most TNC activity originates from the highly

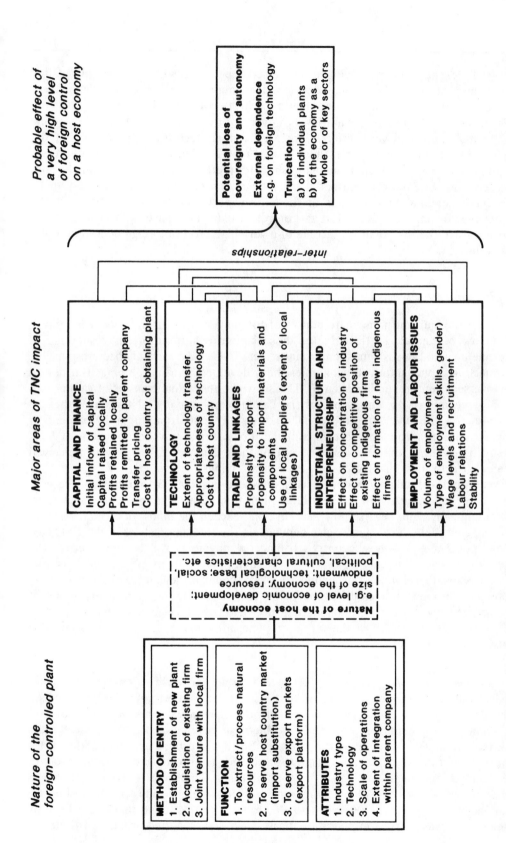

Figure 12.1 Effects of transnational corporations on host economies

Nature of the foreign-controlled plant

METHOD OF ENTRY
1. Establishment of new plant
2. Acquisition of existing firm
3. Joint venture with local firm

FUNCTION
1. To extract/process natural resources
2. To serve host country market (import substitution)
3. To serve export markets (export platform)

ATTRIBUTES
1. Industry type
2. Technology
3. Scale of operations
4. Extent of integration within parent company

Nature of the host economy
e.g. level of economic development; size of the economy; resource endowment; technological base; social, political, cultural characteristics etc.

Major areas of TNC impact

CAPITAL AND FINANCE
Initial inflow of capital
Capital raised locally
Profits retained locally
Profits remitted to parent company
Transfer pricing
Cost to host country of obtaining plant

TECHNOLOGY
Extent of technology transfer
Appropriateness of technology
Cost to host country

TRADE AND LINKAGES
Propensity to export
Propensity to import materials and components
Use of local suppliers (extent of local linkages)

INDUSTRIAL STRUCTURE AND ENTREPRENEURSHIP
Effect on concentration of industry
Effect on competitive position of existing indigenous firms
Effect on formation of new indigenous firms

EMPLOYMENT AND LABOUR ISSUES
Volume of employment
Type of employment (skills, gender)
Wage levels and recruitment
Labour relations
Stability

Inter-relationships

Probable effect of a very high level of foreign control on a host economy

Potential loss of sovereignty and autonomy

External dependence e.g. on foreign technology

Truncation
a) of individual plants
b) of the economy as a whole or of key sectors

industrialized and affluent developed market economies. In so far as the bulk of this activity also flows to these same developed economies the degree of 'dissonance' between a foreign-controlled operation and the host economy is likely to be small.

For less industrialized host countries, however, for those at a lower level of economic development and with very different socio-cultural characteristics from those of the foreign firm's home country, the 'shock waves' may be much greater. In addition, we need to consider not just the immediate, short-term effects of foreign investment on a host economy but, more importantly, the longer-term effects, which are likely to be more fundamental and far-reaching. In the following sections, we look separately at the major areas of impact which a foreign-controlled plant may have on a host economy before considering the possible outcome of a very high level of foreign penetration of a host country.

Capital and finance

The most obvious and immediate impact of foreign investment on a host economy is the inflow of capital. Particularly for those countries suffering from a shortage of investment capital this is a most important effect. TNCs have certainly been responsible for injecting capital into host economies, both developed and developing. But not all new overseas ventures undertaken by TNCs involve the actual transfer of capital into the host economy. At least some of the capital employed by TNCs may either be borrowed on the host country capital markets or arise from the reinvestment of retained earnings from the foreign affiliate. Local firms may be bought with local money. Local firms may be squeezed out of local capital markets by the perceived greater attractiveness of TNCs as a use for local savings.

Even where capital inflow does occur, or where earnings are retained, there will, eventually, be a *reverse* flow as the foreign plant remits earnings and profits back to its parent company. This reverse flow may, in time, exceed the inflow of capital. Any net financial gain to the host country also depends on the trading practices of the TNC (see below, pp. 395–9). A host country's balance of payments will be improved to the extent that the foreign plant exports its output and reduced by its propensity to import. A vital issue, therefore, is the extent to which financial 'leakage' occurs from host economies through the conduit of the TNC. This raises the question of the ability of host country governments to obtain a 'fair' tax yield from foreign firms, many of which are capable of manipulating the terms of their intra-corporate transactions through transfer pricing (see below).

The problem of transfer pricing   TNCs have a strong incentive to engage in transfer pricing; the very large, highly centralized TNC has the greatest potential for doing so. Transfer pricing is a major problem for all host economies, whatever their level of economic development; it is a particular problem for developing countries, which often lack the resources to counteract the practice. But it has proved extremely difficult for governments (and researchers) to gather hard evidence on its actual extent. Most of the evidence is limited to a small number of cases. The most frequently quoted is the study by Vaitsos (1974) of overpricing by firms in four industries in Colombia: pharmaceuticals, rubber, chemicals and electronics. Vaitsos found that overpricing by foreign firms was greatest in pharmaceuticals, such that 'reported profits constitute 3.4

per cent of the effective returns, royalties 14 per cent and overpricing 82.6 per cent' (Vaitsos, 1974, p. 62).

The UNCTC (1988) gives an example of the underpricing of timber exports by a TNC from a Pacific Island country to avoid paying taxes. Concern over the use of transfer pricing to avoid or reduce the liability to pay corporate taxes is not confined to developing countries. In the United States a House of Representatives study in 1990 claimed that more than half of almost forty foreign companies surveyed had paid virtually no taxes over a ten-year period. It was estimated that some $35 billion is being lost through the transfer pricing mechanism. But even in an advanced economy like the United States it is extremely difficult for the Internal Revenue Service to assess the actual extent of transfer pricing.

The costs of attracting foreign investment   The financial gain (or loss) to host countries from foreign investment, therefore, consists primarily of the net balance of capital flows plus the net earnings from trade. But there is also the cost of actually obtaining the foreign investment to consider. Most states, national and subnational, offer very considerable financial and fiscal incentives to attract internationally mobile investment. These costs, together with the provision of the necessary physical infrastructure (as in export-processing zones, for example) may be quite substantial. Hence,

> in the case of EPZs . . . competitive outbidding between sites tends to gradually erode the potential for recovering some of the incurred capital expenses by the government. Indeed, in view of the increasing number of EPZ areas, EPZ authorities tend to offer more and more fiscal incentives to enterprises in order to secure new investments and prevent the relocation of existing plants. For example, lengthy tax holidays which were not needed initially for attracting foreign investment in offshore production, have become an almost indispensable feature of the incentive structure to be offered to potential EPZ investors.
>
> (Maex, 1983, p. 37)

In the United States, states and municipalities have injected very large sums in the Japanese automobile transplants.[2] A similar high level of public investment in incentive packages for foreign companies is common in most Western European countries. It is not uncommon for a considerable proportion of the capital costs of a foreign branch plant to be provided by the host government itself. Indeed, the financial returns may exceed the initial capital subsidies. However, the costs of attracting investment have to be set against the likely alternative. For example, would the labour force otherwise be unemployed and have to be supported by the state? Agencies such as the Irish Industrial Development Authority claim that their subsidies to foreign investors are paid off within four to five years. Much depends, of course, on the 'knock-on' or multiplier effects of the foreign plant on other sectors of the economy.

Technology

> Technology may be embodied in the form of capital goods, such as machinery, equipment and physical structures; or it may be disembodied in such forms as industrial property rights, unpatented know-how, management and organisation, and design and operating instructions for production systems. *Foreign direct investment has traditionally been one of the most important channels of technology transfer as it involves the physical relocation of entire production systems*, combining in a single package capital goods and a number of the forms of disembodied technology.
>
> (UNCTC, 1983b, pp. 162–3, emphasis added)

Three issues are especially important in evaluating the technological impact of foreign enterprises on host economies:

1. the extent to which technology is *transferred* both within the bounds of the TNC and to other – domestically based – enterprises;
2. the *appropriateness* of the technology transferred – both *processes* and *products*;
3. the *costs* to the host economy of acquiring the technology.

Transfer of technology   Each of the modes of foreign involvement discussed in Chapter 7 – foreign direct investment, collaborative ventures, international subcontracting – is a potential channel for technology transfer. Simply by locating some of its operations outside its home country the TNC engages in the geographical transfer of technology. In this respect, technologies are, indeed, spread more widely. There is no doubt, for example, that the newly established foreign branch plants of US firms in the 1960s played a major role in transferring technology to Europe. In so far as a foreign branch plant employs local labour there will be a degree of technology transfer to elements of the local population through training in specific skills and techniques. But the mere existence of a particular technology within a foreign-controlled branch plant does not guarantee that its benefits will be widely diffused through a host economy. The critical factor here is the extent to which the technology is made available to potential users outside the firm either directly, through linkages with indigenous firms, or indirectly via the 'demonstration' effect.

The specific question of linkages is dealt with in the next section of this chapter. The very nature of the TNC, however, may well inhibit the spread of its proprietary technology beyond its own organizational boundaries. Possession and exploitation of technology are a diagnostic feature of the TNC. Such technology – the essential life-blood of the TNC – is not lightly handed over to other firms. Control over its use is jealously guarded: the terms under which the technology is transferred are dictated primarily by the TNC itself in the light of its own overall interests.

A major tendency, as we saw in Chapter 7, is for TNCs to locate the bulk of their technology-creating activities – research and development – either in their home country or in the more advanced industrialized countries. So far, relatively little R & D has been relocated to the developing countries. Thus, 'even if it is admitted that TNCs transfer the best production technology, they do not transfer the capability to generate new technology to affiliates in the Third World. They transfer "know-how" (production engineering) and not "know-why" (basic design, research and development)' (Lall, 1984, p. 10).

Nevertheless, some R & D – mostly lower-level support laboratories – has been located in a few of the newly industrializing countries. In some cases this is a direct result of host government pressure on TNCs to establish R & D facilities in return for entry. Such leverage is probably greatest where the TNC wishes to establish a branch plant to serve the host country market itself:

> Because of the relatively stronger bargaining position of the host country, it can insist on more R & D activity being carried on locally. And because of its interest in local demand and cost conditions and its long-term commitment, the firm has a greater incentive to take advantage of the gains to be obtained from having research activities in intimate association with the host country environment.
>
> (Reuber, 1973, p.188)

In contrast, there is very little incentive for TNCs engaged in export-oriented foreign investment to establish R & D facilities in Third World countries. But it is not only in Third Word countries that the R & D activities of TNCs tend to be under-represented. In Canada, for example, much concern has been expressed at the reluctance of United States TNCs to establish higher-level R & D facilities there. Most of the Canadian R & D laboratories of US companies have been concerned mainly with adapting and modifying US-originated products and processes for the domestic Canadian market. Similar criticisms have been directed at American manufacturing firms in Scotland and at Japanese manufacturing firms in Europe.[3]

Appropriateness of technology   In the case of developing countries a major issue is the appropriateness of the technology transferred via the TNC. Do the processes and products being introduced match local conditions and local needs? New technologies are invariably introduced by TNCs first of all in their home country or in other industrialized countries. Since they reflect the prevailing cost and availability of factors in those countries the technologies tend to be capital- rather than labour-intensive. But in most developing countries the abundant factor tends to be low-skilled labour while capital is relatively scarce. Hence, there is much disagreement about the extent to which TNCs adapt their technology for use in developing countries to make it more appropriate.[4]

Lall and Streeten make a distinction between the *adaptability* of technology and its actual *adaptation*:

(a) As far as *adaptability* goes, much of modern 'high' technology cannot be changed to suit LDCs' endowments: the demands of precision, continuity, scale and complexity are too great. However, some 'low' technologies (for instance, in simple industries such as textiles) and 'peripheral' or 'ancillary' technology (for instance, transport or handling) are more adaptable. The scope of adaptability can be extended, but the cost, in terms of R & D and organisational requirements, may be quite high.

(b) As far as *adaptation* of foreign technology goes, the bulk of *basic or 'core' production technology* transferred by TNCs, both directly and by licensing, is not adapted in any significant way to low-wage conditions, though some *scaling down* of technology seems to be undertaken to adjust to smaller runs than would be appropriate in developed countries.

(Lall and Streeten, 1977, p. 72)

The notion that only a limited degree of adaptation takes place in the technology transferred by TNCs to developing countries is supported by Reuber (1973). He found that, in almost three-quarters of his sample of firms, production technology was transferred intact, with no modification. Yet not all share this view. Lall (1984, p. 9) claims that 'at the micro-level, every new application of a technology entails considerable adaptive effort. The core process may not be significantly altered, but changes in scale, inputs, outputs, automation etc., may constitute between 10–60 percent of total project costs.' Research shows great variation in the relative capital intensity of production by foreign and domestic firms operating in the same industry in developing countries. In some cases, there is no significant difference; in others, foreign firms are significantly more capital-intensive in their operations than domestic firms.[5]

The second major issue in the appropriateness debate relates to the kinds of *products* transferred by TNCs to developing countries. This has become an especially contentious issue largely because of some highly publicized cases. There can be no doubt that, in the drive to create global markets, TNCs have indeed sought to introduce and sell their products throughout the developing world. The creation of particular types of demand and the shaping of consumer tastes and preferences are an intrinsic part of the TNC system. The problem is

> that the use of scarce resources for the production of goods which are over-differentiated, over-packaged, over-promoted, over-specified and within the reach of only a small elite, or, if bought by the poor, at the expense of more essential products, is not conducive to 'national welfare'. This is not to say that all TNC technology is unnecessary in LDCs – clearly, that would be absurd. But the free import of foreign capital and of the sort of technology many TNCs excel in would reproduce the pattern of developed countries and would be undesirable. In other words, a definition of welfare based on meeting *basic social needs* would lead to a *fairly small proportion of TNC technology being regarded as beneficial.*
>
> (Lall and Streeten, 1977, p. 71)

Against those products which are clearly 'inappropriate' to Third World countries, however, there are undoubtedly others that have brought great benefit, including those in agriculture (seeds, fertilizers) and health care.

A final aspect of the appropriateness issue relates to the *environmental and safety dimension* of TNC activity. Do TNCs export technologies to developing countries which are environmentally objectionable or which are less safe than they should be?[6] There have been claims that TNCs systematically shift some of their environmentally noxious or more hazardous operations to developing countries with less stringent environmental and safety standards. Although this may indeed occur in specific cases, there is no evidence to suggest that this is the general practice. Leonard's (1988) study of US companies did not support the claim of US firms engaging in industrial flight to pollution havens. The years following the introduction of stringent environmental regulations in the United States were not characterized by the widespread relocation of pollution-intensive industries to countries with lower regulatory standards. The UNCTC (1988, p. 227) also argued that 'differences in the levels of environmental control have had insignificant impact on the location of TNCs'.

A rather different aspect of the problem relates to safety. Industrial disasters, such as the one at the Union Carbide plant at Bhopal, India, in 1984, focused attention on the safety practices of TNCs. A frequent claim was that TNCs tend to adopt less stringent safety practices in their developing country plants than in their home plants. But, again, there is no *general* evidence to support this:

> While the number of industrial accidents appears to have risen over the last fifteen years, available evidence indicates that transnational corporations have been involved in less than half of them. Many accidents have occurred in purely national firms or in State-owned enterprises. The vast majority of industrial accidents have occurred in the home countries of transnational corporations, or in other developed market economies.
>
> (UNCTC, 1988, p. 228)

Nevertheless, such accidents suggest that both TNCs and host country governments should be far more stringent in their attitudes towards plants which use potentially hazardous processes. In particular, the TNCs should ensure that their plants are built and operated to the same rigorous environmental and safety standards as plants in their home countries. Undoubtedly, many TNCs do this but the lack of information made available by the companies sometimes leaves room for justifiable concern.

Questions about the appropriateness of technology tend to be less relevant in the developed economies for obvious reasons. But one issue of growing importance in such economies is the influence of foreign TNCs on *business organization and practice*. Japanese investment in Britain is a case in point. One argument for encouraging such investment has been that the introduction of highly efficient Japanese business methods will rub off on British industry in general and raise the level of efficiency in the economy. In other words, it is the *demonstration* of the effect of the 'social' innovations of work organization, labour relations, relationships with suppliers and so on which are suggested to be highly appropriate to the needs of the UK economy. Whether such a process of diffusion occurs, and to what extent, only time will tell.

Cost of technology transfer via the TNC   Acquiring technology through the agency of the TNC involves a cost to the host country. But precisely what that cost is, and whether it is a 'reasonable' price to pay, are extremely difficult to determine. First, technology is only one part of the overall package of attributes which the TNC brings to a host country; it is difficult to separate out. Second, assessment of the cost involved assumes that it can be measured against alternative ways of acquiring the same technology. The two major alternatives are (a) to buy or license the technology alone from its owner (the TNC); that is, to 'unbundle' the TNC package; or (b) to produce the technology domestically.

The parallel usually drawn is with Japan, which rebuilt its postwar economy without the introduction of direct foreign investment, mainly by *licensing* technology from Western firms. Although a great deal of technology is licensed by developing (and developed) countries from TNCs it is not always a feasible alternative. A TNC may be unwilling to license the technology or it may charge an exorbitant price. It may be a question of the host country accepting the entire TNC package or getting nothing at all. The possibility of producing the technology domestically may be feasible for some of the more advanced industrial nations but rarely so for developing countries.

Trade and linkages[7]

Two of the most important questions surrounding the impact of TNCs on all host economies, whether developed or developing, concern, first, their role in the host country's *trade* with the outside world, and, second, the extent to which they are *integrated* into the local economy through *linkages with domestic firms*.

The nature of the TNC's behaviour on both counts has far-ranging implications. It affects the host country's balance of payments: exports and import-substituting production by TNCs contribute towards a positive trade balance, imports by TNCs contribute towards a negative trade balance, although this may be offset if the imports are essential for export-producing activities. More significant in the long term, however, is the extent to which TNCs are integrated into the national or local economy. The forging of direct links with indigenous firms is the most significant means by which technology is transferred, additional employment created and opportunities increased for the formation of new local enterprises.

Interfirm linkages are the most important channel through which technological change is transmitted. By placing orders with indigenous suppliers for materials or components which must meet stringent specifications technical expertise is raised. The experience gained in new technologies by local firms may enable them to compete more effectively in broader markets, provided, of course, that they are not tied exclusively to a specific customer. The sourcing of materials locally may lead to the emergence of new domestic firms to meet the demand created, thus increasing the pool of local entrepreneurs. The expanded activities of supplying firms, and of ancillary firms involved in such activities as transport and distribution, will result in the creation of additional employment. But such beneficial spin-off effects will occur only *if* the foreign affiliates of TNCs do become linked to local firms. Where TNCs do not create such linkages they remain essentially as foreign enclaves within a host economy, contributing little other than some direct employment.

Factors influencing linkage formation   How far TNCs actually contribute to the host economy's trade balance and to its deeper and broader economic development through linkages with domestic firms depends on a number of factors. The most obvious is the primary purpose for which the foreign plant is established (Figure 12.1): that is, whether its role is to serve the host market itself or to serve third-country markets using the host economy as an export platform. But the trade and linkage effects of import-substituting and export-oriented plants are more complex than this simple division would suggest. For example, there is considerable disagreement over the extent to which foreign plants export proportionately more or less than domestic firms.

In the East and South East Asian NICs 'the conclusion for most, but not all, countries is that foreign firms do export a higher proportion of their output than their local counterparts . . . The results therefore suggest that MNCs can and do play a positive role in host countries' export-oriented development' (Hill, 1990, p. 35). However, Hill goes on to note (ibid.) 'the possible costs of MNCs in a host country's export drive, namely the generally higher import propensity of MNCs and possible restrictions on their affiliates' exports resulting from the head office's global sales strategy'. This latter point explains the attempts made by some host governments, in both developed and developing countries, to persuade TNCs to grant 'world product mandates' to their plants to give them a wider export market. What export-oriented foreign plants clearly do, however, is to lead to closer integration of the host economy into the international economy.

It is not surprising that TNCs in many developing countries export relatively more than domestic firms because many of them were established explicitly to act as *export platforms*. The same relationship does not necessarily apply in the industrialized countries where most TNCs tend to be oriented to the local market. In the United States, for example, foreign plants export less, but import far more, than domestic firms:

> In 1986, the US had a merchandise trade deficit of $149 billion. That year, foreign affiliates imported $124 billion and exported only $51 billion, for a net deficit of $73 billion . . . leaving aside wholesale operations, foreign firms imported £15.8 billion more than they exported – or nearly 11% of the entire merchandise trade balance.
>
> (Glickman and Woodward, 1989, p. 152)

As far as local linkages are concerned the most significant are *backward* or *supply linkages*. Here, the crucial issue is the extent to which TNCs either import materials and components or procure them from local suppliers. The general consensus, as we have seen, is that the foreign affiliates of TNCs have a much higher propensity to import than domestic firms. But, again, we need to be wary of sweeping generalizations. The actual incidence of local linkage formation by foreign-controlled plants depends upon four major influences:

1. the particular *strategy* being followed by the TNC and the *role* played by the foreign plant in that strategy;
2. the particular *industry* and *technology* involved;
3. the *characteristics* of the *host economy*;
4. *time*.

The influence of TNC strategy on the development of local linkages within host economies has a number of aspects. One is the general corporate policy towards the sourcing of inputs, which will determine the degree of sourcing autonomy granted to individual plants. Those TNCs which are strongly vertically integrated at a global scale are less likely to develop local supply linkages than firms with a lower degree of corporate integration. But even where vertical integration is low the existence of strong links with independent suppliers in the TNC's home country or elsewhere in the corporate network may inhibit the development of local linkages in the host economy. Familiarity with existing supply relationships may well discourage the development of new ones, particularly where the latter are perceived to be potentially less reliable or of lower quality.

A particularly important factor is the role of the foreign plant itself in the TNC's overall strategy: whether it is oriented to the host market and is, therefore, import-substituting, or whether it is an export platform activity. Foreign plants which serve the host market are more likely to develop local supply linkages than are export platform plants. Lall (1978) identifies four types of export-oriented activity in TNCs and suggests that each will have rather different implications for the creation of local supply linkages, particularly in developing countries:

1. *TNCs which began as local market-oriented operations but which have subsequently developed strong export orientation*. Their activities generally utilize relatively stable and unsophisticated technologies and are located in areas where the labour force is skilled but cheap and where there is an established indigenous industrial sector. Such firms may have established significant local linkages during their import-substituting phase which then form the basis of broader export-oriented operations.
2. *TNCs involved in more traditional industries* (e.g. textiles, food processing, sports goods) which employ standardized technology but where product differentiation, marketing or product innovations are important considerations. These activities have a high potential for the creation of local linkages with domestic suppliers to manufacture either components or complete products.
3. *New TNC investments in 'modern' industries which are located in developing country export platforms*. The technologies used are fairly complex and production is directed towards world markets. In general, these activities are very closely controlled from corporate headquarters in the TNCs' home countries and often utilize established supply links. The potential for developing local linkages is fairly limited but not entirely absent.
4. *TNC investments which are, essentially, merely 'sourcing' operations* in which only a particular process (usually highly labour-intensive) is located in developing countries. The clearest example is probably the semiconductor industry (Chapter 10). The potential for developing local linkages is low because of the very tight product specification, the need to keep production costs as low as possible and the dynamic nature of the technology involved.

The third major influence on the establishment of local supply linkages by TNCs is the nature and characteristics of the host economy itself. In general, we would expect to find denser and more extensive networks of linkages between TNCs and domestic enterprises in the developed economies than in the developing economies. Within developing countries such linkages are likely to be greatest in the larger and more industrialized countries than in others. In addition, host country governments may well play a very important role in stimulating local linkages by insisting that TNCs utilize a certain level of locally sourced materials and components. Such *local content* policies have become increasingly widespread in both developing and developed countries and they have undoubtedly had some effect. Indeed, the ILO's view is that, at least in developing countries, 'government intervention in the sourcing choices of MNEs appears to have been the single most powerful determinant for the creation of local linkages of MNEs ... Without such government intervention it is likely that, despite some market pressure, local MNE linkages would be much less developed than they are today in various countries and industries' (ILO, 1981a, p. 94).

But much depends, of course, on the relative strength of the *host country's bargaining power* as regards the TNC, and on the extent to which local supplies are of an appropriate quantity and quality. Again, it tends to be in the larger and the more industrialized developing countries that such local content policies have the greatest impact, and also in those TNC activities serving the local market. Indeed, it could be that the export-oriented industrialization strategies of developing countries actually inhibit the development of local supply linkages.

The fourth influence on the establishment of TNC linkages with local suppliers is time. Particularly in view of the closer relationships between firms and their suppliers which have been emerging (see Chapter 7, pp. 219–220) it should not be expected that a foreign plant, newly established in a particular host country, will immediately develop local supplier linkages. Not only do appropriate suppliers have to be identified but also it takes time for supplier firms to 'tune-in' to a new customer's needs.

Evidence of linkage formation   Factual evidence of local linkage formation by TNCs is extremely variable. Examples can be found at all positions on the spectrum of possibilities, from virtually no local linkages on the one hand to a high level of local linkages on the other. Studies within the smaller developing countries, particularly those with a short history of industrial development, tell a fairly uniform story of shallow and poorly developed supply linkages between local firms and foreign-controlled plants. Hill's (1982) study of the Philippines and Chi's (1981) study of the Penang area of Malaysia both demonstrate this. Hill shows that extensive local procurement of components by foreign plants was the exception rather than the rule. Most of the components used by foreign plants were either imported or produced in-house. Similarly, Chi observed that the supply linkages of foreign plants in the Prai Industrial Area of Malaysia were already predetermined and prearranged and that there was very little interaction with local firms.

Local TNC linkages in Singapore have also tended to be limited although Lim and Fong (1982) showed that TNCs from different national origins tended to have rather different local sourcing policies. Japanese firms tended to be less integrated into the Singapore local economy than American or European firms. On the other hand, 'in the more industrialized Northeast Asian economies ... quite strong linkages have developed, as documented for Taiwan ... [and] ... Korea' (Hill, 1990, p. 41). A common observation is that foreign plants located in export-processing zones (EPZs) are particularly unlikely to develop supplier linkages with the wider economy. This is generally borne out by evidence assembled by ILO (1988), Sklair (1986) and by Warr (1987) although some local linkages are beginning to develop in certain cases.

The situation in the older-industrialized countries is rather different. Often, of course, foreign firms have been in operation for so long that they are an almost indistinguishable part of the landscape and have developed dense networks of local supplier linkages over a long period of time. IBM, for example, claims to have some 11,000 suppliers in the United Kingdom. Even some of the more recent arrivals have gradually built up local supply networks. But, as the experience of Sony in the United Kindom shows, it may take up to ten years to do so. One reason for the generally low level of local supply linkages of Japanese firms in Europe as a whole, therefore, is that many of them are very recent arrivals. On the other hand, it is by no means inevitable that TNCs in industrialized countries will develop strong local linkages. In Canada there have been persistent criticisms of the failure of many US firms to source their major inputs from Canadian suppliers. The extent of local supply links of American electronics plants in Scotland is also small.

Apart from the extent of linkages created by foreign firms in a host economy there is also the question of their *quality* and the degree to which they involve a beneficial transfer of technology (either production or organizational) to supplier firms. A common criticism is that many TNCs tend to procure only 'low-level' inputs from local sources, for example cleaning services and the like. This may be because of deliberate company policy to keep to established suppliers of higher-level inputs or because such inputs are simply not available locally (or are perceived not to be so). Where development of higher-level supply linkages occurs there does seem to be a positive effect on supplier firms. For example, the rigorous specifications laid down by Japanese electronics firms and by American firms such as IBM have undoubtedly had a beneficial technological spin-off on their UK suppliers. In these cases, at least, there is evidence of technological linkages increasing over time.

Industrial structure and entrepreneurship[8]

Although not all TNCs are giant corporations it is often the case that foreign plants are larger than their domestic competitors. Hence, the entry of a foreign plant into a host economy may have a number of repercussions on the structure of domestic industry, particularly on the competitiveness, survival and birth of domestic enterprises. But, as in other aspects of the impact of TNCs, there is no inevitability about such structural effects. Much will depend upon the specific domestic context itself and on the relative size and market power of the TNC affiliate in that context. In general, the difference between a foreign plant and a domestic enterprise in developed market economies, especially those which are themselves sources of TNCs, will be far less than that in developing countries. The industrial structure of most developing economies tends to be much more strongly *dualistic*, with a small, technologically advanced sector (relatively speaking) which is oriented to the more modern urban market and a technologically less advanced sector characterized by traditional production and attitudes. TNCs in developing countries are, by definition, part of the technologically advanced sector of the host economy.

A major *long-term* effect of the entry of TNCs into a host economy – both developed and developing – is likely to be an *increase in the level of industrial concentration*.

The number of firms is likely to be reduced and the dominance of very large firms increased. Lall suggests two reasons;

> First, regardless of the MNC's market conduct, the *attributes* of these enterprises can raise barriers to entry for local firms: MNCs often introduce advanced, usually larger-scale and more capital-intensive technology; they generally produce a wider, more differentiated and better marketed range of products; they utilise newer managerial and organisational skills; they have better access to financial, technical and marketing resources abroad; and they may be more prepared to challenge 'live-and-let-live' rules of the game observed by local oligopolists than local entrants.
>
> Second, their *conduct* may speed up the process of concentration. MNCs are generally the leading forces in the developed countries in diversifying across industries, in effecting takeovers and in lobbying policy makers, and they may be expected to transfer these highly developed strategies to all the host countries in which they operate. Thus, in LDCs, MNCs may purchase local firms on especially favourable terms because of their strong hold over technology or input markets . . . ; they may be able to outlast local competitors in price-cutting wars because of financial staying power; or they may be able to win more favourable concessions from host governments. The market power and tactics of MNCs may also cause higher concentration by inducing defensive mergers among local firms.
>
> (Lall, 1979, p. 328)

Thus, two important effects of TNCs on host economies are, first, the possible squeezing out of existing domestic firms and, second, the suppression of new indigenous enterprises. Both of these fears have been voiced especially strongly in the case of developing countries but there is no reason to believe that they do not also apply to particular sectors in developed economies if high TNC penetration has occurred. But, clearly, the less developed the indigenous sector the greater is the likelihood of its being swamped by foreign entry and of local entrepreneurship being suppressed. But we should beware of assuming that the involvement of TNCs in a particular host economy will inevitably destroy or suppress domestic enterprise. Where substantial local linkages are forged by foreign plants, particularly on the supply side, opportunities for local businesses may well be enhanced. Existing firms may receive a boost to their fortunes or new firms be created in response to the stimulus of demand for materials or components. Another way in which TNCs might contribute to the formation of new enterprises is through the 'spin-off' of managerial staff who set up their own businesses on the basis of experience and skills gained in employment with the foreign firm.

Employment and labour issues[9]

For most ordinary people, as well as for many governments, the most important issue in the debate over the TNC is its effect on jobs (Figure 12.1). Does the entry of a foreign-controlled plant create new jobs? What kinds of jobs are they? Do TNCs pay higher or lower wages than domestic firms? Do TNCs operate an acceptable system of labour relations? Are foreign plants likely to offer stable employment opportunities or are they fly-by-night operations likely to switch from location to location at the merest whim? From our discussion of other aspects of TNC activity it should now be apparent that there is no simple, unequivocal answer to any of these questions which will fit all circumstances. Hence, in examining the possible employment effects of TNCs in host economies we should remember that they are possibilities and not inevitabilities.

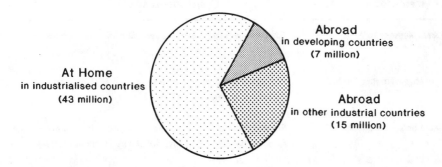

Figure 12.2 Global distribution of direct employment in TNCs, mid-1980s (*Source*: Kreye, Heinrichs and Fröbel (1988), Chart 3)

Volume of employment TNCs employ very large numbers of workers in both developed and developing countries. Estimates for the mid-1980s suggest a total direct world labour force in TNCs of some 65 million, of which 43 million were in TNCs' home countries and 22 million were employed abroad. Of these, roughly 7 million were in developing countries (Figure 12.2). But these figures are not very illuminating in themselves because, as we saw in Chapter 3, production and employment in TNCs are very unevenly distributed geographically. The fact that the 7 million employed in developing countries represents less than 1 per cent of their total labour force is not a very helpful statistic. In some cases TNCs employ between a third and a half of a country's manufacturing labour force; in others the TNC share is very low. A second reason for not relying too heavily on such aggregate figures is that they represent only workers employed *directly* in TNC establishments. To these we must add the *indirect* employment in linked firms which are, themselves, not TNCs.

It is virtually impossible to establish a global figure for the size of the indirect TNC labour force because, as Figure 12.3 shows, there are many ways in which such employment is generated. The number of jobs created directly in a particular TNC plant will depend both on the *size* of its activities and also on the *technological* nature of the operation; particularly on whether it is capital-intensive or labour-intensive. The number of *indirect* jobs created will also depend upon two major factors. First, it will vary according to the extent of local linkages forged by the TNC with domestic firms. As we have seen, such linkage formation may vary considerably in different circumstances. Second, it will depend upon the *amount of income generated* by the TNC and *retained* within the host economy. In particular, the wages and salaries of TNC employees and of those in linked firms will, if spent on locally produced goods and services, increase employment elsewhere in the domestic economy.

A third reason for treating aggregate employment estimates with some caution is they do not show the *net* employment contribution of TNCs to a host economy. In other words, against the number of jobs *created* in and by TNCs we need to set the number of jobs *displaced* by any possible adverse effects of TNCs on indigenous enterprises. For reasons outlined in the preceding section of this chapter, domestic enterprises may be squeezed out by the size and strength of foreign branch plants while new firm formation may be inhibited. In these respects, the effect of the TNC may be to displace existing or potential jobs in domestic enterprises. Hence, the overall

Employment effects	Definition or illustration
Direct employment effects	*Total number of people employed within the MNE subsidiary*
Indirect employment effects	*All types of employment indirectly generated throughout the local economy by the MNE subsidiary*
1. Macro-economic effects	*Employment indirectly generated throughout the local economy as a result of spending by the MNE subsidiary's workers or share-holders*
2. Horizontal effects	*Employment indirectly generated among other local enterprises as a result of competition with the MNE subsidiary*
(a) Narrow horizontal effects	*Employment indirectly generated among local enterprises competing in the same industry as the MNE subsidiary*
(b) Broad horizontal effects	*Employment indirectly generated among local enterprises active in other industries than the MNE subsidiary*
3. Vertical effects	*Employment indirectly generated by the MNE subsidiary among its local suppliers and customers*
(a) Backward effects (or linkages)	*Employment indirectly generated by the MNE subsidiary among its local suppliers (of raw materials, parts, components, services, etc.)*
(b) Forward effects (or linkages)	*Employment indirectly generated by the MNE subsidiary among its local customers (e.g. distributors, service agents etc.)*

Note: The above employment effects, if they could be measured, should be calculated in net terms (i.e. gross employment directly or indirectly generated, minus total employment displacement).

Figure 12.3 The direct and indirect employment-generating effects of TNC subsidiaries (*Source*: International Labour Office (1984a), p. 39)

employment effect of TNCs in host economies depends upon the balance between job–creating and job–displacing forces:

Net employment contribution of TNC to host economy = $(DJ + IJ) - JD$

where: DJ = Number of direct jobs created in TNC
         IJ = Number of indirect jobs in firms linked to TNC and in other sectors
         JD = Number of jobs displaced in other firms

Only specific empirical study can throw real light on this equation and its calculation is fraught with difficulties. What would the employment situation be like in the absence of the TNC? The majority view seems to be that, at least in terms of numbers of jobs, the net employment effect has been favourable to host countries. This view was fairly consistently expressed in the studies of individual host countries commissioned by the

ILO, although few of these studies appear to have made very rigorous calculation of the actual net effects. It may well be that the 'alternative situation' will vary by sector. In the West German study, for example, the ILO argued that domestic firms might well have filled the vacuum in the chemicals industry if TNCs had not been present but in other sectors such as computers such domestic replacement seemed less likely.

Others have taken a different view. Britton and Gilmour (1978) claimed that the result of US investment in the Canadian economy in 1970 was a net loss of 120,600 jobs. This net loss was made up of an estimated gain of 110,000 and an estimated loss of 230,600 jobs. More recently, Glickman and Woodward (1989, p. 128) estimate that in the United States between 1982 and 1986, 'US affiliates . . . eliminated approximately 56,000 more jobs than they created'. This net deficit in employment creation in foreign plants arose from job creation of 386,432 and job loss of 442,295.

Type of employment and wage levels   The number of jobs created by TNCs in host economies is only part of the story. What kind of jobs are they? Do they provide employment opportunities which are appropriate for the skills and needs of the local labour force? The answer to these questions depends very much on the attributes of the foreign plant (Figure 12.1): the type of industry, the nature of the technology used, the scale of operations and the extent to which the foreign plant is integrated into its parent organization. In particular, where the plant 'fits' into the TNC's overall structure and how much decision-making autonomy it has are key factors. In general, the fact that TNCs tend to concentrate their higher-order decision-making functions and their R & D facilities in the developed economies produces a major geographical bias in the pattern of types of employment at the global scale.

In developing countries, the overwhelming majority of jobs in TNC plants are *production jobs*. The ILO estimates that production workers make up between 60 and 75 per cent of the total employment in TNCs in developing countries. Such figures should cause no surprise in view of the nature of most TNC investment in developing countries, particularly in export-platform activities. In export-processing zones, of course, low-level production jobs, especially for young females, are the rule (see Table 6.5). But this partly reflects the types of industry which dominate in EPZs. More generally, the ILO suggests that the proportion of higher-skilled workers employed by TNCs in developing countries has tended to increase over time as has the proportion of local professional and managerial staff. Such changes have progressed furthest in the more advanced industrial countries of Latin America and South East Asia. Even so, the TNC labour force in developing countries remains concentrated in low-skill production and assembly occupations.

Geographical segmentation of high and low-order occupations (white collar and blue collar jobs) in TNCs is not confined solely to that between developed and developing countries. As we saw in Chapter 7, a similar geographical segmentation of TNC functions exists within developed economies. This dichotomy is clearly illustrated in the United Kingdom where the higher-order occupations are disproportionately concentrated in South East England in and around London. A major criticism of the foreign branch plants located in Scotland, Wales, Northern England and Northern Ireland, therefore, is that they employ primarily lower-skilled workers and that they generally lack higher-level employment opportunities. Whether or not better types of job would be available in such areas in the absence of the foreign plants is, of course, an open question.

Related to the type of employment offered by foreign plants is the question of *wages and salary levels*. Two issues are involved here. One relates to the global scale and, especially, to the distinction between developed and developing countries. In so far as TNCs take advantage of geographical differences in prevailing wage rates between countries they do, in fact, 'exploit' certain groups of workers. The exploitation of cheap labour in developing countries at the expense of workers in developed countries is one of the major charges levelled at the TNC by labour unions in Western countries. But another way of looking at the question is to ask whether TNCs pay more or less than local firms in the same industry. If they pay below the prevailing rate then, quite clearly, exploitation is taking place. But the general consensus seems to be that TNCs generally pay either at or above the 'going rate' in the host economy. In developing countries, wages within EPZs – which contain predominantly foreign-controlled plants – tend to be considerably higher than those outside the zones. But there are enormous difficulties in measuring comparative wage levels both within and between countries. Where TNCs do pay above the local rate they may well 'cream off' workers from domestic firms and possibly threaten their survival.

This point relates to the kinds of *recruitment policies* used by TNCs. TNCs tend to operate very careful screening procedures when hiring workers. This may well mean that employees for a newly established foreign plant are drawn from existing firms rather than from the ranks of the unemployed. Another aspect of recruitment, at least in some industries, is the extent to which TNCs recruit particular types of workers to keep labour costs low. In the textiles, clothing and electronics industries, for example, we noted the very strong tendency to prefer females to males in assembly processes and, in some cases, to employ members of minority groups as a means of holding down wage costs and for ease of dismissal. But such practices would appear to be specific to particular industries and should not necessarily be regarded as universally applicable to all TNCs in all industries.[10]

Labour relations   The question of labour relations within TNCs, including the role of labour unions, is a further significant aspect of the employment effects of TNCs in host countries. In most developing countries labour is either weakly organized or labour unions are strictly controlled (or even banned) by the state. Even in developed economies some major TNCs simply do not recognize labour unions in their operations. IBM is a good example of this practice. But most TNCs, however reluctantly, do accept labour unions where national or local circumstances make this difficult to avoid. Whether labour unions are involved or not, the question of the nature of labour relations within TNCs focuses on whether they are 'good' or 'bad', that is, harmonious or discordant. Some studies suggest that TNCs tend to have better labour relations in their plants than domestic firms; others point to a higher incidence of strikes and internal disputes in TNCs. But it is often difficult to separate out the 'transnational' element. In the case of strikes, for example, it may be plant or firm size which is the most important influence rather than nationality of ownership.

One of the most acute concerns of organized labour is that decision-making within TNCs is too remote: that decisions affecting work practices and work conditions, pay and other labour issues are made in some far-distant corporate headquarters which has little understanding or even awareness of local circumstances. Again, only detailed empirical studies can answer these questions for specific cases. Hamill's (1984) study of labour relations in a sample of thirty foreign TNCs operating in the United Kingdom is

Table 12.1 The locus of decision-making within a sample of foreign TNCs in the United Kingdom

Most centralized	Least centralized
Determination of operating budget	Union recognition
Supply of new investment funds	Employer association membership
Cost of wage increases	Collective bargaining
Pension and employee fringe benefits	Structure of collective bargaining
Numbers employed	Wage payment systems
	Settlement of strikes
	Recruitment of managerial staff

(*Source*: based on Hamill (1984), p. 32)

a good example which provides some valuable insights into the issues involved. Some labour relations decisions made by TNCs are far more centralized than others in that they are either made directly at corporate headquarters or require its approval.

As Table 12.1 shows, these areas mainly relate to the operating costs of the subsidiary and reflect the parent company's concern to control financial and labour costs. However, Hamill also found considerable variation between TNCs in their degree of headquarters' involvement in labour relations at their UK plants. Six firms operated highly *centralized* labour relations policies in which virtually every labour relations matter was decided by the company headquarters. Conversely, ten firms operated very *decentralized* systems in which the UK subsidiary had a high degree of autonomy in this area. The remaining fourteen firms fell somewhere between these two extremes.

Such variety in TNC labour relations policies can be explained by several factors:

1. *Degree of intersubsidiary production integration*. Highly integrated production systems require a greater degree of control and coordination.

   > If the overseas subsidiary is a vital part of an integrated production system – a system which could be widely disrupted if this particular key plant was to close down – then labour relations problems at this level could severely affect not only the operational efficiency of the local subsidiary itself, but also that of the whole worldwide network. In such circumstances, therefore, the parent company will become closely involved in labour relations developments at this level.
   >
   > (Hamill, 1984, p. 33)

2. *Nationality of ownership*. Labour relations decision-making was more highly centralized in US companies than in European companies, but this was largely because the US firms were more highly integrated in their operations.
3. *Method of establishment in the United Kingdom*. Labour relations were more highly centralized in newly established (greenfield) foreign plants than in plants acquired from domestic firms.
4. *Performance of the UK subsidiary*. Poor performance tended to result in a significant increase in corporate involvement in labour relations at the plant. This was especially so where such poor performance was perceived to be caused by labour problems.
5. *Relative importance of parent company as a source of funds for the UK subsidiary*. A heavy dependence on corporate finance was generally associated with a fairly centralized labour relations policy.

Hence, Hamill (1984, p. 34) concluded that 'general statements cannot be applied to the organisation of the labour relations function within such firms. Rather, different MNCs adopt different labour relations strategies in relation to the environmental factors peculiar to each firm. In other words, *it is the type of multinational under consideration which is important rather than multinationality itself.*'

Stability of foreign-controlled plants One of the most contentious issues of all in the debate on the employment effects of TNCs in host countries is their long-term stability. Are TNC plants more likely to switch their operations from one country to another? Are they, in the words of one study, 'stickers or snatchers'? A popular view is that TNCs can and do switch and reswitch their operations with great frequency and ease to take advantage of changes in costs or profitability. But, although TNCs certainly do reshuffle their geographical pack through the processes of investment and disinvestment, as previous chapters have demonstrated, the situation is, again, more complex than the popular view would suggest.

> Whilst there is some evidence of production switching to offset strike-bound plants . . . the feasibility of production relocation appears to be much lower . . . It is important to distinguish actual, from threatened, production relocation. Cases of production relocation by MNCs are fairly rare . . . The threat of plant closure is a little more frequent . . . Clearly, there are constraints on the ability of an MNC to switch production, certainly in the short-term. Not only are transference, set-up and production loss costs likely to be substantial, but where the MNC operates a global strategy production relocation may have global restructuring repercussions. This type of consideration is likely to discourage the relocation of production in the event of industrial troubles or the opportunity of lower cost labour. A more likely strategy for the MNC would concern the placement of new investment, not the switching of existing facilities.
>
> (Enderwick, 1979, pp. 228–9)

In other words, the adjustment is often more gradual than is frequently claimed. Employment in established plants in one location may be reduced over a period of time as new investment is developed elsewhere. When closure does occur it is often because the plant no longer has a role to play in the changed corporate circumstances. Unfortunately, there are few empirical studies which examine the comparative stability of foreign-owned and domestic plants. McAleese and Counahan concluded that in Ireland, at least between 1973 and 1977,

> employment loss in MNCs during the recession has not been noticeably different from that of indigenous firms, that MNCs' propensity to close down their plants is similar to that of their domestic counterparts and that employment recovery after the recession was greater in MNCs. The evidence suggests that MNCs do not constitute an unstable element in the economy: they are 'stickers' rather than 'snatchers' as far as the small open host economy like the Irish Republic is concerned.
>
> (McAleese and Counahan, 1979, p. 355)

Broadly similar conclusions were drawn by Townsend and Peck (1986) in their study of employment change in foreign and domestic establishments in Britain. Foreign firms did not display a greater propensity to close their branch operations than did domestic firms.

Of course, the experience over one time period in two small countries proves nothing on its own except to remind us of the need for more detailed research into specific cases. It is all too easy to over-exaggerate the geographical flexibility of TNCs, particularly over a short period. This is not to imply that locational shifts do not occur –

we have seen in several earlier chapters that they certainly do – but rather that gradual shifts are more common than abrupt changes. It is possible, however, that more rapid locational switching may occur among export-platform plants in developing countries. A careful study by Flamm (1984) of the volatility of offshore investment in the semiconductor industry found some substance in the common allegation that semiconductor firms readily shift their offshore assembly operations between Third World locations. Such shifts, he suggests, are only partly a response to changes in relative labour costs in different countries. TNCs in the industry appear to operate a 'diversified portfolio of "risky" LDC production locations' (p. 234).

Dependence, truncation and hollowing out

It should now be clear from our discussion of individual aspects of TNC impact on host economies that no unequivocal *general* evaluation can be made. Whether a foreign plant creates net costs or net benefits will depend on the *specific* context: the inter-action between the attributes and functions of the plant itself within its corporate system and the nature and characteristics of the host economy. It also depends, critically, on the alternatives realistically available. But what if a host economy – or an important sector within the economy – develops a high level of foreign TNC involve-ment? Does the presence of a large foreign-controlled sector tip the balance of evalua-tion? In a long-term sense, the answer to these questions would appear to be 'yes'. Whereas the involvement of some foreign plants in a host economy may well have beneficial effects – not only in creating employment but also in introducing new technologies and business practices – overall dominance by foreign firms is almost certainly undesirable from a host country viewpoint. There are real dangers in acquir-ing the status of a *branch plant economy*.

Precisely what constitutes an undesirable level of foreign penetration is open to debate. It is made more complex by the fact that a country may be dominated by, and dependent upon, external forces even where there is very little *direct* foreign invest-ment in the economy. This may occur, for example, where a large segment of an economy is engaged in subcontracting work for foreign customers. Here, however, our concern is with the effects of a large direct foreign presence. Until recently, most of the debate was focused upon developing countries and formed part of the broader *dependency* debate. But such concern is no longer confined to developing countries. Most obviously, those developed economies which have a very high level of foreign direct investment also share the same kinds of problem. Even in those developed economies where, overall, the degree of foreign penetration is relatively low the dominance of specific sectors by foreign firms is beginning to raise concern. The current controversy over foreign (especially Japanese) penetration of the US economy is a good example of this.

The potential problems of a high level of foreign penetration of a host economy can be summarized under two closely related headings: dependence and truncation or 'hollowing out'.

Dependence   A major consequence of a high level of dependence on foreign enter-prises is a reduction in the host country's *sovereignty* and *autonomy*: its ability to make its own decisions and to implement them. At the heart of this issue are the different – often conflicting – goals pursued by nation states on the one hand and TNCs on the other. Each is concerned to maximize its own 'welfare' (in the broadest sense). Where much of a host country's economic activity is effectively controlled by foreign firms, non-national goals may well become dominant. It may be extremely difficult for the host government to pursue a particular economic policy if it has insufficient leverage over the dominant firms. The tighter the degree of control exercised by TNCs within their own corporate hierarchies and the lower the degree of autonomy of individual plants the greater this loss of host country sovereignty is likely to be. In the *individual* case this may not matter greatly but where such firms *collectively* dominate a host economy or a key economic sector it most certainly does matter.

Perhaps the most significant aspect of dependence upon a high level of foreign direct investment is that of *technological* dependence. This is

> the continuing inability of a . . . country to generate the knowledge, inventions, and innovations necessary to propel self-sustaining growth . . . If a country does not produce its own technology in at least some industries, it is argued, it will suffer slower growth and more disadvantageous terms of trade in the long run . . . Technological dependence may mean slower or 'distorted' growth and reduced economic sovereignty.
>
> (Newfarmer, 1983, pp. 177–8)

Truncation   A second consequence of a very high level of foreign dominance is the likely truncation, or 'hollowing out', of various facets of the economy. Truncation, as Hayter (1982) explains, is an 'umbrella concept' which summarizes the costs of foreign investment to host economies. The term can be interpreted at two levels. The most obvious level is that of the *foreign-controlled plant or firm* itself: 'A truncated firm is one which does not carry out all the functions – from the original research required through to all the aspects of marketing – necessary for developing, producing and marketing goods. One or more of these functions are carried out by the foreign parent' (Government of Canada, 1972, p. 405).

Truncation is implicit in the very nature of the large, geographically extensive TNC. As we have seen, TNCs characteristically subdivide their internal operations and locate specialized units in different types of location. Essentially, a truncated plant is one which concentrates mainly or exclusively on production activities and which, therefore, lacks the higher-level administrative and R & D functions. Again, the problem arises not from this phenomenon in itself but where *most* foreign-controlled plants in a host economy are truncated. One result will be a highly skewed set of occupational opportunities with relatively few openings for higher-skilled and professional, scientific and technical workers. In addition, truncated plants are likely to be deficient in technological dynamism, relying upon their parent company for innovative activity.

The second level at which truncating effects of foreign dominance may apply is that of *an entire industry* or even of a *host economy as a whole*: 'As the proportion under foreign control rises, an industry becomes a shell. In terms of its products, the industry seems to be complete and comprehensive, but large elements of the production system are missing or deficient'(Britton and Gilmour, 1978, p. 98). The result is a 'hollowed-out' economy: the 'shell' referred to in Britton and Gilmour's quotation.

More generally, a high level of foreign penetration will tend to exacerbate the *negative* repercussions of foreign plants discussed in earlier sections of this chapter. In particular it will tend to inhibit or suppress the development of *indigenous* firms either because foreign plants create few local linkages or because indigenous firms are squeezed out by the competitive strength of foreign plants which are backed by much larger corporate resources:

> Thus much of the drive, enthusiasm, and invention that lie at the heart of economic growth is removed, reduced, or at best, suppressed . . . one would expect that the region would not be a leader in developing new products, processes and technologies, which, in turn, suggests that innovation will not be a major force in the local economy, which further implies that there will not be a substantial development of new enterprises or indeed of growth within existing enterprises.
>
> (Firn, 1975, p. 410)

To repeat, dependence and truncation of a host economy do not inevitably arise from the entry of a TNC. But they do seem to be a most likely outcome of a high level of foreign penetration. Hence, it is in such economies that the symptoms are most clearly seen. The best-documented case is that of Canada, where concern over the very high level of foreign – predominantly US – control goes back a good many years. A particularly strong view was expressed by Britton and Gilmour in their 1978 report *The Weakest Link*. They directly attributed Canada's relatively poor industrial and trade performance in manufacturing to the high level of foreign direct investment. In their opinion, 'over the decades an underdeveloped industrial structure typical of satellite or hinterland economies has been generated . . . Canada is industrially backward. This position seems associated with technological underdevelopment . . . There is no avoiding the implication that the indifferent performance of high-technology industry in Canada . . . reflects the high degree of foreign ownership of these activities' (op. cit., pp. 52, 59, 80). It may be argued, of course, that Canada is something of a special case, particularly because of its immediate geographical proximity to the United States. It is certainly significant that US firms are the dominant foreign presence in Canada. But the kind of repercussion resulting from such a very high level of foreign ownership is at least an indication of what *may* happen where similar conditions apply, whether in developed or developing economies.

None of this is to argue that foreign investment should be avoided completely. What should be avoided by host economies is an excessive degree of foreign penetration. The major need is to avoid *technological dependence*, because it is technology which is the seed corn of future economic development. As Lall points out,

> the correct strategy then must be a judicious and careful blend of permitting TNC entry, licensing and stimulation of local technological effort. The stress must always be – as it was in Japan – to keep up with the best practice technology and to achieve production efficiency which enables local producers (regardless of their origin) to compete in world markets. This objective will necessitate TNC presence in some cases but not in others.
>
> (Lall, 1984, p. 11)

The bargaining relationship between TNCs and host economies[11]

In the final analysis, the relationship between TNCs and host countries (actual or potential) revolves around their relative bargaining power. As Nixson (1988, p. 378) points out, 'it is this process that in large part determines the extent, nature and distribution between the participating agents of the costs and benefits that arise as a result of DFI'. For the sake of simplicity we assume that a host country can be regarded as a single entity in the bargaining relationship. In fact, of course, many competing

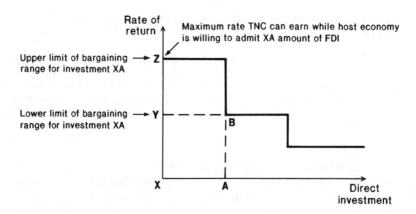

Figure 12.4 A simplified model of the bargaining relationship between a TNC and host country (*Source*: based on Nixson (1988), Figure 1)

interest groups are involved, including domestic business interests which may have varying attitudes towards TNCs.

In Figure 12.4 the bargaining range for a particular level of TNC investment (XA) is shown to vary between (1) a lower limit (XY), which is the minimum rate of return that the TNC is prepared to accept for the amount of investment XA; (2) an upper limit (XZ), which is determined by the cost to the host economy of either developing its own operation, finding an alternative investor or managing without the particular advantages provided by the TNC. XZ is the maximum return the TNC can make for the amount of direct investment (XA) permitted by the host economy. It is in the interests of the TNC to try to raise the upper limit (XZ); conversely, it is in the interests of the host economy to try to lower the upper limit: 'The higher is the cost to the host economy of losing the proposed DFI, the greater are the possibilities for the TNC of setting the bargain near the maximum point' (Nixson, 1988, p. 379). On the other hand, the more possibilities the host economy has of finding alternatives the greater its chances of lowering the upper limit.

The greater the competition between TNCs for the particular investment opportunity the greater are the opportunities for the host country to reduce both the upper and lower limits: 'In addition, the host economy has an interest in lowering the lower limit, through the creation of an advantageous "investment climate" (political stability, constitutional guarantees against appropriation, etc.) which might persuade the TNC to accept a lower rate of return' (Nixson, 1988, p. 380). However, the greater the competition between potential host countries for a specific investment the weaker will be any one country's bargaining position.

Gabriel (1966) describes the TNC–host country bargaining relationship in the following terms:

> In the international investment market, foreign corporations compete for investment opportunities, access to which host governments control. These in turn compete for foreign private investments ... The price which the receiving country will ultimately pay is a function of (1) the number of foreign firms independently competing for the investment opportunity; (2) the recognised measure of uniqueness of the foreign contribution (as against its possible provision by local entrepreneurship, public or private); (3) the perceived degree of domestic need for the contribution. The terms the foreign investor will accept, on

the other hand, depend on (1) his general need for an investment outlet; (2) the attractiveness of the specific investment opportunity offered by the host country, compared to similar or other opportunities in other countries; (3) the extent of prior commitment to the country concerned (e.g. an established market position).

(Gabriel, 1966, p. 114)

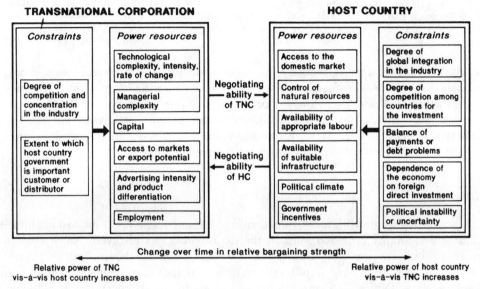

Figure 12.5 Components of the bargaining relationship between TNCs and host countries (*Source*: based on material in Kobrin (1987))

Figure 12.5 implies that the relative bargaining power of TNCs and host countries is a function of three related elements:

1. the *relative demand* by each of the two participants for resources which the other controls;
2. the *constraints* on each which affect the translation of potential bargaining power into control over outcomes;
3. the *negotiating status* of the participants.

Figure 12.5 tends to suggest that host countries are subject to a greater variety of constraints than are TNCs, a reflection of the latter's greater potential flexibility. Nevertheless, the extent to which a TNC can implement a globally integrated strategy is constrained by nation state behaviour. Where a company particularly needs access to a given location and where the host country does have leverage, then the bargain which is eventually struck may involve the TNC in making concessions. It is in this kind of context that the host country's ability to impose performance requirements on foreign firms is greatest. At the extreme, of course, both institutions – TNCs and governments – possess sanctions which one may exercise over the other. The TNC's ultimate sanction is not to invest in a particular location or to pull out of an existing investment. A nation state's ultimate sanction against the TNC is to exclude a particular foreign investment or to appropriate an existing investment.

Most research has concentrated on investment in the natural resource sectors and little work has been done in manufacturing industries. In the natural resource industries, the accepted view seems to be that of the 'obsolescing bargain' in which

> once invested, fixed capital becomes 'sunk', a hostage and a source of bargaining strength. The high risk associated with exploration and development diminishes when production begins. Technology, once arcane and proprietary, matures over time and becomes available on the open market. Through development and transfers from FDI the host country gains technical and managerial skills that reduce the value of those possessed by the foreigner.
> (Kobrin, 1987, pp. 611–12)

In much of manufacturing industry, however – particularly in those sectors in which technological change is frequent and where global integration of operations is common – 'the bargain will obsolesce slowly, if at all, and the relative power of MNCs may even increase over time' (op. cit., p. 636). In so far as such industries are becoming increasingly important in the world economy, the associated shift in bargaining power towards the TNC will pose major problems for host countries.

Summary

The organizing framework for our discussion of the impact of TNCs on host economies, Figure 12.1, simply provided a list of the major processes involved. No attempt was made to indicate the interconnections between each of the individual processes themselves. Yet it is most important to stress that they are all *interlinked* in highly complex ways. Figure 12.6 shows some of the major interconnections. It helps us to appreciate just how complex the relationship can be between the operation of a foreign-controlled unit and the host economy in which it is placed. Most studies of the impact of TNCs on host economies tend to focus on just a small segment of these relationships. There is nothing wrong in this; it is often necessary to do so in order to make the research task feasible. But it is important always to set such partial investigations within the overall picture. Ideally, of course, one would wish to be able to obtain empirical data on every one of the linked processes shown in Figure 12.6 so that an estimate of the *net* impact of a TNC on a host economy can be established. Sadly, that remains a utopian goal.

Transnational corporations and home economies[12]

Most of the arguments over the possible costs and benefits of TNCs have been concerned with their effects on host economies. This is not surprising: the geographical destinations of TNC investments are far more diverse and numerous than their origins. As we saw in Chapter 3, most TNCs originate from a relatively small number of developed countries. Of these, the United States has been until recently by far the most significant and it was there that widespread concern about the possible domestic impact of overseas investment first surfaced. But the issue is one which faces an increasing number of countries as more and more firms have extended their operations across national boundaries. As we have seen at several points throughout this book, a general tendency of the past two decades has been for large firms in particular to expand their overseas operations at a much faster rate than their domestic activities.

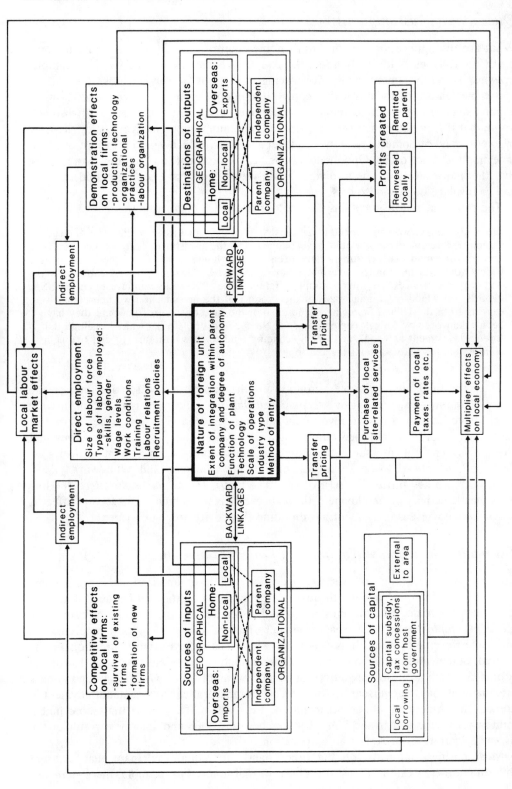

Figure 12.6 The impact of a TNC on a host economy: the total picture

Questions and assumptions

What are the implications for the firm's home country of such increased international production? Does it adversely affect the country's economic welfare by, for example, drawing away investment capital, displacing exports or destroying jobs? Or is it an inevitable feature of today's highly competitive global economy which forces firms to expand overseas in order to remain competitive? Proponents of overseas investment argue that the overall effects on the domestic economy will be positive, raising the level of exports and of domestic activity to a level above that which would prevail if overseas investment did not occur. Opponents of overseas investment, on the other hand, argue that the major effect will be to divert capital which could have been invested at home and to displace domestic exports.

The problem, as Musgrave has pointed out in the US context, is that

> these opposing viewpoints depend in a critical way on the initial assumptions. The first such assumption refers to what would have happened if the investment had not been made abroad: (a) would that investment have been made at home or (b) would these resources which went into the foreign investment have been used in higher levels of consumption and/or public services? Another assumption refers to the effect of foreign investment on US exports: (c) would the foreign sales of the product of the investment have been filled by exports from the United States in the absence of the investment, or (d) would they have been taken over by foreign competitors? . . . Needless to say, studies which conclude that foreign investment is 'good' for the US economy tend to take positions (b) and (d) while contrary studies take positions (a) and (c).
>
> (Musgrave, 1975, p. 137)

Thus, the fundamental problem in assessing the impact of TNCs – this time on the home economy – is, once again, a *counterfactual* one. The need is to establish what would have happened if such investment had not occurred. But, since the alternative situation cannot be established empirically, we have to make assumptions about that alternative. The critical issue is the extent to which domestic investment could realistically be *substituted* for overseas investment. As Glickman and Woodward (1989) point out, two key processes are involved. The first is the *displacement effect*: the loss of domestic activity or employment through overseas investment. The second effect is the *stimulus* overseas investment has on domestic activity and employment.

Estimating the employment impact of outward investment

The act of establishing an overseas operation will have implications for the home country's balance of payments, through its influence on capital and financial flows and its effects on trade. But the most obvious implication for the average citizen is the effect on *employment*.[13] Overseas investment may have very significant repercussions on domestic employment. It is this issue which has received most attention in the United States and which has become increasingly contentious in those European countries which are also the sources of significant numbers of TNCs. Labour interests are adamant that the overseas activity of TNCs dramatically increases unemployment at home. The American labour organization, the AFL–CIO, argued that some half a million jobs were lost in the United States between 1966 and 1969 as a result of the overseas operations of American corporations.

Most business interests, on the other hand, are equally adamant that overseas investment increases, or at least preserves, jobs at home. For example, the Emergency

Committee on American Trade claimed that US overseas investment had a 'substantially positive' effect on domestic employment. The United States Chamber of Commerce described the effect as 'positive'. The Confederation of British Industry similarly regards the effects of British overseas investment on the domestic economy as beneficial, or at least benign. The problem with these diametrically counterpoised claims is that they are based on *totally opposing assumptions*. The labour-interest estimates frequently assume that *all* the activity undertaken overseas could realistically have been retained at home. The business-interest estimates tend towards the opposite assumption: that *none* of the overseas activity could realistically have been retained at home. Quite clearly, more sophisticated approaches are needed to estimate even the approximate impact of overseas investment on domestic employment.

A pioneer study in this respect was that by Hawkins (1972). He subdivided the possible direct employment effects into four categories:

1. *The production-displacement effect (DE)*: employment losses arising from the diversion of production to overseas locations and the serving of foreign markets by these overseas plants rather than by home country plants, that is, the displacement of exports.
2. *The export-stimulus effect (XE)*: employment gains from the production of goods for export created by the foreign investment which would not have occurred in the absence of such investment.
3. *The home office effect (HE)*: employment gains in non-production categories at the company's headquarters made necessary by the expansion of overseas activities.
4. *The supporting firm effect (SE)*: employment gains in other domestic firms supplying goods and services to the investing firm in connection with its overseas activities.

Thus, the net employment effect (NE) of overseas investment on the home economy is:

$$NE = -DE + XE + HE + SE$$

Hawkins calculated the employment-displacement effect for three differing assumptions of the amount of production which could have remained in the United States if the foreign investment had not occurred for the year 1968. His estimates ranged from a possible gain of 279,000 jobs to a possible loss of 322,000 jobs.

Hawkins's three alternative scenarios were based upon somewhat arbitrary substitution values. A more recent and more sophisticated attempt to estimate the employment effects of overseas investment was made by Glickman and Woodward (1989). They used complex formulae to calculate the displacement and stimulus effects of overseas investent by US firms for the period 1977 to 1986. They estimated that such investment resulted in a net loss of 2.7 million domestic jobs (i.e. jobs which, in the absence of the overseas investment, would have been created in the United States). Within this aggregate total, 3.3 million jobs were displaced by overseas investment and 588,000 jobs stimulated by such investment. The adverse employment impact varied substantially from one industry to another, however, being greatest in non-electrical machinery and chemicals, industries in which US overseas investment was particularly high. There was also a differential impact on *different groups of workers*. In general, women workers were harder hit than male workers; black and Hispanic workers more than white workers; blue collar workers more than higher-skilled white collar workers.

These comments apply to *net* job change. But, within the overall displacement through overseas investment there were many particular cases of *employment growth* through the stimulus effect of overseas investment. Often, however, the jobs stimulated are not necessarily in the same industries, occupational groups or geographical locations as the jobs displaced. *The winners and the losers are not the same.* Two quotations make this point very clearly:

> the skill, industry, and location mix of those eliminated jobs will differ from the characteristics of the jobs created through export stimulus and managerial-staff accretions as a result of the foreign operations . . . It becomes a problem of a structural mismatch between the jobs eliminated and the jobs created, even if the latter dominate. One of the obvious results is that the jobs that are eliminated will be almost exclusively production jobs . . . Other staff, and occasionally managerial, jobs may be eliminated, but this is less frequent and the reabsorption of such workers in similar jobs in the same firm is much more likely than in the case of production workers . . . The skill mix [of the jobs created] obviously does not conform closely to the jobs eliminated through substitution of foreign production for exports. Thus one tendency, on a priori grounds, is for an expansion of foreign investment to relatively reduce the demand for production workers and to expand the number of clerical, professional, skilled, and managerial workers within the same industry.
>
> (Hawkins and Jedel, 1975, cited in ILO, 1981b, p. 60)

> On computer tapes, jobs may be interchangeable. In the real world they are not. A total of 250,000 new jobs gained in corporate headquarters does not, in any political or human sense, offset 250,000 old jobs lost on the production line. When Lynn, Massachusetts becomes a ghost of its former self, its jobless citizens find little satisfaction in reading about the new headquarters building on Park Avenue and all the secretaries it will employ. The changing composition of the workforce and its changing geographical location brought about by the globalization of US industry are affecting lives of millions of Americans in serious and largely unfortunate ways.
>
> (Barnet and Muller, 1975, p. 302)

Is overseas investment discretionary or obligatory?

It is virtually impossible to say, with certainty, that overseas investment could equally as well have been made in the firm's home country. We can make various assumptions about what might have happened, but that is all. Ultimately, the key lies with the *motivations* which underlie specific investment decisions. Firms invest overseas for a whole variety of reasons, as we have seen at various points throughout this book: to gain access to markets or to defend positions in those markets; to circumvent trade barriers; to diversify the firm's production base; to reduce production costs, and so on.

It might be argued that foreign investment which is undertaken for *defensive* reasons – to protect a firm's existing markets, for example – is less open to criticism than *aggressive* overseas investment. The argument in the case of defensive investment would be that in the absence of overseas investment the firm would lose its markets and that domestic jobs would be lost anyway. Such investment might be made necessary by the erection of trade barriers by national governments, by their insistence on local production, or by the appearance of competitors in the firm's overseas markets. But, presumably, defensive investment might also include the relocation of production to low-wage countries in order to remain competitive in cost terms. Here, the alternative might be the introduction of automated technology in the domestic plant which would also lead to a loss of jobs.

In this context, the product life-cycle model has obviously formed the basis of much of the argument regarding the probable effects of overseas investment on domestic employment.[14] As a product moves through the stages of the cycle the relocation of production to overseas locations, first, to maintain access to markets and, subsequently, to hold down production costs is depicted as being inevitable (see pp. 139–142). But as we have seen, the reasons for international production are far more varied and complex than the product cycle model suggests. It is, therefore, a matter of fine judgement to draw the line between defensive and aggressive foreign investment. Even the product cycle can be interpreted in both ways.

Although there may well be some clear-cut cases – particularly where access to markets is obviously threatened or where proximity to a localized material is mandatory – there will inevitably be many instances where there is substantial disagreement over the need to locate overseas rather than at home. The various interest groups will have different perceptions of the situation. Thus,

> Multinational management tends to explain investments abroad as reflecting imperative requirements of markets and competition, i.e. global change factors calling for such adjustment by multinational enterprises in the interest of efficient production and ultimate survival. Its critics see the search for profits as the driving factor.
>
> The available evidence is mixed as to the extent to which multinational enterprises do, or can, in fact exercise discretion in decisions to transfer production abroad on a large scale, especially to low-wage developing countries. Instances can be found of multinational enterprises apparently having set up off-shore sourcing subsidiaries, in Hong Kong, Singapore or elsewhere, before they were obliged to do so in self-defence, by competition in home or export markets, either from local firms or from multinational enterprises based in other home countries. On the other hand, examples can be found where such foreign subsidiaries were set up only after substantial competition, for instance from national enterprises in NICs, had been encountered in home or third markets. Still other cases exist – notably in the US television industry – where some MNEs resisted production transfers abroad until they were thrust upon the firm by imminent bankruptcy. The problem for the researcher is that, in all such cases, observable events must be compared with hypothetical, non-observable alternatives. *Thus a clear-cut general answer as to the degree to which MNEs have been wilful, discretionary agents of production transfers abroad, is not possible. For this purpose, the actual decision-making processes of individual multinationals . . . would need to be investigated through detailed case studies.*
>
> (ILO, 1981b, pp. 69–70, emphasis added)

It is this kind of complexity that makes the adoption of national policies towards outward investment so problematical. A comprehensive restriction on overseas investment by domestic firms might make a national economy worse, rather than better, off. On the other hand, wholesale outflow of investment must surely be detrimental to home country interests. If there is to be a national policy in this area, therefore, it must be both well informed and selective. It is worth recalling that Japanese policy changed from that of tight restriction on outward investment to one of carefully encouraging certain types of overseas investment as a matter of national policy.

International regulation of transnational corporations[15]

In Chapter 6 we outlined the ways in which individual national governments attempt to influence the behaviour of TNCs. Figure 6.3 summarized the major policy areas. The fundamental problem facing *individual* nation states is that each contains only part of any TNC's operations. This is so whether the TNC is foreign or domestic. Thus, individual attempts to regulate the behaviour of TNCs are heavily circumscribed by the territorial disparity between the nation state and the TNC. At the same time, it is potentially easy for the TNC, particularly the global corporation, to play off one state against another as competition for scarce internationally mobile investment has intensified. In so far as TNCs have the flexibility to substitute one location for another in choosing the site for new or expanded investment it is often very difficult for an individual government to take a rigorous stance towards TNCs. In so doing the government may lose the potential investment to another, neighbouring, country.

The notion that the relationship between TNCs and nation states is unequal and that, in many cases, the balance of power in economic matters may lie with the TNC led to calls for the *international* regulation of transnational corporations, particularly by developing countries. Such exhortations have been amplified by labour unions in the Western economies, deeply concerned about the effects of TNCs on employment – especially on stability of employment – as TNCs reorganize their operations on a global scale. During the 1970s distinct signs of an international dimension began to appear at the governmental level. The various attempts at international regulation reflect the nature of the dominant interests involved. In general, the labour union lobby has joined the developing countries in urging a *mandatory* regulation of certain aspects of TNC activity while the business lobby, often backed by developed country governments, has argued for *voluntary* codes of conduct for TNCs.

The areas of TNC activity which have received most attention in the attempts at international regulation are:

1. the disclosure of information by TNCs;
2. taxation issues, particularly those relating to tax evasion and transfer pricing;
3. employment protection;
4. industrial relations;
5. antitrust (competition) issues.

However, the actual approach adopted between one set of guidelines and codes and another varies considerably. Four major bodies have attempted to develop international regulations for TNCs:

1. *The OECD Guidelines for Multinational Enterprises*, introduced in 1976, are voluntary. They reflect the concern by the 'rich' country governments to regulate possible excesses of economic (and social) behaviour by TNCs whilst, at the same time, demonstrating the beneficial influence of TNCs. The OECD Guidelines cover all five areas listed above.
2. *The International Labour Organization Declaration of Principles Concerning Multinational Enterprises and Social Policy* of 1977 specifically deals with the employment and industrial relations effects of TNCs. Like the OECD Guidelines, the ILO Declaration is voluntary.
3. *The United Nations Code of Conduct for Transnational Corporations*, initially drafted in 1982, is still under negotiation. The major impetus came from the developing countries. In addition to the five areas listed earlier, the UN Code covers technology transfer, ethical practices in international business transactions, the question of political interference by TNCs in national affairs, and respect for national sovereignty. Although the desire of developing country governments and labour unions was to make the UN Code mandatory, it will, in fact, be voluntary.
4. *The European Community approach to regulation of TNCs* is aimed at specific, legally binding measures rather than at a comprehensive code of conduct. EC

regulations on the publication of accounts for the subsidiary companies of TNCs are being considerably strengthened; the national tax authorities have evolved co-operative procedures which will focus, especially, on corporate tax evasion and transfer pricing. The most publicized and contentious area, however, concerned the proposals to compel TNCs above a certain size to make more information available to their employees on their investment plans and to increase worker participation in the decision-making process. This so-called 'Vredeling directive' ran into considerable opposition and has not been implemented.

Despite these various initiatives, relatively little real progress has been made in introducing regulatory codes of conduct for TNCs. Indeed, the focus has shifted: 'the main emphasis in international negotiations is today placed on formulating standards concerning the treatment (or protection) of TNCs rather than standards regarding their conduct' (UNCTC, 1988, p. 360). It is significant that one of the areas of negotiation in the Uruguay Round of the GATT concerned 'trade-related investment measures' (TRIMs) whose aim was to ensure fair treatment for TNCs.

Conclusion

The aim of this chapter has been to explore the major ways in which TNCs may affect national economies and groups such as labour. Potentially, TNCs affect many aspects of economic life through their channelling of investment into new locations and through their specific application of technology in both processes and products. They have enormous implications for a country's trade because so much of the world's trade is now carried on within the boundaries of the TNC. Through their interrelationships with other firms they have the capacity to influence much broader segments of the economy, both directly and indirectly. Their potential impact on employment – not just on the number of jobs but on their type – is likewise immense. But the direction of these effects – whether positive or negative, beneficial or detrimental to national economies and their populations – is not at all easy to determine. It is particularly dangerous to make broad, sweeping generalizations about the effects of TNCs. Almost certainly, the extreme points of view, those which regard the TNC as inherently and unreservedly 'good' or 'bad', are misguided. A single judgement, applicable at all times and in all places, is simply not possible.

Whether we are concerned with the impact of TNCs on host economies or with their repercussions for home countries the crux of the problem is the same. It is a *counterfactual* problem. We cannot unambiguously establish what the situation would be like *if* the TNC were not involved in a particular case. *Assumptions* have to be made as to what the alternative situation would realistically be. The 'bottom line' is the *net effect* which takes into account the opportunities forgone by the presence or absence of the TNC. But the situation has become vastly more complex in today's highly interconnected global economy. From a host economy's viewpoint, could the particular item of technology, the particular level of employment, and so on, be created without the involvement of the TNC? In some cases, the answer will undoubtedly be 'yes': in others, the answer will just as undoubtedly be 'no'. Similarly, from a home economy's viewpoint, what would be the effect if the country's firms were not involved in overseas activities? Can a firm opt out of international operations in today's increasingly global production environment? Would jobs at home be lost anyway even if the firm did not invest overseas? Would the economy be better or worse off? Everything will depend, as we have seen, on whether the overseas investment is obligatory or discretionary, but even here differences of viewpoint will exist. What is quite clear is that *TNCs tie national and local economies more closely into the global economy*.

In this chapter we have concentrated overwhelmingly on the *economic* impact of TNCs. However, they are not only extremely significant transmitters of economic change but also of *social*, *cultural* and *political* change. This is particularly the case for developing countries. A person employed in a TNC factory or office acquires not just work experience but also a new set of attitudes and expectations. The effect of the employment of women is often to transform, not always favourably, family structures and practices. TNCs introduce patterns of consumption which reflect the preferences of industrialized country consumers. In this respect, the transnational advertising agencies are especially important. They 'are not simply trying to sell specific products in the Third World, but are engineering social, political and cultural change in order to ensure a level of consumption that is "the material basis for the promotion of a standardized global culture"' (Sklair, 1991, p. 149). But it is not just TNCs in advertising and the 'cultural' industries which are important channels for the transmission of a global culture; the production of goods is part of the same process because TNC goods embody particular cultural attributes.

TNCs also exert substantial political influence through the sheer scale of their resources as well as their potential geographical flexibility. However, in the opinion of the UNCTC (1988) the 'undue influence on the domestic policies of host developing countries' is less evident today than during the 1960s and 1970s and the 'instances of political interference appear to have declined noticeably'. The fact remains, nevertheless, that the relationship between TNCs and governments, whether of host or home countries, is a delicate one. A clear and unequivocal judgement of the costs and the benefits is not easy to achieve.

Notes for further reading

1. General accounts of the host country impact of TNCs are provided by Hood and Young (1979), Caves (1982), UNCTC (1988), Glickman and Woodward (1989), Graham and Krugman (1989).
2. Glickman and Woodward (1989), chapter 8, provide a detailed account of state and municipality incentive packages in the United States.
3. For a discussion of this issue see Dunning (1986).
4. The debate is well summarized in Reuber (1973), Lall and Streeten (1977), Stewart (1978), ILO (1984a), Kirkpatrick, Lee and Nixson (1984), Lall (1984), UNCTC (1988).
5. Review of empirical studies from a variety of developing countries are povided by UNCTC (1988), Hill (1990).
6. The literature on the environmental dimension of TNC activity is not large although there has been much popular coverage of major disasters such as at the Union Carbide plant in Bhopal, India, in 1984 and at the Hoffman La Roche plant at Seveso, Italy, in 1976. Overall reviews are provided by Pearson (1987), Leonard (1988), UNCTC (1988).
7. The general trade and linkage effects of TNCs in host countries are discussed by Reuber (1973), Lall (1978), ILO (1981a, b), Caves (1982), UNCTC (1983b), Halbach (1988), Hill (1990).
8. Discussion of the possible effects of TNCs on industrial structure and entrepreneurship in

host economies is provided by Reuber (1973), Lall and Streeten (1977), Hood and Young (1979), Lall (1979), Caves (1982), Dunning (1985), Hill (1990).

9. The most comprehensive treatment of employment effects of TNCs can be found in two reports by the ILO (1981a, b), which deal separately with industrialized and developing countries. Other useful sources include Reuber (1973), Enderwick (1979, 1982), Hamill (1984), Kreye, Heinrichs and Fröbel (1988), UNCTC (1988).

10. The particular case of the employment of women by TNCs is discussed by Elson and Pearson (1981, 1989), ILO (1985).

11. The literature on bargaining relationships between TNCs and host economies is fairly limited. Useful contributions are provided by Gabriel (1966), Kobrin (1987), Nixson (1988).

12. Most of the literature relates to the United States. Glickman and Woodward (1989) and Musgrave (1975) examine the impact on the US economy. Hood and Young (1979) provide a useful general summary.

13. Discussion of the employment effects of overseas investment on the home economy are provided by Enderwick (1979, 1982), ILO (1981a, b), Chaudhuri (1983), Gaffikin and Nickson (1984), Glickman and Woodward (1989).

14. The influence of product cycle thinking on views about the employment displacing effects of overseas investment is discussed by Enderwick (1982).

15. The most recent comprehensive treatment of this topic is UNCTC (1988), chapter XX. See also Robinson (1983), McDermott (1989).

Chapter 13

Making a Living in the Global Economy: Problems of Adjusting to Global Shift

Introduction: where will the jobs come from?

In this final chapter our emphasis moves from describing the patterns of global shifts in economic activity and from explaining the underlying causes of those shifts to their *effects* on people. Such effects are manifold. Towards the end of Chapter 12 we referred briefly to the social and cultural effects of the processes of globalization and of transnational corporations. One suggested effect is the emergence of a global culture and the potential demise of national, regional, local and ethnic cultures in the face of the globalization of many aspects of life.[1] The focus of this chapter, however, is primarily on the impact of the internationalization and globalization of economic activity on *employment opportunities*. There is a very good reason for adopting this specific focus.

The key to an individual's or a family's material well-being is *income*. The major source of income (for all but the exceptionally wealthy) is employment, or self-employment. In turn, there is no doubt that the complex, interlocking processes of internationalization and globalization exert a major influence on the structure of employment opportunities across the world. The position adopted in this book is that 'global shift' (used as a shorthand term to encompass the whole set of internationalization and globalization processes) is a *structural* phenomenon: a deep-seated *secular* trend. But the world economy, and its constituent parts, is also affected by *cyclical* forces: the roller-coaster of booms and slumps; prosperity and recession. The peaks and troughs of such business cycles, through their effects on investment (and disinvestment) decisions, tend to amplify structural change. This makes it very difficult to distinguish unambiguously between structural and cyclical forces.

Since the end of the Second World War the world economy has experienced enormous cyclical variation in economic activity: the unparalleled growth of the long boom which lasted from the early 1950s to the mid-1970s; the deep world recession of the second half of the 1970s and the early 1980s; the impressive economic recovery of the later 1980s; the uncertainty of prospects for the 1990s. Nigel Thrift captures the essence of this uncertainty very well:

> The economic crisis of the 1970s has not gone away: rather, in the 1980s it has been brushed under the carpet. Now it shows signs of emerging again but on a larger and much nastier scale. The world economy in the 1980s has consisted of a remarkable juggling act in which three chief sets of actors on the world economic stage – multinational corporations, banks and the governments of countries – have all striven to gain maximum advantage whilst simultaneously promoting economic recovery. This act has produced a world-economic order which is more and more uncertain and yet which is also more and more inured to that

uncertainty. It is a world-economic order which is addicted to the knife-edge. It is a world-economic order hooked on speed.

(Thrift, 1989, p. 16)

Underlying global cyclical trends, therefore, are *global structural changes* associated with the increasing internationalization and globalization of economic activity. The world economic map has become much more complicated than it was only forty or fifty years ago. Although world production, trade and investment are still dominated by the developed market economies, the position of individual industrial nations has changed dramatically. The United States is no longer the clear and undisputed industrial leader as it was in the immediate postwar period. Its hegemonic position was eroded initially by the resurgent European economies (notably West Germany) but more recently, and comprehensively, by the spectacular rise of Japan as a world industrial power. The global economy is now *multi-polar*.

At the same time new centres of production have emerged in what had been, historically, the periphery of the world economy. The emergence of some newly industrializing countries is clear evidence that there has been a substantial geographical shift of economic activity away from the core, even though the extent of the shift is less than popular opinion tends to believe. The complexity of global change has rendered simplistic notions of 'core' and 'periphery' less useful or capable of capturing the nature of today's global economy. The world is more a 'mosaic of unevenness in a continual state of flux' than a simple dichotomous structure of core and periphery.

All parts of the world face major problems of adjustment to these far-reaching changes. But the *nature* of the problem, and certainly its *perception*, varies according to each country's position in the global system. In this respect, the view from the older-industrialized countries (OICs) is very different from the view from the newly industrializing countries (NICs) and different again from the least industrialized countries (LICs). But position in the global economy is only part of the picture. It is far too simplistic to 'read off' a country's or a region's problems (and solutions) solely from its place in a global division of labour. Internal circumstances – cultural, social, political as well as economic – are of great importance. Nevertheless, there are problems which affect OICs, NICs and LICs in different ways as *groups* of countries as they grapple with the repercussions of global economic change and attempt to adjust to its employment impact.

The question 'Where will the jobs come from?' faces all countries, whatever their position in the global economy. We face a truly desperate employment crisis at the global scale. At the end of 1983 there were approximately 35 million unemployed in the OECD countries, a figure unheard of since the 1930s: 'almost 20,000 net jobs had to be created every day over the five-year period from 1984 to 1989 if OECD unemployment was to be cut back to the 1979 level of 19 million' (*OECD Observer*, September 1984, p. 5). In fact, by 1989, the number unemployed in the OECD countries had fallen to almost 25 million, still a very high figure.

Serious as the unemployment position is in the industrialized nations it pales into insignificance compared with the problems of most developing countries, particularly the least industrialized countries. At least in older-industrialized countries the growth of the labour force is now easing. In the low-income countries of Asia, by contrast, 'the labour force is expected to increase by 250 million between 1975 and the end of the century . . . This is about double the 1950–1975 increase and equal to the level of the labour force in 1950' (Renshaw, 1981, p. 57).

The whole question of employment and unemployment – its dimensions, the theories put forward to explain it, the solutions suggested for its alleviation – is far too complex and large a subject to be covered adequately in the final chapter of an already long book. Such a task is not attempted here. The more limited aim is to show how the *processes of global shift* have contributed towards the employment and unemployment problems facing the OICs, the NICs and the LICs.

Problems of the older-industrialized countries: deindustrialization, employment and unemployment

The deindustrialization debate[2]

Some time during the late 1970s a new 'buzz-word' entered the economic and political vocabulary of the older industrialized economies of the West – *deindustrialization*. Academic conferences were held, books and papers were written; talking and writing about the new phenomenon became a growth industry in itself. Politicians began to use the term in their speeches, though in different ways according to their party labels. Not surprisingly, a countermovement subsequently emerged: a plea for the *reindustrialization* of the older industrial economies. This movement was especially prominent in the United States where a special issue of *Business Week* (30 June 1980, p. 56) sounded the clarion call: 'The US economy must undergo a fundamental change if it is to retain a measure of economic viability let alone leadership in the remaining 20 years of this century. The goal must be nothing less than the re-industrialization of America. A conscious effort to rebuild America's productive capacity is the only real alternative to the precipitous loss of competitiveness of the last 15 years.' More recently Cohen and Zysman (1987) have argued forcefully that 'manufacturing matters' and that the US economy (and, by extension, the other industrialized nations) must create anew an internationally competitive manufacturing sector.

Similar exhortations have been heard in other Western economies. Deindustrialization has been interpreted in a number of different ways:

1. as a *declining share of manufacturing* in total output or employment;
2. as an *absolute decline* in manufacturing output or employment;
3. as an *inability to compete internationally* in the production and export of manufactured goods.

Of these interpretations, the first is the least satisfactory. A well-established feature of economic development is the systematic sectoral shift of employment emphasis from the primary through the secondary to the tertiary and quaternary sectors. Some writers equate this shift with the transition to a post-industrial society, others to a self-service economy. What is indisputable is that the service sector has greatly increased in importance in all industrialized economies. But this need not, in itself, signal a problem. More significant are the changes which have been occurring in the manufacturing sector itself. Some of these changes affect virtually all the OICs, others are more serious in particular countries. In the United Kingdom, for example, there has been both an absolute decline in manufacturing employment and also a dramatic decline in international competitiveness. For the first time, the United Kingdom has become a net importer of manufactured goods.

In an international context, therefore, the most satisfactory definition of deindustrialization is (3) above: *the inability of a country to compete internationally in the production and export of manufactured goods, both for domestic and foreign markets.* The causes of deindustrialization in any particular country are extremely complex and are, at least partly, specific to that country. Historical legacy, political, social and economic structures are all important. Undoubtedly, however, a major contributory factor is the *global shift of economic activity*, and this certainly influences the availability of employment opportunities.

Trends in employment and unemployment in the older-industrialized countries[3]

General trends   The general trend since at least the 1960s has been for employment in services to grow more rapidly than employment in manufacturing. Even so, it was only during the 1970s that manufacturing employment actually declined in absolute terms in the major European economies other than the United Kingdom, where manufacturing employment began its steep fall after 1966. In terms of total employment the experience of the United States and of the European economies has been startlingly different. 'Since 1970 more than 30 million new jobs have been created in the United States, whereas there has been hardly any growth in employment in Europe. In the four biggest European economies (whose combined population is slightly more than that of the United States), the number of jobs stagnated between 1970 and 1987' (*OECD Observer*, 1988, p. 9). Virtually all the net new jobs created in the United States were in the services sector: 29 million of the 30 million jobs. Within the European economies, too, it is the services sector which has been the major source of new employment. In view of the static employment picture, however, such service jobs were, in effect, substituting for jobs lost in manufacturing.

Compared with the 1960s and early 1970s, unemployment rates in the industrialized countries have increased dramatically. Figure 13.1 shows the trends for the 1973–89 period. There was a massive increase in unemployment in virtually all the industrialized countries between 1973 and the early 1980s. In 1973, the average for all the OECD countries was 3.3 per cent; in 1983 the average unemployment rate was 8.6 per cent. By 1989 the average rate had fallen back to 6.7 per cent but this was still double the 1973 average. As Figure 13.1 reveals, however, there was a very wide variation in unemployment rates between individual industrialized countries. Only Japan and Sweden of the countries shown had persistently low rates of unemployment. During the second half of the 1980s the position did improve in most cases but the gap between the situation in the early 1970s and late 1980s remained wide in most countries. A particularly serious problem is the persistence of long-term unemployment. Table 13.1 shows that the percentage of the unemployed who had been out of work for more than a full year increased in every case. The problem is especially acute in Europe, where in several countries around half the unemployed had been out of work for more than a year.

The selective nature of unemployment   Although the general level of unemployment, including long-term unemployment, remains very high in most industrialized countries, its actual incidence is extremely uneven. *Job loss is a selective process.* Two particular dimensions stand out in this respect. First, different social groups experience different

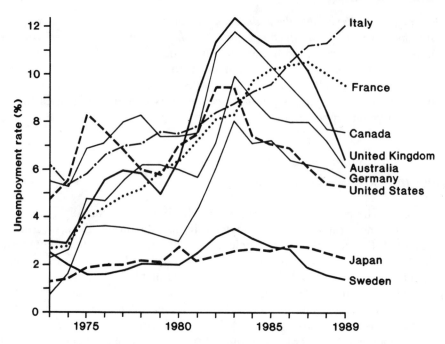

Figure 13.1 Unemployment rates in industrialized countries, 1973–89 (*Source*: OECD (1989) *Quarterly Labour Force Statistics*)

Table 13.1 Long-term unemployment in industrialized countries

| | Percentage of unemployed workers out of work for more than 12 months | | | |
	1973	1979	1983	1986
Belgium	51.0	58.0	62.8	68.9
France	21.6	30.3	42.6	47.8
Germany	8.5	19.9	28.5	32.0
Italy	—	35.8	41.9	56.4
Netherlands	12.8	27.1	43.7	56.3
United Kingdom	26.9	24.5	36.2	41.1
Sweden	—	6.8	10.3	8.0
United States	3.3	4.2	13.3	8.7
Canada	—	3.5	9.5	10.9
Japan	—	16.5	15.5	17.2
Australia	—	18.1	27.5	27.5

— Data unavailable

(*Source*: OECD *Economic Outlook*; OECD *Employment Outlook*, various issues)

levels of unemployment. Second, unemployment tends to be geographically uneven within countries. As far as *social groups* are concerned, 'the persons least exposed to unemployment are men between 25 and 54 years of age with a good education or good training . . . This leaves a large number of people who are more vulnerable to unemployment: women, youth, older workers and minorities . . . Most of these people are unskilled or semi-skilled workers' (ILO, 1984b, p. 42).

The vulnerability of women and young people to unemployment reflects two major

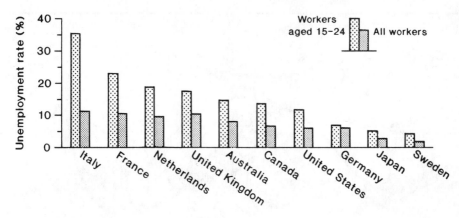

Figure 13.2 Unemployment rates among young workers in industrialized countries, 1987 (*Source*: OECD, *Economic Outlook*, various issues)

features of the labour markets of the OICs. The participation of women in the labour force – particularly married women – has increased dramatically. A large proportion of these are employed as part-time workers in both manufacturing and services, especially the latter. Youth unemployment during the 1980s partly arose from the entry on to the labour market of vast numbers of 1960s 'baby boom' teenagers. In most industrialized countries, therefore, unemployment rates among the young (under twenty-five years) are roughly twice as high as that for the over-twenty-fives. In some cases youth unemployment is three times higher than adult unemployment (Figure 13.2).

Unemployment tends to be especially high among minority groups within the population. In the United States in the 1980s, for example, unemployment among black youths was 150 per cent higher than among white youths. Similarly, unemployment rates among Hispanic youths were at least 50 per cent higher than among white youths. In Western Europe the problem of minority group unemployment reflects the large-scale immigration of labour in the boom years of the 1960s.[4] Relatively easily absorbed – indeed welcomed as 'guestworkers' – in the good times, the migrant workers now face enormous problems both in times of economic recession and also because of longer-term decline in the demand for certain kinds of worker. Figure 13.3 shows the major flows of migrant labour in the world since the 1960s. In continental Europe most of the migrant labour came from the Mediterranean rim – North Africa, Turkey, Greece, Portugal, Spain and southern Italy. In the United Kingdom, where the nature of the immigration was different because of Commonwealth obligations, most migrants came from South Asia (India, Pakistan, Bangladesh) and the Caribbean. In the United States the major sources of new migrants were Mexico and parts of the Caribbean.

As we shall see later, such migration has been of great importance for the countries of origin. For the European host countries, too, the migrants have performed an extremely significant role. In the 1960s there were severe labour shortages as the European economies grew very rapidly. One response was to recruit migrant labour on temporary contracts. The migrant workers were overwhelmingly young, male, unaccompanied – and unskilled. Their numbers grew spectacularly:

> foreign workers in West Germany rose from about 300,000 in 1960 to a peak of about 2.6 million at the beginning of the energy crisis in September 1973 . . . The number then in

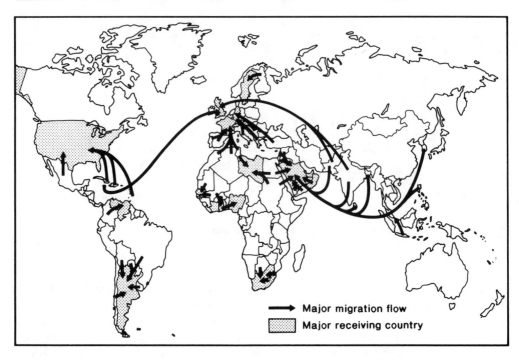

Figure 13.3 Major international labour migration flows (*Source*: Population Information Program (1983), p. M−249)

north-west Europe as a whole (excluding Britain) was about 6 million, and they comprised about 30 per cent of the labour force in Switzerland and Luxembourg, 10 per cent in France and West Germany, 7 per cent in Belgium and 3 per cent in the Netherlands.

(Jones, 1990, p. 246)

Geographical patterns of unemployment Thus, unemployment tends to fall especially heavily on certain sensitive groups within the population of the older industrialized countries. In addition, unemployment tends to be *geographically differentiated*, a reflection of the locational trends in economic activity within individual countries. We have noted examples of such trends at various points in the preceding chapters. In Chapter 7, for example, we discussed the tendency for the specialized units of large multi-plant enterprises, particularly TNCs, to display specific locational preferences and to create a particular map of employment opportunity. In the industry case studies we identified numerous instances of highly localized employment loss associated with the closure or contraction of large plants. Within the OICs, three broad types of geographical change in employment patterns – and, therefore, in unemployment – have been apparent in recent years.[5] First, there have been *broad interregional shifts in employment opportunities*, as exemplified by the relative shift of investment from 'Snowbelt' to 'Sunbelt' in the United States and from north to south within the United Kingdom.

A second locational trend, which may cut across these broad regional shifts, has been the relative decline of the large urban–metropolitan areas as centres of manufacturing activity and the growth of new manufacturing investment in non-metropolitan and rural areas. Such a trend is apparent in both North America and many parts of Western Europe as part of a quite powerful decentralization tendency. Very closely related to this tendency, though predating it, is the decline of employment, especially manufacturing employment, in the *inner cities of the older-industrialized countries*. In virtually every case the inner urban cores have experienced massive employment loss as the focus of economic activity shifted first from central city to suburb and subsequently to less urbanized areas. This inner-city dimension is not confined solely to declining regions though it is more prevalent there. But even cities in the growth regions of the United States Sunbelt or South East England (notably inner London) have suffered. Some of the highest unemployment rates of all, therefore, are to be found in the older inner-city areas.

For particular social groups within the inner cities the rate of unemployment was substantially higher. Among young, minority group members of the population of inner cities unemployment rates of 60 to 70 per cent are not unknown. Thus, the phenomenon of deindustrialization is most dramatically experienced in the older industrial cities of all the developed market economies as well as in those broad regions in which the decline of specific industries has been especially heavy. The physical expression of this deindustrialization is the mile upon mile of industrial wasteland; the human expression is the despair of whole communities, families and individuals whose means of livelihood have disappeared. One outcome of these cataclysmic changes has been the growth of an *informal* or *hidden economy*, a world of interpersonal cash transactions or payments in kind for services rendered, a world much of which borders on the illegal and some of which is transparently criminal.

Japan: an exceptional case? Of the leading industrialized countries only Japan appears not to have experienced the kinds of unemployment problems we have been discussing, or at least not to the same degree. Indeed, while virtually all the other industrialized countries have been suffering from labour surplus, Japan has experienced labour shortages. This reflects Japan's marked success as both a domestic and an export economy. It also reflects some particular characteristics of the Japanese labour market, including the practice of lifetime employment in the larger companies (though not in the myriad of small firms). There is a considerable degree of 'hidden' unemployment in the Japanese economy.

Japan's older industrial regions have certainly been affected by job losses, however. By the end of the 1980s, too, it was being suggested that Japan might well begin to catch the 'Western disease' of higher unemployment in the 1990s. The Japanese unemployment rate, though minuscule compared with that in North America and, especially, Western Europe, is now twice as high as in the early 1970s. Demographic changes of the 1970s – the Japanese 'baby-boom' – will bring more young workers on to the labour market in the 1990s. At the same time, the Japanese economy is becoming increasingly internationalized as more Japanese firms shift some of their operations overseas.

Causal factors: a recapitulation

The profound problem of unemployment which has gripped all the older-industrialized countries (with the notable exception so far of Japan) has no simple explanation (and, therefore, no simple, magic cure). The fact that, within the generally high level of unemployment, there are very substantial differences both between individual countries and also between parts of the same country suggests the operation of both general and specific forces. The most general explanation of an overall high level of unemployment in the OICs between the early 1970s and mid-1980s was the effect of world recession. Recession, whatever its causes, drastically reduces levels of demand for goods and services. Thus, the ILO estimated that the bulk of unemployment in the OICs as a whole is 'demand-deficient' unemployment. But the general force of recession does not explain the spatial variation in unemployment between and within countries. In fact, a whole set of interconnected processes operates simultaneously to produce the changing map of employment and its reverse image, unemployment. In this chapter we are concerned with the structural effect of 'global shift', the processes of internationalization and globalization, on employment. We cannot provide a quantitative assessment of this influence; we can merely suggest the kinds of influence involved. Let us now recapitulate the ways in which processes may help to explain the current situation in the OICs given the specific circumstances (social, cultural, institutional) which exist in individual countries.

Global restructuring   Fundamentally, the highly uneven pattern of employment and unemployment reflects the outcome of a complex industrial restructuring process at the global scale. However, the problems created for particular social groups and particular geographical areas were greatly exacerbated by the deep economic recession and by demographic trends in the size and composition of the labour force. Restructuring of operations by business enterprises, especially the large TNCs, is a continuous process, as we argued in Chapter 7. But its effects are more keenly felt in times of recession than in times of growth. At the same time, recession intensifies the efforts of enterprises to rationalize and restructure their activities to sustain their profitability.

A key feature of the postwar period has been the development and intensification of *global competition* in virtually every industry, both old and new. In such circumstances, business enterprises adopt a variety of strategies to ensure their survival and to grow. Geographical and product diversification have greatly increased as have efforts to cut back production costs through processes of intensification, investment in new, labour-saving technology and rationalization. New, more flexible work practices are being adopted on an extensive scale. Such flexibility takes two major forms:

1. *Numerical flexibility*, whereby firms frequently adjust the number of workers in their labour force to meet changes in market conditions. Often this involves the more extensive use of part-time workers.
2. *Functional flexibility*, in which the kind of work, and the number of tasks, are changed. Rigid divisions between tasks are broken down; a degree of multi-skilling is created.

The transnationalization of production   Most significant, however, is the fact that many of these processes have themselves become *increasingly global* in their extent. Virtually all large firms, and many medium-sized ones, have become *transnational* enterprises. They engage in global, rather than merely domestic, production, whether directly through the establishment of overseas branch plants or indirectly through international alliances and subcontracting. As such, their strategies impinge upon both their home country and also the host countries in which they operate. But the process itself is very complex.

Both the reasons for engaging in international production and also the effects of such operations vary greatly between different types of firm in different industries and according to the characteristics of home and host countries. But there is no doubt that a good deal of the changing level and distribution of employment opportunities in the OICs is associated with the actions of TNCs. As we have seen, the large TNCs in particular have been altering the *relative balance* of their activities between home and overseas locations. Some of this involves the location and relocation of production units in Third World countries but the majority is still within the older-industrialized countries themselves. In general, international production networks have become far more extensive. In one sense, therefore, TNCs can indeed be blamed for some of the loss of manufacturing jobs in their home countries. But as we argued in Chapter 12, the question is very complex. What would have been the realistic alternative to such overseas investment? Could the specific investment have been made at home? In an increasingly global economy is it possible for firms to opt out of international production? Would the absence of overseas investment have increased employment at home or would it have resulted in even greater employment loss?

Sweeping generalizations – pro or con – cannot be made. Nevertheless, where the volume of outward investment is very high in relation to domestic production, there are bound to be adverse effects on the domestic economy and, especially, on employment. The job losses associated with overseas investment have been both geographically and occupationally uneven. The older the plants the more likely they are to be replaced by new plants in different locations. Even relatively new plants may be closed if their efficiency falls below the target set by comparison with other plants in the transnational network. Generally, it has been the basic production activities, especially semi-skilled and unskilled operations, which have shown the greatest propensity to be relocated overseas. Where the displaced plants are either very large individually or where there are substantial localized concentrations the effects are especially severe. Where, as in the case of the automobile industry, linkages with other industries are especially strong the 'knock-on' effects of transnational restructuring can be very severe indeed.

Technological change   Within the general process of international restructuring by TNCs a most important process is that of technological change, both in products and processes. Both have important employment implications. In general, product innovations tend to increase employment opportunities overall as they create new demands. On the other hand, process innovations are generally introduced to reduce production costs and increase productive efficiency. They tend to be labour-saving rather than job-creating. Such process innovations are characteristic of the mature phase of product cycles and became a dominant phenomenon from the late 1960s onwards. The general effect of process innovations, therefore, is to increase labour productivity and to permit the same, or an increased, volume of output from the same, or even a smaller, number of workers. Each of our industry case studies demonstrated this tendency. But, again, the impact of such technological change on jobs tends to be uneven. In most cases, it is the semi- and unskilled workers who are displaced in the largest numbers. It is manual workers rather than professional, technical and supervisory workers whose numbers have been reduced most of all, although the spread of the new information technologies may well change this situation.

There can be little doubt that changes in process technology, in conjunction with the locational shifts in production units overseas, have adversely affected the employment opportunities of less-skilled members of the population. The most vulnerable have been the young, the old and the members of minority groups in the OICs. These two forces – technological change and locational shift – are important basic causes of structural unemployment. They are made more intractable in particular circumstances by other obstacles to adjustment such as institutional rigidities on either the management or labour side (or both) and low levels of occupational and geographical mobility. The particular industrial history of an area is a most important contributory factor, as the varying experience of different nations shows.

There has been much debate about the contribution of technological change to unemployment. Since this is a topic of great significance for future patterns of industrialization we shall return to it again. Against the arguments that technology is a major destroyer of jobs in general, Tobin (1984, p. 83) has argued that 'mass unemployment is not technological in origin. New technologies do, of course, displace particular workers and work hardships upon them, and upon whole industries and regions. But human labour in general has never yet become obsolete.'

The influence of government policies   More significant at the macro-level are the actions of national governments in stimulating or depressing the overall demand for goods and services and, hence, in affecting the level of demand-deficient unemployment. As Tobin (1984, p. 99) states, 'The severe decline in world economic activity since 1979 was the result of restrictive macroeconomic policies deliberately and consciously adopted in almost all major countries. The common goal was to overcome the price acceleration accompanying the second oil crisis.'

One of the most contentious issues in the debate over unemployment and deindustrialization in the OICs is the contribution of *imports of manufactured goods from developing countries*, notably the NICs. The rapid development of manufacturing production in a small number of NICs and their accelerating involvement in world trade has been a major theme of this book. It is one of the most striking aspects of global shift in the world economy. In some cases, notably Singapore and Taiwan, a major driving force has been the involvement of foreign TNCs. In other NICs, such as South Korea and Hong Kong, the direct participation of TNCs has been relatively small. The common element in virtually all the NICs, however, is a very strong *government involvement* in guiding or directing the economy. The adoption and pursuit of vigorous export-oriented industrial strategies were discussed in detail in Chapter 6 and in the industry case studies. By enhancing their initial comparative advantage of a large and cheap labour force through the provision of basic physical infrastructure (including EPZs), by substantial financial and tax incentives and, most of all, by their 'guarantee' of a malleable labour force, the NIC governments created a powerful new force on the world economic scene.

Effects of import penetration from developing countries   The basic question is: how far has the industrialization of these fast-growing economies contributed towards the *de*industrialization of the older industrialized countries? In overall terms, the answer would seem to be 'only to a limited extent'. We saw in Chapter 2 that manufactured imports from developing countries to OECD countries amounted to only 13.4 per cent of total manufactured imports in 1985. The four leading East and South East Asian NICs were responsible for almost three-quarters of these imports. A 1982 UNIDO study asserted that 'for the time being and for the industrialised countries as a group jobs lost due to imports of manufactures from developing countries are more than compensated by jobs created as a result of exports of manufactured goods to the developing countries' (UNIDO, 1982, p. 70). Since 1982 the trade surplus which the industrialized countries had with the NICs has been transformed into a deficit. Most of this is explained by the deterioration in the United States' trade position.

Up-to-date estimates of the employment impact of NIC trade with the industrialized countries are not available. The calculations made for the 1970s give some idea, however, of the difficulties involved in making such estimates and the varying results which may emerge. The UNIDO study cites estimates by Balassa that in 1976 the OECD countries actually gained 1.5 million jobs as a result of their trade in manufactures with developing countries. But for trade with just nine NICs another study cited suggested a net gain of only 81,000 jobs for the developed OECD countries (made up of a gain of 1.34 million jobs through exports to the NICs and loss of 1.26 million jobs from NIC imports). A study of the effects of trade between the EC countries and NICs during the 1970s claims a small negative employment effect but 'the employment forgone owing to the increase in imports between 1970 and 1977 may be estimated to amount to 0.6% of the total employment in the six (EEC) countries together' (Schumacher, 1984, p. 339).

Schumacher goes on to point out, however, that the effects on employment are extremely uneven between different groups of workers and in different industries. Women tend to be affected more than men; lower-paid workers more than higher-paid workers. These differential occupational effects reflect the sectoral bias of NIC imports noted in Chapter 2. Import penetration was highest initially in clothing and textiles, footwear and leather goods, electrical goods, toys and a variety of miscellaneous consumer goods. More recently, import penetration has increased in technologically more sophisticated industries including electrical machinery, telecommunications, data processing, motor vehicles and components.

Since several of the industries most affected by NIC imports are also highly localized within individual OICs, the effects on jobs are highly concentrated geographically, as well as occupationally and socially. Thus, although the direct impact of NIC industrialization on the deindustrialization of the OICs is small so far, it is a far more important factor in certain industries and in specific geographical areas. But even where the direct impact is apparently small the *indirect effects* may be far greater. The threat of such competitive imports has surely been a significant factor in encouraging firms to invest in labour-saving technologies and to rationalize and restructure their operations, both domestically and globally.

For most individual industrialized countries, however, a far greater employment impact occurs through trade with other industrialized countries. This is even true of some sectors in which NIC imports are very significant, notably clothing. A good deal of the employment loss in the UK textiles and clothing industries which could be regarded as being trade-related (as opposed to productivity-related) was the outcome of import penetration from countries such as Italy, West Germany and the United States. Table 13.2 shows some estimates of the relative loss or gain in employment experienced by the six major EC countries as a result of their trade with developed and developing countries respectively.

Each of the six EC countries had a positive employment balance from their trade with developing countries as a whole although all except one (Belgium) had negative employment balances from trade with six South East Asian NICs. But compared with the effects of trade with other industrialized countries the impact was relatively small.

Table 13.2 Effects on employment arising from trade in industrial products: major EC countries, 1977

Trade with industrialized countries

	With Western countries (incl. EC) (thousands)		With Japan (thousands)		With the United States (thousands)	
	Employment generated by exports	Employment forgone due to imports	Employment generated by exports	Employment forgone due to imports	Employment generated by exports	Employment forgone due to imports
West Germany	2,831.2	2,112.2 (+719.0)	42.8	113.8 (−71.0)	273.6	188.8 (+84.8)
France	1,380.0	1,742.7 (−362.7)	17.9	62.8 (−44.9)	116.8	172.4 (−55.6)
Italy	1,599.0	1,067.8 (−531.2)	25.2	32.6 (−7.4)	159.7	111.1 (+48.6)
United Kingdom	2,250.3	2,492.5 (−242.2)	47.3	136.1 (−88.8)	309.9	373.7 (−63.8)
Netherlands	595.4	817.9 (−222.5)	5.0	27.9 (−22.9)	26.9	63.9 (−37.0)
Belgium	699.2	729.1 (−29.9)	4.2	15.2 (−11.0)	40.1	44.9 (−4.8)

Trade with developing countries

	With developing countries (thousands)		With South East Asia* (thousands)	
	Employment generated by exports	Employment forgone due to imports	Employment generated by exports	Employment forgone due to imports
West Germany	956.8	358.0 (+598.8)	54.0	107.3 (−53.3)
France	679.9	233.6 (+446.3)	24.5	39.0 (−14.5)
Italy	685.4	181.1 (+504.3)	21.1	25.0 (−3.9)
United Kingdom	1,124.9	362.7 (+762.2)	98.6	129.0 (−30.4)
Netherlands	124.8	69.1 (+55.7)	6.7	27.0 (−20.3)
Belgium	121.3	62.6 (+58.7)	11.0	10.7 (+0.3)

* Singapore, Taiwan, Philippines, South Korea, Hong Kong, Malaysia

(*Source*: based on D. Schumacher (1982) A comparative analysis of the impact of trade in industrial products on the employment pattern in six EEC countries: report on a research project, in UNIDO (1982), Tables 1–6)

For the EC countries the biggest adverse effect on employment was the result of the very large trade deficit with Japan. There can be little doubt that, in those sectors on which the Japanese have targeted their energies, Japanese import penetration has increased unemployment in EC countries. The situation is much the same in the United States. For four of the six major EC countries overall domestic employment was also adversely affected by the trade balance with the United States.

Alternative forms of adjustment to global shift

Removing obstacles to adjustment   An overriding concern with the problems of adjustment to intensified global competition and global shift has come to affect all the older industrialized countries. The problem is intensified by the speed of adjustment needed to the internationalization of labour markets (Gray, 1987). Domestic macroeconomic policies vary from one country to another. For example, there has been a basic difference of approach between the United States, which has financed economic growth – and employment growth – through running a huge budget deficit, and most European countries, which have pursued strongly deflationary policies. In both cases, however, the focus has been on *removing obstacles* to the efficient production or supply of goods and services in the context of keeping down inflation. In Europe, especially, cutting back on public expenditure has been a major strategy, an approach carried to its greatest extreme in the United Kingdom. In the United Kingdom, too, much emphasis has been placed on removing perceived inefficiencies in the labour market; reducing the power of the trade unions, urging the unemployed to 'price themselves into jobs'.

On the whole, labour is occupationally and geographically less mobile in Europe than in North America, for reasons which lie deep in social and economic history and in cultural attitudes. But there are many real obstacles to the geographical mobility of labour. Not only must deep community and family ties be broken but also the nature of the housing market may make relocation extremely difficult. For example, selling a house in an area of industrial decline may be virtually impossible; getting accommodation at a realistic price in an area of growth may be equally difficult. Rigidities in the public housing sector may also inhibit geographical mobility. The whole question of the geographical mobility of labour in the OICs is more complex than is often supposed, especially by politicians.

Quite apart from efforts by governments to remove what they see as obstacles to economic expansion without increasing inflation, four types of policy have received a great deal of attention in the older industrialized countries:

1. the promotion of small firms;
2. the development of new technologies;
3. the attraction of foreign investment;
4. the protection of certain industries against imports.

Each of these merits a chapter in itself; here our concern is simply to outline very briefly their implications for employment creation.

Promoting small firms   There has been a remarkable swing of the pendulum in attitudes towards firms of different sizes. In the 1960s the key to economic (and employment) growth was seen to be the very large firm and the very large plant. Only in such organizations, it was held, could economies of scale be achieved to enhance competitiveness. Many governments encouraged mergers between enterprises towards this end. By the 1970s, in complete contrast, disillusionment with the large enterprise had set in and the employment panacea was seen to be the small firm with its supposed dynamism and lack of rigidity. This view was strongly reinforced by Birch's (1987) claim that 88 per cent of all net new jobs created in the United States between 1981 and 1985 had been in firms employing fewer than twenty workers.

A number of writers make the connection between the development of flexible production methods, new ways of organizing production (which we discussed in Chapters 4 and 7) and the dynamism of small firms. Such a combination, it is argued, is leading to the development of 'new industrial districts' and 'new industrial spaces'. Thus, 'at the very time when the deindustrialization of core regions was reaching its peak in the late 1970s and early 1980s . . . a number of other areas were experiencing rapid growth of industrial output and employment. This growth . . . has in many cases been posited on the development of alternative ways of organizing production systems and labour markets' (Storper and Scott, 1989, p. 21). However, as was argued in Chapter 7 (pp. 224–5), although such new industrial spaces are, indeed, an important development, their significance has probably been exaggerated and over-generalized on the basis of a small number of cases.

Reservations have also been expressed about the job-creating significance of small firms. It is certainly true that a large number of jobs are created in small firms in most industrialized countries. Researchers such as Storey and Johnson (1987) argue, however, that Birch's estimates for the United States (referred to above) are exaggerated. In their view, the total national impact of jobs created in small firms will continue to be rather modest. It would take many thousands of small firms to replace the jobs lost through the rationalization processes of even a few large firms, let alone to create additional jobs. There is no doubt that small firms are an important source of growth in an economy. But it should not be forgotten that the majority of small firms are far from dynamic, that most depend upon large firms for their markets and that the failure rate of small firms is very high. Thus, the 'small firm fix' is likely to be limited in its impact on unemployment. A policy for small firms should not be at the expense of a policy towards large firms.

Developing new technologies   A similar caution is necessary towards the 'technological fix' often advocated as the salvation of areas in economic distress. Certainly innovation is the life-blood of any economy. Not only must large firms 'innovate or die' but so must entire nations. As we have seen at several points in this book, governments have become universally involved in attempting to stimulate technological developments within their own national territories, either directly or indirectly. But controversy has always raged over the employment implications of introducing new technology.[6] Does new technology create or destroy jobs? If it creates new jobs are they different from the old jobs either in terms of skills required or in their geographical location? Historical experience seems to show that, *in aggregate*, new technologies create more jobs than they destroy, at least over the longer term. This seems to occur because they create new demands for goods and services, many of which could not have been foreseen.

Petrella (1984), however, warns against too ready adoption of what he terms the technological fallacy, that simply by injecting a large dose of technological innovation into the economies of a region or country, one will inevitably recover a high growth rate and, consequently, a fresh growth in employment, increase in purchasing power and new rising standards of living. There are three reasons why this is a fallacy. First, ... the interpretations made of the relationships between technology and employment are vague and insubstantial in terms of theory and analysis. The real world is not quite as simple and linear as many would like it to be. Second – and this applies particularly to Western Europe – more technology does not always mean more growth and more growth does not necessarily mean more employment ... Third, the current technological and industrial changes are taking place in Europe against a background of national 're-industrialisation' processes activated by countries in competition with one another, the result being that even the modest benefits that technology can bring in terms of employment are being eroded and lost.

(Petrella, 1984, p. 353)

As we observed earlier, technological innovations in production processes tend to reduce rather than to increase employment. Thus, the new wave of information technologies may well not create the number of jobs some of its advocates suggest. According to studies in the United States,

'the number of high-tech jobs created over the next decade (1983–1993) will be less than half of the 2 million jobs lost in manufacturing in the past three years ... While high-tech industries will generate 10 times the number of jobs expected from the rest of industry, it will still amount to only 730,000 to 1 million jobs ...' Few jobs will be created in high-tech industries for three reasons: (1) They have a relatively small base ... (2) Their productivity will rise rapidly ... (3) There is a tendency to take the non-skilled jobs abroad, given the constant lure of cheaper foreign labour.

(Petrella, 1984, p. 355)

Hence, much of the growth associated with the new technologies may well be 'jobless growth'. The new information technologies are also having a major effect on many service industries and may well reduce the capacity of the services sector to absorb employees displaced from manufacturing as they have done in the OICs since the 1960s.

What is certain is that new technologies *redefine* the nature of the jobs performed, the skills required and the training and qualifications needed. They alter the balance of the labour force between different types of worker. To some writers the outcome is the *de-skilling* of the labour force but this is by no means a universal outcome. *Reskilling* and *multi-skilling* are also significant outcomes of technological change. The new technologies and the industries based upon them will not, however, inevitably emerge in the same geographical locations as the old industries. The terms 'Sunrise' and 'Sunset' industries themselves imply a geographical distinction (the sun does not rise and set in the same place!).

This geographical dimension of technological change is apparent at both the international and the intranational scales. At the international scale, for example, Western Europe is lagging behind both the United States and Japan in a number of key sectors, not least in the new information technologies. Within Europe, too, there are considerable technological gaps between individual nations, although these tend to vary by industry. In all cases, however, it is clear that much of the new economic activity based on new technologies has a different geographical profile from that of the obsolescent industries which are being displaced (and relocated overseas). As Hall (1985) points out, the 'anatomy of job creation' is rather different from the 'anatomy of job loss'.

Attracting foreign investment   The third kind of adjustment policy pursued by the OICs in an attempt to cope with employment decline has been one of attracting foreign investment. As we saw in Chapter 6, rivalry for the investment favours of TNCs has become intense and often pursued at the highest governmental and state level. Prime ministers and presidents exert their influence either overtly or covertly to persuade the major TNCs to locate new, job-creating investment in their particular countries. Such efforts have been especially notable in industries such as motor vehicles, because of the large scale of the investments, and electronics, because of its high-tech nature. In recent years Japanese firms have become the prime target for both the US and Western European enticement efforts. Whether such a policy is beneficial in the long term is a matter of debate, as we discussed in Chapter 12. From a political point of view, of course, the 'foreign investment fix' has the advantages of having a high profile and being relatively quick. Undoubtedly new jobs are created by inward investment but whether there is a gain in *net* terms depends on its impact on existing firms and on the effect of foreign investment on the country's technological development. In any case, there is a limited amount of internationally mobile investment to go round and since much of it will be in higher-technology industries anyway the number of jobs created may well be far less than in the past.

Protecting domestic industries   Running in parallel with attempts to attract foreign investment have been increasing measures to restrict imports of manufactured goods by most OICs. This fourth type of policy – the 'protectionist fix' – is especially contentious not only in terms of its likely effects on the OICs themselves but also globally because of its implications for the economic well-being of developing countries. The kinds of trade-protection measures which governments can pursue were discussed in some detail in Chapter 6. Trade frictions have undoubtedly increased across the entire spectrum of countries within most of the OICs. 'Managed' trade or 'strategic' trade policies have become more widespread in most of the older-industrialized countries.[7] Quite apart from the frictions between the United States and the EC in certain industries, and even between individual members of the EC, there have been two major targets for this new wave of protectionism. One is Japan, and it is not difficult to see why this should be so given Japan's rapid emergence as a global industrial power. Japan has become the world's leading exporter in certain very important industries. Japan's targeted penetration of the domestic markets of the OICs has led to the negotiation of 'voluntary' agreements to hold down the level of Japanese imports. At the same time, much resentment has built up over the alleged difficulty of foreign firms in selling products within Japan itself even though, in terms of overt trade barriers, Japan is now a relatively open economy. But allegations continue to be made by Western firms and governments of non-tariff and bureaucratic obstacles to entry into the Japanese market.

The question of trade between Japan and the Western industrial nations has become a highly politicized issue. As Table 13.2 shows, the impact of Japanese exports on employment in EC countries lends substance to this concern. What is less understandable, in terms of the magnitudes involved, is the strongly protectionist stance taken towards imports from the NICs. Table 13.2 suggests that, overall, NIC imports have not been a significant direct cause of unemployment in the EC countries. Yet, in many ways, the relationship with the NICs is the crux of the adjustment debate in the OICs.[8] The problem is industry-specific rather than general. How far should industries in which the OICs no longer have a comparative advantage be allowed to run down and become the exclusive preserve of the NICs, whether first or second generation? The 'rational' economic view would regard the answer to such a question as self-evident. The OICs should move out of what are, for them, obsolete activities which can be manufactured more cheaply in developing countries and move to higher-technology products and into the sophisticated service industries. But in the real world such a 'rational' view is more difficult to sustain.

NIC imports have become such a contentious issue for a number of reasons. The most general is the *context* in which they have occurred. The very rapid growth of NIC imports of manufactured goods came at a time when the OICs had begun to experience severe problems of recession. Although the actual level of NIC imports may be relatively low (even in the industries most affected) it is the very rapid *growth* of such imports which causes particular problems of adjustment. It also so happens that the most rapid growth of NIC imports has occurred in industries which are especially sensitive in the OICs.

Protectionist sentiment in the OICs is further reinforced by the very keen price competitiveness of NIC imports. This tends to produce a reaction among both employers and employees in the OICs that the competition is 'unfair'.

> The sense of injustice is based partly on the fact that wages and labour standards are known to be lower in the supplying countries and partly on the belief that low prices are made possible by export subsidies, and is reinforced by the fact that the supplying countries themselves often restrict imports very severely.
>
> (Renshaw, 1981, p. 42)

The basic argument for protection against imports from Japan and the NICs is that it is necessary to give domestic industry time to adjust: it provides a *breathing space*. There can be no doubt of the justification of such temporary measures in those sectors where international competition has intensified very rapidly. But if the breathing space is used, as it should be, to restore an industry's competitiveness or to shift into new activities, it will not necessarily preserve employment. As we have seen, new investment is likely to be labour-saving; new activities may be located in different places. It may also be the case that protection against imports in the sectors most sharply affected by NIC growth will not prevent an inevitable decline in employment in the OICs.

Many would argue that the highly labour-intensive industries are intrinsically at a disadvantage in the high-wage countries. Renshaw summarizes the position in the following way:

> In principle there is a good case for temporary protection, in conjunction with appropriate adjustment assistance policies, as a means of facilitating structural adjustment to foreign competition. In practice, profound conflicts emerge. Protection discourages the necessary contraction of the industry. The extent to which the competitiveness of the industry can be restored by new investment is frequently exaggerated. There is a direct conflict between restoring competitiveness and maintaining employment, and in quantitative terms the importance of competition from imports and as a source of job losses is usually over-estimated.
>
> (Renshaw, 1981, p. 54).

Clearly, the older-industrialized countries continue to face considerable difficulties in adjusting to the intensified competitive environment of a global economy. Preserving, let alone creating, jobs for their active populations has become infinitely more complex in today's highly interconnected world. The imperatives which drive nation states to strive to enhance their international competitiveness will not always be job-creating. In particular, all the industrialized economies face the task of alleviating the adjustment problems of declining industries and declining areas. However, the problems facing the affluent industrialized economies pale into insignificance when set beside the problems facing the world's developing countries.

Problems of the Third World in a global economy

Heterogeneity of the Third World

In large part, though by no means entirely, the economic progress and well-being of Third World countries are linked to what happens in the developed market economies. A continuation of buoyant economic conditions in the industrialized economies, with a general expansion of demand for both primary and manufactured products, would undoubtedly help Third World countries. The impact of the oil crises of 1973 and 1979 was even more serious for the oil-importing developing countries than it was for the advanced industrial nations, even though some of the NICs managed to maintain higher growth rates than the OICs throughout the 1970s and 1980s (see Chapter 2). But the notion that 'a rising tide will lift all boats', while containing some truth, ignores the enormous variations that exist between countries. The shape of the 'economic coastline' is highly irregular; some economies are beached and stranded way above the present water level. For such countries there is no automatic guarantee that a rising tide of economic activity would, on its own, do very much to refloat them.

For the Third World as a whole, the basic problem is one of *poverty* together with a *lack of adequate employment opportunities*. In a 1976 report the ILO summarized the situation in stark terms:

> More than 700 million people live in acute poverty and are destitute. At least 460 million persons were estimated to suffer from a severe degree of protein-energy malnutrition even before the recent food crisis. Scores of millions live constantly under a threat of starvation. Countless millions suffer from debilitating diseases of various sorts and lack access to the most basic medical services. The squalor of urban slums is too well known to need further emphasis. The number of illiterate adults has been estimated to have grown from 700 million in 1962 to 760 million towards 1970. The tragic waste of human resources in the Third World is symbolised by nearly 300 million persons unemployed or underemployed in the mid-1970s.
>
> (ILO, 1976, p. 3)

A decade and a half later, the precise numbers may have changed but the basic dimensions of the problem surely have not.

Apart from the yawning gap between developed and developing countries as a whole, however, there are enormous disparities within the Third World itself. Indeed, it may well be that the term 'Third World' is now redundant[9] because it suggests a uniformity of conditions which do not exist. Krugman (1989), for example, suggests a fourfold division:

1. inward-looking primary exporters;
2. wards of the international community (the countries suffering the deepest poverty);
3. new manufacturing exporters;
4. problem debtors.

From our viewpoint one of the most important differences is in the level of industrialization. As we saw in Chapter 2, only a very small number of developing

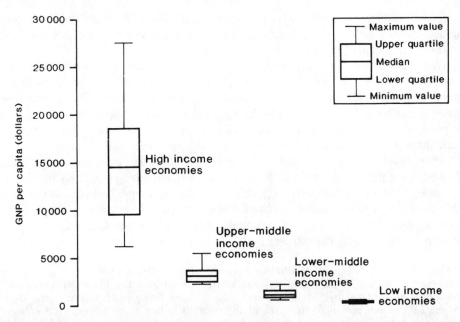

Figure 13.4 Variations in GNP per capita between groups of countries, 1988 (*Source*: World Bank (1990), Table 1)

countries have developed even a modest industrial capacity. A mere ten countries produce more than two-thirds of all manufacturing output in the Third World.

Something of the broader variation between developing countries is shown in Figure 13.4 and Table 13.3. Following the World Bank scheme, developing countries are divided into three categories on the basis of income (GNP per capita). The median income in the thirty-five lowest-income countries in 1988 was a mere $290, approximately one-eleventh that of the upper-middle-income countries and one-fiftieth that of the high-income economies. Within each category, however, there were very large variations in income levels. The thirty-five low-income countries represent the world's hard-core poverty areas. With some exceptions they are also the least-industrialized

Table 13.3 Measures of well-being in groups of developing countries

Country group		Adult literacy rate (%)	Life expectancy at birth (yrs)	Population per physician	Daily per capita calorie supply
Low income (35 countries)	mean	44	60	5,580	2,384
	range	13–88	47–71	1,000–78,970	1,595–2,609
Lower-middle income (34 countries)	mean	27	65	2,520	2,733
	range	6–67	47–72	490–15,580	1,880–3,336
Upper-middle income (14 countries)	mean	24	68	1,220	3,117
	range	4–50	53–77	310–2,790	2,494–3,688
High income (25 countries)	mean	—	76	470	3,376
	range	—	64–78	170–1,310	1,880–3,336

— Data unavailable

(*Source*: based on World Bank (1990), Tables 1, 28)

countries in the world. Conversely, the NICs are predominantly upper-middle-income countries, although both Singapore and Hong Kong, as well as some Middle East oil countries, are included by the World Bank in the high-income group.

Such wide variations in per capita income are associated with a whole variety of social indicators which reflect major disparities in material well-being. Table 13.3 summarizes just some of these and reveals an absolutely clear picture: the lowest-income countries suffer from a multidimensional problem of poverty and deprivation while the upper-middle-income countries are better off in almost every respect. But, again, there are substantial variations within each income group, variations which are far greater than any experienced within the industrial market economies.

Not only are the variations in well-being between developing countries much greater than those between industrialized countries but also variations between *different parts of the same developing country* tend to be much greater. In particular, the differential between urban and rural areas is especially great. Gilbert and Gugler summarize the situation very concisely. On the one hand,

> Urban poverty in the Third World is on a scale quite different to that in the developed countries . . . In the Third World city the relative poverty of the black Baltimore slum-dweller is accentuated by absolute material deprivation. Some poor people in the United States suffer from malnutrition, most of the poor in Indian cities fall into this category. Overcrowded tenement slums and too few jobs are abhorrent, but the lack of fresh water, medical services, drainage, and unemployment compensation adds to this problem in most Third World cities.
>
> (Gilbert and Gugler, 1982, pp. 23, 25)

On the other hand, in the Third World

> Cities are centres of power and privilege . . . Certainly, many urban dwellers live in desperate conditions . . . [but] . . . even those in the poorest trades reported that they were better off than they had been in the rural areas . . . The urban areas, and especially the major cities, invariably offer more and better facilities than their rural hinterlands.
>
> (op. cit., pp. 50, 52)

Development or dependency? An outline of conflicting viewpoints

The starkness of the facts of the global development gap and the nature of the conditions in most Third World countries leave little room for dispute regarding the extent of the problem facing many countries. Where controversy does exist is over the *causes* of underdevelopment (and, therefore, in the policies advocated to deal with it). Expressed in the baldest of terms, the debate has tended to polarize around two opposed schools of thought. One is that the development process is one of diffusion or spread over time from developed to underdeveloped countries. The counterposed view is that the development of 'peripheral' countries is virtually impossible in a capitalist world economy; that the very nature of the world economic system is one in which the core exploits the periphery and inhibits its development. In fact, of course, the debate is far more complex than this.[10]

The notion that development is a more or less linear or stage-like progression was expressed most emphatically in Rostow's *The Stages of Economic Growth* (1960). According to Rostow (1960, p. 4), 'It is possible to identify all societies, in their economic dimensions, as lying within one of five categories: the traditional society, the preconditions for take-off, the take-off, the drive to maturity, and the age of high mass-consumption.' Progress through the sequence was largely a matter of overcoming various obstacles to development. The critical stage was the 'take-off' period in which, in particular, productive investment increased beyond a critical level, leading economic sectors (usually manufacturing) developed, and the institutional structure became amenable to economic growth. Once these conditions were achieved 'take-off into sustained growth' would occur.

Rostow's grand design achieved wide currency. It formed a central pillar in the Western-based view of *development as modernization*, in which the so-called traditional economies of the Third World would be transformed into modern versions of the Western model. Whether such a development path was appropriate – or desirable – was rarely questioned by its advocates. Whether such a process was likely to occur in reality was also little questioned:

> Modernisation theory . . . had as its central essence a belief in the inevitable development of the Third World . . . The well-worn metaphor was of course that of developing countries sitting as aeroplanes at the end of a runway, about to take off into a process of self-sustained growth and fuelled by the diffusion of knowledge, capital, and culture from the developed world. The failure of most states to taxi to the runway, let alone get airborne, saw the onset of a bout of pessimism in which the central tenets of modernisation theory came under challenge.
>
> (Higgott, 1984, p. 59)

A major weakness of the stages theory of growth was that it

> failed to take into account the crucial fact that contemporary Third World nations are part of a highly integrated and complex international system in which even the best and most intelligent development strategies can sometimes be nullified by external forces beyond their control . . . One simply cannot claim, as many economists did in the 1950s and 60s, that development is simply a matter of 'removing obstacles' and supplying various 'missing components' like capital, foreign exchange, skills and management.
>
> (Todaro, 1989, p. 61)

One major reaction was to argue that Third World development was impossible in the prevailing economic system. This was the essence of the body of *dependency theory*. According to Frank, who coined the phrase 'the development of underdevelopment',

> Economic development and underdevelopment are the opposite faces of the same coin. Both are the necessary result and contemporary manifestation of internal contradictions in the world capitalist system. Economic development and underdevelopment are not just relative and quantitative, in that one represents more economic development than the other; economic development and underdevelopment are relational and qualitative in that each is structurally different from, yet caused by its relation with, the other. Yet development and underdevelopment are the same in that they are the product of a single, but dialectically contradictory, economic structure and process of capitalism . . . One and the same historical process of the expansion of capitalism throughout the world has simultaneously generated – continues to generate – both economic development and structural underdevelopment.
>
> (Frank, quoted in Brookfield, 1975, p. 163)

In its most dogmatic form, dependency theory does not allow for the industrial development of peripheral countries within a capitalist world economy. They are locked into a stagnationist position as suppliers of primary products for the core economies. The problem was that, as in the case of the stages of growth theory, the real world did not fit the theory too well. In particular, a number of Third World countries – the NICs – emerged as significant industrial forces at the global scale during the 1960s and 1970s. The general claim that capitalism could not generate industrial development in the peripheral dependent economies was shown to be invalid. Not surprisingly, there have been many different interpretations of these 'anomalous' or 'deviant' cases.[11]

A major weakness of dependency theory was its virtual disregard of the *internal* conditions in different countries. The emphasis was on *external* factors in development with little or no room for the operation of specific historical, cultural and political forces within countries. Yet when one looks at the NICs in detail it is clear that specific internal circumstances have indeed been extremely important in the development process. It is more realistic, therefore, to regard the development process as an interplay between external and internal forces. There is no doubt that the state of the world economy in the 1960s and early 1970s was a crucial element in the development of the NICs. The massive expansion of world trade, the rapid growth of TNCs, the developments in the enabling technologies were, as we have seen, of vital importance. But on their own these do not explain particular NIC development. It was the way in which these various forces coincided with internal conditions in particular countries which explains why certain countries industrialized and others did not.

Studies of industrial development in South Korea and Taiwan, for example, both demonstrate the importance of 'concrete' experiences: the role of particular historical, cultural and political factors. Both countries had approximately fifty years as Japanese colonies, during which time the imperial power built up the physical infrastructure to facilitate the countries' role as suppliers of primary products to Japan. Much of this infrastructure remained after the Japanese left in 1945. During the colonial period a substantial educational system had been developed in both rural and urban areas. After the war both countries received massive aid from the United States to act as bulwarks against communism. Both had experienced substantial change in their agricultural and land tenure system which reduced the power of an entrenched landholding class. In both countries there is a strong sense of national identity and purpose wedded to a strong centralized government. (A feature of all the NICs, as we saw in Chapter 6, is a very strong state involvement in economic affairs.)

Thus, the nature of the international economic environment, together with the particular conditions existing in individual countries, explains the progress of industrialization in specific Third World countries. In some cases, the local circumstances include a natural resource endowment although in cases such as Singapore and Hong Kong this is clearly not so. Industrialization of these two states owes much to their particular historical evolution as ports and commercial centres under British rule together with their geographical position and their cultural and political make-up. Hence, although there are undoubtedly degrees of dependence within the world economy the actual situation is not as simple and deterministic as much of dependency theory has claimed.

Equally, however, the nature of the interconnections within the world economy *has* tended to favour the core industrial nations at the expense of the rest. In many cases, this asymmetrical relationship has indeed retarded development but the experience of the NICs shows that this is not an inevitable consequence. *The crucial point is the particular combination of external and internal forces.* Recognition of this interplay is vital not just to understand what has happened in the past but also in terms of likely future developments. Extrapolating the experience of the NICs to other less industrialized countries without an appreciation of the specific nature and importance of

Table 13.4 Structure of the labour force in developing countries, 1960 and 1980*

Country group		Percentage of the labour force in					
		Agriculture		Industry[†]		Services	
		1960	1980	1960	1980	1960	1980
Low income	mean	77	70	9	15	14	15
	range	54–95	46–93	1–18	2–20	3–37	5–38
Lower-middle income	mean	71	55	11	17	18	28
	range	39–93	20–87	2–26	4–33	5–39	9–60
Upper-middle income	mean	49	30	20	28	31	42
	range	8–67	2–42	9–52	16–57	19–69	32–62
Industrial market economies	mean	18	6	38	38	44	56
	range	4–42	2–18	25–50	29–46	27–57	44–66

*Latest available data
[†] Includes, mining, manufacturing, construction, electricity, water, gas

(*Source*: based on World Bank (1983) *World Development Report, 1983*, New York: Oxford University Press, Table 21)

domestic circumstances is extremely misleading. There is an inevitability neither about industrial development nor about its absence in today's highly interdependent world.

Employment, unemployment and underemployment in the Third World[12]

The basic problem: labour force growth outstrips the growth of jobs Although the employment structure of Third World countries has undergone marked change (Table 13.4) the fact remains that most developing countries are predominantly agricultural economies. As Table 13.4 shows, an average of 70 per cent of the labour force in the lowest-income countries was employed in agriculture in 1980 (in some countries the figure was above 90 per cent) compared with only 6 per cent in the industrial market economies and a mere 2 per cent in the United Kingdom. Even in the upper-middle-income group (in which most industrial development has occurred) agriculture employed almost one-third of the labour force. In each category the relative import-ance of agriculture has declined even though in absolute terms the numbers employed in agriculture continued to grow. The balance of employment has shifted towards the other sectors in the economy: industry and services.

These broad sectoral changes in employment in Third World countries have to be seen within the broader context of growth in the size of the labour force. The contrast with the experience of the industrialized countries in the nineteenth century is espec-ially sharp. During that earlier period the European labour force increased by less than 1 per cent per year on average; in today's developing countries the labour force is growing at more than 2 per cent every year. Thus, the labour force in the Third World doubles roughly every thirty years compared with the ninety years taken in the nine-teenth century for the European labour force to double. Hence, it is very much more difficult to absorb the exceptionally rapid growth of the labour force into the economy. The problem is not likely to ease in the near future because 'labour force growth is determined mainly by past population growth with a lag of about 15 years. Con-sequently, the high, and in some countries increasing rates of population growth of the late 1960s and 1970s will not be reflected in labour force growth rates until the 1980s and 1990s' (World Bank, 1979, p. 48).

There is, therefore, an enormous difference in labour force growth between the older industrialized countries on the one hand and the developing countries on the other. But the scale of the problem also differs markedly between different parts of the Third World itself. By far the greatest problem exists in low-income Asian countries, where the projected increase of 250 million in the labour force between 1975 and 2000 is twice that of the region's labour force growth rate between 1950 and 1975. Of course, pressure on the labour market is lessened where lower population growth rates occur.

The basic dilemma facing most Third World countries, therefore, is that the growth of the labour force vastly exceeds the growth in the number of employment opportunities available. Harris and Rashid's (1986, p. 286) study of employment trends in developing countries claimed that the sectoral and occupational distribution of the labour force showed signs of improvement during the 1970s. However, 'the rate of growth of the labour force outstripped the rate of creation of jobs . . . Employment conditions for the employed improved during the 1970s but the numbers of unemployed rose sharply . . . The increasing numbers entering the labour force – from the larger population, from agriculture, and through the school system – mean that the unemployed are going to spend longer periods without work.'

The Third World labour market: formal and informal sectors   It is extremely difficult to quantify the actual size of the unemployment problem in Third World countries. There are three main reasons for this. One is the simple lack of accurate statistics. A second is the nature of the unemployment itself, which tends to be somewhat different from that in the developed economies. A third reason is the structure of most Third World economies, particularly their division into two distinctive, though closely linked, sectors: *formal* and *informal*.[13] Published figures in Third World countries tend to show a very low level of unemployment, in some cases lower than those recorded in the industrial countries. But the two sets of figures are not comparable. Unemployment in developing countries is not the same as unemployment in industrial economies. To understand this we need to appreciate the nature of the labour market in Third World countries.

The *formal* sector is the sector in which employment is in the form of wage labour, where jobs are (relatively) secure and hours and conditions of work clearly established. It is the kind of employment which characterizes the majority of the workforce in the developed market economies. But in most Third World countries the formal sector is not the dominant employer, even though it is the sector in which the modern forms of industry are found. As Stewart points out, the formal sector employs a minority of the labour force:

> many more work in traditional modes, working directly for themselves in subsistence activities, self-employed on marketed activities, and working for others on a traditional rather than a wage-employed basis, such as share cropping. Traditional forms of activity are not confined to agriculture; traditional crafts are practised on this basis. Services too . . . are also largely outside the wage employment nexus. With growing urbanization, following industrialization, new forms of urban activity have developed – the so-called 'informal' sector – in which people perform a variety of functions, manufacturing, services and construction for themselves and for the modern sector in and around rapidly growing towns.
>
> (Stewart, 1978, p. 34)

This *informal* sector, which encompasses both legal and illegal activities, is not totally separate from the formal sector: the two are interrelated in a variety of complex ways. The informal sector is especially important in urban areas; some estimates suggest that between 40 and 70 per cent of the urban labour force may work in this sector (Gilbert and Gugler, 1982). But measuring its size accurately is virtually impossible. By its very nature, the informal sector is a floating, kaleidoscopic phenomenon, continually changing in response to shifting circumstances and opportunities.

In a situation where only a minority of the population of working age are 'employed' in the sense of working for wages or salaries, defining unemployment is a very different issue from that in the developed economies. In fact, the major problem in Third World countries is *underemployment*,

> a state of low labour productivity, sporadic employment and depressed earnings... the underemployed are usually understood to comprise most of the rural landless, many small farmers, many of the urban self-employed and most employees of small-scale urban enterprises. While some are not fully employed because of such factors as seasonal variations in the demand for agricultural labour, others work long hours throughout the year but earn very little from their low productivity jobs. Their common characteristic – low income – identifies them as the core of the poverty problem... The poor... cannot afford to be unemployed; they are obliged to accept underemployment.
>
> (World Bank, 1979, pp. 46–7)

The urban–rural dimension Underemployment and a general lack of employment opportunities are widespread in both rural and urban areas in Third World countries. There is a massive underemployment and poverty crisis in rural areas arising from the inability of the agricultural sector to provide an adequate livelihood for the rapidly growing population and from the very limited development of the formal sector in rural areas. Some industrial development has occurred in rural areas, notably in those countries with a well-developed transport network. Mostly this is subcontracting work to small workshops and households in industries such as garment manufacture. But the bulk of the modern industries are overwhelmingly concentrated in the major cities. It is in the big cities that the locational needs of manufacturing firms are most easily satisfied. Yet despite the considerable growth of manufacturing and service industries in the cities the supply of jobs in no way keeps pace with the growth of the urban labour force. Not only is natural population increase very high in many Third World cities but also migration from rural areas has reached gigantic dimensions. The pull of the city for rural dwellers is directly related to the fact that urban employment opportunities, scarce as they are, are much greater than those in rural areas.

In complete contrast to the OICs, therefore, where a growing *counter*-urbanization trend was evident for some years, urban growth in most Third World countries has continued to *accelerate*.[14] The highest rates of urban growth are now in developing countries where the number of very large cities has increased enormously. The sprawling shanty towns of the Third World city are the physical expression of this explosive growth. In the OICs, as we have seen, most industrial growth now occurs away from the major urban centres; in the Third World the reverse is the case. Virtually all industrial growth is in the big cities. More generally,

> the features of contemporary urbanisation in developing countries differ markedly from those of historical experience. Whereas urbanisation in the industrialised countries took many decades, permitting a gradual emergence of economic, social and political institutions to deal with the problems of transformation, the process in developing countries is occurring

far more rapidly, against a background of higher population growth, lower incomes, and fewer opportunities for international migration. The transformation involves enormous numbers of people: between 1950 and 1975, the urban areas of developing countries absorbed some 400 million people; between 1975 and 2000, the increase will be close to one billion people . . . The rate of urban population growth in these countries is likely to decline after 1975, but it is expected to remain three to four times as high as the urban growth rates of the industrialised countries in this period.

(World Bank, 1979, p. 72)

Labour migration as a 'solution'   Despite its considerable growth in at least some Third World countries, manufacturing industry has made barely a dent in the unemployment and underemployment problem of most developing countries. Only in the very small NICs, such as Hong Kong and Singapore – essentially city states with a minuscule agricultural population – has manufacturing growth absorbed large numbers of people. Indeed, Singapore has experienced a labour shortage and has had to resort to controlled in-migration from time to time. In all other cases, however, the problem is not so much that large numbers of people have not been absorbed into employment – they have – but that the *rate of absorption* cannot keep pace with the growth of the labour force.

One commonly adopted solution has been to 'export' labour to foreign countries. As Figure 13.3 showed, there have been massive flows of international labour migration in the past three decades, encouraged by many Third World governments. But the effects of such out-migration on the countries of origin are not necessarily all beneficial. It is certainly true that out-migration helps to reduce pressures in local labour markets. It is also true that the remittances sent home by migrant workers make a very important contribution to the home country's balance of payments position and to its foreign exchange situation (as well as to the individual recipients and their local communities). Indeed, in many cases the value of foreign remittances is equivalent to a very large share of the country's export earnings (Table 13.5). Other supposed benefits of out-migration include the learning of skills which, when the migrant returns, will help to upgrade the home country's economic and technological base.

The other side of the coin is less attractive for the labour-exporting countries. The

Table 13.5 Remittances of migrant workers as a percentage of export earnings

Home country	Percent of home country's earnings from merchandise exports	
	1981	1988
Pakistan	71.4	46.3
Portugal	69.8	33.1
Egypt	67.5	75.3
Turkey	53.2	15.1
Bangladesh	48.8	60.0
Morocco	45.2	36.0
Yugoslavia	37.1	38.3
Greece	27.4	31.0
Sri Lanka	22.2	24.3

(*Source*: World Bank (1983), *World Development Report, 1983*, Tables 9, 14; World Bank (1990), Tables 14, 18)

migrants are often the young and most active members of the population. Further, as Jones points out:

> growing familiarization with foreign consumption styles leads to disdain for domestic products and a growing dependence on expensive foreign imports . . . returning migrants are rarely bearers of initiative and generators of employment. Only a small number acquire appropriate vocational training – most are trapped in dead-end jobs – and their prime interest on return is to enhance their social status. This they attempt to achieve by disdaining manual employment, by early retirement, by the construction of a new house, by the purchase of land, a car and other consumer durables, or by taking over a small service establishment like a bar or taxi business; there is also a tendency for formerly rural dwellers to settle in urban centres. There is thus a reinforcement of the very conditions that promoted emigration in the first place. It is ironic that those migrants who are potentially most valuable for stimulating development in their home area – the minority who have acquired valuable skills abroad – are the very ones who, because of successful adaptation abroad, are least likely to return. There are also problems of demographic imbalance stemming from the selective nature of emigration. Many villages in Southern Europe have been denuded of young men, with consequences not only for family formation and maintenance but also for agricultural production.
>
> (Jones, 1990, p. 250)

There is no question, therefore, that the magnitude of the employment and unemployment problem in developing countries is infinitely greater than that facing the older-industrialized countries, serious as their problem undoubtedly is. The biggest problem in Third World countries is *underemployment* and its associated *poverty*. The high rate of labour force growth in many developing countries continues to exert enormous pressures on the labour markets of both rural and urban areas. Such pressures are unlikely to be alleviated very much by the development of manufacturing industry alone. With one or two exceptions among the NICs, industrial growth has done little to reduce the severe problems of unemployment and underemployment – with their resulting poverty – in Third World countries.

Sustaining growth and ensuring equity in the newly industrializing countries

The spectacular industrial growth of a group of Third World countries – the NICs – has been one of the most significant developments in the world economy in recent years. Indeed, some would argue that the four leading Asian NICs should no longer be regarded as Third World or developing countries at all. Certainly, there can be no doubting the remarkable industrial progress of this group of countries although these 'industrial miracles' are not without their serious internal difficulties. Three kinds of problem can be identified here:

1. sustaining economic growth;
2. ensuring that such growth is achieved with equity for the countries' own people; and
3. the problem of foreign debt which faces some, though not all, newly industrializing countries.

Sustaining economic growth

Economic growth in the NICs has been based primarily upon an aggressive *export-oriented* strategy. This depends fundamentally on the continued growth and openness of overseas markets, especially in the industrialized countries. During the 1960s the conditions were indeed favourable; in the 1990s the omens are far less propitious as the OICs have reduced their demands for NIC exports, partly through the deliberate operation of protectionist trade measures. From the NICs' viewpoint, therefore, the macroeconomic expansion of the industrialized economies is vital. But this will be effective only if trade barriers are also removed or at least lowered. The present political climate in the OICs makes both possibilities somewhat remote.

At present there are growing *trade frictions*, particularly between the United States, their biggest export market, and the leading Asian NICs – especially South Korea and Taiwan, which have a substantial trade surplus. They are also regarded by the United States as countries that are less open to industrialized country imports than they might be. Consequently, the NICs will need to develop further their own domestic market although if this were to be done by raising trade barriers which are already high, the likely result would be to make the OICs even more reluctant to modify their own protectionist stance. There is also the problem that the smaller NICs have very limited domestic markets; this was the major reason for adopting an export-oriented strategy in the first place. Expansion of domestic demand in some NICs is also constrained by the immense debt burden which a number of them now carry (see below, pp. 452–454).

A second problem facing the leading NICs in sustaining economic growth arises from the *growing competition from other developing countries* – the 'proto-NICs'. In Chapter 2 we focused mainly on the leading NICs, but there is a further group of potential NICs which have also been growing substantially as manufacturing centres. This 'next tier' includes such countries as Malaysia, the Philippines, Thailand, Pakistan, Indonesia, Colombia, Chile, Peru, Egypt, Turkey and, potentially the most serious competitor of all, the People's Republic of China. Competition from these lower-wage countries has intensified as labour costs in the first tier of NICs have risen. The competition is obviously most severe in the lower-skill, labour-intensive activities on which NIC industrialization was originally based.

Thus, the development of competition from other developing countries, together with trends in the automation of some labour-intensive processes in the industrialized countries, has added to the pressures on the leading NICs to shift to more skill-intensive and capital-intensive products and processes. The southern European NICs – Spain, Portugal and Greece – face an additional complication. Their new status as full members of the EC considerably alters their development prospects. On the one hand, their products will eventually have unfettered access to the entire EC market; on the other hand, their own economies will become open to the full force of competition from the industrialized economies of the EC. They also now face increasing competition from the opening up of Eastern Europe (as, of course, do the Third World NICs in general).

Whether or not a second, third or fourth division of NICs really will emerge to threaten the 'super league' is, however, a matter of considerable argument. Cline (1982) puts forward the 'fallacy of composition' argument that what is possible for a small number of cases is not possible for all, or even the large majority of cases. He argues that if the East Asian model of export-led development were to become characteristic of all developing countries 'it would result in untenable market penetration into industrialized countries' (Cline, 1982, p. 88). Conversely, the World Bank argues against this viewpoint:

First, the capacity of industrial nations to absorb new imports may be greater than supposed . . . Second, the idea that a large number of economies might suddenly achieve export-to-GDP ratios for manufactures like Hong Kong, Korea or Singapore is highly implausible . . . Third, export-oriented countries would produce different products, and intra-industry trade is likely to be important. Finally, the first wave of newly industrializing countries is already providing markets for the labour-intensive products of the countries that are following.

(World Bank, 1987, p. 81)

Ensuring economic growth with equity

Sustaining economic growth is only one of the difficulties facing the NICs (both existing and potential) in today's less favourable global environment. Sustaining growth *with equity* for the populations of the NICs themselves is also a major problem. A widely voiced criticism of industrialization in developing countries has been that its benefits have not been widely diffused to the majority of the population. There is indeed evidence of highly uneven income distribution within many developing countries, as Table 13.6 reveals. In countries such as Brazil, Mexico, Turkey and Malaysia, for example, the share of total household income owned by the top 10 and 20 per cent of households was very much higher than that in the industrial market economies.

This pattern does not apply in all cases. South Korea, for example, has a household income distribution very similar to that of the industrial market economies (though at much lower levels, of course):

South Korean real per capita incomes have more than tripled since the early 1960s and real incomes have been increasing in both the agricultural and industrial sectors . . . In fact, a number of quality of life indicators show substantial improvement . . . including health, housing, employment, and education indicating that the well-being of South Korea has been improving . . . Income distribution in South Korea does not bear out the pattern of middle and upper class privilege so typical in the Third World . . . On the contrary, South Korea's

Table 13.6 Distribution of income by household group within some developing countries

Country (year)	Percentage share of household income in:		
	Lowest 20% of households	Highest 20% of households	Highest 10% of households
Brazil (1983)	2.4	62.6	46.2
Turkey (1973)	3.5	56.5	40.7
Malaysia (1987)	4.6	51.2	34.8
Mexico (1977)	2.9	57.7	40.6
Philippines (1985)	5.5	48.0	32.1
South Korea (1976)	5.7	45.3	27.5
Hong Kong (1980)	5.4	47.0	31.3
India (1983)	8.1	41.4	26.7
Singapore (1982–3)	5.1	48.9	33.5
Average for the industrial market economies	6.3	39.7	24.0

Note: The figures are estimates from a variety of sources and for different dates. They should be treated with care.

(*Source*: based on World Bank (1990), Table 30)

distribution of income is more equal than most Third World countries and is comparable to the distribution in the United States . . . This distribution pattern is the result of asset levelling that was brought about by land reform in the late 1940s and the destruction of physical capital during the Korean War.

(Barone, 1983, p. 46)

A broadly similar conclusion regarding income distribution in the case of Taiwan is suggested by Barrett and Whyte (1981). Although the contrast between the Asian NICs and those in Latin America is very marked, however, this does not necessarily prove that there is a direct relationship between export-oriented industrialization and improvements in the distribution of income. Kirkpatrick (1990) shows that, of the four leading Asian NICs, only Taiwan experienced a clear improvement in income distribution over the period from 1965 to 1983.

Income distribution is one aspect of the 'growth with equity' question. Another is the broader social and political issue of democratic institutions, civil rights and labour freedom. Although the degree of repression and centralized control in these countries may sometimes be exaggerated, the fact is that such conditions do exist in a number of cases. The very strong state involvement in economic management in most NICs has brought with it often draconian measures to control the labour force. Labour laws tend to be extremely stringent and restrictive; in many instances strikes are banned. In certain cases the regime is one of tight military rule. How far the success of NICs – particularly in attracting foreign investment – really depends on the use of strongly authoritarian measures is difficult to ascertain. But until such repressive behaviour is relaxed, the achievements of some of the NICs in the strictly economic sphere must be regarded with some reservations. In this respect, it is significant that in the late 1980s both South Korea and Taiwan began to move along the democratization path. As the persistence of labour disturbances in South Korea shows, however, the transition is not proving to be easy.

The debt problem

A third serious problem for the NICs in the 1990s is the burden of *financial debt* which many have incurred, particularly since the late 1970s (Figure 13.5).[15] Much of the industrial growth of the NICs has been financed by overseas borrowing (as was that of the industrializing countries in the nineteenth century). Before the 'second oil shock' of 1979 there were no major problems. Exports from NICs continued to hold up well despite the onset of recession in the OICs. Capital was needed for investment in the NICs; after the 1973 oil crisis the huge volume of petrodollars had to be *recycled* by the commercial banks to prevent the world financial system from seizing up. The banks were only too ready to lend to the more successful Third World countries in order to achieve this.

The debt crisis broke with great suddenness in the early 1980s when the first, and most spectacular, incident was the financial collapse of Mexico in August 1982. It soon became clear that a number of Third World countries were in deep financial difficulty. The problem was particularly concentrated in the middle-income group of developing countries which had come to depend most heavily on commercial lending. The fact that the low-income countries were not involved in no way indicates their lack of financial difficulty – on the contrary. But commercial banks have generally been unwilling to

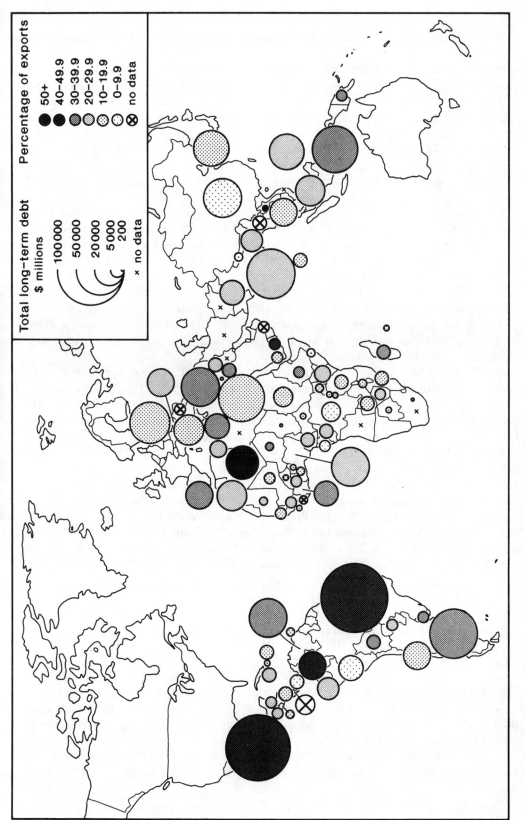

Figure 13.5 The map of developing country debt (*Source*: World Bank (1990), Table 23)

lend to the very poor countries so that most of their borrowing is in the public sector aid programmes.

In the early 1980s, therefore, the problem of recycling funds was suddenly displaced by that of *rescheduling* the massive debts of some Third World countries. An increasing number of countries now find themselves unable to repay the interest on the sums borrowed let alone reduce the basic sum. Some are having difficulty paying the interest on the interest as they have had to borrow more simply to avoid going under completely. The emergence of this very serious problem reflects a whole host of factors. Some are internal to the countries involved: undoubtedly there has been some profligate spending on unnecessary prestige projects. But this is not the fundamental cause.

The most important factors relate to developments within the global economy since the late 1970s. As we have seen, demand for manufactured goods and materials in the OICs declined as recession deepened and as protectionist measures intensified. The market for NIC exports slackened very substantially. At the same time, the governments of the industrialized countries began to pursue very tight financial and fiscal policies which, while reducing price inflation in their own economies, forced up interest rates. In particular, the need for the United States to finance its enormous budget deficit has kept interest rates especially high. Thus, floating interest-rate debts now cost far more to service than when they were initially incurred. New loans taken out became extremely expensive.

The largest debt problems are in Latin America. Brazil, Mexico and Argentina in particular have been continuously in the financial headlines since 1982 as each country seemed to be on the brink of financial collapse. Each of the major debtor nations has been forced to seek the co-operation of the international financial institutions – the IMF and the major commercial banks – in rescheduling their debts. So far, total breakdown has been averted but the very severe conditions the IMF places on the debtor countries as the price of rescue create serious social and political problems for the countries themselves. Whether the lid can be kept on such a volatile pot is by no means certain. As yet, the East and South East Asian countries have not experienced such problems, although in 1983 the Philippines was hovering on the brink of financial chaos. But the relative buoyancy of the Asian NICs has helped to keep them out of the grip of the moneylenders. Thus, although South Korea has borrowed very heavily, its continuing high-level export performance has enabled it to service the debt.

Ensuring survival and reducing poverty in the least industrialized countries

Although the NICs certainly face problems, they are not of the same magnitude or seriousness as those facing the least industrialized, low-income countries. As Figure 13.4 and Table 13.3 demonstrated, the poorest thirty or so Third World countries are poor not just in terms of income but also in virtually every other aspect of material well-being. They are the countries of the deepest poverty, several of which face mass starvation. For large numbers of people in the low-income countries (and in some of the higher-income countries, too) life is of the lowest material quality.

The causes of low income levels

Poverty, then, is the most crushing burden of all. As Todaro argues,

> low levels of living (insufficient life-sustaining goods and inadequate or nonexistent education, health, and other social services) are all related in one form or another to low incomes. These low incomes result from the low average productivity of the entire labour force, not just those working. Low labour force productivity can result from a variety of factors, including on the supply side poor health, nutrition, and work attitudes; high population growth, and high unemployment and underemployment. On the demand side inadequate skills, poor managerial talents, and overall low levels of worker education may, along with the importation of developed country labour-saving techniques of production, result in the substitution of capital for labour in domestic production. The combination of low labour demand and large supplies results in the widespread underutilisation of labour. Moreover, low incomes lead to low savings and investment which also restrict the total number of employment opportunities. Finally . . . low incomes are also thought to be related to large family size and high fertility since children provide one of the few sources of economic and social security in old age for very poor families . . . The important point to remember . . . is that *low productivity, low incomes, and low levels of living are mutually reinforcing phenomena*. They constitute what Myrdal has called a process of 'circular and cumulative causation' in which low incomes lead to low levels of living (income plus poor health, education etc.) that keeps productivity low, which in turn perpetuates low incomes, and so on.
>
> (Todaro, 1989, p. 93)

Dependence on a narrow economic base

A most important contributory factor in the poverty of low-income countries (and of some of the lower-middle-income countries too) is their dependence on a very narrow economic base together with the nature of the conditions of trade. We saw earlier (Table 13.4) that the overwhelming majority of the labour force in low-income countries is employed in agriculture. This, together with the extraction of other primary products, forms the basis of these countries' involvement in the world economy. Approximately four-fifths of the exports of developing countries are of primary products compared with less than one-quarter for the developed economies.

In the classical theories of international trade, based upon the comparative advantage of different factor endowments, it is totally logical for countries to specialize in the production of those goods for which they are well endowed by nature. Thus, it is argued, countries with an abundance of particular primary materials should concentrate on producing and exporting these and import those goods in which they have a comparative disadvantage. This was the rationale underlying the 'old' international division of labour in which the core countries produced and exported manufactured goods and the countries of the global periphery supplied the basic materials. According to traditional trade theory all countries benefit from such an arrangement. But such a neat sharing of the benefits of trade presupposes some degree of equality between trading partners, some stability in the relative prices of traded goods and an efficient mechanism – the market – which ensures that, over time, the benefits are indeed shared equitably.

In the real world – and especially in the trading relationships between the industrialized countries and the low-income, primary-producing countries – these conditions do not hold. In the first place, there is a long-term tendency for the composition of demand to change as incomes rise (Figure 5.3). Thus, growth in demand for manu-factured goods is greater than the growth in demand for primary products. This immediately builds a bias into trade relationships between the two groups of countries, favouring the industrialized countries at the expense of the primary producers.

Over time, these inequalities tend to be reinforced through the operation of the *cumulative* processes of economic growth. The prices of manufactured goods tend to increase more rapidly than those of primary products and, therefore, the *terms of trade* for manufactured and primary products tend to diverge. (The terms of trade are simply the ratio of export prices to import prices for any particular country or group of countries.) As the price of manufactured goods increases relative to the price of primary products, the terms of trade move against the primary producers and in favour of the industrial producers. For the primary producers it becomes necessary to export a larger quantity of goods in order to buy the same, or even a smaller, quantity of manufactured goods.

Although the terms of trade do indeed fluctuate over time, there is no doubt that they have recently been deteriorating for the primary producing countries:

> Prices for primary commodities . . . fluctuate sharply with global supply and demand. During the 1980s prices for many primary commodities fell to their lowest levels since World War II. Non-oil commodity prices declined for most of the decade . . . By 1989 average com-modity prices were still 33% lower than in 1980 . . . The decline in the terms of trade during the 1980s has been most pronounced in Sub-Saharan Africa and Latin America.
>
> (World Bank, 1990, p. 13)

Calls for a 'New International Economic Order'

The severity of the situation facing the developing countries in general and the low-income, least industrialized countries in particular, has led to demands for a radical change in the workings of the world economic system. The demand by a group of Third World countries has been for a *New International Economic Order* (NIEO), a demand set out formally in a United Nations Declaration in 1974. The United Nations Declaration

> called for the replacement of the existing international economic order, which was charac-terised by inequality, domination, dependence, narrow self-interest and segmentation, by a new order based on equity, sovereign equality, interdependence, common interest and co-operation among States irrespective of their economic and social systems.
>
> (United Nations, 1982, p. 3)

Table 13.7 summarizes the major objectives of the NIEO. Of particular interest for our present purposes are those relating to trade and to industry (including technology). Overall, the demand is for better access to world markets for both primary products and manufactured goods for the Third World. In the case of primary products a key demand is for stabilization schemes to remove the serious fluctuations in both demand and prices for them, fluctuations which have an especially severe impact on many Third World countries. In the case of manufactured goods, the demand is for a reduction in protection of developed country markets, for the OICs to adopt more positive adjustment policies in their home economies to permit the expansion of developing country exports and for measures to ease the transfer of technology.

The demand for a fairer distribution of world industrial production was formulated more precisely in the 1975 Lima Declaration and Plan of Action for Industrial Co-operation. The Lima Declaration set a target for 25 per cent of world industrial

Table 13.7 Major objectives of the United Nations Declaration for a New International Economic Order

International trade

1. Improvement of the terms and conditions of trade, including the removal of tariff and non-tariff barriers.
2. Adoption of an integrated approach to trade in commodities, including the stabilization of demands and prices.
3. Development of an international food programme.
4. Strengthened co-operation between developing countries, including regional integration.

Industrialization and technology

1. Redeployment of industrial productive capacity to developing countries.
2. Establishment of mechanisms for the transfer of technology to developing countries.
3. Regulation of transnational corporations.
4. Better use of natural resources in order to reduce waste.

International finance and aid

1. Reform of the IMF, including a more stable flow of development assistance.
2. Participation of developing countries in IMF decision-making.
3. Increased transfer of resources through the World Bank and the IMF, particularly to make extra capital available to the poorest countries on favourable terms.
4. Renegotiation of the debts of developing countries; transformation into longer-term 'soft' loans; more favourable interest rates.
5. Attainment of UN development assistance targets: each developed country to progressively increase its development assistance to a minimum of 0.7 per cent of its GNP.

Social issues

1. Achieving a more equitable distribution of income and raising the level of employment.
2. Providing health services, education, higher cultural standard and qualification for the workforce, and assuring the well-being of children and the integration of women in development.
3. Assuring the sovereignty of states.
4. Compensation for adverse effects on the resources of states due to foreign occupation, colonial domination or apartheid.
5. Restructuring the economic and social sections of the UN.

(*Source*: based on Todaro (1989), pp. 609–12)

production to be located in the developing countries by the year 2000. It is notable that the necessary

> rapid and accelerated industrialisation in developing countries should be directed towards the development primarily of domestic self-generated and self-sustaining growth. Industrialisation should seek to meet the internal demand for goods and services, but it should also be the means whereby exports would yield the resources necessary for the purchase of capital goods and other inputs for sustained development.
>
> (United Nations, 1982, p. 30)

Global problems require global solutions

Very little progress has been made on most of these demands. Renshaw (1981) suggests three reasons for the lack of a comprehensive response from the developed economies of the North. First, they are increasingly preoccupied with their own internal problems of inflation and unemployment and, as we have seen, the pressures to increase protection against imports from developing countries have increased rather than decreased. Second, the comprehensive nature of the NIEO requires a coherent and collective response involving the major international institutions – the GATT, the IMF, the World Bank and the United Nations. There are immense practical difficulties in organizing such collective action, not the least of which is the lack of political will. Third, there is no consensus among the countries of the North on the kinds of reform required, whether such reforms are feasible and, most of all, whether they are desirable. To many in the North there is little wrong with the existing international economic order which cannot be put right simply by leaving it alone and allowing market forces to reign.

The very poor developing countries, those at the bottom of the well-being league table, have benefited least from the internationalization and globalization of economic activities. Both their present and their future are dire. Ways have to be found to solve the problems of poverty and deprivation. The 1990 *World Bank World Development Report* reveals that although substantial progress has been made in raising levels of material well-being there are no fewer than 1 billion people living in the developing world on an income of less than $370 a year (around $1 per day!). 'No task should command a higher priority for the world's policy makers than that of reducing global poverty. In the last decade of the twentieth century it remains a problem of staggering dimensions' (World Bank, 1990, p. 5).[16]

In such a highly interconnected world as we now inhabit, and as our children will certainly inhabit, it is difficult to believe that anything less than global solutions can deal with such global problems. But how is such collaboration to be achieved? Can a new international economic order be created or is the likely future one of international economic disorder?[17] There can be no doubt that the present immense global inequalities are a moral outrage. The problem is one of reconciling what many perceive to be conflicting interests. For example, one of the biggest problems facing the older-industrialized countries is unemployment. As we have seen, the causes of unemployment are complex but if one factor is perceived to be import penetration by developing country firms then the pressure to adopt restrictive trade policies becomes considerable.

The alternative is to adopt policies which ease the adjustment for those groups and areas adversely affected and to stimulate new sectors. But this requires a more positive attitude than most Western governments have been prepared to adopt. For the poorer developing countries it is unlikely, however, that industrialization will provide the solution to their massive problems. Despite its rapid growth in some developing countries, manufacturing industry has made barely a small dent in the unemployment and underemployment problems of developing countries as a whole. For most, the answer must lie in other sectors, particularly agriculture, but the seriousness of these countries' difficulties necessitates concerted international action.

Notes for further reading

1. 'Global culture' is discussed by Sklair (1991) and in the set of essays edited by Featherstone (1990).
2. Important contributions to this debate are by Blackaby (1979), Bluestone and Harrison (1982), Cohen and Zysman (1987), Rowthorn and Wells (1987), Harrison and Bluestone (1988), Radice (1989).
3. Employment and unemployment trends in the industrialized countries are documented regularly in OECD publications, e.g., *OECD Economic Outlook, OECD Observer*. A review of the situation at the end of the 1980s can be found in OECD (1989).
4. International labour migration is discussed by Cohen (1987), ILO (1984b), Jones (1990). King (1990) analyses the situation in Western Europe.
5. Spatial shifts within the United States are described by Bluestone and Harrison (1982), Sawers and Tabb (1984), and within Western Europe by Clout *et al.* (1989) and Pinder (1990).

6. See Freeman, Clark and Soete (1982), Petrella (1984), Standing (1984).
7. As part of their policy recommendations to deal with the deindustrialization of the US economy both Cohen and Zysman (1987) and Harrison and Bluestone (1988) advocate some form of managed trade policy.
8. For a comprehensive discussion of this issue see Helleiner (1980), Renshaw (1981), Turner and McMullen (1982), Cable (1983), Schumacher (1984).
9. Harris (1986).
10. Brookfield (1975), Higgott (1984), Corbridge (1986), Todaro (1989), Peet (1991) provide broad-ranging reviews of the 'development of development thinking'.
11. Critiques of dependency and related theories of development can be found in the sources in note 10 and in Palma (1981), Seers (1981), Sklair (1991). Hamilton (1983) surveys the position of the four Asian NICs, Limqueco (1983) deals with the ASEAN countries. The 'deviant' cases of South Korea and Taiwan are analysed by Barone (1983) and by Barrett and Whyte (1981) respectively.
12. Detailed studies of employment changes in developing countries include those by World Bank (1979), ILO (1984b), Harris and Rashid (1986).
13. Stewart (1978), ILO (1984b) discuss problems of measuring and defining unemployment in Third World countries. The characteristics of the formal and informal sectors are described by Hart (1973), Roberts (1978), Gilbert and Gugler (1982), ILO (1984b), Armstrong and McGee (1985).
14. There is a large literature on urbanization in the Third World, including World Bank (1979), Gilbert and Gugler (1982), Armstrong and McGee (1985), Drakakis-Smith (1990).
15. Strange (1986, 1988), Corbridge (1988) provide substantial discussions of the debt problem in developing countries.
16. The 1990 World Bank *World Development Report* is devoted entirely to an analysis of poverty in developing countries.
17. Helleiner (1980) coined the term 'international economic disorder'. Thrift (1989) discusses the geography of international economic disorder.

Epilogue: 'Through a Glass, Darkly'

> Time present and time past
> Are both perhaps present in time future,
> And time future contained in time past.
>
> T. S. Eliot, *Four Quartets*

What of the future? Although 'time present' and 'time past' are indeed 'present in time future' we cannot simply extrapolate current patterns and processes into the future. The future is an amalgam of the probable and the unpredictable. Change occurs within an existing context but it also transforms that context, usually gradually, occasionally rapidly. What does seem certain is that the tendency towards an increasingly highly interconnected and interdependent global economy will intensify. The fortunes of nations, regions, cities, neighbourhoods, families and individuals will continue to be strongly influenced by their position in the global network. In a rapidly shrinking and interconnected world there is no hiding place.

A major question mark must hang over the future form of the international financial system. It is clear that the production of goods and services in the global economy is highly interlocked with the global financial system of foreign exchange rates and international currencies; buying and selling of shares in international and domestic companies; scheduling, rescheduling and securitization of debts of both business firms and entire countries; fluctuating interest rates.

> The problem with the international geopolitics of international finance is that the international financial system has become, in effect, a separate state . . . we are left with a strange world, one in which money capital flows freely, and is becoming less and less regulated, while movement of goods in the 'productive' economy has become more and more negotiated and regulated. Yet, this productive economy is increasingly susceptible, in principle if not in practice, to the institutions of money capital.
>
> (Thrift, 1990, p. 1136)

Susan Strange (1986, p. 1) has coined the term 'casino capitalism' to describe the international financial system: 'every day games are played in this casino that involve sums of money so large that they cannot be imagined. At night the games go on at the other side of the world . . . (the players) . . . are just like the gamblers in casinos watching the clicking spin of a silver ball on a roulette wheel and putting their chips on red or black, odd numbers or even ones.'

Back in the 'real world' of goods and services, what will the future global map of economic activity look like? Where will the new industries and the new technologies be located? At present, *the global economy appears to be polarizing around three major regional blocs* – North America, Western Europe and East and South East Asia (focused on Japan and the four leading NICs) – the *global triad*. Will such polarization intensify in the future? Much may depend upon the future shape of the global trading system and, especially, the outcome of the Uruguay Round of the GATT. At the time of writing this appeared to have foundered on the intransigence of the major countries on a variety of issues, notably agriculture and services. Will, as many suggest, the global economy become fragmented into protectionist blocs, inward- rather than outward-looking, with all this would mean particularly for developing countries?

A major question at the global scale is how far the centre of gravity of the global economy will shift towards the Pacific Basin. Will the twenty-first century be the 'Pacific Century' as some forecasters maintain? Much will depend not only on developments in the Pacific Basin itself (in which, of course, the United States and Canada are also involved) but also on whether the European economies can recover their economic momentum. Japan has been the major force in the global shift of the centre of economic gravity. But if a more general and powerful shift is to occur in the industrial balance of power from Atlantic to Pacific, other industrial centres must continue to develop.

The current harbingers of such developments are the East and South East Asian NICs. Often these have been termed the 'new Japans' although only South Korea has really tried to replicate the Japanese model of industrialization. Will the 'gang of four' continue to grow? Will a second generation of NICs emerge fully? We have seen some evidence of such a second tier, but it is by no means certain that these will be able to grow at anything like the rate of the earlier generation of Asian NICs. The world is now a different place in terms both of the potential for export-led growth and technologically. In particular, the increasing automation of labour-intensive processes may be reducing the incentive for TNCs to relocate these activities to developing countries. If this tendency should develop more widely then it would greatly alter the prospects for export-based industrialization in Third World countries. A growing domestic market – possibly regional in scale – is becoming more necessary. But we should also beware of simply transferring the experience of one country to that of another; internal conditions have been extremely important in the growth of the first generation of NICs.

In the Pacific context, of course, the great unknown is the likely extent of China's involvement in the global economy. Will its 'open-door' policy, initiated in 1979 but greatly set back by the massacres in Tianenmen Square in 1989, continue to develop? Some quite remarkable progress has been made in opening up China economically. But immense problems – many of which are political and ideological – remain in transforming China into a modern economy. The internal power struggles continue. There have been huge difficulties in modernizing both the production and the distribution systems and in generating a sufficient flow of foreign exchange and foreign capital to finance construction projects and to pay for necessary imports of intermediate and capital goods. A key issue will be the manner in which Hong Kong is reabsorbed into China after 1997. Will the 'one country, two systems' formula actually work? Hong Kong is clearly highly intertwined with China economically and not just in its traditional role as a gateway for trade. A most significant development has been the spectacular growth of Hong Kong-owned factories being established in southern China. In effect, this region now acts as an extension of Hong Kong's labour market. A very high proportion of the manufacturing and assembly of several Hong Kong industries is now performed in China with Hong Kong acting as the co-ordinator.

Elsewhere in the Pacific Basin, what does the future hold for countries such as Australia? In recent years the Australian economy has undergone a marked reorientation away from its traditional links with the United Kingdom. Initially the shift seemed to be towards the United States; more recently Japan has come to dominate Australia's external relationships. Australia has been particularly prominent in trying to encourage more formal links between countries in the region in the form of a Pacific Basin Initiative. What role will the ASEAN countries perform in a Pacific community? Will they continue to be dominated by Japan or can they build a less dependent, more regionally balanced, economic system? So far, the six ASEAN countries have not operated economically as a coherent unit. They have competing, rather than complementary, economic structures.

Within the North American context there are signs, as we have seen, of increasing regional integration. Canada and the United States are already implementing their free trade agreement. The United States and Mexico are talking about a similar bilateral agreement – much to the apprehension of many Canadians, some of whom are, in any case, not very happy with their free trade agreement with the United States. The potential implications of a broader regional agreement, eventually perhaps encompassing the countries of Latin America as well, are huge. More specifically, what does the future hold for economic development in Latin America? Can Brazil and the other major industrial countries break free from their internal and external constraints? What will their relationship be with the United States and with the changing international division of labour as a whole?

Of all the major world regions, economic integration has proceeded furthest in Western Europe. The EC is by far the most important regional economic grouping in the world. But despite its origins in the late 1950s, the European Community remains fragmented along national lines in many important respects. Concern that such 'balkanization' was inhibiting the Community's competitive position within the global economy led to the Single European Act of 1985, whose aim was the completion of the single market by 31 December 1992. Whether or not this occurs by then its potential development has stimulated business firms and national governments, both inside and outside the Community, to re-evaluate their strategies. To those outside there is a real fear of the development of a 'fortress Europe', a prospect strongly denied by the EC itself. Even when all the internal legislative barriers have been removed, however, there will still be significant national and local differences – in taste, culture, language and the like – which will not disappear. To talk of a single, *uniform* European market is nonsense.

Of course, when the Single European Act was passed in 1985 nobody foresaw the enormous political changes which were to occur in Eastern Europe within four years. It is still unclear what the outcomes will be. The former German Democratic Republic has become the eastern region of a newly united Germany. This, in itself, will create enormous opportunities – and problems – for the German economy. Given Germany's dominant economic position in Western Europe, the effects on the other European economies of the costs of German integration will also be very considerable. It is far less easy to see what will happen in the other Eastern European economies as they endeavour to make the unheard-of transition from state socialism to market capitalism. As with the developing countries elsewhere in the world, much will depend on the buoyancy of the global economy as a whole. Perhaps the greatest question mark of all hangs over the Soviet Union. Will the move towards *glasnost* and *perestroika*, initiated by Mikhail Gorbachev, continue? Can the Soviet economy be rebuilt whilst, at the same time, the internal nationalist divisions are resolved?

The future, then, is a land of many questions to which there are no certain answers. In trying to understand the processes of global economic change in this book we have emphasized the importance of two major kinds of institution – transnational corporations and nation states – operating in a complex and volatile technological environment. Through their strategies and interactions TNCs and states have reshaped the global economic map and contributed towards the increasing globalization of economic activities. There is every reason to believe that they will continue to be the primary forces even though their particular behaviour, and the interrelationships between them, will certainly change. The key point is that each is both a political and an economic institution. Nation states, whilst essentially political institutions, have become increasingly involved in economic matters. Transnational corporations, though fundamentally economic in function, have become increasingly political in their actions and impact. Hence, the 'topography' of tomorrow's global economic map, like that of yesterday's and today's, will be the outcome of both economic and political forces. The two cannot properly be separated. In the end, however, the issues are not merely academic. The global economy and all its participants, from transnational corporations and national governments to individual citizens, face a major global challenge: to meet the material needs of the world community as a whole in ways which reduce, rather than increase, inequality and which do so without destroying the environment.

Bibliography

ABLER, R. (1975) Effects of space-adjusting technologies on the human geography of the future, in R. Abler, D. Janelle, A. Philbrick and J. Sommer (eds) *Human Geography in a Shrinking World*, Duxbury Press, North Scituate, Mass, chapter 3.

ALLEN, J. (1988) Service industries: uneven development and uneven knowledge, *Area*, Vol. 20, pp. 15–22.

ALTSHULER, A., ANDERSON, M., JONES, D., ROOS, D. and WOMACK, J. (1984) *The Future of the Automobile*, Allen & Unwin, London.

AMIN, A. and ROBINS, K. (1990) The re-emergence of regional economies? The mythical geography of flexible accumulation, *Environment and Planning D: Society and Space*, Vol. 8, pp. 7–34.

AMSDEN, A. H. (1989) *Asia's Next Giant: South Korea and Late Industrialization*, Oxford University Press, New York.

ANSON, R. and SIMPSON, P. (1988) *World Textile Trade and Production Trends*, Economist Intelligence Unit Special Report 1108, London.

ARMSTRONG, W. and McGEE, T. G. (1985) *Theatres of Accumulation: Studies in Asian and Latin American Urbanization*, Methuen, London.

ASHOFF, G. (1983) The textile policy of the European Community towards the Mediterranean countries: effects and future options, *Journal of Common Market Studies*, Vol. XXII, pp. 17–45.

BAILLY, A., MAILLAT, D. and REY, M. (1987) *Nouvelles Articulations des Systèmes de Production et Rôle des Services: Une Analyse Comparative Internationale et Inter-régionale*, CEAT, Lausanne.

BAIROCH, P. (1982) International industrialisation levels from 1750 to 1980, *Journal of European Economic History*, Vol. 11, pp. 269–333.

BAKIS, H. (1987) Telecommunications and the global firm, in F. E. I. Hamilton (ed.) *Industrial Change in Advanced Economies*, Croom Helm, London, pp. 130–60.

BARNET, R. J. and MULLER, R. E. (1975) *Global Reach: The Power of the Multinational Corporation*, Cape, London.

BARONE, C. A. (1983) Dependency theory, Marxist theory and salvaging the idea of capitalism in South Korea, *Review of Radical Political Economics*, Vol. 15, pp. 41–67.

BARR, K. (1981) On the capitalist enterprise, *Review of Radical Political Economics*, Vol. 12, pp. 60–70.

BARRETT, R. E. and WHYTE, M. K. (1981) Dependency theory and Taiwan: analysis of a deviant case, *American Journal of Sociology*, Vol. 87, pp. 1064–89.

BARTLETT, C. A. (1986) Building and managing the transnational, in Porter (1986) chapter 12.

BARTLETT, C. A. and GHOSHAL, S. (1987) Managing across borders: new strategic requirements, *Sloan Management Review*, Vol. 7, Summer, pp. 7–17.

BARTLETT, C. A. and GHOSHAL, S. (1989) *Managing Across Borders: The Transnational Solution*, Harvard Business School Press, Boston.

BEHRMAN, J. N. and FISCHER, W. A. (1980) *Overseas R & D Activities of Transnational Companies*, Oelgeschlager, Gunn & Hain, Cambridge, Mass.

BENSON, I. and LLOYD, J. (1983) *New Technology and Industrial Change*, Kogan Page, London.

BERTRAND, O. and NOYELLE, T. (1988) *Human Resources and Corporate Strategy: Technological Change in Banks and Insurance Companies*, OECD, Paris.

BESSANT, J. and HAYWOOD, B. (1988) Islands, archipelagos and continents: progress on the road to computer-integrated manufacturing, *Research Policy*, Vol. 17, pp. 349–62.

BEYERS, W. B. (1989) Speed, information exchange, and spatial structure, in H. Ernste and C. Jaeger (eds) *Information Society and Spatial Structure*, Belhaven Press, London, chapter 1.

BHAGWATI, J. N. (1988) *Protectionism*, MIT Press, Cambridge, Mass.

BHAGWATI, J. N. (1989) United States trade policy at the crossroads, *The World Economy*, Vol. 12, pp. 439–79.

BHASKAR, K. (1980) *The Future of the World Motor Industry*, Kogan Page, London.

BIRCH, D. L. (1987) *Job Creation in America*, Free Press, New York.

BLACKABY, F. (ed.) (1979) *De-Industrialization*, Heinemann, London.

BLACKBURN, P., COOMBS, R. and GREEN, K. (1985) *Technology, Economic Growth and the Labour Process*, Macmillan, London.

BLOOMFIELD, G. T. (1978) *The World Automotive Industry*, David and Charles, Newton Abbot.

BLUESTONE, B. and HARRISON, B. (1982) *The Deindustrialization of America*, Basic Books, New York.

BRADFORD, C. I. JR. (1987) Trade and structural change: NICs and next tier NICs as transitional economies, *World Development*. Vol. 15, pp. 299–316.

BRAUDEL, F. (1984) *Civilisation and Capitalism. 15th–18th Centuries*, 3 Vol., Collins, London.

BRITTON, J. N. H. and GILMOUR, J. M. (1978) *The Weakest Link*, Science Council of Canada, Ottawa.

BRITTON, S. (1990) The role of services in production, *Progress in Human Geography*, Vol. 14, pp. 529–46.

BROOKFIELD, H. (1975) *Interdependent Development*, Methuen, London.

BROOKS, H. E. and GUILE, B. R. (1987) Overview, in B. R. Guile and H. Brooks (eds) *Technology and Global Industry: Companies and Nations in the World Economy*, National Academy Press, Washington, DC, pp. 1–15.

BUCKLEY, P. J. and CASSON, M. (1976) *The Future of Multinational Enterprise*, Macmillan, London.

BUCKLEY, P. J. and MIRZA, H. (1986) Pacific Asia's multinationals, *Euro–Asia Business Review*, Vol. 5, no. 4, pp. 39–43.

BUCKLEY, P. J. and MIRZA, H. (1988) The strategy of Pacific Asian multinationals, *The Pacific Review*, Vol. 1, pp. 50–62.

BUSINESS INTERNATIONAL (1987) *Competitive Alliances: How to Succeed at Cross-Regional Collaboration*, Business International Corporation, New York.

BYLINSKY, G. (1983) The race to the automatic factory, *Fortune*, 21 Feb. pp. 52–64.

CABLE, V. (1983) *Protectionism and Industrial Decline*, Hodder & Stoughton, London.

CABLE, V. and CLARKE, J. (1981) *British Electronics and Competition with Newly Industrializing Countries*, Overseas Development Institute, London.

CALLINGAERT, M. (1988) *The 1992 Challenge from Europe: Development of the European Community's Internal Market*, National Planning Association, Washington, DC.

CARTER, C. (ed.) (1981) *Industrial Policy and Innovation*, Allen & Unwin, London.

CASSON, M. (ed.) (1983) *The Growth of International Business*, Allen & Unwin, London.

CAVES, R. E. (1971) International corporations: the industrial economics of foreign investment, *Economica* (New Series), Vol. XXXVIII, pp. 1–27.

CAVES, R. E. (1982) *Multinational Enterprise and Economic Analysis*, Cambridge University Press.

CAWSON, A. (1989) European consumer electronics: corporate strategies and public policy, in Sharp and Holmes (1989), chapter 3.

CECCHINI, P. (1988) *The European Challenge 1992; The Benefits of a Single Market*, Wildwood House, Aldershot.

CHAUDHURI, A. (1983) American multinationals and American employment, in C. P. Kindleberger and D. B. Andretsch (eds) *The Multinational Corporation in the 1980s*, MIT Press, Cambridge, Mass, chapter 12.

CHEN, C. H. (1986) Taiwan's foreign direct investment, *Journal of World Trade Law*, Vol. 20, pp. 639–64.

CHEN, E. K. Y. (1981) Hong Kong Multinationals in Asia: characteristics and objectives, in K. Kumar and M. G. McLeod (eds) *Multinational Corporations from Developing Countries*, Lexington Books, Lexington, Mass, chapter 5.

CHEN, E. K. Y. (1983) Multinationals from Hong Kong, in Lall (1983), chapter 3.

CHERRY, C. (1978) *World Communication: Threat or Promise?*, Wiley, Chichester.

CHESNAIS, F. (1986) Science, technology and competitiveness, *Science Technology Industry Review*, Vol. 1, pp. 85–129.

CHI, SECK CHOO (1981) Industrial organisations and industrial estates: a case study, *Tijdschrift voor Economische en Sociale Geografie*, Vol. 72, pp. 235–41.

CHU, YUN-HAN (1989) State structure and economic adjustment of the East Asian newly industrialising countries, *International Organisation*, Vol. 43, pp. 647–72.

CLAIRMONTE, F. F. and CAVANAGH, J. H. (1981) *The World in their Web: Dynamics of Textile Multinationals*, Zed Press, London.

CLAIRMONTE, F. F. and CAVANAGH, J. H. (1983) Transnational corporations and the struggle for the global market, *Journal of Contemporary Asia*, Vol. 13, pp. 446–80.

CLAIRMONTE, F. F. and CAVANAGH, J. H. (1984) Transnational corporations and services: the final frontier, *Trade and Development*, Vol. 5, pp. 215–73.

CLINE, W. R. (1982) Can the East Asian model of development be generalised?, *World Development*, Vol. 10, pp. 81–90.

CLINE, W. R. (1987) *The Future of World Trade in Textiles and Apparel*, Institute for International Economics, Washington, DC.

CLOUT, H., BLACKSELL, M., KING, R. and PINDER, D. (1989) *Western Europe: Geographical Perspectives*, 2nd edn, Longman, London.

COASE, R. (1937) The nature of the firm, *Economica*, Vol. 4, pp. 386–405.

COHEN, B. J. (1990) The political economy of international trade, *International Organisation*, Vol. 44, pp. 261–81.

COHEN, R. B. (1981) The new international division of labour, multinational corporations and the urban hierarchy, in M. Dear and A. J. Scott (eds) *Urbanisation and Urban Planning in Capitalist Society*, Methuen, London, chapter 12.

COHEN, R. B. (1987) *The New Helots: Migrants in the International Division of Labour*, Gower, Aldershot.

COHEN, S. S. and ZYSMAN, J. (1987) *Manufacturing Matters: The Myth of the Post-Industrial Economy*, Basic Books, New York.

CONTRACTOR, F. and LORANGE, P. (eds) (1987) *Co-operative Strategies in International Business*, Lexington Books, Lexington.

COOMBS, R. and JONES, B. (1989) Alternative successors to Fordism, in H. Ernste and C. Jaeger (eds) *Information Society and Spatial Structure*, Belhaven Press, London, chapter 8.

CORBRIDGE, S. (1986) *Capitalist World Development: A Critique of Radical Development Geography*, Macmillan, London.

CORBRIDGE, S. (1988) The debt crisis and the crisis of global regulation, *Geoforum*, Vol. 19, pp. 109–30.

COWLING, K. and SUGDEN, R. (1987) Market exchange and the concept of a transnational corporation, *British Review of Economic Issues*, Vol. 9, pp. 57–68.

CRAIG, L. C. (1981) Office automation at Texas Instruments Incorporated, in M. L. Moss (ed.) *Telecommunications and Productivity*, Addison-Wesley, Reading, Mass.

CURZON, G., DE LA TORRE, J., DONGES, J., MACBEAN, A., WAELBROECK, J. and WOLF, M. (1981) *MFA Forever? The Future of the Arrangement for Trade in Textiles*, Trade Policy Research Centre, London.

CZINKOTA, M. R., RIVOLI, P. and RONKAINEN, I. A. (1989) *International Business*, Dryden Press, Chicago.

DANIELS, J. D. and RADEBAUGH, L. H. (1989) *International Business: Environments and Operations*, 5th edn, Addison-Wesley, Reading, Mass.

DANIELS, P. W. (1991) A world of services? Inaugural Lecture, Portsmouth Polytechnic.

DANIELS, P. W., THRIFT, N. J. and LEYSHON, A. (1989) Internationalization of professional producer services: accountancy conglomerates, in P. Enderwick (ed.) *Multinational Service Firms*, Routledge, London, chapter 4.

DAVIS, E., GEROSKI, P. A., KAY, J. A., MANNING, A., SMALES, C., SMITH, S. R. and SZYMANSKI, S. (1989) *1992: Myths and Realities*, London Business School Centre for Business Strategy, London.

DAVIS, E. and SMALES, C. (1989) The integration of European financial services, in Davis *et al.* (1989), chapter 5.

DAVIS, S. M. (1976) Trends in the organisation of multinational corporations, *Columbia Journal of World Business*, Vol. 11, pp. 59–71.

DE MEZA, D. and VAN DER PLOEG, F. (1987) Production flexibility as a motive for multinationality, *Journal of Industrial Economics*, Vol. 35, pp. 343–51.

DICKEN, P. (1987) Japanese penetration of the European automobile industry: the arrival of Nissan in the United Kingdom, *Tijdschrift voor Economische en Sociale Geografie*, Vol. 79, pp. 94–107.

DICKEN, P. (1988) The changing geography of Japanese foreign investment in manufacturing industry: a global perspective, *Environment and Planning A*, Vol. 20, pp. 633–53.

DICKEN, P. (1991) Europe 1992 and strategic change in the international automobile industry, *Environment and Planning A*, Vol. 23.

DICKEN, P. and LLOYD, P. E. (1981) *Modern Western Society: A Geographical Perspective on Work, Home and Well-Being*, Paul Chapman, London.

DICKEN, P. and LLOYD, P. E. (1990) *Location in Space: Theoretical Perspectives in Economic Geography*, 3rd edn, Harper & Row, New York.

DIEBOLD, W. JR. (1982) Past and future industrial policy in the United States, in J. Pinder (ed.) *National Industrial Strategies and the World Economy*, Croom Helm, London, chapter 6.

DIZARD, W. P. (1966) *Television: A World View*, Syracuse University Press.

DONAGHU, M. T. and BARFF, R. (1990) Nike just did it: international subcontracting and flexibility in athletic footwear production, *Regional Studies*, Vol. 24, pp. 537–52.

DORE, R. (1986) *Flexible Rigidities: Industrial Policy and Structural Adjustment in the Japanese Economy, 1970–1980*, Stanford University Press.

DOSI, G. (1983) Semiconductors: Europe's precarious survival in high technology, in Shepherd, Duchene and Saunders (1983), chapter 9.

DOSI, G., FREEMAN, C., NELSON, R., SILVERBERG, G. and SOETE, L. (eds) (1988) *Technical Change and Economic Theory*, Pinter, London.

DOZ, Y. (1986) *Strategic Management in Multinational Companies*, Pergamon, Oxford.

DRAKAKIS-SMITH, D. (ed.) (1990) *Economic Growth and Urbanization in Developing Areas*, Routledge, London.

DRUCKER, P. (1946) *The Concept of the Corporation*, John Day, New York.

DRUCKER, P. (1986) The changed world economy, *Foreign Affairs*, Vol. 64, pp. 768–91.

DUNNING, J. H. (1973) The determinants of international production, *Oxford Economic Papers*, New Series, Vol. 25, pp. 289–336.

DUNNING, J. H. (1977) Trade, location of economic activity and the MNE: a search for an eclectic approach, in B. Ohlin, P. O. Hesselborn and P. M. Wijkman (eds) *The International Allocation of Economic Activity*, Macmillan, London, chapter 12.

DUNNING, J. H. (1979) Explaining changing patterns of international production: in defence of the eclectic theory, *Oxford Bulletin of Economics and Statistics*, Vol. 41, pp. 269–96.

DUNNING, J. H. (1980) Towards an eclectic theory of international production: some empirical tests, *Journal of International Business Studies*, Vol. 11, pp. 9–31.

DUNNING, J. H. (1983) Changes in the level and structure of international production: the last 100 years, in Casson (1983), chapter 5.

DUNNING, J. H. (ed.) (1985) *Multinational Enterprises, Economic Structure and International Competitiveness*, Wiley, Chichester.

DUNNING, J. H. (1986) *Japanese Participation in British Industry*, Croom Helm, London.

DUNNING, J. H. (1988a) *Explaining International Production*, Unwin Hyman, London.

DUNNING, J. H. (1988b) *Multinationals, Technology and Competitiveness*, Unwin Hyman, London.

DUNNING, J. H. (1989) Transnational corporations and the growth of services: some conceptual and theoretical issues, *UNCTC Current Studies*, no. 9.

DUNNING, J. H. and CANTWELL, J. (1987) *IRM Directory of Statistics of International Investment and Production*, Macmillan, Basingstoke.

DUNNING, J. H. and NORMAN, G. (1983) The theory of the multinational enterprise: an application to multinational office location, *Environment and Planning A*, Vol. 15, pp. 675–92.

DUNNING, J. H. and NORMAN, G. (1987) The location choice of offices of international companies, *Environment & Planning A*, Vol. 19, pp. 613–31.

DUNNING, J. H. and RUGMAN, A. M. (1985) The influence of Hymer's dissertation on the theory of foreign direct investment, *American Economic Review, Papers & Proceedings*, Vol. 75, pp. 228–32.

ECONOMIST INTELLIGENCE UNIT (1983) The economic balkanisation of Europe, *European Trends*, Vol. 2, pp. 31–5.

EDWARDS, G. (1982) Four sectors: textiles, man-made fibres, shipbuilding, aircraft, in J. Pinder (ed.) *National Industrial Strategies and the World Economy*, Croom Helm, London, chapter 4.

ELSON, D. (1988) Transnational corporations in the new international division of labour: a critique of 'cheap labour' hypotheses, *Manchester Papers on Development*, Vol. IV, pp. 352–76.

ELSON, D. (1989a) Bound by one thread: the restructuring of UK clothing and textile multinationals, in A. MacEwan and W. K. Tabb (eds) *Instability and Change in the World Economy*, Monthly Review Press, New York, pp. 187–204.

ELSON, D. (1989b) The cutting edge: multinationals in the EEC textiles and clothing industry, in Elson and Pearson (1989), chapter 5.

ELSON, D. (1990) Marketing factors affecting the globalization of textiles, *Textiles Outlook International*, March, pp. 51–61.

ELSON, D. and PEARSON, R. (1981) 'Nimble fingers make cheap workers': analysis of women's employment in Third World export manufacturing, *Feminist Review*, Vol. 7, pp. 87–107.

ELSON, D. and PEARSON, R. (eds) (1989) *Women's Employment and Multinationals in Europe*, Macmillan, Basingstoke.

ENDERWICK, P. (1979) Multinationals and labour: a review of issues and responses, *Management Bibliographies and Reviews*, Vol. 5, pp. 219–34.

ENDERWICK, P. (1982) Labour and the theory of the multinational corporation, *Industrial Relations Journal*, Vol. 13, pp. 32–43.

ENDERWICK, P. (1989) Multinational corporate restructuring and international competitiveness, *California Management Review*, Vol. 32, pp. 44–58.

ERNST, D. (1985) Automation and the worldwide restructuring of the electronics industry: strategic implications for developing countries, *World Development*, Vol. 13, pp. 333–52.

ERNST, D. (1987) US–Japanese competition and worldwide restructuring of the electronics industry: a European view, in Henderson and Castells (1987), chapter 3.

ERZBERGER, H. R. and SONDEREGGER, U. (1989) Satellite offices for computer specialists at Crédit Suisse, in H. Ernste and C. Jaeger (eds) *Information Society and Spatial Structure*, Belhaven Press, London, chapter 11.

EUH, J. D. and MIN, S. H. (1986) Foreign direct investment from developing countries: the case of Korean firms, *The Developing Economies*, Vol. XXIV, pp. 149–68.

FARRANDS, C. (1982) The political economy of the Multi-Fibre Arrangement, in C. Stevens (ed.) *EEC and the Third World: A Survey*, Vol. 2, Hodder & Stoughton, London, chapter 6.

FEATHERSTONE, M. (ed.) (1990) *Global Culture: Nationalism, Globalization and Modernity*, Sage, London.

FERNANDEZ-KELLY, M. P. (1989) International development and industrial restructuring: the case of garment and electronics industries in southern California, in A. MacEwan and W. K. Tabb (eds) *Instability and Change in the World Economy*, Monthly Review Press, New York, pp. 147–65.

FINGER, J. M. and LAIRD, S. (1987) Protection in developed and developing countries – an overview, *Journal of World Trade Law*, Vol. 21, pp. 9–23.

FIRN, J. R. (1975) External control and regional development: the case of Scotland, *Environment and Planning A*, Vol. 7, pp. 393–414.

FLAMM, K. (1984) The volatility of offshore investment, *Journal of Development Economics*, Vol. 16, pp. 231–48.

FONG, P. E. and KOMARAN, R. V. (1985) Singapore multinationals, *Columbia Journal of World Busines*, Vol. 20, no. 2, pp. 35–43.

FORESTER, T. (ed.) (1985) *The Information Technology Revolution*, Blackwell, Oxford.

FORESTER, T. (1987) *High-Tech Society: The Story of the Information Technology Revolution*, Blackwell, Oxford.

FRANKO, L. G. (1976) *The European Multinationals: A Renewed Challenge to American and British Big Business*, Harper & Row, London.

FRANKO, L. G. (1983) *The Threat of Japanese Multinationals: How the West Can Respond*, Wiley, Chichester.

FRANSMAN, M. (1990) *The Market and Beyond; Co-operation and Competition in Information Technology in the Japanese System*, Cambridge University Press.

FREEMAN, C. (1982) *The Economics of Industrial Innovation*, Pinter, London.

FREEMAN, C. (1987) The Challenge of New Technologies, in OECD, *Interdependence and Co-operation in Tomorrow's World*, OECD, Paris, pp. 123–56.

FREEMAN, C. (1988) Introduction, in Dosi *et al.* (1988).

FREEMAN, C., CLARK, J. and SOETE, L. (1982) *Unemployment and Technical Change*, Pinter, London.

FREEMAN, C. and PEREZ, C. (1988) Structural crises of adjustment, business cycles and investment behaviour, in Dosi *et al.* (1988) chapter 3.

FRIEDMANN, J. (1986) The world city hypothesis, *Development and Change*, Vol. 17, pp. 69–83.

FRÖBEL, F., HEINRICHS, J. and KREYE, O. (1980) *The New International Division of Labour*, Cambridge University Press.

FUENTES, A. and EHRENREICH, B. (1983) *Women in the Global Factory*, South End Press, Boston.

FUJITA, M. and ISHIGAKI, K. (1986) The internationalization of Japanese commercial banking, in Taylor and Thrift (1986), chapter 7.

GABRIEL, P. (1966) The investment in the LDC: asset with a fixed maturity, *Columbia Journal of World Business*, Vol. 1, no. 3, pp. 113–20.

GAFFIKIN, F. and NICKSON, A. (1984) *Jobs Crisis and the Multinationals*, Trade Union Resource Centre, Birmingham.

GATT (1989) Services in the domestic and global economy, in *International Trade 1988–1989*, GATT, Geneva, Part III.

GERMIDIS, D. (ed.) (1980) *International Subcontracting: A New Form of Investment*, OECD, Paris.

GERTLER, M. (1988) The limits to flexibility: comments on the post-Fordist vision of production and its geography, *Transactions of the Institute of British Geographers*, Vol. 13, pp. 419–32.

GIBBS, D. C. (1988) Restructuring in the Manchester clothing industry: technical change and inter-relationships between manufacturers and retailers, *Environment and Planning A*, Vol. 20, pp. 1219–33.

GIBBS, M. (1985) Continuing the international debate on services, *Journal of World Trade Law*, Vol. 19, pp. 199–218.

GIDDY, I. H. (1978) The demise of the product cycle model in international business theory, *Columbia Journal of World Business*, Vol. 13, pp. 90–7.

GIDDY, I. H. and YOUNG, S. (1982) Conventional theory and unconventional multinationals: do new forms of multinational enterprise require new theories?, in A. M. Rugman (ed.) *New Theories of the Multinational Enterprise*, Croom Helm, London, chapter 4.

GILBERT, A. and GUGLER, J. (1982) *Cities, Poverty and Development: Urbanisation in the Third World*, Oxford University Press.

GLICKMAN, N. J. and WOODWARD, D. P. (1989) *The New Competitors: How Foreign Investors Are Changing the US Economy*, Basic Books, New York.

GOVERNMENT OF CANADA (1972) *Foreign Direct Investment in Canada*, Information Canada, Ottawa.

GRAHAM, E. M. and KRUGMAN, P. R. (1989) *Foreign Direct Investment in the United States*, Institute for International Economics, Washington, DC.

GRAY, H. P. (1987) The internationalization of global labour markets, *Annals of the American Academy of Political and Social Science*, Vol. 492, pp. 96–108.

GRAY, S. J. and McDERMOTT, C. (1989) *Mega-Merger Mayhem: Takeover Strategies, Battles and Contests*, Paul Chapman, London.

GROSSE, R. (1982) Regional offices in multinational firms, in A. M. Rugman (ed.) *New Theories of the Multinational Enterprise*, Croom Helm, London, chapter 6.

GRUBEL, H. G. (1989) Multinational banking, in P. Enderwick (ed.) *Multinational Service Firms*, Routledge, London, chapter 3.

GRUNWALD, J. and FLAMM, K. (1985) *The Global Factory*, The Brookings Institution, Washington, DC.

GWIAZDA, A. (1982) Some characteristic trends in East–West economic relations, *National Westminster Bank Quarterly Review*, August, pp. 53–61.

GWYNNE, R. N. (1979) Oligopolistic reaction, *Area*, Vol. 11, pp. 315–20.

HACHTEN, W. A. (1974) Mass media in Africa, in A. Wells (ed.) *Mass Communications: A World View*, National Press Books, Palo Alto.

HAIG, R. M. (1926) Toward an understanding of the metropolis, *Quarterly Journal of Economics*, Vol. 40, pp. 421–33.

HÅKANSON, L. (1979) Towards a theory of location and corporate growth, in F. E. I. Hamilton and G. J. R. Linge (eds) *Spatial Analysis, Industry and the Industrial Environment, Volume 1. Industrial Systems*, Wiley, Chichester, chapter 7.

HALBACH, A. J. (1988) Multinational enterprises and subcontracting in the Third World: A study of inter-industrial linkages, *International Labour Office Multinational Enterprises Programme, Working Paper 58.*

HALL, P. (1985) The geography of the Fifth Kondratiev, in P. Hall and A. Markusen (eds.) *Silicon Landscapes*, Allen & Unwin, Boston, chapter 1.

HALL, P. and PRESTON, P. (1988) *The Carrier Wave: New Information Technology and the Geography of Innovation, 1846–2003*, Unwin Hyman, London.

HAMILL, J. (1984) Labour relations decision-making within multinational corporations, *Industrial Relations Journal*, Vol. 15, pp. 30–5.

HAMILL, J. (1989) *Mediterranean Textiles and Clothing*, Economist Intelligence Unit Special Report 1121, London.

HAMILTON, A. (1986) *The Financial Revolution*, Penguin, Harmondsworth.

HAMILTON, C. (1983) Capitalist industrialisation in East Asia's four little tigers, *Journal of Contemporary Asia*, Vol. 13, pp. 35–73.

HARRIS, G. (1984) The globalization of advertising, *International Journal of Advertising*, Vol. 3, pp. 223–34.

HARRIS, G. T. and RASHID, Z. BIN ABDUL (1986) The employment performance of developing countries during the 1970s, *The Developing Economies*, Vol. 24, pp. 272–87.

HARRIS, N. (1986) *The End of the Third World: Newly Industrialised Countries and the Decline of an Ideology*, Penguin, Harmondsworth.

HARRISON, B. and BLUESTONE, B. (1988) *The Great U-Turn: Corporate Restructuring and the Polarizing of America*, Basic Books, New York.

HART, K. (1973) Informal income opportunities and urban employment in Ghana, *Journal of Modern African Studies*, Vol. 11, pp. 61–81.

HARVEY, D. (1982) *The Limits to Capital*, Blackwell, Oxford.

HARVEY, D. (1987) Three myths in search of a reality in urban studies, *Environment & Planning, D: Society and Space*, Vol. 5, pp. 367–76.

HAWKINS, R. G. (1972) Job displacement and the multinational firm: a methodological review, *Occasional Paper 3, Center for Multinational Studies*, Washington, DC.

HAYTER, R. (1982) Truncation, the international firm and regional policy, *Area*, Vol. 14, pp. 277–82.

HEENAN, D. A. (1979) The regional headquarters decision: a comparative analysis, *Academy of Management Journal*, Vol. 22, pp. 410–15.

HEENAN, D. A. and PERLMUTTER, H. V. (1979) *Multinational Organizational Development: A Social Architecture Perspective*, Addison-Wesley, Reading, Mass.

HELLEINER, G. K. (1973) Manufactured exports from less-developed countries and multinational firms, *Economic Journal*, Vol. 83, pp. 21–47.

HELLEINER, G. K. (1980) *International Economic Disorder; Essays in North–South Relations*, Macmillan, London.

HELLEINER, G. K. and LAVERGNE, R. (1979) Intra-firm trade and industrial exports to the United States, *Oxford Bulletin of Economics and Statistics*, Vol. 41, pp. 297–312.

HENDERSON, J. (1989) *The Globalization of High Technology Production*, Routledge, London.

HENDERSON, J. and CASTELLS, M. (eds) (1987) *Global Restructuring and Territorial Development*, Sage, London.

HEPWORTH, M. (1989) *Geography of the Information Economy*, Belhaven, London.

HESSELMAN, L. (1983) Trends in European industrial intervention, *Cambridge Journal of Economics*, Vol. 7, pp. 197–208.

HEWINGS, G. J. D. (1977) *Regional Industrial Analysis and Development*, Methuen, London.

HIGGOTT, R. (1984) Export-oriented industrialisation, the new international division of labour and the corporate state in the Third World: an exploratory essay on conceptual linkages, *Australian Geographical Studies*, Vol. 22, pp. 58–71.

HILL, H. (1982) Vertical inter-firm linkages in LDCs: a note on the Philippines, *Oxford Bulletin of Economics and Statistics*, Vol. 44, pp. 261–71.

HILL, H. (1990) Foreign investment and East Asian economic development, *Asian-Pacific Economic Literature*, Vol. 4, pp. 21–58.

HIRSCH, S. (1967) *Location of Industry and International Competitiveness*, Clarendon, Oxford.

HIRSCH, S. (1972) The United States electronics industry in international trade, in L. T. Wells (1972), pp. 39–54.

HOBSBAWM, E. J. (1979) The development of the world economy, *Cambridge Journal of Economics*, Vol. 3, pp. 305–18.

HOFFMAN, K. (1985) Clothing, chips and competitive advantage: the impact of microelectronics on trade and production in the garment industry, *World Development*, Vol. 13, pp. 371–92.

HOFFMAN, K. and RUSH, H. (1988) *Microelectronics and Clothing: The Impact of Technical Change on a Global Industry*, Praeger, New York.

HOLMES, J. (1986) The organization and locational structure of production subcontracting, in A. J. Scott and M. Storper (eds) *Production, Work and Territory: The Geographical Anatomy of Industrial Capitalism*, Allen & Unwin, London, chapter 5.

HOLMES, J. (1990) The globalization of production and the future of Canada's mature industries: the case of the automobile industry, in D. Drache and M. S. Gertler (eds) *The New Era of Global Competition: State Policy and Market Power*, McGill, Queens Press, Montreal, chapter 7.

HOLMES, J. (1991) The continental integration of the North American automobile industry: from the Auto Pact to the FTA, *Environment and Planning A*, Vol. 23.

HONE, A. (1974) Multinational corporations and multinational buying groups: their impact on the growth of Asia's exports of manufactures – myths and realities, *World Development*, Vol. 2, pp. 145–9.

HOOD, N. and YOUNG, S. (1979) *The Economics of Multinational Enterprise*, Longman, London.

HOOD, N. and YOUNG, S. (1982) US Multinational R & D: corporate strategies and policy implications for the UK, *Multinational Business*, Vol. 2, pp. 10–23.

HOWELLS, J. R. L. (1984) The location of research and development: some observations and evidence from Britain, *Regional Studies*, Vol. 18, pp. 13–29.

HOWELLS, J. R. L. (1990) The internationalization of R&D and the development of global research networks, *Regional Studies*, Vol. 24, pp. 495–512.

HUERTAS, T. F. (1990) US multinational banking: history and prospects, in G. Jones (ed.) *Banks as Multinationals*, Routledge, London, chapter 13.

HYMER, S. H. (1972a) The efficiency (contradictions) of multinational corporations, in G. Paquet (ed.) *The Multinational Firm and the Nation-State*, Collier-Macmillan, Don Mills, pp. 49–65.

HYMER, S. H. (1972b) The multinational corporation and the law of uneven development, in J. N. Bhagwati (ed.) *Economics and World Order*, Macmillan, London, pp. 113–40.

HYMER, S. H. (1976) *The International Operations of National Firms: A Study of Direct Foreign Investment*, MIT Press, Cambridge, Mass.

INOGUCHI, T. and OKIMOTO, D. I. (eds) (1988) *The Political Economy of Japan*, Stanford University Press.

INTERNATIONAL INSTITUTE OF COMMUNICATIONS (1990) *The Global Telecommunications Traffic Boom*, IIC, London.

INTERNATIONAL LABOUR OFFICE (1976) *Employment, Growth and Basic Needs: A One World Problem*, ILO, Geneva.

INTERNATIONAL LABOUR OFFICE (1981a) *Employment Effects of Multinational Enterprises in Developing Countries*, ILO, Geneva.

INTERNATIONAL LABOUR OFFICE (1981b) *Employment Effects of Multinational Enterprises in Industrialised Countries*, ILO, Geneva.

INTERNATIONAL LABOUR OFFICE (1984a) *Technology Choice and Employment Generation by Multinational Enterprises in Developing Countries*, ILO, Geneva.

INTERNATIONAL LABOUR OFFICE (1984b) *World Labour Report*, Vol. 1, ILO, Geneva.

INTERNATIONAL LABOUR OFFICE (1985) *Women Workers in Multinational Enterprises in Developing Countries*, ILO, Geneva.

INTERNATIONAL LABOUR OFFICE (1988) *Economic and Social Effects of Multinational Enterprises in Export Processing Zones*, ILO, Geneva.

ISARD, W. (1956) *Location and Space Economy*, MIT Press, Cambridge, Mass.

JACQUEMIN, A. (ed.) (1984) *European Industry: Public Policy and Corporate Strategy*, Clarendon, Oxford.

JAMES, B. G. (1985) Alliance: the new strategic focus, *Long Range Planning*, Vol. 18, pp. 76–81.

JANELLE, D. (1969) Spatial reorganisation: a model and a concept, *Annals of the Association of American Geographers*, Vol. 59, pp. 348–64.

JENKINS, R. (1984a) Divisions over the international division of labour, *Capital and Class*, Vol. 22, pp. 28–57.

JENKINS, R. (1984b) *Transnational Corporations and the Latin American Automobile Industry*, Macmillan, London.

JOHNSON, C. (1982) *MITI and the Japanese Economic Miracle: The Growth of Industrial Policy, 1925–1975*, Stanford University Press.

JOHNSON, C. (1985) The institutional foundations of Japanese industrial policy, *California Management Review*, Vol. XXVII, pp. 59–69.

JOHNSTON, R. J. (1982) *Geography and the State*, Macmillan, London.

JONES, H. R. (1990) *A Population Geography*, 2nd edn, Paul Chapman, London.

JULIUS, DeANNE (1990) *Global Companies and Public Policy: The Growing Challenge of Foreign Direct Investment*, Pinter, London.

KAPLINSKY, R. (1988) Restructuring the capitalist labour process: some lessons from the car industry, *Cambridge Journal of Economics*, Vol. 12, pp. 451–70.

KENNEDY, P. (1987) *The Rise and Fall of the Great Powers*, Random House, New York.

KINDLEBERGER, C. P. (1969) *American Business Abroad*, Yale University Press, New Haven.

KINDLEBERGER, C. P. (1988) The 'new' multinationalization of business, *Asean Economic Bulletin*, Vol. 5, pp. 113–24.

KING, R. (1990) The social and economic geography of labour migration from guestworkers to immigrants, in Pinder (1990), chapter 10.

KIRKPATRICK, C. H. (1990) Export oriented industrialization and income distribution in the Asian newly industrializing countries, in M. P. Van Dijk and H. S. Marcussen (eds) *Industrialization in the Third World*, Frank Cass, London, chapter 3.

KIRKPATRICK, C. H., LEE, N. and NIXSON, F. I. (1984) *Industrial Structure and Policy in Less Developed Countries*, Allen & Unwin, London.

KNICKERBOCKER, F. T. (1973) *Oligopolistic Reaction and Multinational Enterprises*, Harvard Business School, Boston.

KNOX, P. and AGNEW, J. (1989) *The Geography of the World Economy*, Edward Arnold, London.

KOBRIN, S. J. (1987) Testing the bargaining hypothesis in the manufacturing sector in developing countries, *International Organization*, Vol. 41, pp. 609–38.

KOBRIN, S. J. (1988) Strategic integration in fragmented environments: social and political assessments by subsidiaries of multinational firms, in N. Hood and J. E. Vahlne (eds) *Strategies in Global Competition*, Croom Helm, London, chapter 4.

KOJIMA, K. (1990) *Japanese Direct Investment Abroad*, International Christian University, Tokyo.

KOJIMA, K. and OZAWA, T. (1985) *Japan's General Trading Companies: Merchants of Economic Development*, OECD Development Centre, Paris.

KOO, B.-Y. (1985) Korea, in Dunning (1985) chapter 9.

KOSTECKI, M. (1987) Export-restraint arrangements and trade liberalization, *The World Economy*, Vol. 10, pp. 425–53.

KREYE, O., HEINRICHS, J. and FRÖBEL, F. (1988) Multinational enterprises and employment, *International Labour Office Multinational Enterprises Programme, Working Paper 55*.

KRUGMAN, P. R. (1989) Developing countries in the world economy, *Daedalus*, Vol. 118, pp. 183–203.

LALL, S. (1973) Transfer pricing by multinational manufacturing firms, *Oxford Bulletin of Economics and Statistics*, Vol. 35, pp. 173–95.

LALL, S. (1978) Transnationals, domestic enterprises and industrial structure in host LDCs: a survey, *Oxford Economic Papers*, Vol. 30, pp. 217–48.

LALL, S. (1979) Multinationals and market structure in an open developing economy: the case of Malaysia, *Weltwirtschaftliches Archiv*, Vol. 115, pp. 325–50.

LALL, S. (1983) *The New Multinationals: The Spread of Third World Enterprises*, Wiley, Chichester.

LALL, S. (1984) Transnationals and the Third World: changing perceptions, *National Westminster Bank Quarterly Review*, May, pp. 2–16.

LALL, S. and STREETEN, P. (1977) *Foreign Investment, Transnationals and Developing Countries*, Macmillan, London.

LANGDALE, J. V. (1989) The geography of international business telecommunications: the role of leased networks, *Annals of the Association of American Geographers*, Vol. 79, pp. 501–22.

LEONARD, H. J. (1988) *Pollution and the Struggle for World Product: Multinational Corporations, Environment and International Comparative Advantage*, Cambridge University Press.

LEONTIADES, J. (1971) International sourcing in the LDCs, *Columbia Journal of World Business*, Vol. VI, no. 6, pp. 19–28.

LEWIS, M. K. and DAVIS, J. T. (1987) *Domestic and International Banking*, Philip Allan, Oxford.

LEYSHON, A., THRIFT, N. J. and TOMMEY, C. (1988) The rise of the British provincial financial centre, *Progress in Planning*, Vol. 31, pp. 151–229.

LI, K. T. (1988) *The Evolution of Policy Behind Taiwan's Development Success*, Yale University Press, New Haven.

VAN LIEMT, G. (1988) *Bridging the Gap: Four Newly Industrializing Countries and the Changing International Division of Labour*, ILO, Geneva.

LIM, L. Y. C. (1980) Women workers in multinational corporations: the case of the electronics industry in Malaysia and Singapore, in K. Kumar (ed.) *Transnational Enterprises: Their Impact on Third World Societies and Cultures*, Westview Press, Boulder, pp. 109–36.

LIM, L. Y. C. and FONG, P. E. (1982) Vertical linkages and multinational enterprises in developing countries, *World Development*, Vol. 10, pp. 585–95.

LIM, M. H. and TEOH, K. F. (1986) Singapore corporations go multinational, *Journal of South East Asian Studies*, Vol. XVII, pp. 336–65.

LIMQUECO, P. (1983) Contradictions of development in ASEAN, *Journal of Contemporary Asia*, Vol. 13, pp. 283–302.

LIN, V. (1987) Women electronics workers in South East Asia: the emergence of a working class, in Henderson and Castells (1987), chapter 6.

LOVERING, J. (1990) Fordism's unknown successor: a comment on Scott's theory of flexible accumulation and the re-emergence of regional economies, *International Journal of Urban and Regional Research*, Vol. 14, pp. 159–74.

MACBEAN, A. I. and SNOWDEN, P. N. (1981) *International Institutions in Trade and Finance*, Allen & Unwin, London.

MCALEESE, D. and COUNAHAN, M. (1979) 'Stickers' or 'snatchers'? Employment in multinational corporations during the recession, *Oxford Bulletin of Economics and Statistics*, Vol. 41, pp. 345–58.

MCDERMOTT, M. C. (1989) *Multinationals: Foreign Divestment and Disclosure*, McGraw-Hill, London.

MCHALE, J. (1969) *The Future of the Future*, George Braziller, New York.

MCMULLEN. N. (1982) *The Newly Industrialising Countries: Adjusting to Success*, British–North American Committee, Washington, DC.

MAEX, R. (1983) Employment and multinationals in Asian export processing zones, *International Labour Office Multinational Enterprise Programme, Working Paper 26*.

MAGAZINER, I. C. and HOUT, T. M. (1980) *Japanese Industrial Policy*, Policy Studies Institute, London.

MAIR, A., FLORIDA, R. and KENNEY, M. (1988) The new geography of automobile production: Japanese transplants in North America, *Economic Geography*, Vol. 64, pp. 352–73.

MAJARO, S. (1982) *International Marketing: A Strategic Approach to World Markets*, Allen & Unwin, London.

MALECKI, E. (1980) Corporate organisation of R & D and the location of technological activities, *Regional Studies*, Vol. 14, pp. 219–34.

MANDEL, E. (1980) *Long Waves of Capitalist Development*, Cambridge University Press.

MENSCH, G. (1979) *Stalemate in Technology. Innovations Overcome the Depression*, Ballinger, New York.

MILES, R. E. and SNOW, C. C. (1986) Organizations: new concepts for new forms, *California Management Review*, Vol. XXVIII, pp. 62–73.

MODY, A. (1990) Institutions and dynamic comparative advantage: the electronics industry in South Korea and Taiwan, *Cambridge Journal of Economics*, Vol. 14, pp. 291–314.

MODY, A. and WHEELER, D. (1987) Towards a vanishing middle: competition in the world garment industry, *World Development*, Vol. 15, pp. 1269–84.

MORGAN, K. and SAYER, A. (1983) The international electronics industry and regional development in Britain, *University of Sussex School of Urban and Regional Studies Working Paper 34*.

MORGAN, K. and SAYER, A. (1988) *Microcircuits of Capital: 'Sunrise' Industry and Uneven Development*, Polity, Cambridge.

MORICI, P. (1989) The Canada–US Free Trade Agreement, *The International Trade Journal*, Vol. III, pp. 347–73.

MORRIS, D. and HERGERT, M. (1987) Trends in international collaborative agreements, *Columbia Journal of World Business*, Vol. XXII, no. 2, pp. 15–21.

MOSS, M. C. (1987) Telecommunications, world cities and urban policy, *Urban Studies*, Vol. 24, pp. 534–46.

MUKERJEE, D. (1986) *Lessons from Korea's Industrial Experience*, Institute of Strategic and International Studies, Kuala Lumpur.

MUSGRAVE, P. B. (1975) *Direct Investment Abroad and the Multinationals: Effects on the United States Economy*, USGPO, Washington, DC.

NAKASE, T. (1981) Some characteristics of Japanese-type multinationals today, *Capital and Class*, Vol. 13, pp. 61–98.

NEGRINE, R. and PAPATHANASSOPOULOS, S. (1990) *The Internationalization of Television*, Pinter, London.

NELSON, K. (1986) Labour demand, labour supply and the suburbanization of low-wage office work, in A. J. Scott and M. Storper (eds) *Production, Work and Territory; The Geographical Anatomy of Industrial Capitalism*, Allen & Unwin, Boston, chapter 8.

NEWFARMER, R. S. (1983) Multinationals and marketplace magic in the 1980s, in C. P. Kindleberger and D. B. Andretsch (eds) *The Multinational Corporation in the 1980s*, MIT Press, Cambridge, Mass, chapter 8.

NIXSON, F. (1988) The political economy of bargaining with transnational corporations: some preliminary observations, *Manchester Papers on Development*, Vol. IV, pp. 377–90.

NOGUÉS, J. J., OLECHOWSKI, A. and WINTERS, L. A. (1986) The extent of nontariff barriers to industrial countries' imports, *The World Bank Economic Review*, Vol. 1, pp. 181–99.

NUÑEZ, W. P. (1990) *Foreign Direct Investment and Industrial Development in Mexico*, OECD Development Centre Studies, Paris.

O'BRIEN, P. (1980) The new multinationals: developing country firms in international markets, *Futures*, Vol. 12, pp. 303–16.

OECD (1979) *The Impact of the Newly Industrializing Countries on Production and Trade in Manufactures*, OECD, Paris.

OECD (1987) *International Investment and Multinational Enterprises*, OECD, Paris.

OECD (1988) *The Newly Industrializing Countries: Challenge and Opportunity for OECD Industries*, OECD, Paris.

OECD (1989) *Economies in Transition: Structural Adjustment in OECD Countries*, OECD, Paris.

OECD Observer (1989) The great American job machine, Vol. 152, June–July, pp. 9–12.

OHLIN, B. (1933) *Inter-regional and International Trade*, Harvard University Press, Cambridge, Mass.

OHMAE, K. (1985) *Triad Power: The Coming Shape of Global Competition*, Free Press, New York.

OKIMOTO, D. I. (1989) *Between MITI and the Market: Japanese Industrial Policy for High Technology*, Stanford University Press.

OMAN, C. (1989) *New Forms of Investment in Developing Country Industries: Mining, Petroleum, Automobiles, Textiles, Food*, OECD, Paris.

OZAWA, T. (1979) *Multinationalism, Japanese Style*, Princeton University Press.

PALIWODA, S. J. (1986) *International Marketing*, Heinemann, London.

PALLOIX, C. (1975) The internationalization of capital and the circuit of social capital, in Radice (1975), chapter 3.

PALLOIX, C. (1977) The self-expansion of capital on a world scale, *Review of Radical Political Economics*, Vol. 9, pp. 1–28.

PALMA, G. (1981) Dependency and development: a critical overview, in Seers (1981), chapter 1.

PARDEE, S. E. (1988) Globalization of financial markets, *Economic Impact*, Vol. 1, pp. 21–4.

PARK, Y. S. (1989) Introduction to international financial centers: their origin and recent development, in Y. S. Park and M. Essayyad (eds) *International Banking and Financial Centers*, Kluwer Academic, Boston, chapter 1.

PEARCE, R. D. (1990) *The Internationalization of Research and Development by Multinational Enterprises*, Macmillan, London.

PEARSON, C. S. (ed.) (1987) *Multinational Corporations, Environment and The Third World*, Duke University Press, Durham.

PECK, J. A. (1990) Circuits of capital and industrial restructuring: adjustment in the Australian clothing industry, *Australian Geographer*, Vol. 21, pp. 33–52.

PEET, R. (ed.) (1987) *International Capitalism and Industrial Restructuring*, Allen & Unwin, London.

PEET, R. (1991) *Global Capitalism: Theories of Social Development*, Routledge, London.

PELKMANS, J. and WINTERS, L. A. (1988) *Europe's Domestic Market*, Routledge, London.

PEREZ, C. (1985) Microelectronics, long waves and world structural change, *World Development*, Vol. 13, no. 3, pp. 441–63.

PERLMUTTER, H. V. and HEENAN, D. A. (1986) Co-operate to compete globally, *Harvard Business Review*, March–April, pp. 136–52.

PERRY, M. (1990) The internationalization of advertising, *Geoforum*, Vol. 21, pp. 35–50.

PETRELLA, R. (1984) Technology and employment in Europe: problems and proofs, *Science and Public Policy*, Vol. 11, pp. 352–9.

PHILLIPS, D. R. (1986) Special Economic Zones in China's modernization: changing policies and changing fortunes, *National Westminster Bank Quarterly Review*, February, pp. 37–48.

PHILLIPS, D. R. and YEH, A. G. O. (1990) Foreign investment and trade: impact on spatial structure of the economy, in T. Cannon and A. Jenkins (eds) *The Geography of Contemporary China: The Impact of Deng Xiaoping's Decade*, Routledge, London, chapter 9.

PINDER, D. (ed.) (1990) *Western Europe: Challenge and Change*, Belhaven, London.

PIORE, M. J. and SABEL, C. F. (1984) *The Second Industrial Divide: Possibilities for Prosperity*, Basic Books, New York.

PITELIS, C. and SUGDEN, R. (eds) (1991) *The Nature of the Transnational Firm*, Routledge, London.

PLANT, R. (1981) *Industries in Trouble*, ILO, Geneva.

POMFRET, R. (1982) Trade preferences and foreign investment in Malta, *Journal of World Trade Law*, Vol. 16, pp. 236–50.

POPULATION INFORMATION PROGRAM (1983) Migration, population growth and development, *Population Reports*, M–7, Johns Hopkins University, Baltimore.

PORTER, M. E. (1985) *Competitive Advantage: Creating and Sustaining Superior Performance*, Free Press, New York.

PORTER, M. E. (ed.) (1986) *Competition in Global Industries*, Harvard Business School Press, Boston.

PORTER, M. E. (1990) *The Competitive Advantage of Nations*, Macmillan, London.

RADICE, H. (ed.) (1975) *International Firms and Modern Imperialism*, Penguin, Harmondsworth.

RADICE, H. (1989) British capitalism in a changing global economy, in A. MacEwan and W. K. Tabb (eds) *Instability and Change in the World Economy*, Monthly Review Press, New York, pp. 64–81.

RADICE, H. (1991) Capital, labour and the state in the world economy, *Society and Space*, Vol. 9.

RAWSTRON, E. M. (1958) Three principles of industrial location, *Transactions of the Institute of British Geographers*, Vol. 25, pp. 125–42.

REDFERN, P. (1987) Infamy! Infamy! They've all got it in for me!, *Environment & Planning D: Society and Space*, Vol. 5, pp. 413–17.

REED, H. C. (1989) Financial centre hegemony, interest rates, and the global political economy, in Y. S. Park and N. Essayyad (eds) *International Banking and Financial Centers*, Kluwer Academic, Boston, chapter 16.

REES, J. (1978) On the spatial spread and oligopolistic behaviour of large rubber companies, *Geoforum*, Vol. 9, pp. 319–30.

REICH, R. B. and MANKIN, E. D. (1986) Joint ventures with Japan give away our future, *Harvard Business Review*, March–April, pp. 78–86.

REICH, S. (1989) Roads to follow: regulating direct foreign investment, *International Organiza-*

tion, Vol. 43, pp. 543–84.

REID, N. (1990) Spatial patterns of Japanese investment in the US automobile industry, *Industrial Relations Journal*, Vol. 21, pp. 49–59.

RENSHAW, G. (1981) *Employment, Trade and North–South Co-operation*, ILO, Geneva.

REUBER, G. L. (1973) *Private Foreign Investment in Development*, Clarendon, Oxford.

RHYS, D. G. (1989) Smaller car firms – will they survive?, *Long Range Planning*, Vol. 22, pp. 22–9.

RICHARDSON, J. D. (1990) The political economy of strategic trade policy, *International Organization*, Vol. 44, pp. 107–35.

RIDDLE, D. I. (1986) *Service-Led Growth: The Role of the Service Sector in World Development*, Praeger, New York.

ROBERT, A. (1983) The effects of the international division of labour on female workers in the textile and clothing industries, *Development and Change*, Vol. 14, pp. 19–37.

ROBERTS, B. (1978) *Cities of Peasants: The Political Economy of Urbanization in the Third World*, Edward Arnold, London.

ROBINSON, J. (1983) *Multinationals and Political Control*, Gower, Aldershot.

ROGERS, E. M. and LARSEN, J. K. (1984) *Silicon Valley Fever: Growth of High Technology Culture*, Basic Books, New York.

ROOT, F. R. (1990) *International Trade and Investment*, 6th edn, South Western Publishing, Cincinnati.

ROSS, R. and TRACHTE, K. (1983) Global cities and global classes: the peripheralisation of labour in New York City, *Review*, Vol. VI, pp. 393–431.

ROSTOW, W. W. (1960) *The Stages of Economic Growth*, Cambridge University Press.

ROTHWELL, R. (1981) Pointers to government policies for technical innovation, *Futures*, Vol. 13, pp. 171–83.

ROTHWELL, R. (1982) The role of technology in industrial change: implications for regional policy, *Regional Studies*, Vol. 16, pp. 361–9.

ROWTHORN, R. and WELLS, J. R. (1987) *Deindustrialization and Foreign Trade: Britain's Decline in a Global Perspective*, Cambridge University Press.

RUGMAN, A. M. (1981) *Inside the Multinationals*, Croom Helm, London.

SABEL, C. F. (1989) Flexible specialization and the re-emergence of regional economies, in P. Hirst and J. Zeitlin (eds) *Reversing Industrial Decline? Industrial Structure and Policy in Britain and her Competitors*, Berg, Leamington, pp. 17–70.

SALIH, K., YOUNG, M. L. and RASIAH, R. (1988) The changing face of the electronics industry in the periphery: the case of Malaysia, *International Journal of Urban and Regional Research*, Vol. 12, pp. 375–403.

SAWERS, L. and TABB, W. K. (eds) (1984) *Sunbelt/Snowbelt: Urban Development and Regional Restructuring*, Oxford University Press, New York.

SAYER, A. (1986) New developments in manufacturing: the just-in-time system, *Capital and Class*, Vol. 30, pp. 43–72.

SAYER, A. (1989) Postfordism in question, *International Journal of Urban and Regional Research*, Vol. 13, pp. 666–95.

SAXENIAN, A. (1985) The genesis of Silicon Valley, in P. Hall and A. Markusen (eds) *Silicon Landscapes*, Allen & Unwin, Boston, chapter 2.

SCAMMELL, W. M. (1980) *The International Economy Since 1945*, Macmillan, London.

SCHILLER, D. (1982) Business users and the telecommunications network, *Journal of Communication*, Vol. 32, pp. 84–96.

SCHOENBERGER, E. (1988a) From Fordism to flexible accumulation: technology, competitive strategies and international location, *Environment & Planning D: Society and Space*, Vol. 6, pp. 245–62.

SCHOENBERGER, E. (1988b) Multinational corporations and the new international division of labour: a critical appraisal, *International Regional Science Review*, Vol. 11, pp. 105–19.

SCHOENBERGER, E. (1989) Thinking about flexibility: a response to Gertler, *Transactions of the Institute of British Geographers*, Vol. 14, pp. 98–108.

SCHONBERGER, R. J. (1982) *Japanese Manufacuring Techniques: Nine Hidden Lessons in Simplicity*, Free Press, New York.

SCHROEDER, T. G. (1989) *Operations Management: Decision Making in the Operations Function*, 3rd edn, McGraw-Hill, New York.

SCHUMACHER, D. (1984) North–South trade and shifts in employment: a comparative analysis of six European Community countries, *International Labour Review*, Vol. 123, pp. 333–48.

SCHUMPETER, J. (1939) *Business Cycles: A Theoretical, Historical and Statistical Analysis of the Capitalist Process*, McGraw-Hill, New York.

SCHUMPETER, J. (1943) *Capitalism, Socialism and Democracy*, Allen & Unwin, London.

SCOTT, A. J. (1987) The semiconductor industry in South East Asia: organization, location and the international division of labour, *Regional Studies*, Vol. 21, pp. 143–60.

SCOTT, A. J. (1988) Flexible production systems and regional development, *International Journal of Urban and Regional Research*, Vol. 12, pp. 171–85.

SCOTT, A. J. and ANGEL, D. P. (1987) The US semiconductor industry: a locational analysis, *Environment and Planning A*, Vol. 19, pp. 875–912.

SCOTT, A. J. and ANGEL, D. P. (1988) The global assembly operations of US semiconductor firms: a geographical analysis, *Environment and Planning A*, Vol. 20, pp. 1047–67.

SCOTT-QUINN, B. (1990) US investment banks as multinationals, in G. Jones (ed.) *Banks as Multinationals*, Routledge, London, chapter 14.

SEERS, D. (ed.) (1981) *Dependency Theory: A Critical Reassessment*, Pinter, London.

SERVAN-SCHREIBER, J.-J. (1968) *The American Challenge*, Hamish Hamilton, London.

SHARP, M. and HOLMES, P. (eds) (1989) *Strategies for New Technology*, Philip Allan, London.

SHARP, M. and SHEARMAN, C. (1987) *European Technological Collaboration*, Routledge, London.

SHARPSTON, M. (1975) International subcontracting, *Oxford Economic Papers, New Series*, Vol. 27, pp. 94–135.

SHEARD, P. (1983) Auto production systems in Japan: organizational and locational features, *Australian Geographical Studies*, Vol. 21, pp. 49–68.

SHEPHERD, G. (1983) Textiles: new ways of surviving in an old industry, in Shepherd, Duchene and Saunders (1983), chapter 2.

SHEPHERD, G., DUCHENE, F. and SAUNDERS, C. (eds) (1983) *Europe's Industries. Public and Private Strategies for Change*, Pinter, London.

SIEGEL, L. (1980) Delicate bonds: the global semiconductor industry, *Pacific Research*, Vol. 11, pp. 1–26.

SKLAIR, L. (1986) Free zones, development and the new international division of labour, *The Journal of Development Studies*, Vol. 22, pp. 753–9.

SKLAIR, L. (1989) *Assembling for Development: The Maquila Industry in Mexico and the United States*, Unwin Hyman, London.

SKLAIR, L. (1991) *Sociology of the Global System, Social Change in Global Perspective*, Harvester Wheatsheaf, Hemel Hempstead.

DE SMIDT, M. (1990) Philips: A global electronics firm restructures its home base, in M. de Smidt and E. Wever (eds) *The Corporate Firm in a Changing World Economy*, Routledge, London, chapter 3.

SMITH, D. M. (1981) *Industrial Location: An Industrial–Geographical Analysis*, 2nd ed, Wiley, New York.

DE SOLA POOL, I. (1981) International aspects of telecommunications policy, in M. L. Moss (ed.) *Telecommunications and Productivity*, Addison-Wesley, Reading, Mass, chapter 7.

SPERO, J. E. (1990) *The Politics of International Economic Relations*, 3rd ed, Allen & Unwin, London.

STANDING, G. (1984) The notion of technological unemployment, *International Labour Review*, Vol. 123, pp. 127–47.

STEED, G. P. F. (1981) International location and comparative advantage: the clothing industries and developing countries, in F. E. I. Hamilton and G. J. R. Linge (eds) *Spatial Analysis, Industry and the Industrial Environment, Volume 2: International Industrial Systems*, Wiley, Chichester, chapter 7.

STEELE, P. (1988) *The Caribbean Clothing Industry*, Economist Intelligence Unit Special Report 1147, London.

STEGEMANN, K. (1989) Policy rivalry among industrial states: what can we learn from models of strategic trade policy?, *International Organization*, Vol. 43, pp. 73–100.

STEWART, F. (1978) *Technology and Underdevelopment*, Macmillan, London.

STOPFORD, J. M., DUNNING, J. H. and HABERICH, K. O. (1981) *The World Directory of Multinational Enterprises*, Macmillan, London.

STOPFORD, J. M. and TURNER, L. (1985) *Britain and The Multinationals*, Wiley, Chichester.

STOREY, D. J. and JOHNSON, S. (1987) *Job Generation and Labour Market Change*, Macmillan, London.

STORPER, M. (1985) Oligopoly and the product cycle: essentialisim in economic geography, *Economic Geograhy*, Vol. 61, pp. 260–82.

STORPER, M. and SCOTT, A. J. (1989) The geographical foundations and social regulation of flexible production complexes, in J. Wolch and M. Dear (eds) *The Power of Geography; How Territory Shapes Social Life*, Unwin Hyman, Boston, chapter 2.

STORPER, M. and WALKER, R. (1983) The theory of labour and the theory of location, *International Journal of Urban and Regional Research*, Vol. 7, pp. 1–41.

STORPER, M. and WALKER, R. (1984) The spatial division of labour: labour and the location of industries, in Sawers and Tabb (1984), chapter 2.

STORPER, M. and WALKER, R. (1989) *The Capitalist Imperative: Territory, Technology and Economic Growth*, Blackwell, Oxford.

STRANGE, S. (1986) *Casino Capitalism*, Blackwell, Oxford.

STRANGE, S. (1988) *Till Debt Us Do Part*, Penguin, Harmondsworth.

SWANN, D. (1988) *The Economics of the Common Market*, 6th edn, Penguin, London.

TAYLOR, M. J. (1986) The product cycle model: a critique, *Environment and Planning A*, Vol. 18, pp. 751–61.

TAYLOR, M. J. and THRIFT, N. J. (eds) (1982) *The Geography of Multinationals*, Croom Helm, London.

TAYLOR, M. J. and THRIFT, N. J. (eds) (1986) *Multinationals and the Restructuring of the World Economy*, Croom Helm, London.

TAYLOR, P. J. (1990) *Political Geography: World Economy, Nation State and Locality*, 2nd edn, Longman, London.

TEECE, D. J. (1980) Economies of the scope and the scope of the enterprise, *Journal of Economic Behavior and Organization*, Vol. 1, pp. 223–47.

TEULINGS, A. W. M. (1984) The internationalization squeeze: double capital movement and job transfer within Philips worldwide, *Environment and Planning A*, Vol. 16, pp. 597–614.

THRIFT, N. J. (1987) The fixers: the urban geography of international commercial capital, in Henderson and Castells (1987), chapter 9.

THRIFT, N. J. (1989) The geography of international economic disorder, in R. J. Johnston and P. J. Taylor (eds) *A World in Crisis? Geographical Perspectives*, 2nd edn, Blackwell, Oxford, chapter 2.

THRIFT, N. J. (1990) The perils of the international financial system, *Environment and Planning A*, Vol. 22, pp. 1135–6.

THRIFT, N. J. AND LEYSHON, A. (1988) 'The gambling propensity': banks, developing country debt exposures and the new international financial system, *Geoforum*, Vol. 19, pp. 55–69.

TICKELL, A. T. (1992) Capital cities? Rethinking financial services, *Environment and Planning A*, 24.

TOBIN, J. (1984) Unemployment in the 1980s: macroeconomic diagnosis and prescription, in A. J. Pierre (ed.) *Unemployment and Growth in the Western Economies*, Council on Foreign Relations, New York, pp. 79–112.

TODARO, M. P. (1989) *Economic Development in the Third World*, 4th edn, Longman, New York.

TOFFLER, A. (1971) *Future Shock*, Pan, London.

TOLCHIN, M. and TOLCHIN, S. (1988) *Buying into America: How Foreign Money is Changing the Face of our Nation*, Times Books, New York.

TOWNSEND, A. and PECK, F. (1986) The role of foreign manufacturing in Britain's great recession, in Taylor and Thrift (1986), chapter 10.

TOYNE, B., ARPAN, J. S., BARNETT, A. H., RICKS, D. A. and SHIMP, T. A. (1984) *The Global Textile Industry*, Allen & Unwin, London.

TURNER, L. (1982) Consumer electronics: the colour television case, in Turner and McMullen (1982), chapter 4.

TURNER, L. (1987) *Industrial Collaboration with Japan*, Routledge, Andover.

TURNER, L. and McMULLEN, N. (eds) (1982) *The Newly Industrializing Countries: Trade and Adjustment*, Allen & Unwin, London.

UNCTAD (1985) *Production and Trade in Services: Policies and their Underlying Factors Bearing upon International Service Transactions*, UNCTAD, Geneva.

UNCTAD (1988) Services in the world economy, in *Trade and Development Report, 1988*, United Nations, New York, Part Two.

UNCTC (1983a) *Transnational Corporations in the International Auto Industry*, United Nations, New York.

UNCTC (1983b) *Transnational Corporations in World Development: Third Survey*, United Nations, New York.

UNCTC (1986) *Transnational Corporations in the International Semiconductor Industry*, United Nations, New York.

UNCTC (1988) *Transnational Corporations in World Development: Trends and Prospects*, United Nations, New York.

UNCTC (1989a) *Foreign Direct Investment and Transnational Corporations in Services*, United Nations, New York.

UNCTC (1989b) The process of transnationalization and transnational mergers, *UNCTC Current Studies Series A*, no. 8.

UNIDO (1980) Export Processing Zones in developing countries, *UNIDO Working Paper on Structural Changes*, no. 19, UNIDO, Vienna.

UNIDO (1981) *Restructuring World Industry in a Period of Crisis – The Role of Innovation; An Analysis of Recent Developments in the Semiconductor Industry*, UNIDO, Vienna.

UNIDO (1982) *The International Industrial Restructuring Process: The EEC, The European Periphery and Selected Developing Countries*, UNIDO, Vienna.

UNITED NATIONS (1982) *Towards the New International Economic Order*, United Nations, New York.

UTTERBACK, J. M. (1987) Innovation and industrial evolution in manufacturing industries, in B. R. Guile and H. Brooks (eds) *Technology and Global Industry: Companies and Nations in the World Economy*, National Academy Press, Washington, DC, pp. 16–48.

VAITSOS, C. (1974) *Inter-Country Income Distribution and Transnational Enterprises*, Clarendon, Oxford.

VAN DUIJN, J. J. (1983) *The Long Wave in Economic Life*, Allen & Unwin, London.

VAUPEL, J. W. and CURHAN, J. P. (1969) *The Making of Multinational Enterprise*, Harvard University Press, Cambridge, Mass.

VAUPEL, J. W. and CURHAN, J. P. (1973) *The World's Multinational Enterprises: A Sourcebook of Tables Based on a Study of the Largest US and Non-US Manufacturing Corporations*, Harvard University Press, Cambridge, Mass.

VERNON, R. (1966) International investment and international trade in the product cycle, *Quarterly Journal of Economics*, Vol. 80, pp. 190–207.

VERNON, R. (1971) *Sovereignty at Bay. The Multinational Spread of US Enterprises*, Basic Books, New York.

VERNON, R. (1974) The location of economic activity, in J. H. Dunning (ed.) *Economic Analysis and the Multinational Enterprise*, Allen & Unwin, London, p. 89–114.

VERNON, R. (1979) The product cycle hypothesis in a new international environment, *Oxford Bulletin of Economics and Statistics*, Vol. 41, pp. 255–68.

WALKER, R. (1988) The geographical organization of production systems, *Environment & Planning D: Society and Space*, Vol. 6, pp. 377–408.

WALKER, R. (1989) A requiem for corporate geography: new directions in industrial organization, the production of place and uneven development, *Geografiska Annaler*, Vol. 71B, pp. 43–57.

WALKER, R. and STORPER, M. (1981) Capital and industrial location, *Progress in Human Geography*, Vol. 5, pp. 473–509.

WALLERSTEIN, I. (1979) *The Capitalist World Economy*, Cambridge University Press.

WALLERSTEIN, I. (1984) *The Politics of the World Economy*, Cambridge University Press.

WARF, B. (1989) Telecommunications and the globalization of financial services, *Professional Geographer*, Vol. 41, pp. 257–71.

WARR, P. G. (1984) Korea's Masan Free Export Zone: benefits and costs, *The Developing Economies*, Vol. 22, pp. 169–84.

WARR, P. G. (1987) Malaysia's industrial enclaves: benefits and costs, *The Developing*

Economies, Vol. XXV, pp. 30–55.

WELLS, A. (1972) *Picture-Tube Imperialism?*, Orbis, New York.

WELLS, L. T. Jr. (ed.) (1972) *The Product Life Cycle and International Trade*, Harvard Business School, Boston.

WELLS, L. T. Jr. (1983) *Third World Multinationals: The Rise of Foreign Investment from Developing Countries*, MIT Press, Cambridge, Mass.

WELLS, P. E. and COOKE, P. N. (1991) The geography of international strategic alliances in the telecommunications industry: the cases of Cable and Wireless, Ericsson, and Fujitsu, *Environment and Planning A*, Vol. 23, pp. 87–106.

WHICHARD, O. G. and SHEA, M. A. (1985) 1982 benchmark survey of US direct investment abroad, *United States Department of Commerce Survey of Current Business*, Vol. 65, no. 12, pp. 37–57.

WILKINS, M. (1970) *The Emergence of Multinational Enterprise: American Business Abroad from the Colonial Era to 1914*, Harvard University Press, Cambridge, Mass.

WILKINS, M. (1974) *The Maturing of Multinational Enterprise: American Business Abroad from 1914–1970*, Harvard University Press, Cambridge, Mass.

WILLIAMSON, O. E. (1975) *Markets and Hierarchies*, Free Press, New York.

WILSON, B. D. (1978) Foreign divestment in the multinational investment cycle: the US experience, *Multinational Business*, Vol. 2, pp. 1–11.

WINTERS, L. A. (1985) *International Economics*, 3rd edn, Allen & Unwin, London.

WOMACK, J. R., JONES, D. T. and ROOS, D. (1990) *The Machine that Changed the World*, Rawson Associates, New York.

WONG, K. Y. and CHU, D. K. Y. (eds) (1985) *Co-ordination in China: The Case of the Shenzhen Special Economic Zone*, Oxford University Press, Hong Kong.

WORLD BANK (1979) *World Development Report, 1979*, Oxford University Press, New York.

WORLD BANK (1987) *World Development Report, 1987*, Oxford University Press, New York.

WORLD BANK (1990) *World Development Report, 1990*, Oxford University Press, New York.

YANNOPOULOS, G. N. (1983) The growth of transnational banking, in Casson (1983) chapter 9.

YOSHINO, M. Y. and LIFSON, T. B. (1986) *The Invisible Link: Japan's Sogo Shosha and the Organization of Trade*, MIT Press, Cambridge, Mass.

YOUNG, A. K. (1979) *The Sogo Shosha: Japan's Multinational Trading Companies*, Westview Press, Boulder.

YOUNG, S., HOOD, N. and HAMILL, J. (1988) *Foreign Multinationals and the British Economy*, Croom Helm, London.

Subject Index